T0230755

CRC Handbook of Flowering

Volume V

Editor

Abraham H. Halevy
Professor
Department of Ornamental Horticulture
The Hebrew University of Jerusalem
Rehovot
Israel

CRC Press is an imprint of the
Taylor & Francis Group, an **informa** business

First published 1986 by CRC Press
Taylor & Francis Group
6000 Broken Sound Parkway NW, Suite 300
Boca Raton, FL 33487-2742

Reissued 2018 by CRC Press

Library of Congress Cataloging-in-Publication Data

(Revised for volume 5)
Main entry under title:

CRC handbook of flowering.

Includes bibliographies and indexes.
1. Plants, Flowering of—Handbooks, manuals, etc.
2. Plants, Cultivated—Handbooks, manuals, etc.
I. Halevy, A. H. (Abraham H.), 1927-
II. Title: C.R.C. handbook of flowering.
III. Title: Handbook of flowering.
SB126.8.C73 1985 635.9 83-21061
ISBN 0-8493-3910-3 (set)

Publisher's Note
The publisher has gone to great lengths to ensure the quality of this reprint but points out that some imperfections in the original copies may be apparent.

Disclaimer
The publisher has made every effort to trace copyright holders and welcomes correspondence from those they have been unable to contact.

ISBN 13: 978-1-315-89347-1 (hbk)
ISBN 13: 978-1-351-07257-1 (ebk)

Visit the Taylor & Francis Web site at http://www.taylorandfrancis.com and the
CRC Press Web site at http://www.crcpress.com

FOREWORD

Life would be awfully dull without flowers. The love of flowers is common to all people all over the world, no matter how different their cultural patterns may be. *Artistically,* flowers play an important role in poetry and in decorative arts for their attractive structure, color and odor in endless variation: for their eternal beauty.

Agriculturally, and perhaps still more so *horticulturally,* flowers play an essential part in the production of all crops. This is, of course, most evident in floriculture, where the flower is the final product. In all fruit crops, whether deciduous or evergreen, and in crops like grains and the fruit-vegetables, the fruit is the final product, but flowers are the indispensable introductory step to fruit formation. For the many crops which are grown for their seeds, including all breeding projects, flowers and fruits cannot be missed for seed formation.

On the other hand, there are crops where flowers and fruits are undesirable — for instance, those which are grown for their roots, stems, or leaves. However, in seed growing and breeding of these crops, flowers again are indispensable.

The above implies that *control of flowering,* whether positively promoting or negatively preventing flowering, is a cultural measure which has to be applied in the growing of most, if not all, crops. No wonder that in several cases more or less sophisticated methods for controlling flowering are rather well known. Quite generally, such empirically developed methods are genetical and ecological, but an endless variation occurs in the requirements for flowering among different species and even among the cultivars of one species.

The human mind is not satisfied by answering the question as to which factors influence flowering — as approached empirically — it also wants to know the *physiological background* of the actions of these factors. This is a much more difficult problem, and disciplines like biophysics, biochemistry, and molecular biology enter the picture. It is self-evident that this approach has made much less progress than the empirical one. However, it has become clear that also in this respect an endless variation occurs among different species and that a general mechanism of action — if it exists — has not been found yet. Be that as it may, the study of flowering, whether empirical or — on a higher level — physiological, contributes to the development of the human mind by the attempt to understand some of the mysteries of Life.

In the foregoing I used "endless variation" no less than three times, and students of flowering, who concentrate on their own plant(s), run the risk of becoming one-sided. It is, therefore, highly important to provide them with a catalogue of the flowering behavior of as many other plants as possible. The present *Handbook* fills the existing gap by bringing together our knowledge of the individual cases. The "Table of Contents" illustrates that it was no simple effort to compose this edition. A. H. Halevy, "Abe" to his friends, has had the courage and the energy to undertake this task. All those who are interested in flowering in some or other way owe him their gratitude.

S. J. Wellensiek

PREFACE

Rise up, my love, my fair one, and come away.
For, lo, the winter is past,
The rain is over and gone;
The flowers appear on the earth;
The time of singing is come,
And the voice of the turtledove is heard in our land;
The fig tree puts forth her green figs,
And the vines in blossom give forth their fragrance.

(Song of Songs 2:10—13)

From antiquity, poets have expressed humanity's association of flowering with spring, renewal, singing, beauty, fragrance, and love. This book deals with the more prosaic aspects of flowering: flower formation and development and the environmental and physiological factors which regulate them.

Several excellent reviews and books on flowering have been published in the last 25 years. These include Lang's[6] and Schwabe's[9] chapters in general encyclopedias of plant physiology, Evan's opening and concluding chapters in his book,[4] the books by Salisbury,[8] Vince-Prue,[11] and Bernier et al.[1,13] and the several review articles in the *Annual Reviews of Plant Physiology*.[2,3,5,7,10,12] With the exception of Evans,[4] these authors have presented a general review of the flowering process and attempted to integrate the data into a unifying theory. Such unifying theories have generally suffered from the disadvantages noted by Evans in the preface to his book;[4] they deal primarily with the earliest events of the flowering process, and they are based on data obtained from a small number of "model" plants. Evans' book, which contains flowering "case histories" of the majority of these species, includes chapters on only 20 plants, of which only one is a woody plant, two are monocotyledons, and none is a gymnosperm. A great wealth of data on the regulation of flower formation and development can be found in the practical literature of agriculture, horticulture, and forestry. Much of this has been often ignored by flowering physiologists. The flowering process is indeed much more diverse than that revealed in the 20 "model" plants presented in Evans' book. It was my aim in planning this book to make a more comprehensive view of the flowering process possible by presenting relevant data from as many plants as possible. This includes the majority of the cultivated plants on which such information is available: field crops, fruits, vegetables, ornamentals, industrial plants, and forest trees, not only of the temperate regions, but also of subtropical and tropical climates. To accomplish this goal I have invited scientists from all over the world to contribute chapters on specific plants or groups of plants. Many of the authors have not only reviewed the available literature, but have also included previously unpublished data. Many of the chapters present the first general review of the flowering process in their specific subject area.

The book deals with all aspects of flowering, including juvenility and maturation, flower morphology, flower induction, and morphogenesis to anthesis. Flower morphogenesis has been taken to include also development of individual flower parts, sex expression, and flower malformations. When possible the authors have attempted to present information on all stages of the flowering process. In many cases, however, this has not been feasible, since little or nothing is known about some of the stages. In most cases the "flowering story" is terminated at anthesis. In some plants, however, flower structure and anthesis are directly related to pollination, and in these cases pollination is also included. In some commercial food crops the description is also extended somewhat beyond anthesis to include important factors in crop production. In many cultivated plants, mostly ornamentals, practical methods for manipulation of flowering are included.

The length of the individual chapters and the emphasis on specific aspects depends in most cases on the availability of experimental information and not on the importance of the plant as a crop or the significance of the physiological stage described. Some important economic plants are absent from this book since little or no information is available on their flowering. In some cases most of the chapter is devoted to a single aspect of flowering, such as juvenility, flower induction, flower development, sex expression, cleistogamy, development of certain flower parts, or flower opening. Some chapters concentrate on physiological aspects, others on ecological, morphological or genetic ones. Other aspects are covered only briefly or even absent, not because they are not important but because they have not been studied in detail.

I am well aware that this book is far from being a comprehensive encyclopedia of flowering, I would greatly appreciate comments from readers on errors found in articles, missing information, and plants not included in the book, whose flowering process have been documented.

The handbook consists of six volumes:

- Volume I — contains general chapters on groups of plants, and individual chapters on plants beginning with the letter A
- Volume II — contains plants of letters B to E
- Volume III — contains plants of letters F to O
- Volume IV — contains plants of letters P to Z
- Volume V — is a supplementary volume and includes chapters on plants not included in Volumes I to IV, listed in alphabetical order.
- Volume VI — will be a second supplementary volume

The merit of the book rests upon the work of the individual authors. I am grateful to them for their efforts, cooperation, and forbearance. I would like to thank my colleagues in the Editorial Board who helped me to select the authors and to review the chapters, and the many other colleagues who helped in reviewing (and sometimes rewriting) specific chapters.

I hope that this handbook will serve as a reference and source book for scientists interested in the flowering process of particular plants, and will draw their attention to the lack of information on important aspects of the flowering process in many important plant species. I also hope that the wealth of information accumulated here will be useful in future attempts to synthesize general theories of the physiology of flowering.

REFERENCES

1. **Bernier, G., Kinet, J. M., and Sachs, R. M.,** *The Physiology of Flowering*, CRC Press, Boca Raton, Fla., 1981, Vol. 1, 149; Vol. 2, 231.
2. **Chouard, P.,** Vernalization and its relation to dormancy, *Annu. Rev. Plant Physiol.*, 11, 191—238, 1960.
3. **Doorenbos, J. and Wellensiek, S. J.,** Photoperiodic control of floral induction, *Annu. Rev. Plant Physiol.*, 10, 147—184, 1959.
4. **Evans, L. T., Ed.,** *The Induction of Flowering. Some Case Histories*, Macmillan, Melbourne, 1969, 328.
5. **Evans, L. T.,** Flower induction and the flowering process, *Annu. Rev. Plant Physiol.*, 22, 365—394, 1971.

6. **Lang, A.,** The physiology of flower initiation, in *Encyclopedia of Plant Physiology,* Vol. 15 (Part I), Ruhland, W., Ed., Springer-Verlag, Berlin, 1965, 1380—1563.
7. **Salisbury, F. B.,** *The Flowering Process,* Pergamon Press, Oxford, 1963, 234.
8. **Salisbury, F. B.,** Photoperiodism and the flowering process, *Annu. Rev. Plant Physiol.,* 12, 293—326, 1961.
9. **Schwabe, W. W.,** Physiology of vegetative reproduction and flowering, in *Plant Physiology, A Treatise,* Steward, F. C., Ed., Vol. VI-A, Academic Press, New York, 1971, 233—411.
10. **Searle, N. E.,** Physiology of Flowering, *Annu. Rev. Plant Physiol.,* 16, 97—118, 1965.
11. **Vince-Prue, D.,** *Photoperiodism in Plants,* McGraw Hill, London, 1975, 444.
12. **Zeevaart, J. A. D.,** Physiology of flower formation, *Annu. Rev. Plant Physiol.,* 27, 321—348, 1976.
13. **Kinet, J. M., Sachs, R. M., and Bernier, G.,** *The Physiology of Flowering,* Vol. 3, CRC Press, Boca Raton, Fla., 1985, 274.

Abraham H. Halevy

PREFACE

Volume V

This supplementary volume of the *Handbook of Flowering* contains chapters that were originally planned for Volumes I to IV, but failed to reach me before the final deadline for these volumes, as well as new chapters commissioned at later dates. In addition, we added two complementary chapters on plants described in previous volumes (Pinaceae and *Rudbeckia*), when it became evident that important information was missing from the earlier chapters.

Among the chapters of this volume are two major contributions on *Rudbeckia* and *Hyoscyamus* by two of the most distinguished authorities on flowering physiology in our generation — Professors Mikhail Chailakhyan and Anton Lang.

We have included a page of errata for Volumes I to IV. I thank my colleagues who drew my attention to the errors, and would appreciate being informed by readers about any further errors that might be detected in the five volumes of the *Handbook.*

While working on the present volume, it became clear to me that this encyclopedic *Handbook* would not be completed with Volume V, and an additional supplementary volume would be needed. Volume VI will be the final volume and will contain chapters on plants not included in Volumes I to V. It will also contain a taxonomic index of all six volumes.

The tentative Table of Contents of Volume VI is as follows: *Acacia, Aeschynanthus, Aglaonema, Agrostemma, Allium* — Ornamental Species, *Alopecurus pratensis, Anigozanthus, Anona, Arisarum, Arum,* Aster pansies, *Atriplex, Banksia, Bellevalia, Bidens radiata, Boronia, Bouteloua, Bromus inermis, Calamintha officinalis, Camellia sinensis, Campanula poscharskyana, Carissa, Caryopteris, Cassia, Chamaecyparis obtusa, Chondrilla juncea, Chrysanthemum segetum, Cirsium, Colchicum tunicatum, Conyza, Coreopsis, Cryptomeria, Cuscuta, Cyanella, Cynoglossum, Datura, Deschampsia, Diospyros, Epilobium, Eustoma grandiflorum, Exacum, Ficus, Gentiana, Gymnaster savatierii, Hieracium, Hydrocharis, Iris ensata, Ixora, Jasminum sambac* and *J. auriculatum, Jasminum grandiflorum, Kleinia, Lagenaria siceraria, Luffa, Lychnis, Lysichitum, Meconopsis, Mimulus, Nicotiana, Nopalxochia, Oliveranthus elegans, Origanum syriacum, Orobanche, Panax, Passiflora, Pentas, Peperomia, Phleum pratense, Picea, Platycodon grandiflora, Poa, Rottboellia exaltata,* Rosaceae — Ornamental Species, *Salvia occidentalis, Schizanthus, Schismus arabicus, Sicyos angulatus, Stokesia laevis, Sternbergia clusiana, Symplocarpus foetidus, Telopea, Themeda australis, Thlaspi, Thuja, Trachelium, Trifolium repens, T. subterraneum, Utricularia, Vicia sativa, Viscaria candida, Xanthorrhoea, Zantedeschia,* and Taxonomic Index to Volumes I through VI.

THE EDITOR

Abraham H. Halevy, Ph.D., is Wolfson Professor of Horticulture and Plant Physiology at the Hebrew University of Jerusalem, Rehovot, Israel.

Dr. Halevy obtained his M.Sc. and Ph.D. degrees from the Hebrew University in 1953 and 1958, respectively.

Dr. Halevy was a Research Fellow at the U.S. Department of Agriculture Research Center at Beltsville, Md. (1958 to 1959) and has been a Visiting Professor at Michigan State University (1964 to 1965) and the University of California, Davis (1970 to 1971, 1976, 1982 to 1984, and 1986).

He has twice received the Alex Laurie award of the American Society of Horticultural Science, and was recently (1983) nominated as a fellow of the society.

Dr. Halevy has published over 200 scientific papers in international journals and has been a guest invited lecturer in numerous symposia around the world. He is currently the Editor-in-Chief of the Israel *Journal of Botany*.

His current research interests are the regulation of flowering and of flower senescence.

CONTRIBUTORS

Paulo de Tarso Alvim Ph.D.
Cocoa Research Center
Scientific Directory of CEPLAC
Itabuna, Bahia
Brazil

Allan M. Armitage, Ph.D.
Department of Horticulture
College of Agriculture
University of Georgia
Athens, Georgia

B. O. Bergh, Ph.D.
Department of Botany and Plant Science
University of California, Riverside
Riverside, California

Jody N. Booze-Daniels, M.S.
Department of Horticulture
Virginia Polytechnic Institute and State University
Blacksburg, Virginia

Rolf Borchert, Ph.D.
Department of Physiology and Cell Biology
University of Kansas
Lawrence, Kansas

K. L. Chadha, Ph.D.
Indian Institute of Horticultural Research
Banglalore, Karnataka State
India

Mikhail Kh. Chailakhyan, Ph.D.
Laboratory of Plant Growth and Development
K. A. Timiryazev Institute of Plant Physiology
U.S.S.R. Academy of Sciences
Moscow
U.S.S.R.

R. M. Davison, Ph.D.
Division of Horticulture and Processing
Department of Scientific and Industrial Research
Auckland
New Zealand

Jerry D. Eastin, Ph.D.
Department of Agronomy
University of Nebraska
Lincoln, Nebraska

C. L. M. van Eijnatten, Dr.ir.
Department of Agronomy
Brawijaya University
Malang
Indonesia

A. R. Ferguson, Ph.D.
Division of Horticulture and Processing
Department of Scientific and Industrial Research
Auckland, New Zealand

Abraham H. Halevy, Ph.D.
Department of Ornamental Horticulture
Hebrew University of Jerusalem
Rehovot
Israel

A. J. Joubert, M.Sc.
Department of Agriculture and Water Supply
Citrus and Subtropical Fruit Research Institute
Nelspruit
South Africa

Avishag Kadman-Zahavi, Ph.D.
Department of Ornamental Horticulture
Agricultural Research Organization
Bet Dagan
Israel

V. G. Kochankov, Ph.D.
Laboratory of Plant Growth and Development
K. A. Timiryazev Institute of Plant Physiology
U.S.S.R. Academy of Sciences
Moscow
U.S.S.R.

Ross E. Koning, Ph.D.
Department of Biological Sciences
Rutgers University
Piscataway, New Jersey

Anton Lang, Dr.Nat.Sci.
Plant Research Laboratory
Michigan State University
East Lansing, Michigan

Luis Carlos Lopes, Ph.D.
Department of Plant Science
Federal University of Viçosa
Viçosa, Minas Gerais
Brazil

Elizabeth M. Lord, Ph.D.
Department of Plant Sciences
University of California, Riverside
Riverside, California

Jon Lovett Doust, Ph.D.
Department of Biology
Amherst College
Amherst, Massachusetts

Robert E. Lyons, Ph.D.
Department of Horticulture
Virginia Polytechnic Institute and State University
Blacksburg, Virginia

Anna Mayers, Ph.D.
Department of Plant Sciences
University of California, Riverside
Riverside, California

Bastiaan J. D. Meeuse, Dr. Tech. Sci.
Botany Department
University of Washington
Seattle, Washington

Teresa C. Minter, M.S.
Department of Botany and Plant Sciences
University of California, Riverside
Riverside, California

Lenis A. Nelson, Ph.D.
Department of Agronomy
Panhandle Extension and Research Center
Scottsbluff, Nebraska

R. N. Pal, Ph.D.
Senior Plant Physiologist
Division of Plant Physiology and Biochemistry
Indian Institute of Horticultural Research
Hessaraghatta Lake
Bangalore, Karnataka
India

R. M. Pandey, D.Phil.
Division of Fruits and Horticultural Technology
Indian Agricultural Research Institute
New Delhi
India

K. Paranjothy, Ph.D.
Biology Division
Palm Oil Research Institute of Malaysia
Selangor
Malaysia

Richard P. Pharis, Ph.D.
Department of Biology
University of Calgary
Calgary, Alberta
Canada

Stephen D. Ross, Ph.D.
Research Laboratory
British Columbia Ministry of Forests
Victoria, British Columbia
Canada

W. W. Schwabe, D.I.C., Ph.D., D.Sc., F.I. Biol.
Department of Horticulture
Wye College
Ashford, Kent
England

Karen G. Shedron
Department of Public Parks and Recreation
Ft. Wayne, Indiana

Dennis P. Stimart, Ph.D.
Department of Horticulture
University of Maryland
College Park, Maryland

Atsushi Takimoto, Ph.D.
Laboratory of Applied Botany
Faculty of Agriculture
Kyoto University
Kyoto
Japan

C. Vonk Noordegraaf, Dr.ir.
Department of Crop Science
Research Station for Floriculture
Aalsmeer
The Netherlands

Thomas C. Weiler, Ph.D.
Department of Floriculture and Ornamental Horticulture
Cornell University
Ithaca, New York

S. J. Wellensiek, Dr. Hort. Sci.
Department of Horticulture
Agricultural University
Wageningen
The Netherlands

Harold F. Wilkins, Ph.D.
Department of Horticulture
University of Minnesota
St. Paul, Minnesota

Herut Yahel, M.Sc.
Department of Ornamental Horticulture
Agricultural Research Organization
Bet Dagan, Israel

Jan A. D. Zeevaart, Ph.D.
MSU-DOE Plant Research Laboratory
Michigan State University
East Lansing, Michigan

TABLE OF CONTENTS

Volume V

*ACTINIDIA DELICIOSA**

En. Kiwifruit; Ch. Mihoutao; Fr. Kiwifruit; Ge. Kiwifruit; It. Actinidia

A. R. Ferguson and R. M. Davison

INTRODUCTION

The kiwifruit of cultivation are large-fruited selections of *Actinidia deliciosa*, little removed in selection terms from wild populations. The genus *Actinidia* (family Actinidiaceae) contains about 50 species of perennial, mostly deciduous vines, occurring mainly in the mountains and hills of southwestern China, although species are found from Siberia and Japan through China to Indochina, India, Malaya, and Indonesia.[45,49] Several of the species have long been used for their edible fruits when collected from the wild — the most important of these are *A. deliciosa* (the kiwifruit) and the closely related *A. chinensis*.

The name "kiwifruit" was originally devised in New Zealand for cultivated plants belonging to the species *A. deliciosa*, but use of the name may become more general. Just as "citrus" and "grape" are collective names which group together cultivars from different species and, even, different genera, so too "kiwifruit" may eventually be used for a wide range of cultivated plants all belonging to the genus *Actinidia*.

Essentially all the available information on the flowering of kiwifruit concerns the cvs grown at present. Most conclusions are based on limited studies in the Auckland and Bay of Plenty districts of New Zealand. It is becoming clear that patterns of flowering can vary considerably in different growing areas and that more comprehensive studies are required in each area.

SYSTEMATICS

Until recently, the kiwifruit has been given the name *Actinidia chinensis* Planch. It was believed that the species, found over a very wide area in China, consisted of a number of variants that were morphologically distinct and occupied separate geographical areas.[13,14,26,34,45-48] Three varieties were described:

1. *Actinidia chinensis* Planch. var. *chinensis*. Fruit almost spherical and covered with soft hairs, often shed when the fruit is ripe, leaving it almost smooth. Three forms had been described.
2. *A. chinensis* var. *deliciosa* A. Chev. (syn. *A. chinensis* var. *hispida* C. F. Liang). Fruit usually long and large and covered with long, persistent hairs. Three forms were recognized, one of which, f. *deliciosa*, included the kiwifruit in cultivation throughout the rest of the world.
3. *A. chinensis* var. *setosa* Li. A variety from Taiwan with hard, spiny hairs on leaf and fruit.

The differences between the varieties *A. chinensis* var. *chinensis* and *A. chinensis* var. *deliciosa* are now considered sufficient to justify treating them as separate species.[34,50,71] The soft-haired variant with smooth-skinned fruit thus retains the name of *A. chinensis* Planch.; the other, with long, stiff hairs, generally persistent on the fruit, becomes *A. deliciosa* (A. Chev.) C. F. Liang et A. R. Ferguson.[50] Thus the kiwifruit takes the name

* Literature search completed August 1, 1984.

A. deliciosa var. *deliciosa*. The taxonomic status of the variant from Taiwan, described by Li[45] as *A. chinensis* var. *setosa* is as yet uncertain, but it should probably be associated with *A. deliciosa*.

ORIGIN AND DISTRIBUTION

The kiwifruit has emerged only relatively recently as a cultivated orchard crop although fruit from wild vines in China have been used for centuries for a variety of purposes. Now there is considerable interest among Chinese horticulturists in the potential of the extensive wild resources of *A. chinensis* and *A. deliciosa*. Annual production in China of fruit from the two species is currently approaching 200,000 t.[71]

Outside of China, all commercial plantings of pistillate kiwifruit throughout the world originated from New Zealand, and all plantings in New Zealand can themselves be traced back to a single introduction of seed from China in 1904.[25,27] This means that only a very small part of the gene pool has so far been exploited. The main pistillate cvs originally identified by Mouat[57] were selected about 50 years ago from very small numbers of seedlings. Of these, 'Hayward' has become by far the predominant cv both in New Zealand and elsewhere because of its markedly superior fruits — they are bigger, of better appearance and flavor, and store better.

New Zealand is still the world center of commercial kiwifruit cultivation, producing in 1983 about 60,000 t of fruit of which 80% would have been exported. However, the crop is now becoming widespread throughout the world, with the most extensive plantings in countries bordering the Mediterranean, notably France and Italy, and in Japan and the U.S. (California). The area and distribution of plantings are changing very rapidly. Furthermore, most kiwifruit orchards, both in New Zealand and in other countries, are less than 10 years old. Vines do not bear heavily until they are 4 or 5 years old and cropping capacity increases thereafter for a number of years.[65] Worldwide production of kiwifruit is increasing rapidly and will continue to do so even if no further plantings are established. Kiwifruit industries will probably develop in other temperate countries, and it is likely that a wide range of cvs and even other *Actinidia* species will eventually be used.

CLIMATIC REQUIREMENTS

A. chinensis and *A. deliciosa* occur naturally in southcentral China, where they are found at forest margins or under the tree canopy in humid, wet, mountainous regions. The two species are separated geographically: *A. chinensis* is found in the warmer districts along the coast and in the eastern provinces, while *A. deliciosa* is found inland in colder areas. When the two species occur together in the one area, they are separated vertically, with *A. chinensis* being found at lower, warmer altitudes.[26,45-48,70]

The preferred climate for kiwifruit cultivation is a comparatively mild, humid one, without extremes of winter and summer temperatures. The cvs of *A. deliciosa* now grown to any extent were originally selected under such climatic conditions in the northern regions of New Zealand. Although it is only recently that kiwifruit plantings have been established in other temperate countries, it is already clear that kiwifruit can acclimatize to conditions that are much more severe than occur in the growing regions of New Zealand. In China itself, in the areas where *A. deliciosa* grows wild, winter temperatures fall well below zero and the top layers of soil can be frozen.[16,40] The main climatic limitations to horticultural production are likely to be damaging winter cold, early autumn or late spring frosts, strong winds, and hot dry summers. Adequate shelter is considered essential for kiwifruit cultivation.[65] Excessively mild winters can delay defoliation and inadequate winter chilling can reduce budbreak and hence the number of flowers.[2,5,51]

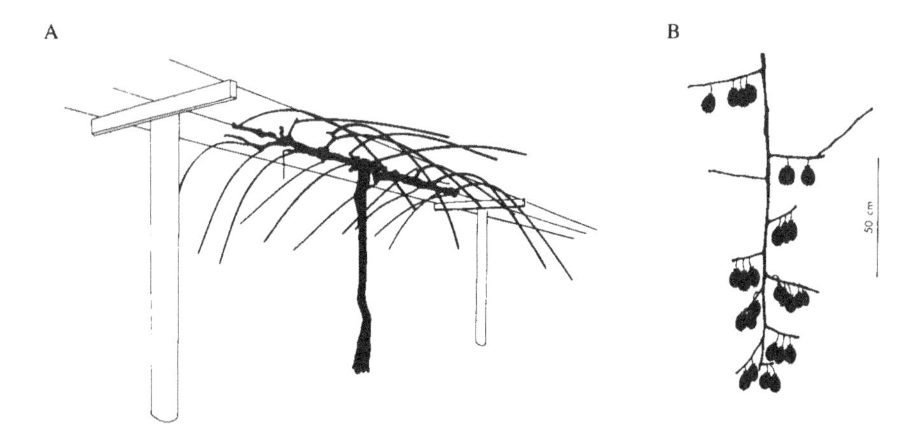

FIGURE 1. (A) Diagram of the structure, after winter pruning, of a mature pistillate vine on a T-bar support system. The long canes tied down on the wires become fruiting arms in the following season; the weight of the fruit and foliage weighs down these arms so that the canopy forms a tunnel. T-bar supports are about 1.8 m high and 5 m apart. (B) Fruiting arm at the end of the second year's growth. Fruit (i.e., flowers) are borne on lateral shoots of the current season's growth. For clarity, the leaves have not been shown.

Under cultivation, kiwifruit flower at about the same time as or slightly later than grapes. This is later in the season than many other fruiting plants and any damage from spring frosts is likely to be on developing flower buds rather than on the flowers themselves. Dormant canes can withstand temperatures down to -10°C without harm, but once spring growth has started, developing flower buds show a tolerance of no more than 1.5 to 2°C frost before death occurs.[37] Susceptibility of the buds to freezing damage does not appear to change during development.

Another consequence of flowering occurring so late in the season is that temperatures are then favorable for bee activity. Pollen transfer from staminate to pistillate vines is a crucial aspect of production.

GROWTH HABIT

In the wild, kiwifruit are usually found growing up through trees or on the edges of forests. Climbing is assisted by the more vigorous shoots twisting tightly around supports and by the interlacing of the shorter shoots.[26] The stem of the young plant is weak, and even mature, cropping plants in cultivation do not become self-supporting, but still require strong structures on which to grow.

Under most systems of orchard management the vine has a main trunk of about 1.8 m and a permanent framework of branches from which fruiting "arms" or canes develop at right angles (Figure 1A). These long, 1-year-old canes are the typical productive units of kiwifruit vines. The axillary buds along the length of such canes break to produce lateral flowering shoots (Figure 1B) and these, in the subsequent year, can in turn produce second-order lateral flowering shoots. Fruiting arms are replaced annually or every 2 or 3 years.

The canopy of a typical 'Hayward' vine occupies a surface area of between 20 and 30 m^2 with some 2000 to 3000 leaves which are large — up to 20 cm wide. A vine would usually carry between 1500 and 2000 fruits: yields of 25 t fruit per hectare are commonly obtained from well-managed orchards, but yields can exceed 50 t/ha.[65]

Shoots of the current season's growth develop, in the main, from buds in leaf axils of shoots formed the previous season. Buds on weaker wood or from the interior of the vine give rise to "terminating" shoots, the tips of which wither and die, leaving a shoot with three to six full-sized leaves (Figure 2C). There is a gradation in length between these and "nonterminating" shoots, which continue to grow for much of the season.[26] If left unpruned, vines rapidly degenerate into a vast tangled mass.

FLOWER MORPHOLOGY

Dioecism

Kiwifruit, like probably all other members of the genus *Actinidia,* are functionally dioecious. Flowers of pistillate plants appear perfect (Figure 2A) but their stamens produce nonviable pollen; flowers of staminate plants (Figure 2B) have a greatly reduced ovary with residual styles. In both types of plant, pollen development is identical up to microspore formation, but in pistillate plants, the nucleus and cytoplasm subsequently atrophy and degenerate.[62]

Dioecism is not absolute. Pistillate plants producing viable pollen have not been detected, but this is not surprising, as any comprehensive screening would be very time consuming. It is much easier to notice the occasional staminate plants that bear fruit. A number of such plants have been collected and bagging experiments indicate that the flowers are self-pollinating and self-setting. The ovaries are only small and stylar development is limited (Figure 3), varying in different parts of the one plant and, it seems, from year to year. The fruit are much too small to be of any commercial value, possibly because the amount and distribution of vascular tissue limits growth;[67] however, they do indicate that development of a hermaphrodite cv might be possible. This would have considerable consequences for kiwifruit cultivation. At present, adequate pollination requires coincident flowering of both staminate and pistillate plants, but different clones vary in time and duration of flowering,[8] and in the apparent viability of the pollen produced.[38] It is therefore necessary to select staminate and pistillate plants that flower at the same time, and plant these in the appropriate proportions and layout. In addition, any factors that affect bee behavior will influence crop yields.

Flowering Shoot

In flowering shoots, both leaves and flowers develop from the one bud. Flowers are borne in small dichasia in the lower leaf axils of flowering shoots. In the following year, flowering shoots are in turn produced from buds in the axils of leaves distal to the last flowering axil. Shoots arising from buds on old wood generally remain purely vegetative for the first year. Flowers occur on both terminating and nonterminating shoots, but are never formed on the shoot apex, which remains vegetative.

The flower dichasium potentially consists of a terminal flower and successive orders of lateral flowers. Frequently, however, these lateral flowers do not develop. Thus 'Hayward' is less floriferous than staminate plants or many other pistillate cvs because there is usually only one flower at each axil, as well as fewer flowering axils per flowering shoot and, in turn, fewer flowering shoots per cane.[8] In most staminate plants and some pistillate cvs (e.g., 'Monty'), flowers occur in groups of three or more at each axil.

Flower Structure

Flowers (Figure 2) are actinomorphic, hypogynous, and usually pentamerous.[45] Pistillate flowers are generally larger than staminate flowers and often contain more perianth parts.[13,35,67] In 'Hayward', flowers at the base of the shoot are generally larger than those distal and also have more perianth parts.[6]

Flower peduncles are about 1.5 cm long, and the pedicels 1 to 2 cm long. The sepals (usually 5) are persistent but develop little after pollination. They are covered by a brownish tomentum. The flower is cup-shaped with 5 (sometimes more) petals about twice the length

A
B
C

FIGURE 3. Flowers of (from left) staminate, fruiting hermaphrodite, and pistillate clones, with petals and stamens removed to show differences in the development of the ovary and styles. The fruiting hermaphrodite was derived from a staminate clone.

and breadth of the sepals. They are broadly obvate but the disposition of the flower parts and the shape of the petals varies with cv[3,4] and age of the flower. When first opened, flowers are white but eventually they become a dirty gold color.

In both staminate and pistillate flowers there are large numbers (up to 200) of hypogynous, free stamens arranged in spirals around the ovary. The long, slender, greenish-white filaments are attached dorsally to the middle of the versatile, yellow anthers which open by longitudinal slits. Anthers appear to be very similar in both staminate and pistillate flowers, but in the latter the filaments are generally much shorter and the stamens, instead of being in an untidy mop, are tucked neatly under the radiating styles.

The pollen is dry and shed in clumps[58,67] which are difficult for pollinating insects, such as bees, to pack into their pollen baskets.[53,63] Pollen from staminate plants is a bright yellow, and that from pistillate plants generally a paler yellow. Pollen grains from staminate plants are prolate to prolate-spherical and tricolpate with crassi-marginate colpi. The exine is thin and without any marked patterning. Grains are binucleate.[24,44,67] Pollen grains from pistillate plants are of roughly the same size, but shriveled in appearance[67] and nonviable, lacking nuclei and protoplasm.

The most obvious difference between staminate and pistillate flowers is in the ovary and styles.[55] In pistillate flowers[33,39,67] the lower parts of the numerous carpels (usually more than 30) fuse to form a syncarpous, superior ovary. Each carpel bears two rows of 10 to 20 ovules attached to the central axis. There are as many styles as there are carpels and these are free, persistent, and radiating, although their disposition varies in different cultivars.[3,4] The distinctive arrangement of the styles (Figure 2A) is the reason for the generic name *Actinidia* (from *actin,* Gk. "ray"). The lower part of the style resembles a tube, the lumen of which (the stylar canal), is loosely filled with transmitting cells. The distal part becomes broadened into a V-shaped stigma, bearing papillae, and with a conspicuous central furrow. The outer epidermis of the ovary is covered by large numbers of multicellular hairs terminating in an elongated cell. The lower parts of the style carry smaller and finer hairs; the upper parts are glabrous.

In staminate flowers, the ovary is much smaller and the ovules, if present, are only rudimentary. The styles are hardly developed at all, forming only a very small tuft on top of the ovary.

No nectar is produced by staminate and pistillate flowers.[15,58,67] Both types of flower have a distinctive fragrance, possibly stronger in staminate flowers.[67]

FLOWER DEVELOPMENT

Flowering Shoot Formation

The primordia that eventually give rise to flowering shoots first develop at the time of budbreak; i.e., a whole year before the differentiated shoot actually emerges. By midsummer, the bud has largely completed development and it enters winter dormancy consisting of 3 to 4 budscales, 2 to 3 transitional leaves, and up to 15 leaf primordia.[8] During winter, the dormant shoot bud is well hidden in the cork of the swollen leaf bases.

Flower Evocation

Some shoots end up purely vegetative, others develop into flowering shoots. Evocation of flowers of pistillate cvs is thought to occur in late summer — in New Zealand, in February or early March. This timing has been determined by sequential deleafing of canes on vines in the field.[18] Generally it does not result in any visible changes, and the potential flowering shoot enters dormancy as a bud containing differentiated vegetative structures and only undifferentiated floral meristems in some leaf axils.[59] Occasionally, the initial stages of differentiation are detectable in autumn.[43] There is thus an unusually long delay between evocation and the development of the flower. In most other deciduous fruiting plants, floral structures are differentiated before buds enter into winter dormancy.[41]

There appear to be no experimental results to indicate the factors that cause buds to become dormant in autumn. As with most deciduous plants, day length and temperature are likely to be important.

Budbreak and Shoot Growth

Development of the overwintering buds recommences in spring, slightly before (European conditions[12,61,72]) or at about the same time (New Zealand conditions) as grape vines. The buds swell and, after 5 to 6 days, begin to open. The leaves unfurl 10 to 15 days later, and, then, after a short lag, rapid shoot growth commences.[9]

Flower Differentiation

Flower differentiation commences in spring as the new shoot emerges.[10,59] The relationship between shoot growth and flower development is shown in Figure 4. The floral meristem starts increasing in size about 10 days before shoot budbreak, and by budbreak (late September/early October, New Zealand), the primordia that develop into the lateral flowers have been formed. Subsequent development of lateral flowers is similar to that of the terminal flower and, although delayed, takes a shorter time. Even so, lateral flowers always open after the terminal flower.

Flower differentiation proceeds rapidly in acropetal direction. At budbreak, the meristematic dome of the terminal flower has already differentiated sepal primordia, and by 10 days after budbreak, when the leaves are unfurling from the shoot bud, the petal and stamen initials have been formed and the stylar lobes start developing. Up to this stage, the flower primordium acts as if it were hermaphroditic but subsequent development results in the production of pistillate or staminate flowers. The ovary and then the locules of pistillate flowers develop and the ovules are formed by about 45 days after budbreak. The stamens differentiate into anthers and filaments. When flower bud opening has reached the stage at

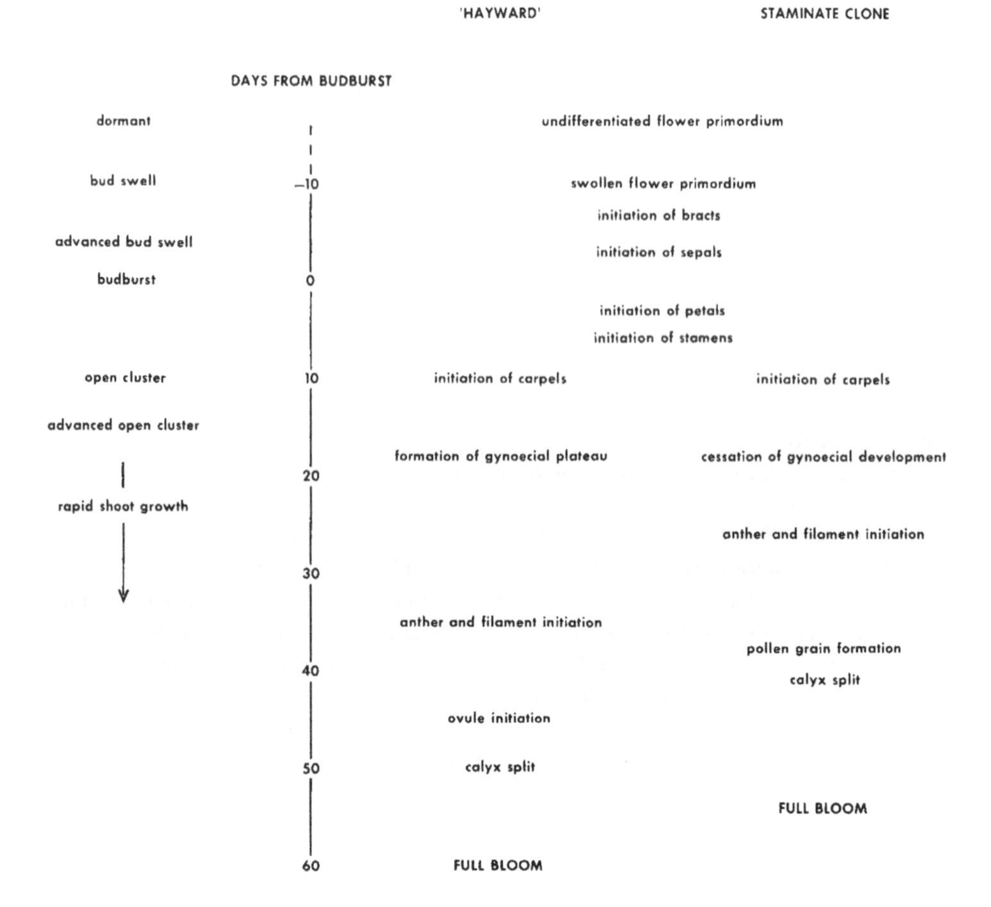

FIGURE 4. Sequence of development of the terminal flower buds of "Hayward" and an unnamed staminate clone (modified from Brundell[10]). The date of budburst varies with cultivar,[9] and the time from budburst to full bloom is also likely to vary.

which the petals are just visible, the stylar lobes separate one from another, the pollen grains separate, the stamen filaments expand rapidly, and the stigmata flatten open. In staminate flowers the ovary and styles develop to only a limited extent.

Flowering

Cultivars differ in their time of flowering. Of the main pistillate cvs named in New Zealand, 'Abbott' and 'Allison' flower earliest, followed by 'Bruno', and then 'Monty' and 'Hayward', with the peak of 'Hayward' flowering 6 to 8 days later than that of 'Abbott'.[28,65] Similar sequences of flowering have been observed in Germany[52] and Italy.[1] Under New Zealand conditions, 'Hayward' flowers at the end of November/beginning of December, about 2 months after shoot budbreak. The main flowering period of an individual vine occupies between 7 and 10 days.

Of the two named staminate clones from New Zealand, 'Matua' usually flowers about a week earlier than 'Tomuri'. As kiwifruit are dioecious, it is necessary for adequate pollination (and hence good cropping) to select staminate clones which coincide in flowering time with pistillate vines. 'Matua' and 'Tomuri' were originally chosen to match early and late flowering pistillate cvs, but 'Tomuri' is often too late for 'Hayward'. The 'Chico male' or

'California male' was selected[5] in California to coincide with 'Chico' (which is now accepted as being 'Hayward').

Pistillate flowers can be pollinated and can set fruit for up to 7 to 9 days after opening.[64,68] This is despite the petals and stamens having turned brown and started to wither within 2 to 3 days. Once a flower has been pollinated, the stigmata also turn brown and wither, whereas unpollinated flowers generally retain unchanged, white stigmata.[68] Staminate flowers produce viable pollen for only the first 2 or 3 days after opening.[64]

Failures in Flower Development

In many parts of New Zealand, fewer than half the overwintering shoot buds of 'Hayward' open in spring. Many on long, vigorous canes or underneath canes remain dormant.[9] This failure of buds to break dormancy is one of the most important factors limiting yield and can result in large parts of the vine being unproductive.

Even if the buds do break, the shoots produced may be only vegetative. In 'Hayward', 10 to 30% of the shoots along a cane normally remain vegetative, but in some instances,[8] more than half are vegetative. This may be due to a failure in evocation or to a failure in subsequent flower differentiation. Most dormant buds along such 1-year-old canes appear to be capable of producing flowering shoots, but whether an individual bud actually does so is thought to depend on the time of budbreak relative to that of other buds.[31]

Many flower buds fail to develop properly, and successive abortion reduces the number of flowers surviving to anthesis. In 'Hayward' almost 40% of flower buds abort during development.[10] In the proximal axils of flowering shoots, buds often stop developing without forming petals. A few of these aberrant primordia do develop and set fruit. The flowers are usually fasciated, owing to the fusion of terminal and lateral flowers, and give rise to fasciated fruit.[10] The incidence of such fruit varies with the season — some vines produce consistently high proportions of fasciated fruit.

FACTORS AFFECTING FLOWERING

All flowers that are set are capable of being carried through to maturity. The most common cause of undersized, unmarketable fruit, usually containing only small numbers of seeds, is inadequate pollination. Most flowers set fruit of some sort unless pollination conditions are very poor or the developing fruit are killed by disease. Furthermore, there is no natural shedding of fruitlets. To achieve the full cropping capacity of a vine, it is therefore important that the number of flowers produced is not limited. Under growing conditions that encourage a much more abundant flowering of 'Hayward', or with cultivation of some of the other pistillate cvs that are more floriferous, the size of the crop may have to be controlled by fruit thinning or more severe winter pruning.[66]

It should be emphasized that we still know very little of the various factors likely to affect flowering in kiwifruit, and any conclusions can be only tentative.

Juvenility

Kiwifruit seedlings may take 3 to 4 years in the field before they overcome a juvenile phase and start to flower. Grafts onto rootstocks may take as little as 18 months to flower heavily (i.e., in the second leaf after grafting), depending on the age and size of the stock and the particular scion cultivar. Staminate clones tend to flower earlier and more heavily than pistillate clones, and the main cv 'Hayward' tends to be less precocious than others such as 'Monty' and 'Abbott'.

Previous Cropping Level

Crop load in one season can influence flower production in the following season. Mature vines can fall into a cycle of alternate heavy and light crops if overcropped in the heavy-bearing years. The phenomenon of biennial cropping occurs in many fruit trees, and the reduced flower production in the off years has been variously attributed to a depletion of nutrients by a previous heavy crop, and/or to the specific inhibition by growth substances such as gibberellic acid produced by the developing seed.[56] In kiwifruit, such factors might be expected to have a strong influence, especially when vines are carrying heavy crops. Evocation occurs about midway between flowering and fruit maturity, at a time when the fruit is still developing. Nevertheless, this tendency towards biennial flowering and cropping does not appear to be deeply entrenched in kiwifruit. A comparison of flower numbers per cane between vines that had borne either heavy or light crops the previous season showed that light-cropping vines produced about 30% more flowers.[19]

Vegetative Growth

In many fruit trees there is likewise a competitive balance between vegetative growth and flowering. It is therefore not surprising that there is evidence of such interactions also occurring in kiwifruit. Removal of the young leaves during the early stages of growth of the flowering shoot resulted in more flowering primordia completing normal development. Young expanding leaves are a major metabolic sink in growing shoots and it was suggested that, after defoliation, nutrients were diverted to the developing flower buds, some of which would otherwise have aborted.[7]

Growth Inhibitors

The possibility of an inverse relationship between vegetative growth and flowering encourages the use of procedures that will reduce vegetative growth; e.g., summer pruning or the use of growth retardants. Sprays of daminozide within the concentration range 1000 to 2000 ppm reduce vegetative growth of kiwifruit.[17] Such treatments increase the number of flowers the following season: autumn applications after the cessation of shoot growth similarly increase flower numbers.[21] The use of such chemical growth retardants is a promising method for achieving an appropriate balance between vegetative growth and flowering in kiwifruit, a vine whose growth under some conditions is extraordinarily vigorous. Daminozide can also be used to delay the flowering of male vines and thus achieve a greatly improved overlap of flowering with that of 'Hayward'.[22]

Girdling

Girdling or cincturing of stems is often used to increase flower production of fruit trees or vines. Kiwifruit vines respond readily to girdling: flower numbers the following season are increased and the time of flowering is advanced.[20] Girdling is effective when carried out on vine stems at times throughout the season from about 1 month after flowering. The procedure will encourage young vines into earlier, heavier flowering and fruiting, but there is always the risk of overcropping. An excess fruit load coupled with a carryover effect of the wounding can reduce cane growth and delay development of the fruiting potential of the vine. The use of girdling treatments therefore requires careful monitoring.

Pruning

Kiwifruit cultivars vary in the effects that different pruning systems have on flower production. With cvs such as 'Abbott' or 'Monty', vine shape can be maintained by relatively severe pruning in which canes are shortened drastically but still flower freely. 'Hayward' responds poorly to such pruning regimes. The selection of fruiting wood is more important with 'Hayward' than with the other cvs; for example, in 'Hayward', "water shoots" (the strong, upright new shoots) seldom flower until 2 years old, whereas in the "standard" cultivars this type of shoot flowers more readily.[28-30]

Shortening of replacement fruiting canes in 'Hayward' markedly reduces the number of flowers produced. Greater numbers of flowers are borne on the flowering shoots that develop at the distal end of the replacement cane and any shortening of the cane reduces even further the number of flowers produced by the remaining part of the cane.[23,42] Long replacement canes are therefore necessary.[36] This requirement has led to the development of training systems such as the T-bar and pergola which allow the retention of long canes, either unpruned or merely tipped, to provide the fruiting framework of the vine. Such techniques are often extended to include a pruning cycle in which there is virtually a complete replacement of fruiting canes each year.[65]

Winter Chilling

In tropical and subtropical regions that have mild winters, vines show poor budbreak extending over a long period, leading to low flower numbers and a much extended blossoming period. As with other plants that require a period of winter cold, winter chilling of kiwifruit increases both the percentage and the uniformity of budbreak, and the number of flowers which develop.

Such field observations on the effects of winter chilling on budbreak and flower development have been confirmed by exposing segments of dormant, 1-year-old canes to various periods of chilling at 4°C.[11] Little chilling was required for budbreak, but as chilling was increased, budbreak was advanced and became more uniform. There was also an increase in the number of flower buds, owing to a reduction in the proportion of floral structures that abort early in development.

South African studies[51,73] suggest that kiwifruit require some 750 to 800 Richardson chilling units[60] for satisfactory flower production. In the kiwifruit-growing regions of New Zealand, there is generally an increase in flower production from north to south (i.e., warmer to colder) within a range of chilling conditions of some 750 to 2000 Richardson chilling units.[54] It is recognized, however, that there are difficulties in quantifying the physiological significance of winter chilling units.

Temperature

It might be expected that higher temperatures would assist evocation of kiwifruit flowers, but no information of any such direct effects has been reported. Temperatures during spring are likely to affect the rate of shoot growth and flower bud development.

Flowering times in the field can vary by as much as 7 days over a relatively short distance. In the Bay of Plenty district of New Zealand, for example, flowering is consistently up to 7 days later at the top of a slope going from about 30 m above sea level to a height of 250 to 300 m over a distance of 8 to 9 km. These differences in flowering time are presumably largely a result of cumulative temperature differences. There can also be differences of some days in flowering time between different parts of a single small orchard.

Light

Maximum flower production is encouraged by exposure of canes to light during the period of evocation, and it has been observed that canes growing in the shade under the canopy bear fewer flowers. Such shaded canes have higher proportions of dead, dormant, or purely vegetative buds and the flowering shoots that are produced have fewer flowers.[32] Flowering was reduced by 50% in a field experiment in which canes were bagged with a black cloth (50% transmission of light) and kept under these shaded conditions from midsummer of the previous season until dormancy.[19] This result is consistent with the conclusion, based on the effects of leaf removal, that flower evocation occurs during mid to late summer. Shading to a similar extent (56 to 60% transmission of light) of unrooted segments of canes over the period of budbreak and flower differentiation had no effect on flower numbers, whereas complete darkness strongly inhibited floral development.[7] Otherwise, there do not appear to have been any more detailed studies of the effects of light quality or intensity on kiwifruit flowering.

Differences between Cultivars

Growers, particularly inexperienced growers, often comment that 'Hayward' is a "difficult" cultivar to manage, and that it is easier to achieve good cropping levels with cvs such as 'Abbott' or 'Monty'. Although these other cvs are inherently more floriferous than 'Hayward',[8] it does seem that they can also adapt more readily to a range of climates; e.g., there is an indication that 'Monty' requires fewer hours of winter chilling.[11,69] These differences, together with the very different pruning regime required to ensure adequate cropping in 'Hayward',[28-30] probably explain why growers are initially more successful with the other cvs. The fruit of 'Hayward' is so clearly superior, however, that it is at present the only cv of any commercial importance.

Other Factors

There are obviously many other factors that are likely to affect flower production in kiwifruit. These cannot be quantified at this stage because the crop has been developed so recently that there is very little definitive information. Moisture deficits can reduce flowering in trees and, probably, similarly affect the flowering of kiwifruit. Mineral deficiencies, especially of nitrogen, could also be expected to reduce flower production. These and other factors that may have an influence on the flowering and, hence, fruit production of kiwifruit require further investigation.

REFERENCES

1. **Alberghina, O.,** Ulteriori osservazioni sull'actinidia nel Catanese, in *Incontro Frutticolo l'Actinidia*, Eynard, I., Ed., Università di Torino, Torino, 1978, 89—92.
2. **Allan, P.,** Cultivation practices for kiwifruit fruit with special reference to South African conditions, *Citrus Suptrop. Fruit J.*, 545, 4—6, 11—13, 1979.
3. **Allan, P.,** Kiwifruit cultivar characteristics, *Citrus Suptrop. Fruit J.*, 586, 16—18, 1982.
4. **Astridge, S. J.,** Cultivars of Chinese gooseberry (*Actinidia chinensis*) in New Zealand, *Econ. Bot.*, 29, 357—360, 1975.
5. **Beutel, J. A. and Costa, G.,** The kiwifruit situation in California, in *Atti del II Incontro Frutticolo sull'Actinidia*, Udine, Cassa di Risparmio di Udine e Pordenone, Italy, 1983, 93—118.
6. **Brundell, D. J.,** Flower Development of the Chinese Gooseberry (*Actinidia chinensis* Planch.) and Some Factors Influencing It, M. Hort. Sc. thesis, Massey University, Palmerston North, New Zealand, 1973.
7. **Brundell, D. J.,** Effect of defoliation, shading, and tipping of shoots on flower bud development and vegetative growth of Chinese gooseberry (*Actinidia chinensis* Planch.), *N.Z. J. Agric. Res.*, 18, 365—370, 1975.
8. **Brundell, D. J.,** Quantitative aspects of flowering in the Chinese gooseberry (*Actinidia chinensis* Planch.), *N.Z. J. Agric. Res.*, 18, 371—374, 1975.
9. **Brundell, D. J.,** Flower development of the Chinese gooseberry (*Actinidia chinensis* Planch.). I. Development of the flowering shoot, *N.Z. J. Bot.*, 13, 473—483, 1975.
10. **Brundell, D. J.,** Flower development of the Chinese gooseberry (*Actinidia chinensis* Planch.). II. Development of the flower bud, *N.Z. J. Bot.*, 13, 485—496, 1975.
11. **Brundell, D. J.,** The effect of chilling on the termination of rest and flower bud development of the Chinese gooseberry, *Sci. Hortic.*, 4, 175—182, 1976.
12. **Camponogara, F.,** Analisi delle esperienze italiene sulla coltivazione dell'Actinidia chinensis Planch., *Agric. Venezie*, 33, 249—256, 1979.
13. **Chevalier, A.,** Sur des lianes fruitières intéressantes: les Actinidia, *Rev. Bot. Appl. Agric. Trop.*, 20, 10—15, 1940.

14. **Chevalier, A.,** Un Actinidia à fruits comestibles intéressant pour la France (A. chinensis Planch. var. deliciosa Chev.), *Rev. Bot. Appl. Agric. Trop.*, 21, 240—244, 1941.
15. **Clinch, P. G. and Palmer-Jones, T.,** Effect on honey bees of azinphos-methyl applied as a pre-blossom spray to Chinese gooseberries, *N.Z. J. Exp. Agric.*, 2, 205—207, 1974.
16. **Cui, Z.-X.,** [Cultivation of kiwifruit in China], in *Mihoutaode Zaipei he Liyong*, Qu, Z.-Z., Ed., Nongye Chubanshe, Beijing, 1981, 95—104.
17. **Davison, R. M.,** Experiments with Alar on Chinese gooseberries, *Orchardist N.Z.*, 44, 443—444, 1971.
18. **Davison, R. M.,** Flowering of kiwifruit, in *Proc. Kiwifruit Semin.*, Tauranga, New Zealand Ministry of Agriculture and Fisheries [Tauranga], 1974, 13—16.
19. **Davison, R. M.,** Some factors affecting flowering and cropping in kiwifruit, in *Proc. Kiwifruit Semin.*, Tauranga, New Zealand Ministry of Agriculture and Fisheries, [Tauranga], 1977, 23—27.
20. **Davison, R. M.,** Girdling of young kiwifruit vines to increase cropping, *Orchardist N.Z.*, 53, 5, 7, 1980.
21. **Davison, R. M.,** unpublished data, 1981.
22. **Davison, R. M.,** Delaying the flowering time of male kiwifruit vines with Alar, a plant growth inhibitor, *Orchardist N.Z.*, 57, 133, 1984.
23. **Davison, R. M.,** unpublished data, 1984.
24. **Erdtman, G.,** *Pollen Morphology and Plant Taxonomy. Angiosperms (An Introduction to Palynology. I,* Hafner, New York, 1971, 34—35. (Corrected reprint of 1952 edition.)
25. **Ferguson, A. R.,** E. H. Wilson, Yichang and the kiwifruit, *Arnoldia (Boston)*, 43(4), 24—35, 1983.
26. **Ferguson, A. R.,** Kiwifruit: a botanical review, *Hortic. Rev.*, 6, 1—64, 1984.
27. **Ferguson, A. R. and Lay Yee, M.,** Kiwifruit (*Actinidia chinensis* var. *hispida*), in *Plant Breeding in New Zealand*, Wratt, G. S. and Smith, H. C., Eds., Butterworths (N.Z.), Wellington, 1983, 111—116.
28. **Fletcher, W. A.,** Growing Chinese Gooseberries, Bull. 349, New Zealand Department of Agriculture, Government Printer, Wellington, 1971.
29. **Ford, I.,** Chinese gooseberries — success lies in pruning, *N.Z. J. Agric.*, 122(4), 43—45, 1971.
30. **Ford, I.,** Training and pruning Chinese gooseberries, *N.Z. Gardener*, 28(1), 18—20, 1971.
31. **Grant, J. A. and Ryugo, K.,** Influence of developing shoots on flowering potential of dormant buds of *Actinidia chinensis, HortScience*, 17, 977—978, 1982.
32. **Grant, J. A. and Ryugo, K.,** Influence of within-canopy shading on photosynthesis and fruit characteristics of kiwifruit (*Actinidia chinensis* Planch.), in *Atti del II Incontro Frutticolo sull'Actinidia*, Udine, Cassa di Risparmio di Udine e Pordenone, Italy, 1983, 369—387.
33. **Guédès, M. and Schmid, R.,** The peltate (ascidiate) carpel theory and carpel peltation in *Actinidia chinensis* (Actinidiaceae), *Flora (Jena)*, Abt. B 167, 525—543, 1978.
34. **Gui, Y.-L.,** [Comparative morphological observations of (*Actinidia chinensis* Planch. var. *chinensis* and *A. chinensis* Planch. var. *hispida* C. F. Liang], *Acta Phytotaxon. Sin.*, 19, 304—307 + 2 plates, 1981.
35. **Habart, J. L.,** La baie de l'*Actinidia chinensis* Planch. var. *chinensis, Fruits*, 29, 191—207, 1974.
36. **Hawes, L. and Lupton, T.,** Setting out a kiwifruit orchard, *Proc. Kiwifruit Pollination Semin.*, Hamilton and Tauranga, New Zealand Ministry of Agriculture and Fisheries, Hamilton, 1982, 11—20.
37. **Hewett, E. W. and Young, K.,** Critical freeze damage temperatures of flower buds of kiwifruit (*Actinidia chinensis* Planch.), *N.Z. J. Agric. Res.*, 24, 73—75, 1981.
38. **Hopping, M. E.,** Kiwifruit pollination: influence of male clones, *Proc. Kiwifruit Semin.*, Tauranga, New Zealand Ministry of Agriculture and Fisheries, [Tauranga, N.Z.], 1981, 21—25.
39. **Hopping, M. E. and Jerram, E. M.,** Pollination of kiwifruit (*Actinidia chinensis* Planch.): stigma-style structure and pollen tube growth, *N.Z. J. Bot.*, 17, 233—240, 1979.
40. Hunan Agricultural Institute, [Initial survey reports on kiwifruit resources in Hunan], in *Mihoutaode Zaipei he Liyong*, Qu, Z.-Z., Ed., Nongye Chubanshe, Beijing, 1981, 105—133.
41. **Jackson, D. I. and Sweet, G. B.,** Flower initiation in temperate woody plants, *Hortic. Abstr.*, 42, 9—24, 1972.
42. **Lamiani Mignani, I., Poma Treccani, C., and Cattaneo, D.,** Influenza del tipo di potatura sul germogliamento dell' Actinidia chinensis Pl., in *Atti del II Incontro Frutticola sull'Actinidia*, Udine, Cassa di Risparmio di Udine e Prodenone, Italy, 1983, 241—248.
43. **Lamiani Mignani, I., Poma Treccani, C., and Youssef, J.,** Aspetti della differenziazione a fiore nella Actinidia chinensis Pl., in *Atti del II Incontro Frutticolo sull'Actinidia*, Udine, Cassa di Risparmio di Udine e Pordenone, Italy, 1983, 333—341.
44. **Lechner, S.,** Anatomische Untersuchungen über die Gattungen Actinidia, Saurauia, Clethra und Clematoclethra mit besonderer Berücksichtigung ihrer Stellung im System, *Bot. Zentralbl.*, 32, 431—467, 1915.
45. **Li, H.-L.,** A taxonomic review of the genus Actinidia, *J. Arnold Arbor. Harv. Univ.*, 33, 1—61, 1952.
46. **Liang, C.-F.,** [Classification of *Actinidia chinensis* Planch.], *Acta Phytotaxon. Sin.*, 13, 32—35 + plate, 1975.
47. **Liang, C.-F.,** [New taxa of the genus *Actinidia* Lindl.], *Guihaia*, 2, 1—6, 1982.
48. **Liang, C.-F.,** [An addition to the infraspecific taxa of *Actinidia chinensis* Planch.], *Acta Phytotaxon. Sin.*, 20, 101—104, 1982.

49. **Liang, C.-F.,** [On the distribution of *Actinidia* species], *Guihaia*, 3, 229—248, 1983.
50. **Liang, C.-F. and Ferguson, A. R.,** Emendation of the Latin name of *Actinidia chinensis* Pl. var. *hispida* C. F. Liang, *Guihaia* 4, 181—182, 1984.
51. **Lötter, J. de V.,** The suitability of the different areas in Southern Africa for the commercial production of Hayward kiwifruit, in *Atti del II Incontro Frutticolo sull'Actinidia*, Udine, Cassa di Risparmio di Udine e Pordenone, Italy, 1983, 119—138.
52. **Maurer, K. J.,** Actinidia, deren Arten, Sorten und Anbaupraxis, *Mitt. Rebe Wein Obstbau Fruchteverwertung*, 21, 298—403, 1971.
53. **Maurer, K. J.,** Neue blütenbiologische Erkenntnisse bei Kiwi, *Mitt. Obstau*, 20, 71—73, 76, 77, 1976.
54. **McAneney, K. J. and Kerr, J. P., Eds.,** Environmental Inputs to Agronomic Research: Guidelines, Agricultural Research Division, New Zealand Ministry of Agriculture and Fisheries, Wellington, 1984.
55. **McGregor, S. E.,** Insect pollination of cultivated crop plants, *U.S. Dep. Agric. Agric. Handb.*, 496, 146—148, 1976.
56. **Monselise, S. P. and Goldschmidt, E. E.** Alternate bearing in fruit trees, *Hortic. Rev.*, 4, 128—173, 1982.
57. **Mouat, H. M.,** New Zealand varieties of yang-tao or Chinese gooseberry, *N.Z. J. Agric.*, 97, 161, 163, 165, 1958.
58. **Palmer-Jones, T. and Clinch, P. G.,** Observations on the pollination of Chinese gooseberries variety 'Hayward', *N.Z. J. Exp. Agric.*, 2, 455—458, 1974.
59. **Polito, V. S. and Grant, J. A.,** Initiation and development of pistillate flowers in *Actinidia chinensis*, *Sci. Hortic.*, 22, 365—371, 1984.
60. **Richardson, E. A., Seeley, S. D., and Walker, D. R.,** A model for estimating the completion of rest for 'Redhaven' and 'Elberta' peach trees, *HortScience*, 9, 331—332, 1974.
61. **Rivals, P.,** Notes biologiques et culturales sur l'Actinidia de Chine (Actinidia sinensis Planchon), *J. Agric. Trop. Bot. Appl.*, 11, 75—83, 1964.
62. **Rizet, G.,** Contribution a l'étude biologique et cytologique de l' *Actinidia chinensis*, *C.R. Seances Soc. Biol. Paris*, 139, 140—142, 1945.
63. **Sale, P. R.,** Subtropical fruit: pollination: an introduction, *N.Z. Min. Agric. Fish. Aglink*, HPP 97, 1978.
64. **Sale, P. R.,** Kiwifruit pollination. Male to female ratios and bee activity, *N.Z. Min. Agric. Fish. Aglink*, HPP 233, 1981.
65. **Sale, P. R.,** Kiwifruit Culture, Williams, D. A., Ed., Government Printer, Wellington, New Zealand, 1983.
66. **Sale, P. R.,** Management of kiwifruit for optimum performance, in *Atti del II Incontro Frutticolo sull'Actinidia*, Udine, Cassa di Risparmio di Udine e Pordenone, Italy, 1983, 139—148.
67. **Schmid, R.,** Reproductive anatomy of Actinidia chinensis (Actinidiaceae), *Bot. Jahrb. Syst. Pflanzengesch. Pflanzengeogr.*, 100, 149—195, 1978.
68. **Stevens, I. and Forsyth, M.,** Pollination, fertilisation, and fruit development, in *Proc. Kiwifruit Pollination Semin.*, Hamilton and Tauranga, M. Reid, Ed., New Zealand Ministry of Agriculture and Fisheries, Hamilton, [1982], 2—10.
69. **Weet, C. S.,** Variety selection important to kiwi production, *Avocado Grower*, 2(10), 44—47, 51, 1978.
70. **Zhang, R.-M.,** [Selected varieties of kiwifruit], in *Mihoutaode Zaipei he Liyong*, Qu, Z.-Z., Ed., Nongye Chubanshe, Beijing, 1981, 134—141.
71. **Zhu, H.-Y.,** *Zhonghua Mihoutao Zaipei*, Shanghai Kexue Jishu Chubanshe, Shanghai, 1983.
72. **Zuccherelli, G.,** Avversità dell'Actinidia, *Inf. Fitopatol.*, 29(5), V/29—V/22, 1979.
73. **Zyl, H. J. van, Bester, G. W. J., and Joubert, A. J.,** Cultivation of kiwifruit, *Decid. Fruit Grower*, 29, 18—24, 1979.

ANACARDIUM OCCIDENTALE

En. Cashew; Fr. Anacardier; Ge. Kaschubaum; Sp. Maranon

C. L. M. van Eijnatten

INTRODUCTION

Cashew trees are grown for their kernels, the embryo of the seed contained in the nut. Minor products are derived from the shell and from the fruitlike swollen pedicel. Cashew originates in South America, but has been distributed widely to such an extent that the major producing areas now occur in eastern Africa and south India. World production is estimated at 535 thousand tons of which only 16% originate in South America.

Cashew (*Anacardium occidentale* L.) belongs to the family Anacardiaceae, and the genus *Anacardium* contains several tropical American species.[1] The cashew develops into a dome-shaped tree with an evergreen canopy up to 15 m high; the stem often develops in a tortuous way, especially under suboptimal conditions. The crop is cultivated mainly between 15°N and 15°S, although the limits of cultivation apparently are around 26°N or S. The elevation should be low, although in exceptional circumstances cultivation succeeds at up to 1000 m of altitude (west Zambia, south Tanzania).

Optimal temperatures for cashew are around 27°C, and rainfall ranging from 1000 to 2000 mm is adequate, providing a period of 5 to 6 dry months interrupts the rains. However, because of its hardiness and its extensive root system cashew is often cultivated in more marginal areas, particularly in respect of rainfall.

FLORAL MORPHOLOGY

The inflorescence is a drooping panicle or thyrsus of varying shape. Its main axis is indeterminate, but secondary and higher order axes are monochasial cymes.

On appearance, lateral buds of the main axis produce 5 to 11 branches from which the cymes emanate. Some weeks later, flower buds are produced. At anthesis of the first flower, the panicle is about 5 weeks old. The flowers develop and open gradually over a period of 5 or 6 weeks. One inflorescence may produce many flowers during this period. This number may rise to 1100 flowers per inflorescence.

The flowers of cashew are of 2 types as the plant is andromonoecious. On each of the cymes making up the panicle the first flower to open is often hermaphrodite, and lateral buds develop into male flowers. For this reason, hermaphrodite flowers tend to appear during the early part (first 3 weeks) of the flowering period of any panicle. The proportion of hermaphrodite flowers is, therefore, higher in the early part of the flowering season. Considered over time, the proportion of hermaphrodite flowers often is in the order of 12 to 16%. This may, however, drop to as low as 3%.

The flowers are small, white in color and aromatic just after opening. After a few days, the flower turns pink and finally turns brown and dries up. The flower has a short pedicel of 2 mm. It is pentamerous, as a rule. Five narrow, green, pubescent sepals 3 to 5 mm long alternate with 5 lanceolate petals 1 to 1.5 cm in length (Figure 1). After anthesis the petals are reflexed. Their pale green to white color with red stripes gradually changes to red after some days. The stamens are upright and have parallel filaments crowned with globular anthers. These are basifixed bilobed and dehiscent lengthwise; the color is red, but turns grey at the time of dehiscence. Usually one stamen is much larger than the others; sometimes

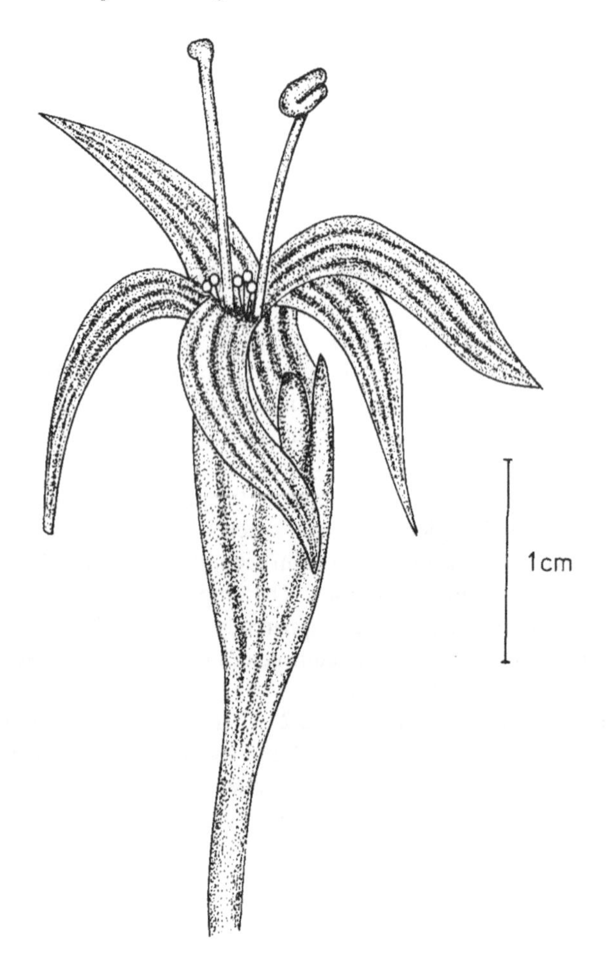

FIGURE 1. Diagram of hermaphrodite flower of cashew.

a large stamen is absent, or there may be 2, 3, or even 4 of such stamens, well exerted above the corolla. The other stamens, usually 9, are short, their anthers remaining within the flower. The large, stout stamen carries an anther, which is also larger in comparison with those of the remaining stamens. Both types, apparently, produce viable pollen.[3,4] The ovary is superior, monocarpellary and unilocular. In the hermaphrodite flowers it contains one ascending, anatropous and parietal ovule. At the distal end of the ovary the style emerges, which ends in a slightly swollen pistil. In the hermaphrodite flowers the gynoecium is of the same size as the large stamen or longer (6 mm or more). In male flowers the gynoecium remains smaller than the large stamen, i.e., 0.1 to 6.0 mm.

GROWTH AND FLOWERING PATTERN

The seedling develops vegetatively for 2 or 3 years. Under favorable conditions the young plants may produce their first inflorescence during the 2nd year, but usually this happens in the 3rd year of growth. However, even 1st-year seedlings have been reported to produce inflorescences at the age of 6 months.

Shoot growth may occur throughout the year, especially when rainfall is well distributed. In most areas, however, one or two distinct dry periods occur, which also lead to a periodicity in growth. Subsequent to a pronounced dry season many young shoots start to develop. The apical bud of these flushes gives rise to an inflorescence some 3 to 4 months onwards, terminating the extension growth of the shoots. Subsequent flushing occurs much more partially and is less pronounced. Such flushes arise from lateral buds just below the inflorescence of the earlier flush.

Flowers may occur on these additional shoots, but often they remain vegetative, probably because of the presence of inflorescences and/or developing fruits elsewhere in the tree canopy. If they do not produce flowers at the end of the extention growth of these shoots, such apical buds may produce an inflorescence during a subsequent flowering season without further extension growth of that flush. It is noteworthy that the occurrence of two pronounced dry seasons may lead twice to a burst of flowering, provided these periods are well-spaced in time.

Flowering starts with anthesis of the first flowers on a first panicle. Gradually the number of flowering panicles increases, as well as the number of simultaneously open flowers per inflorescence. The flowering intensity of a tree, therefore, increases. Of the flowers, 50% are produced during a period of 4 to 6 weeks from about a month after appearance of the first flower. The whole process of anthesis from the first to the last flowers takes some 3 months. The peak period varies from one tree to another and the main flowering season of cashew in any one area usually is spread over 4 months.

Despite the regular flowering season individual trees may come into bloom at different times of the year. This "out of season" production of flowers is likely to be related more to microclimatical/ecological conditions, particularly access to water, than to a different genetic make-up of the trees.

Individual flowers tend to open in the morning hours, continuing to some extent into the early hours of the afternoon. The stigma appears receptive from before anthesis till the following day. Anther dehiscence occurs more rapidly at higher than at lower temperatures. Longer time between anthesis and anther dehiscence reduces the probability of self pollination. The pollen kernels are sticky and insect pollination is, therefore, likely to be required. Also, the aromatic scent of cashew flowers points in this direction.

REFERENCES

1. **Ohler, J. G.,** Cashew, Communication 71, Department of Agricultural Research, Royal Institute for the Tropics, Amsterdam, 1969, 260 pp.
2. **Agnoloni, M. and Giuliani, F.,** Cashew Cultivation, Library of Tropical Agriculture, Ministry of Foreign Affairs, Istituto Agronomico, Florence, 1969.
3. **Northwood, P. J.,** Some observations on flowering and fruitsetting in the cashew, *Anacardium occidentale* L., *Trop. Agric. Trinidad,* 43(1), 35—42, 1966.
4. **Copeland, H. F.,** Observations on the reproductive structures of *Anacardium occidentale, Phytomorphology (India),* 11, 314—325, 1961.

AQUILEGIA × HYBRIDA

En. Columbine; Fr. Colombine; Ge. Akelie; Sp. Columbino

Thomas C. Weiler and Karen G. Shedron

INTRODUCTION

The Ranunculaceae offers a number of ornamental flowering plants. Among them is *Aquilegia × hybrida* Sims. About 70 *Aquilegia* species are native to the temperate and mountain regions of the northern hemisphere.[1] *Aquilegia caerulea, A. longissima, A. chrysantha, A. canadensis,* and *A. vulgaris* have been interbred to create the hybrids commonly grown in gardens today.[6] Cultivars with hooked spurs usually indicate hybridization of the European *A. vulgaris,* and those with long, straight spurs, hybridization of the North American *A. caerulea* and *A. chrysantha.*[1] The derivation of the common name columbine comes from "columba", Latin for dove, which the spurred flowers resemble. The genus name, *Aquilegia,* is derived from the Latin "aquilla", for eagle, alluding to the claw-like curve at the end of each spur of *A. vulgaris.*[13]

Columbine flowers consist of 5 petaloid sepals and 5 true petals, each with a long, hollow backward-projecting spur which varies in length from 1.5 cm to as much as 15 cm depending on cv.[5] The sepals may be the same color as the petals or a contrasting color. Flowers are 4 to 10 cm across, and have many stamens and 5 pistils.[1] Their colors, mostly pastel, include many tones of red, pink, yellow, blue, and lavender as well as white. The leaves are two to three times ternately divided, and the fruit is a many-seeded follicle.

'McKana's Giant' (Figure 1) is perhaps the most popular of the hybrid types due to its large flowers. Plants may grow 0.8 to 1 m tall. Cultivars such as 'Dragonfly' and 'Fairyland' are dwarf, reaching only to 0.4 to 0.5 m tall.[12] Seeds sown in the summer will produce flowering-sized plants by the next spring. Otherwise they rarely flower the first year.[5,7] Plants are propagated from seed or by division in the spring. In the northern hemisphere columbines often bloom May to June, depending upon cultivar and location. Plants tolerate full sun although light shade prolongs the flowering season.[6] The seed pods are removed as soon as the flowers fade. A moist, well-drained soil is ideal. Columbines often are short-lived and may need to be replaced every few years.[5,7] Ignateva[8] described morphogenesis of the plant, noting deterioration of parts of the vegetative tissue with age.

PLANT AGE AND VERNALIZATION

Shedron and Weiler[12] studied means of regulating *A. hybrida* development. Of young plants with 12 or more leaves, 83 to 100% were induced to flower by cold treatment (4.5°C) whereas only 0 to 50% of the plants with 7 to 8 leaves flowered. For induction of mature plants (Figure 2), 10 to 12 weeks of cold were required, and this varied somewhat between cultivars. For example, after 12 weeks of cold storage, 50% of 7-leaf 'McKana's Giant' plants flowered, while no plants of 'Fairyland' or 'Crimson Star' flowered. For 12- to 15-leaf plants receiving 12 weeks of cold storage, 100% of 'McKana's Giant' and 'Crimson Star' plants flowered while 83% of the 'Fairyland' plants flowered.

As in other vernalized plants, the presence of leaves was not necessary during cold treatment. In fact, leaves present during storage abscised when growth resumed at warmer temperatures.

While the effects of temperatures between 13 and 18°C on growth rate were minor, minimally vernalized plants flowered 100% only at 13°C, indicating either possible devernalization by 18°C or vernalization by 13°C.

FIGURE 1. *Aquilegia* × *hybrida*.

PHOTOPERIOD AND GROWTH

During growth after various storage treatments, long photoperiods of 18 hr created by 7 to 18 μmol m^{-2} sec^{-1} incandescent lighting at night did not replace cold induction; rather, they promoted growth to the visible bud stage. After 12 weeks of vernalization, plants grown in long photoperiods required only 79% the time to reach the visible bud stage required by short-photoperiods-grown plants. The negative consequence of this treatment was elongated leaf petioles. Daminozide (2500 to 5000 mg/ℓ) or ancymidol (50 to 100 mg/ℓ) treatments counteracted the effects of incandescent light while chlormequat (1500 to 5000 mg/ℓ) was ineffective. Chemical growth-retarding treatments did not affect rate of flowering. LaRoche[11] was able to suppress leaf enlargement by using small containers for plant growth. Thus plant flowering and growth may be regulated for specific gardening, botanical display, and potted plant use.

IN VITRO CULTURE OF FLOWER BUDS

In vitro differentiation of excised *A. formosa* flower buds has been studied extensively.[2-4,14,15] Although anthesis was not achieved when buds were excised before or just after carpel initiation, stamens, stamenodia, and carpels initiated and partially differentiated on a basal medium of mineral nutrients, sucrose, vitamins, and amino acids as well as the hormones IAA, GA, and kinetin. Coconut milk was an integral part of the medium in early studies,[2-4,14] and was eventually replaced by altering the medium and notably raising the level of kinetin to 10^{-6} to 10^{-7} *M*.[15] Thus in vitro floral bud differentiation noted in earlier studies also was achieved in a fully defined medium. Further, kinetin was found to be essential for carpel initiation; GA for sepal, petal, and carpel differentiation; and GA and kinetin for stamen differentiation. At 10^{-5} *M* kinetin, primordia destined to become stamenodia differentiated into carpels, an example of hormonally altered floral development.[15]

FIGURE 2. Cold-stored *Aquilegia* × *hybrida* 'Crimson Star' (left) and untreated (right).

In other studies[9,10] bisected floral buds regenerated the damaged apex after an initial 2 days of starch accumulation. Stored starch was depleted days 3 to 7 after bisection as apex integrity was reestablished.

REFERENCES

1. **Bailey Hortorium Staff,** *Hortus Third,* Macmillan, New York, 1976.
2. **Bilderback, D. E.,** The effect of amino acids upon the development of excised floral buds of *Aquilegia, Am. J. Bot.,* 58, 203—208, 1970.
3. **Bilderback, D. E.,** The effects of hormones upon the development of excised floral buds of *Aquilegia, Am. J. Bot.,* 59, 525, 1972.
4. **Bilderback, D. E., Karpoff, A. J., and Tepfer, S. S.,** Development of excised floral buds of *Aquilegia:* the coconut-milk problem, *Am. J. Bot.,* 55, 1042—1046, 1968.
5. **Crockett, J. U.,** *Perennials,* Time-Life, New York, 1972.
6. **Cumming, R. W. and Lee, R. E.,** *Contemporary Perennials,* Macmillan, New York, 1960.
7. **Davis, K. C.,** Aquilegia, in *Cyclopedia of Horticulture,* Bailey, L. H., Ed., Macmillan, New York, 339—343, 1942.

8. **Ignateva, I. D.,** Morphogenesis in *Aquilegia vulgaris,* Bot. Zh. *(Leningrad),* 49, 358—371, 1964.
9. **Jensen, L. C. W.,** Experimental bisection of *Aquilegia* floral buds cultured *in vitro.* I. The effect on growth, primordia initiation and apical regeneration, *Can. J. Bot.,* 49, 487—493, 1971.
10. **Jensen, L. C. W.,** Experimental bisection of *Aquilegia* floral buds cultured *in vitro.* II. Cytological changes following bisection, *Can. J. Bot.,* 50, 1611—1615, 1972.
11. **LaRoche, G.,** The effects of restricting root growing space, decreasing nutrient supply, and increasing water stress on the phenetics of *Aquilegia canadensis* L. (Ranunculaceae), *Bull. Torrey Bot. Club,* 107, 220—231, 1980.
12. **Shedron, K. G. and Weiler, T. C.,** Regulation of growth and flowering in *Aquilegia* × *hybrida* Sims., *J. Am. Soc. Hortic. Sci.,* 107, 878—882, 1982.
13. **Shosteck, R.,** *Flowers and Plants,* Quadrangle, New York, 1974.
14. **Tepfer, S. S., Greyson, R. I., Craig, W. R., and Hindman, J. L.,** *In vitro* culture of floral buds of *Aquilegia, Am. J. Bot.,* 50, 1035—1045, 1963.
15. **Tepfer, S. S., Karpoff, A. J., and Greyson, R. I.,** Effects of growth substances on excised floral buds of *Aquilegia, Am. J. Bot.,* 53, 148—157, 1966.

ASCLEPIAS TUBEROSA

En. Butterfly Weed, Pleurisy Root, Milkweed; Fr. Asclepiade tubereuse; Ge. Schwalbenschwanz

Robert E. Lyons

INTRODUCTION

The genus *Asclepias* falls within the tribe Asclepiadeae, subfamily Cyanchoidae, and family Asclepiadaceae, one of the most advanced of the dicotyledonous families.[6] The Asclepiadaceae has approximately 220 genera and 2000 species distributed worldwide, primarily in tropical and subtropical regions.[1] Although the ornamental plant industry has developed the potential of many genera within this family (e.g., *Stephanotis, Hoya* and *Ceropegia*), the genus *Asclepias* is better known for its weedy characteristics. However, one member of this reputably problematic genus, *A. tuberosa,* is gaining interest as an outdoor garden ornamental,[8] and, recently, as a cut flower.

A. tuberosa L. is native to eastern Canada, northern Mexico, and three fourths of the U.S. With the exception of the state of Arizona, the area west of the Rocky Mountains is sparsely populated by this species, especially the Pacific coastlines of California, Oregon, and Washington.[5] Seed is shed in the fall and requires chilling temperatures prior to germination the following May or June, coinciding with shoot emergence from dormant buds on the roots of established plants. The clumps of deep-set, tuberous roots most frequently grow in well-drained prairies, meadows, and land which has been disturbed and unreclaimed by man. Although full sunlight favors vigorous growth, the species has also adapted to the partial shade of woodland borders.

The type species has an erect, pubescent stem which attains a height of 0.5 to 1 m and branches profusely to not at all. Unlike many other species within this family, the sap is clear rather than milky. Leaves are alternate, linear to lanceolate or oblanceolate, hairy, 5 to 10 cm long, and sessile to very short petioled.[1,5] The apex of the main stem often divides into a triplet of branches during floral initiation.[13]

Bailey[1] recognizes two varieties of the species: *A. tuberosa* var. *tuberosa,* which has abruptly pointed leaves due to a distal widening of the blade and grows chiefly east of the Appalachian Mountains; and *A. tuberosa* var. *interior* (Woodson) Shinners, which occurs in the western range of the species and possesses leaves with a distal blade widening to give a tapered appearance. According to Kartesz and Kartesz[7] the above varieties have been elevated to subspecies status. The same authors recognize two other subspecies: *A. tuberosa* ssp. *terminalis* Woods. and *A. tuberosa* ssp. *rolfsii* (Britt. ex Vail) Woods.

The most notable characteristic of *A. tuberosa,* and the entire genus, is the occurrence of pollen grains in a sticky mass called the pollinium. This feature is paralleled only by the Orchidaceae, the most advanced of the monocot families.[6]

FLORAL MORPHOLOGY

Bookman[3] has provided the most recent and comprehensive account of *Asclepias* floral morphology. Her study examined *A. speciosa,* a widespread species in the western U.S. with a floral morphology representative of the entire genus. The evolution of *Asclepias* to such an advanced level has resulted in an extremely complex flower structure which displays much organ fusion and specialization. Each flower consists of 5 sepals surrounding 5 petals to compose the perianth. The stamens have fused to the head of the stigma to form a conspicuous central structure called the gynostegium. The cucullus is a unique floral structure composed of two distinct outgrowths of the staminal filament, the hood and horn, and each resembles its namesake (Figure 1). The horn is partially and incompletely wrapped by the hood and frequently functions to obstruct the footwork of visiting pollinators. This usually results in pollinium removal or insertion as the visitor awkwardly maneuvers in search of nectar. The 5 cuculli which surround the gynostegium are collectively known as the corona.

The gynoecium consists of 2 distally fused pistils which terminate in a single stigmatic head, which is not the site of pollen reception (Figure 1). Instead, each of the 5 true stigmatic surfaces is partially enclosed within a fissure (Figure 2) created by the lateral wing-like projections of adjacent anthers.[3]

The androecium is composed of 5 fused stamens. Each bears a bilocular anther which encloses 2 separate pollinia, each containing a mass of approximately 500 pollen grains. The translator arm is a highly specialized appendage which connects 2 pollinia from separate but adjacent anthers. Both pollinium-translator arm complexes are joined by a common bead-like body of tissue, the corpusculum, which is situated above each true, receptive stigmatic surface (Figure 2). The entire collection of 2 pollinia, 2 translator arms, and corpusculum is called the pollinarium (Figure 3). This complex set of floral parts creates a unique relationship between pollinator and pollen donor/acceptor.[12]

The inflorescence is umbelliform and is located at terminal and axillary positions along the stem.[5]

JUVENILITY AND PHOTOPERIODISM

Very little can be said about the flowering mechanism of *A. tuberosa,* in part due to the scant attention given the species in research efforts. The effect of photoperiod has been given some attention.

A. tuberosa is a day-neutral plant which commonly remains vegetative for at least 1 year following germination. However, despite the inability of photoperiod to cause premature flower initiation, vegetative and tuberous root development depend greatly upon light duration. Under SD, leaf production and stem elongation cease, while lateral branching becomes rapid, yet never fully develops (Figure 4). Older leaves drop from an approximate perpendicular position relative to the stem to one which is very oblique. This ultimately gives the appearance of wilting, despite full tissue turgidity. Furthermore, apical leaves do not fully expand under SD and remain closely bunched together due to the cessation of stem elongation. By contrast, LD promote vigorous shoot extension (Figure 5), continuous leaf production, typical leaf blade development, and a perpendicular position along the stem. Apical dominance is enhanced by LD as evidenced by an almost complete inhibition of lateral branching[9] (Figure 6).

These contrary effects due to different daylengths may in part be due to a photoperiodic control of photosynthate partitioning. The shoot to root dry-weight ratio was significantly greater when *A. tuberosa* was cultivated under LD than under SD.[9]

FLORAL DEVELOPMENT

Much research has focused on *A. tuberosa* floral development as it relates to pollinia removal and pollination. In an extensive study of this species in Virginia and North Carolina, Wyatt[13] reported that flower number per umbel and umbel number per plant were extremely variable. In fact, there was no correlation between the number of flowers on each umbel and umbel number per branch, branch number per stem, or stem number per plant. The total flower number per plant, however, was positively correlated with the number of stems per plant and the number of umbels per plant.

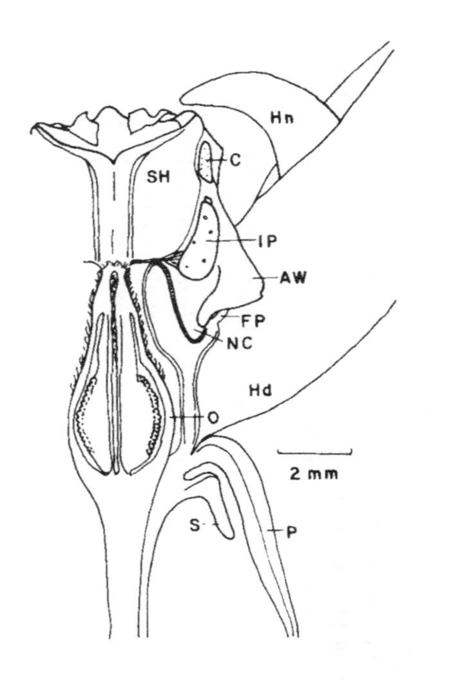

FIGURE 1. Longitudinal section through flower showing inserted pollinium (IP) above nectar cavity. AW, anther wing; C, corpusculum; FP, fleshy pad; Hd, hood; Hn, horn; IP, inserted pollinium; O, ovary; P, petal; S, sepal; SH, stigmatic head. (From Bookman, S. S., *Am. J. Bot.*, 68, 675—679, 1981.

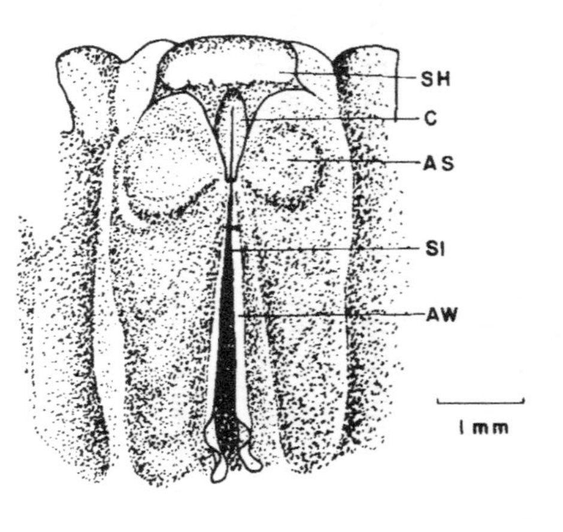

FIGURE 2. Side view of gynostegium. AS, anther sac; AW, anther wing; C, corpusculum; Sl, slit; SH, stigmatic head. (From Bookman, S. S., *Am. J. Bot.*, 68, 675—679, 1981. With permission.)

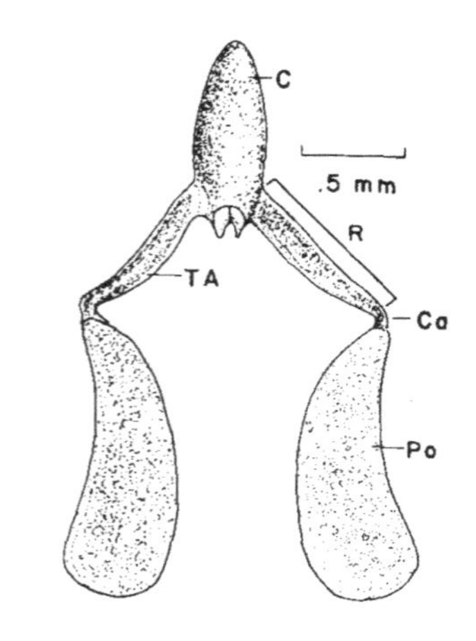

FIGURE 3. Pollinarium as oriented in flower. Ca, caudicle; C, corpusculum; Po, pollinium; R, retinaculum; TA, translator arm. (From Bookman, S. S., *Am. J. Bot.*, 68, 675—679, 1981. With permission.)

FIGURE 4. *A. tuberosa* exhibits little stem elongation, poor leaf production and development, and extensive branch initiation when maintained under SD. (Photograph by R. E. Lyons.)

Wyatt[13] also stated that the occurrence of larger inflorescences favored an increased rate of visitation by potential pollinators. Ironically, natural selection has encouraged the perpetuation of smaller, moderately sized inflorescences. *A. tuberosa* has a strong self-incompatibility system which would negate many of the pollinations by an overabundance of visitors, since the probability of selfing would be greater.[12-14]

FIGURE 5. A comparison of vegetative morphology of *A. tuberosa* grown under LD and SD. (Photograph by R. E. Lyons.)

A. tuberosa flowers are receptive to pollination for only 7 to 8 days following anthesis, but flowering time within and among umbels on a single plant can span 4 to 5 weeks.[14] Pollen tube germination occurs on the convex side of the pollinium, and when the tube is in contact with the receptive stigmatic surface, leads to corolla withering and ovary swelling in 3 to 4 days.[12]

BEHAVIOR OF RELATED SPECIES

Other *Asclepias* species have been studied more extensively since they are often invasive weeds within agronomic crops. Much attention has been given to *A. syriaca*, which often grows alongside *A. tuberosa* and has been shown to share similar juvenility and photoperiodic characteristics.[2,4,9,10] For these reasons, both species may share additional developmental

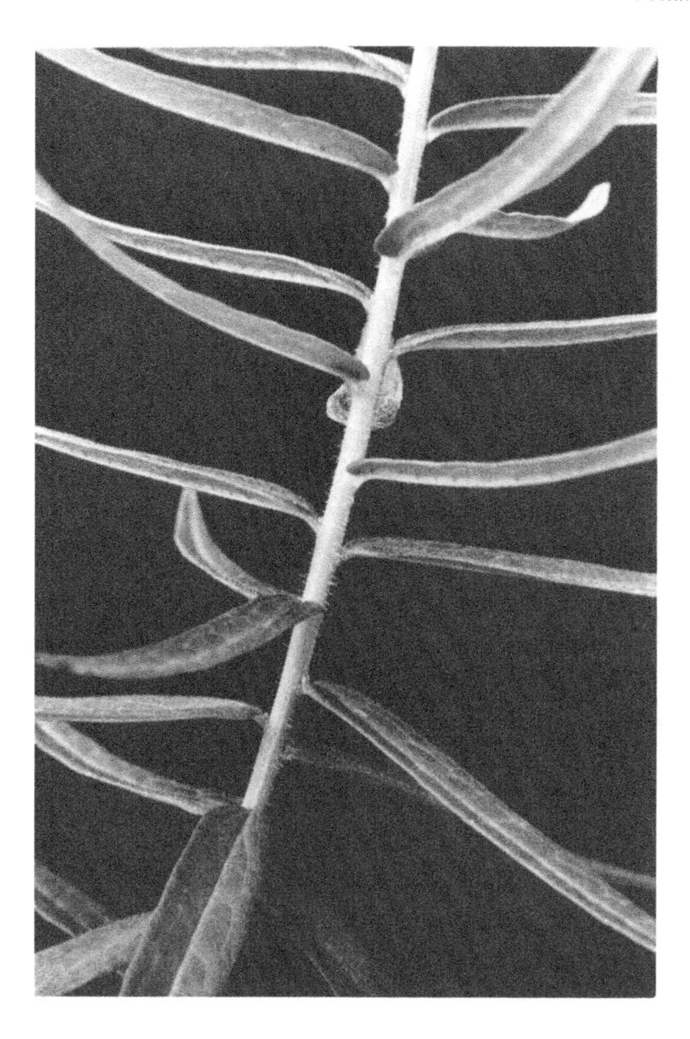

FIGURE 6. Strong inhibition of branching on *A. tuberosa* grown under LD. (Photograph by R. E. Lyons.)

characteristics. *A. syriaca* will regenerate new plants from fragmented root sections, but will not flower the first year unless the fragments are a minimum of 30 to 45 cm. Hence, there has been a restoration of the juvenile state. Peak flowering of *A. syriaca* generally occurs the third year after germination and gradually declines in subsequent years.[10]

Willson and Price[11] have reported that larger inflorescences aid in attracting greater numbers of potential pollinators, but make no mention of an evolutionary trend. They do reason, however, the the significance of inflorescence size, in contrast to *A. tuberosa,* is probably not a function of pollen compatibility. Since *A. syriaca* is highly self-compatible, seed production is theoretically limited only by the number of successful pollinations. Hence, larger inflorescences function to provide a greater quantity of available pollinaria, which ultimately enhance the chances of self- and cross-pollinations. As a result, a large number of fruits begin to develop. Yet this system, which apparently guarantees a highly successful rate of species repopulation, is actually antagonized by the resource limitations of the plant since many fruits abort prematurely.

REFERENCES

1. **Bailey, L. H.,** *Manual of Cultivated Plants,* Macmillan, New York, 1977, 814—815.
2. **Bhowmik, P. C. and Bandeen, J. D.,** The biology of Canadian weeds. XIX. *Asclepias syriaca* L., *Can. J. Plant Sci.,* 56, 579—589, 1976.
3. **Bookman, S. S.,** The floral morphology of *Asclepias speciosa* (Asclepiadaceae) in relation to pollination and clarification in terminology for the genus, *Am. J. Bot.,* 68(5), 675—679, 1981.
4. **Cramer, G. L.,** The physiology of common milkweed, *Proc. Annu. Meet. N. Central Weed Cont. Conf.,* 32, 105—106, 1977.
5. **Gleason, H. A. and Cronquist, A.,** *Manual of Vascular Plants of Northeastern U.S. and Adjacent Canada,* D. Van Nostrand, New York, 1968, 556—557.
6. **Heywood, V. H.,** *Flowering Plants of the World,* Mayflower Books, New York, 1978, 225—226.
7. **Kartesz, J. T. and Kartesz, R.,** *A Synchronized Checklist of the Vascular Flora of the U.S. and Greenland.* Volume 2, *The Biota of North America,* University of North Carolina Press, Chapel Hill, 1980.
8. **Lyons, R. E.,** An orange wildling for your garden, *Garden,* 7(5), 16—19, 1983.
9. **Lyons, R. E. and Booze, J. N.,** Effect of photoperiod on first year growth of 2 *Asclepias* species, *HortScience,* 18(4) (Abstr.), 575, 1983.
10. **Minshall, W. H.,** The biology of common milkweed, *Proc. Annu. Meet. N. Central Weed Cont. Conf.,* 32, 101—104, 1977.
11. **Willson, M. F. and Price, P. W.,** The evolution of inflorescence size in *Asclepias* (Asclepiadaceae), *Evolution,* 31, 495—511, 1977.
12. **Wyatt, R.,** Pollination and fruit-set in *Asclepias:* a reappraisal, *Am. J. Bot.,* 63(6), 845—851, 1976.
13. **Wyatt, R.,** The reproductive biology of *Asclepias tuberosa.* I. Flower number, arrangement and fruit-set, *New Phytol.,* 85, 119—131, 1980.
14. **Wyatt, R.,** The reproductive biology of *Asclepias tuberosa.* II. Factors determining fruit-set, *New Phytol.,* 88, 375—385, 1981.

ASTER NOVI-BELGII

En. Michaelmas daisy, New York aster; Fr. Aster de la Nouvelle-Belgique; Ge. Glattblattaster

W. W. Schwabe

DESCRIPTION

The Michaelmas daisy (*Aster novi-belgii* L., Compositae; 2n = 54) is a native of North America widely grown in gardens and has become naturalized in Britain, frequently as a garden escape. It is distinguished from the New England aster (*A. novi-angliae*) by its smooth leaves and blue flowers. Numerous other species and their hybrids are commonly grown in gardens, and these include many others: *A. acris, A. alpinus, A. amellus, A. cordifolius, A. ericoides, A. frikartii, A. tradescantii,* and *A. yunnanensis.*

A. novi-belgii is an autumn-flowering species with normally blue flowers, but there is a range of cultivars and unnamed variants. The cv used in much of the experimental work described is a white-flowered one and was named 'Wimbledon' for its garden origin. It reaches approximately 30 to 50 cm in height and is a vigorous grower. Dwarf cvs (approximately 15 cm) used in some of the experiments included a blue cv 'Victor' and a red-flowering 'Little Pink Boy'.

The highly synchronized flowering in the Michaelmas daisy suggested that there must be a close environmental control mechanism. Initially this was suspected to be merely a daylength effect. However, subsequent work has revealed a much more complex mechanism involving both temperature and photoperiod.

PROPAGATION

In nursery practice, Michaelmas daisy is propagated by division of the rootstock with its spreading basal shoots and rhizomes. For experimental purposes, single shoots with roots or cuttings can be used.

A. novi-belgii is relatively undemanding and grows well in most soils and composts which are reasonably well drained. For experimental purposes a soilless compost with added mineral nutrients has been found to give uniform growth.

VEGETATIVE GROWTH

The vegetative growth made by *A. novi-belgii* is very much affected by the environment, and in Table 1 some data are recorded for height growth in different daylength sequences. It is also clear that leaf numbers are affected very greatly (Table 1) by those conditions.

Moreover, the vernalization (chilling) treatment required to promote flowering (see below) has a very great effect on this plant. The differences between the extremes in one experiment indicated an almost 100-fold difference in height alone.

Even in fully vernalized plants, also given long photoperiods (LD) for 4 weeks, the temperature-daylength interaction effect on height growth is striking (Table 2), and higher temperatures raise heights very greatly, especially in combination with "relatively longer" short photoperiods, though 29°C is supraoptimal.

The Michaelmas daisy produces (often in large numbers) rhizomes from the lateral stem buds, which grow just below the soil surface and then, by changing their diageotropic growth, turn into short, almost-rosetted shoots under the main canopy of the flowering shoots. Apart from the effects on flowering, daylength also influences the pattern of veg-

Table 1
HEIGHT GROWTH AND LEAF NUMBERS PRODUCED IN CV 'WIMBLEDON' EXPOSED TO DIFFERENT DURATIONS OF LD (CONTINUOUS LIGHT) TREATMENT BEFORE TRANSFER TO DIFFERENT FINAL SHORTER DAYLENGTHS, WITH AND WITHOUT INITIAL VERNALIZATION; FIVE REPLICATES (6 WEEKS, 5°C)

	No. of LD						
	Unvernalized			Vernalized			
Length of photoperiod (hr)	0	14	24	0	14	24	LSD ($p = 0.05$)
	Heights (cm)						
8	0.4	1.2	8.1	3.1	5.7	29.7	10.6
12	0.8	1.8	21.7	5.4	15.8	37.2	
16	35.9	46.1	53.1	56.6	62.2	72.7	
	Leaf number per plant						
8	9.0	12.0	19.0	15.0	18.0	20.0	3.1
12	10.0	15.0	21.0	18.0	18.0	19.0	
16	18.0	22.0	22.0	21.0	22.0	26.0	

Table 2
EFFECT OF TEMPERATURE DURING FINAL SD TREATMENT ON HEIGHTS (cm) OF 'WIMBLEDON' PLANTS PREVIOUSLY VERNALIZED (6 WEEKS, 5°C) AND GIVEN 4 WEEKS OF LD (16 hr)

	Temperature (°C)			
Daylength	15	22	29	Mean
8 hr	22.9	35.0	38.2	32.1
11 hr	41.1	55.5	55.6	50.9
14 hr	86.0	82.9	70.9	80.0

Note: LSD (interaction, $p = 0.05$) 12.2. LSD (main effect of daylength, $p = 0.05$) 7.0.

etative growth in modifying the production of lateral branches on the stem and at the plant base. Thus, in an 8-hr photoperiod at 20°C, fewer main stem laterals (1.1) were produced compared with 3.9 in 13 hr (LSD for $p = 0.05$ being 1.5); by contrast, basal shoot numbers were higher in 8 than 13 hr, i.e., 5.2 vs. 2.7 (LSD for $p = 0.01$ being 1.7), highly significant differences, probably related to the early entry into dormancy in 8-hr days. These basal shoots ensure the survival of the plant as a perennial, for, if they all elongated and became floral, there would be no viable shoots left for the next season. In the chrysanthemum a

Table 3
EFFECTS OF VERNALIZATION IN INTERACTION WITH FINAL DAYLENGTH ON 'WIMBLEDON'

Final daylength (hr)	8	12	14	8	12	14	8	12	14
	No. of flowers produced			Inverse of days to flower (days^{-1} × 10^3)			Final height (cm)		
Unvernalized	0.2	1.9	6.0	1.3	2.6	5.3	1.5	3.8	20.1
6 Weeks cold	1.0	2.5	17.9	8.3	9.0	8.3	7.6	13.0	33.4
	LSD (p = 0.001)		5.6	LSD (p = 0.001)		1.5	LSD (p = 0.01)		4.1

Note: After chilling, plants were exposed to LD for 0, 2, or 4 weeks, but these data have been pooled and only the first-order interaction for final daylength and vernalization is shown.

similar situation obtains,[1] and here it has been shown that the production of such basal shoots is the result of devernalization by low-intensity light or darkness on the young shoots developing in the shade of the elongated and flowering main shoots. Tests as to whether a similar situation exists in the Michaelmas daisy are described below.

FLOWERING

A series of experiments indicated that 'Wimbledon' required not only a period of chilling, but also appropriate daylength treatment to follow on the chilling treatments. Interestingly, "normal" flowering resulted only if the sequence of treatments given actually included vernalization by exposure to low temperature for several weeks, and a subsequent exposure to LD conditions (e.g., 16-hr photoperiods) corresponding to the approximate length of the longest summer day at the latitude of southern England. Again a period of some weeks of such exposure was required, and this needed to be followed by SD treatment, with the SD itself being of an appropriate photoperiod, neither too long nor too short.

Since virtually all the results described have not been published previously, a series of data on the vegetative growth and flowering behavior is given in the tables below to illustrate the effects quantitatively, and these indicate the extraordinary degree of environmental control in *A. novi-belgii* ('Wimbledon' and others).

The Need for Vernalization

As for the LD and SD requirements, the evidence for the chilling requirement is best seen from the tabulated data which involve also the photoperiodic treatments. However, a few results may be given here. In an experiment with cv 'Wimbledon' in which plants were chilled at 5°C for 6 weeks or left unchilled (20°C), vernalization (or its absence) was followed by LD exposure for 0, 2, or 4 weeks (data here pooled), and a final series of short daylengths of 8, 12, or 14 hr.

Effects on flowering are seen in the data for the vernalization × final daylength interactions (Table 3).

In an experiment with dwarf cv 'Victor', an interesting difference is seen between 4 and 8 weeks of chilling in terms of flower numbers. In this cv a longer final daylength is required for optimum flowering (Table 4).

The highly beneficial effects of a longer period of vernalization are also clear from these data and confirm the chilling requirement for 'Victor'. Of course, there were also other effects on leaf numbers to flowering, etc. but only the mean heights of the plants vernalized

Table 4
EFFECT OF 4 AND 8 WEEKS OF VERNALIZATION (4°C) ON FLOWERING IN CV 'VICTOR', FOLLOWED BY 4 OR 8 WEEKS OF CONTINUOUS LIGHT AND FINAL DAYLENGTHS OF 10, 14, AND 16 HR

Vernalization weeks	No. LD weeks	Final daylength (hr) 10	14	16
4	4 weeks	0	0	0
	8 weeks	1.0[a]	1.0[a]	12.6
8	4 weeks	1.0[a]	1.0[a]	4.8
	8 weeks	1.0[a]	1.0[a]	38.6

Note: Five replicates per treatment. No statistics required.

[a] All terminal flowers aborted.

for 4 and 8 weeks may be referred to (i.e., 5.8 and 12.2 cm, respectively) to illustrate the magnitude of the effects.

Substitution of Vernalization and LD Treatment by Gibberellic Acid

It is known for several species of plants that a chilling and/or long photoperiod requirement may be satisfied by exogenous application of GA_3. This was also tested in 'Wimbledon' in an experiment in which 1.3 mℓ of 50 ppm GA_3 was applied as drops to the shoot apex.

The results clearly indicated that in this species GA_3 was effective in causing flowering in appropriate short daylengths when the plants had neither been chilled nor exposed to LD previously.

Table 5 shows the heights of plants at the end of the experiment and the proportion of shoots induced to form floral buds in various daylengths after GA_3 treatment. The controls failed to flower in any daylength and have therefore been summed over no-GA_3 treatments. Interestingly, the abortion of flower buds caused by too short a final photoperiod is not modified by GA_3 pretreatment and is determined by photoperiod alone (cf. lower part of Table 5).

The Long-Day Requirement

The necessity for LD treatment in order to induce flowering in *A. novi-belgii* became obvious at an early stage in the investigation and the effect is displayed in virtually all the factorial experiments described above and below (see Tables 4, 7, and 9).

Even in the absence of chilling treatment the promotive effect of long photoperiods is clear in 'Little Red Boy' held for 7, 14, or 28 days in continuous light, or for 7, 14, 28, or 35 days in 16-hr days, even though the degree of flowering was very low. The flower numbers per plant were 0.8, 1.0, and 1.5 for continuous light, and 0.7, 0.5, 0.3, and 3.0 for 16-hr days, respectively. It is interesting, too, that with 24-hr LD only 28 days were required to give 1.5 flowers per plant while in 16-hr days 35 cycles were required to give 3.0.

The evidence is that in all cvs a period of several weeks of LD is needed to give its full promotive effect.

Table 5
EFFECTS OF 1.3 mℓ OF 50-PPM GA_3[a] ON FLOWERING AND INFLORESCENCE DEVELOPMENT AND ABORTION IN 5 FINAL DAYLENGTHS

	GA_3					Controls (summed over all 5 daylengths)
Daylength (hr)	**11**	**12**	**12½**	**13**	**14**	
Height (cm) at budding (controls at end of expt.)	15.9	2.12	19.7	32.6	27.5	2.1
No. of shoots vegetative	14	9	9	2	6	179
No. of shoots budded	22	27	27	34	30	1
No. of budded shoots flowering	7	18	14	20	30	1
No. of budded shoots aborted	15	9	13	14	0	0

Note: Unvernalized 'Wimbledon' plants previously exposed to LD were used with 36 replicate shoots per treatment. Data for non-GA_3-treated plants showed no daylength differences and have been pooled. Also, plants were either held at a constant day/night temperature of 21°C or in a 25°C day/18°C night temperature regime. As there were no significant differences, the two sets of data were also pooled.

[a] Applied on 13 occasions as 0.1-mℓ drops per shoot.

Table 6
FLOWERING IN 'WIMBLEDON' CHILLED FOR 6 WEEKS FOLLOWED BY 4 WEEKS EXPOSURE TO LD (16 HR) BEFORE EXPOSURE TO DIFFERENT FINAL DAYLENGTHS

	No. of shoots		
Final daylength (hr)	**Vegetative**	**Flowering**	**Inflorescence aborted**
16	16	0	0
14	4	17	0
12	0	3	16
8	0	1	22

The requirement for LD pretreatment is seen particularly clearly in the combined (factorial) experiments in which the need for SD as the final daylength is equally obvious. Thus, while recognized as very important, the LD effect can be regarded as established and need not be detailed further.

The Short-Day Requirement

The need for a final short photoperiod to give normal flowering was, of course, also seen in numerous factorial experiments. Moreover, the actual length of the SD is highly significant (see the section titled "Flower Abortion" below). However, data obtained in an experiment comparing four final daylengths showed the effects so clearly that no statistical confirmation was required (Table 6).

The optimum final daylength varies somewhat with cv; thus in an experiment with 'Little Red Boy', the number of flowers per plant in a 14-hr final daylength reached 148 while in a 12-hr day it was only 8 (the LSD $p = 0.001$ being 36.2). By contrast, in 'Victor' a 16-

Table 7
PERCENTAGE OF 'WIMBLEDON' PLANTS INITIATING FLOWER BUDS AFTER EXPOSURE TO 8, 16, OR 12 LD INTERRUPTED AFTER EVERY 4, 2, OR 1 CYCLES BY 1, 2, OR 3 SHORT PHOTOPERIODS FOLLOWED BY RETURN TO A FINAL 8-HR SD

Total no. of LD	Uninterrupted controls	Mean of all interruption treatments	Grouping of LD			No. consecutive SD at each interruption		
			4	2	1	1	2	3
8	20	11	20	6.7	6.7	13	6.7	13
16	60	53	87	67	6.7	53	60	47
24	100	64	100	67	27	80	60	53
Mean	60	43	69	47	13	49	42	38

Note: All three first-order interactions are shown. There were 5 replicate plants in each of the 30 treatments.

hr photoperiod was not yet too long to permit flowering; thus there are cv differences in the maximum tolerable photoperiod for the SD requirement. The lower limit is, of course, equally important, in that daylengths below this strongly induce the onset of dormancy and cause flower abortion (see below).

Interruption of LD Treatment by Intercalated SD, and Vice Versa

It is known that in such SD-requiring species as *Kalanchoe blossfeldiana, Perilla ocymoides,* and *Glycine soja* (see chapters in volumes 3, 4, and 1, respectively), interruption of SD treatment by intercalated LD reduces or abolishes the effect of such flower-inductive treatment.[2] Similarly, it is known that in LD-requiring species, an LD induction period may be interrupted by SD without the induction effect being eliminated.[3] It seemed of interest, therefore, to study this situation in *A. novi-belgii.* In this experiment, unvernalized 'Wimbledon' plants were used and the following treatment administered: three total numbers of LD given (8, 16, and 24 cycles at 16 hr), interrupted after every 4, 2, or 1 cycles by either 3, 2, or 1 SD (8 hr), making 30 sequences when uninterrupted controls are included. All plants were placed in a final 8-hr SD after LD treatment was completed. For reasons of space only a brief summary of the data can be given in Table 7 in which data on the percentages of plants flowering are shown in interactions of paired factors.

It is clear from these percentage data that the number of LD cycles given is most important, and only with 24 LD did all plants flower. The interruption by SD was highly detrimental, on average, even with 24 LD. The numbers of LD given consecutively (grouping) was also important, and with groups of four the intercalated SD could be tolerated, whereas interruptions after each LD were highly detrimental. Interestingly, the numbers of SD given consecutively had a variable but much smaller effect.

In another test, the effect of interrupting the final SD treatment in the sequency by intercalated LD was tested. The data are again presented as percentages of plants flowering, but as the total period of SD given had to be kept to a fixed number, the maximum flowering did not reach 100% even in the controls — as would be expected. All 'Wimbledon' plants were given an initial period of 21 cycles of 16-hr LD. After every 1, 2, 5, or 10 SD, the intercalated LD were given either singly or in pairs (Table 8).

The data suggest that in *A. novi-belgii* LD interruption of SD is inhibitory to induction and that it matters greatly how many SD are given in any one sequence, the deleterious LD

Table 8
PERCENTAGE OF 'WIMBLEDON' PLANTS INITIATING FLOWER BUDS AFTER AN INITIAL PERIOD OF 21 LD (16 HR), FOLLOWED BY 10, 20, OR 30 SD WHOSE SEQUENCE WAS INTERRUPTED BY EITHER SINGLE OR PAIRED LD AFTER 1, 2, 5, OR 10 SD CYCLES

Total no. SD	Uninterrupted controls	Mean of all interruption treatments	Grouping of SD given consecutively				No. of intercalated LD given together		Mean
			1	2	5	10	1	2	
10	17	2.8	0	0	8.3	17	5.6	0	4.8
20	33	15	0	17	25	17	17	12.5	17
30	17	33	33	25	42	33	33	33	31

Note: First-order interactions are shown.

Table 9
EFFECTS OF SEQUENCE OF CHILLING, LD, AND SD TREATMENTS ON FLOWERING OF 'WIMBLEDON'

Treatment sequence	SD/ cold/ LD	SD/ LD/ cold	LD/ cold/ SD	LD/ SD/ cold	Cold/ SD/ LD	Cold/ LD/ SD
No. of flowers per plant	0.7	7.0	10.2	3.6	1.8	7.4
Final height per plant (cm)	28.8	29.1	30.6	20.7	31.0	30.7

Note: Cold = 4 weeks at 5°C in 8 hr light; LD = 3 weeks in $16^1/_2$-hr days; SD = 3 weeks in 12-hr days.

effect becoming less the more SD are grouped together. Also, of course, there is an interaction with the total number of SD given, and clearly 30 was still suboptimal. Two intercalated LD given together are more harmful than one, but only with a shorter SD induction period.

The basic qualitative conclusion from these two tests is that both LD and SD periods should be given as block treatments and cannot be given in any form of alternation without their effects becoming diminished.

Sequence of Chilling and Long- and Short-Day Treatments

Although it would seem logical that the optimum sequence of environmental treatments should be in the order of their occurrence under outdoor conditions (i.e., winter cold, spring and summer LD, late summer and autumn SD), it seemed worthwhile to test whether this was in fact correct. The experiment with 'Wimbledon' combined 4 weeks of chilling at 4°C with 3 weeks at 20°C in a $16^1/_2$-hr LD, or a 12-hr SD. All three conditions were given either in the beginning, middle, or final treatment period resulting in six treatments. Unfortunately, a number of plants died in some treatments, making a statistical analysis inappropriate, but the main effects on flowering are nonetheless obvious from Table 9.

Interestingly, the sequence between LD and cold does not seem to matter greatly as long as both occur in the first two slots of the treatment cycle, but flowering was fully effective only if the triple sequence ended either with SD or with cold (itself combined also with SD). It would have been desirable to have also had a cold treatment combined with LD, but this proved technically impossible.

Interaction of Photoperiod with Temperature

In this experiment on the effects of temperature during the final SD period all plants were given 4 weeks vernalization in 8-hr days, followed by 4 weeks in 16-hr LD at 20°C. Subsequently they were divided into nine treatment groups; i.e., three temperatures (15, 22, and 29°C) combined with three SD photoperiods (8, 11, and 14 hr). The striking effects on plant growth and floral development are seen in Figure 1.

The quantitative effects on inflorescence numbers per plant are given in Table 10 and illustrate the magnitude of the temperature/photoperiod interaction, both increasing temperature and longer photoperiods (though still effectively SD) raising flower numbers, especially when high levels of both factors are combined. Also the marked effect of high temperature at that stage on floral abortion (see below) is clear from the data and from Figure 1.

DEVERNALIZATION

In a perennial plant, an annual vernalization requirement can only arise for one of two reasons: either (1) not all meristems subjected to the original chilling become vernalized, perhaps acquiring competency to become vernalized only at some later stage, or (2) a process of devernalization occurs. In the chrysanthemum it was found[1,4] that all meristems are susceptible to chilling and that devernalization by exposure of basal shoots to low-intensity light under the canopy of the main flowering shoots caused devernalization, especially at raised temperatures. In the Michaelmas daisy, too, no evidence was found that there were unvernalizable meristems (e.g., on low laterals).

In an experiment in which the behavior of the basal shoots was tested, the treatments included decapitation of the main shoots and also exposure to severe shading (an effective method for devernalization of the chrysanthemum).

All 'Wimbledon' plants had been vernalized for 6 weeks and were given 16-hr LD for 4 weeks before transfer to a final daylength of 13 hr.

When decapitated to the base, after chilling, the plants produced a mean of 1.2 basal shoots which subsequently flowered with 6.0 flowers per shoot, and the same number of flowering shoots were produced when decapitation was delayed for 4 weeks, i.e., until after the LD treatment (with 8.2 flowers). If, however, the main shoot had merely had its tip cut off and all main-stem lateral branches removed, only three of six plants produced flowering basal shoots (0.5, with 1.2 flowers). Severe shading after detipping and disbudding of the main shoot resulted in even fewer flowering basals developing (i.e., 0.3 with 1.2 flowers).

In another experiment, in which plants were exposed to 6 weeks chilling and subsequently to a 16-hr photoperiod for 4 weeks, five of five control flowered with five flowers per plant when transferred to their final 12-hr photoperiod, while only two out of five plants did so with 0.4 flowers per plant when they were severely shaded.

Clearly severe shading appears detrimental to flowering, and while the data are not adequate, they suggest a devernalizing effect such as has been established for the chrysanthemum. Data for a further shading/devernalization experiment are shown in Table 11.

It is clear from this that only partial devernalization was achieved, especially when the shading treatment was given during a 16-hr LD. This LD treatment did, of course, have itself the usual promoting effect which accounts for the higher values for flower numbers.

In an experiment to test whether devernalization is possible by high-temperature treatment (32°C) a clear-cut result was obtained. High temperature does not on its own devernalize, but *per contra* accelerates the progress towards flowering, which takes place sooner and at a lower leaf number and stem height (Table 12).

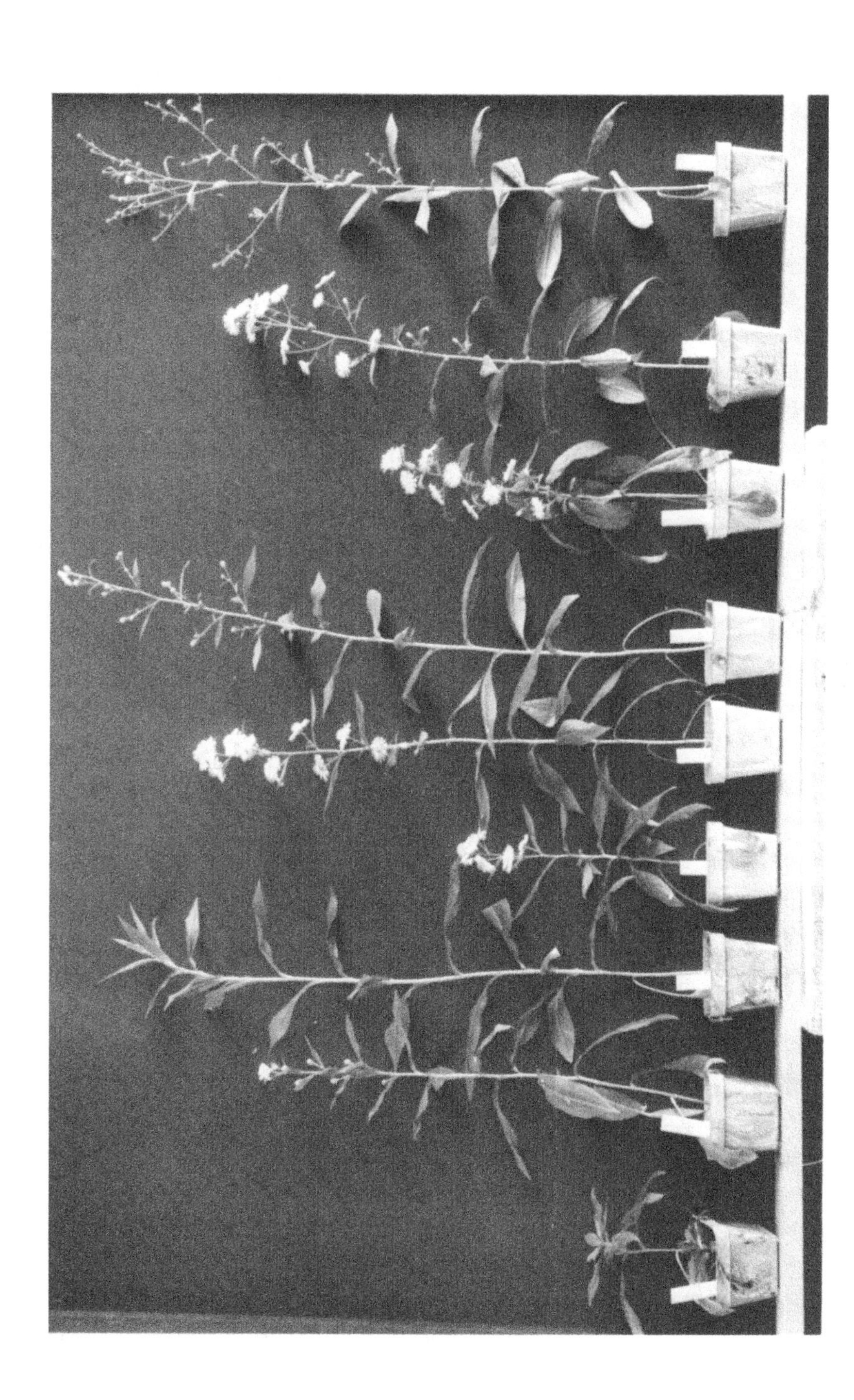

Table 10
EFFECT OF TEMPERATURE DURING FINAL SHORT-PHOTOPERIOD TREATMENT ON INFLORESCENCE NUMBERS IN 'WIMBLEDON' PREVIOUSLY CHILLED FOR 4 WEEKS AT 5°C AND GIVEN LD (16 HR) FOR 4 WEEKS (SEVEN REPLICATES)

	Temperature (°C)			
Daylength	**15**	**22**	**29**	**Mean**
8 hr	1.3	7.3	12.2	6.9
11 hr	7.2	13.8	18.0	13.0
14 hr	11.3	24.5	48.1	28.0
	6.6	15.2	26.1	

Note: LSD (interaction, $p = 0.05$) 4.8. LSD (main effects, $p = 0.05$) 2.8.

Table 11
DEVERNALIZATION BY SEVERE SHADING OF 'WIMBLEDON'

	Daylength during shading								
	SD (10 hr)				LD (16 hr)				
Period of shading (weeks):	**0**	**3**	**4**	**6**	**0**	**3**	**4**	**6**	**LSD ($p = 0.05$)**
No. of flowers per plant	4.5	0.9	0.8	0.9	4.5	9.0	4.8	4.8	3.18
Days to flowering	61	65	70	85	66	84	87	110	9.6
Final leaf nos.	24.5	22.5	21.0	21.0	24.3	29.0	25.5	27.8	2.2
Final height (cm)	19.6	11.8	9.3	9.2	38.8	56.5	40.9	46.4	9.6

Note: Plants were vernalized for 4 weeks at 4°C, then exposed to 12°C for 4 days before transfer to either 10- or 16-hr photoperiods at 20°C and in severe shading (approximately 250 lux) for 0, 3, 4, or 6 weeks, after which all plants were returned to full light at 20°C in an SD of 13 hr. Since some of the 6 replicates died during the low-intensity light treatment, the statistical evaluation shows rather large LSD.

Table 12
EFFECT ON FULLY VERNALIZED 'WIMBLEDON' PLANTS (6 WEEKS AT 4°C) OF SUBSEQUENT EXPOSURE TO 13-HR SD AT 32°C

	20°C controls	**4 days at 32°C**	**14 days at 32°C**	**LSD ($p = 0.05$)**
Time to visible budding (days)	116.5	120.2	99.0	12.17
Final leaf no. at flowering	23.5	21.7	18.8	2.08
Final height at flowering (cm)	23.4	19.7	13.4	6.42

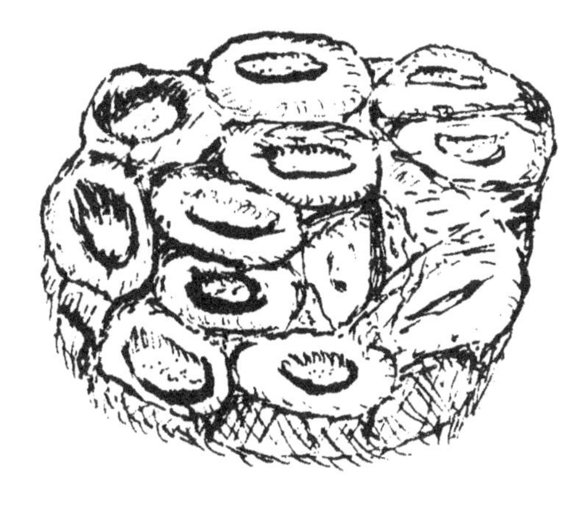

FIGURE 2. Appearance of terminal apex of inflorescence aborted in very short days.

FLOWER ABORTION

One of the surprising observations made in plants exposed to the 8-hr SD treatments, following previous vernalization and LD exposure, was the fact that not only height growth ceased, but also that plants showing the first macroscopic signs of flowering failed to produce normal inflorescences. When these were dissected, it was clear that commonly the progress to flowering had come to a halt when the large composite receptacle had developed and when floret initials had formed on it, reaching the early "cup-stage". According to the interval between the onset of 9-hr photoperiods and the dissection, these were alive but apparently "inactive", or they had turned brown and the whole young inflorescence was dead. This abortion of inflorescences in 8-hr SD contrasted with continuous development in "longer" SD, e.g., 12 or 13 hr.

Since abortion coincided with the cessation of vegetative growth of other plant parts (basal shoots) it clearly represents the onset of dormancy, i.e., *another,* and *different,* response of the plant to the environment regulating its life cycle. However, with very short photoperiods (e.g., 8 hr) the dormancy response predominated over the flowering response (Table 6 and Figure 2).

Another experiment with variety 'Victor' suggested that very prolonged exposure to 16-hr LD reduced the tendency to abortion; i.e., with 0, 7, 14, or 24 LD all plants initiated and aborted all flowers in subsequent 8-hr SD, but after 84 LD subsequent flowers developed normally, the mean heights of the latter treatment being 12.6 cm compared with the mean of all other treatments of 4.8 cm. In this cv even a 12-hr final daylength was found to be too short to permit normal flowering development. In 'Wimbledon' a 12-hr day still permitted normal flower development.

However, there is a strong interaction with temperature, and discrepancies are explained by the fact that high temperatures encourage normal development, while low temperatures promote the onset of dormancy (see Table 13 and also Figure 1).

Abortion of inflorescences by too short a SD could be simulated by auxin application; 500 ppm IAA in lanolin paste applied to the terminal main shoot just below the tip caused 100% abortion in vernalized plants, also given a 4-week period of continuous illumination before transfer to 20°C in a 12-hr final daylength. By contrast, all the controls flowered normally. Though suggestive, this is not, of course, proof that abortion in very short photoperiods is caused by too high an endogenous auxin level.

In view of the favorable effects of GA_3 in preventing flower abortion in 8-hr SD, an

Table 13
INTERACTION OF DAYLENGTH AND TEMPERATURE ON INFLORESCENCE DEVELOPMENT OF 'WIMBLEDON'

Photoperiod (hr)	Temperature (°C)		
	15	22	29
8	1 veg (6 ab)	3 fl (4 ab)	5 fl (2 ab)
11	1 fl (6 ab)	5 fl (2 ab)	7 fl
14	5 fl (2 ab)	2 fl (5 ab)	4 fl (2 veg)

Note: Seven plants per treatment; veg = vegetative plant, fl = plant with normal flowers, ab = plant with aborted inflorecences.

Table 14
EFFECT OF GA_3 AND ABA ON INFLORESCENCE NUMBERS AND SIZE, AS WELL AS NUMBER OF LIGULATE FLORETS PER INFLORESCENCE AND PLANT HEIGHT, IN 'WIMBLEDON' PLANTS VERNALIZED AND EXPOSED TO 24 PHOTOPERIODS BEFORE TRANSFER TO 8-HR DAYS (12 REPLICATES)

Treatment	Days to flowering from start of 8-hr SD	Plant height (cm)	Inflorescence no. per plant	Terminal inflorescence size (diameter, cm)	No. of ligulate florets per inflorescence
Control	26.0	21.1	1.7	3.0	46.3
GA_3 (25 ppm; applied as drops to the tip 2 × per week)	24.0	42.1	2.8	3.2	38.4
ABA (25 ppm; applied as drops to the tip 2 × per week)	25.0	19.2	1.2	2.6	22.5
LSD (p = 0.05)	1.3	3.7	1.0	0.4	6.8

attempt was made to study the effects of ABA under very short daylengths to discover whether abortion would be enhanced. However, using 'Wimbledon' (vernalized at 5°C in 8 hr for 4 weeks and then held at 15°C in 24-hr days for 3 days followed by 24 days at 20°C in 24-hr days before transfer to 8-hr days at 20°C), ABA had virtually no effect on abortion. Nevertheless there were effects on number and size of inflorescences and also numbers of ligulate florets, as well as heights of plants (Table 14), with ABA reducing all characters measured, while GA_3 promoted or was equal to the controls. An attempt to restimulate arrested inflorescences into completing their growth by chilling the plants at 5°C for 4 weeks and further LD treatment failed totally.

CONCLUSIONS

The Michaelmas daisy thus represents one of the most accurately regulated species in its flowering behavior, with detailed adaptations to the environment, so that each phase of climatic change has become a *necessary* factor in the sequence of those needed to induce flowering — and also dormancy. A period of winter cold needs to be followed by the long days of late spring and early summer, and flowering is initiated only when the shortening days of late summer follow this sequence, while the short days of late autumn induce cessation of growth and flowering leading to dormancy of all shoots. Altogether this represents a remarkable example of environmental dependence and adaptation.

ACKNOWLEDGMENTS

I am much indebted to Miss P. Kite for help with much of the experimental work reported.

REFERENCES

1. **Schwabe, W. W.,** Factors controlling flowering in the chrysanthemum. V. De-vernalisation in relation to high temperature and low light intensity treatments, *J. Exp. Bot.*, 6, 435—450, 1955.
2. **Schwabe, W. W.,** Studies of long-day inhibition in short-day plants, *J. Exp. Bot.*, 10, 317—329, 1959.
3. **Snyder, W. E.,** Mechanism of the photoperiodic response of *Plantago lanceolata* L., a long-day plant, *Am. J. Bot.*, 35, 520—525, 1948.
4. **Schwabe, W. W.,** Factors controlling flowering in the chrysanthemum. VI. De-vernalisation by low-light intensity in relation to temperature and carbohydrate supply, *J. Exp. Bot.*, 8, 220—234, 1957.

ASTER PILOSUS

Avishag Kadman-Zahavi and Herut Yahel

INTRODUCTION

Aster pilosus Willd. (Compositae, Asteraceae), a native of North America, is a perennial herb with two distinct phases of growth[1] (Figure 1). The winter phase is a rosette, with relatively large leaves (up to 18 cm long and 1.8 cm wide). Stem elongation starts in spring, and stem height may reach 1 m. Secondary, tertiary, and higher orders of branching bring the plant to its summer phase, which is a bush, bearing tiny spiny leaves (0.5 × 0.1 cm). Flower buds are initiated in late summer and the flowers open in September,[2] covering the bushes with a cascade of small white flowers (Figure 1D). Ligulate florets are white and the tubular florets are yellow, turning beige before senescence. The stems subsequently dry up and break off during the winter. Sessile regeneration buds break out concurrently with stem drying.

PHOTOPERIODIC RESPONSES

Like the *Aster novi-belgii*,[3] *A. pilosus* plants readily respond to daylength conditions: long days induce stem elongation, which is a prerequisite for flowering, but flower initiation is possible only under SD conditions. Allard and Garner[2] considered *A. pilosus* to be a SDP, probably because their SD treatments commenced in April, after the stems already started elongating. Unlike *A. novi-belgii*,[3] vernalization is not required for stem elongation, at least in the plants used in our experiments that are grown in the gardens in Israel; and at any time these plants may be induced to elongate under LD conditions. This is also the case for cvs of *A. novi-belgii* cultivated in Israel.[4]

Duration of Long-Day Treatment

Unless otherwise mentioned, 4 hr of night-break with 70 to 100 lx of incandescent light was used to create LD conditions. The number of LD needed for stem elongation was tested under field conditions in October 1981. Young rosette plants were illuminated for 30, 40, or 50 nights, and subsequently continued to develop under the natural winter daylength (10 hr, 40 min to 11 hr, 30 min). It was found that 30 LD induced stem elongation in only 25% of the plants; these plants subsequently flowered 80 days after being returned to natural SD. A period of 40 LD was enough to induce 100% elongation; these plants flowered at the same time as those in the first group, i.e., 70 days after being transferred to SD. Plants illuminated in 50 nights flowered 16 days later, but more flowering branches were produced.

Day Extensions vs. Night-Break Illuminations

In another experiment it was found that a higher total of light energy was needed to induce elongation when the light was applied in 5-hr day extensions than in 4-hr NB (using either incandescent, high pressure sodium, or Gro-lux fluorescent lamps).

Cyclic Lighting

It was also found that there was no difference in the effect of incandescent light whether the same total light energy was applied continuously or in cyclic flashes over one-third of the time (2 min light — 4 min dark) at triple the intensity, applied over the same period at night.

FIGURE 1. *Aster pilosus*, different phases of development. (A) Plants grown in 10-hr SD since January 1981; (B) plants grown in natural daylength, stems with open flowers; (C)

D

C
LD→
14.9.81→
ND
19.10.81

Table 1
RELATIONSHIP BETWEEN DURATION OF LD TREATMENT AND NUMBER OF NATURAL SD TO VISIBLE BUDS AND TO ANTHESIS

Location[a]	End of LD treatment	No. LD	No. SD[b] to visible buds	No. SD[b] to anthesis
Besor	Oct. 21, 1981	30	50	68
	Nov. 1, 1981	40	40	68
	Nov. 11, 1981	50	35	74
Lachish	Nov. 18, 1982	62	45	100
	Nov. 18, 1982	90	25	80
Bet Dagan	Sept. 9, 1980	167	27	34
	Oct. 4, 1980	187	17	28
	Oct. 18, 1980	201	21	43

[a] Experiments were conducted in three locations in Israel. Besor is the most southern and warmest location.
[b] Natural short days.
[c] Only 25% of the plants elongated and flowered.

Type of Lamp

The number of light quanta (mmol m^{-2}) needed for the induction of elongation was not consistently different when different types of lamps were used (high pressure sodium, cool white fluorescent, Gro-lux fluorescent, or incandescent).

Branching and Stem Development under LD Conditions

When the plants were kept in LD for long periods, the apical meristems did not continue to produce new leaf primordia indefinitely, but after some time they entered a period of rest. Apical dominance was lost and secondary and higher orders of branching commenced. Leaf size on upper order branches was reduced drastically and the plants appeared very bushy.

Induction of Flowering

Flowering is inhibited under LD, although after prolonged periods (6 to 7 months) under LD, mature stems may eventually produce a few flowers even under LD. The rate of flower initiation in SD depends on the maturity of the stems (Table 1). In one experiment, plants were placed in 10 hr SD for 7, 14, or 21 days and then returned to LD. Seven SD did not induce flower formation; 14 SD caused some flower initiation in all the plants, but subsequent flower development was slow. Under SD, flower buds became visible after 21 days. Plants returned to LD at this stage flowered almost simultaneously with similar plants kept continuously in SD.

The number of flowers produced depends on the number of available sites. The flowers develop on the terminal apices of branches regardless of their size. The number of such branches depends on the duration of the LD period prior to the start of SD and on the total photosynthetic light energy.

Critical Daylength for Flowering and for Stem Elongation

From Table 1 it may be seen that about 40 SD are needed for flowering of mature stems. In Israel, flowering of *A. pilosus* usually starts late in September. This means that natural daylength in August (about 13 hr 40 min) allows flower initiation; this should be considered the critical daylength for flower induction under summer temperature conditions (see also Reference 2).

Critical daylength for stem elongation is much shorter. Plants kept continuously under 10 hr SD did not elongate, but plants held during winter under natural daylength in a heated greenhouse (minimum temperature 17°C) started to elongate already in February, i.e., at a daylength slightly longer than 11 hr. This means that there is a large overlap in the daylength for stem elongation and flowering. In areas where the winter is warm, as at the Besor Research Station, or in heated greenhouses, where elongation starts in winter, the plants may start flowering in spring before the days become too long. This flowering is then followed by the elongation of the basal buds and a second flowering flush in autumn.

Effects of Gibberellic Acid

In winter, GA_3 sprays (25 to 50 ppm) caused some stem elongation; these stems branched early and remained very short. Higher doses of GA_3 produced poor spindly stems. GA thus cannot entirely replace the LD, but GA_3 sprays do advance stem elongation under LD conditions.

Regeneration Buds

These buds are always present at the base of the stem. As long as the stems continue active growth, the buds develop very slowly. Their development is enhanced concurrently with flower development in autumn (Figure 1C). This is not an SD effect, because the development of the basal buds may be enhanced at any time by the excision of the whole stem. If pruning is not complete, buds on the stem itself become active. These buds usually immediately start to elongate.

Vegetative Propagation

The plants may be propagated either by division of the rosettes or by rooting of the regeneration buds which are found at the base of the stems and even on the roots. Rooting can be accomplished easily either under mist or high humidity conditions. Sometimes even detached leaves may root and regenerate new buds. It is also possible to use stem cuttings, but such cuttings usually elongate prematurely and produce poor plants.

REFERENCES

1. **Peterson, D. L. and Bazzaz, F. A.,** Life cycle characteristics of *Aster pilosus* in early successional Habitats, *Ecology*, 59(5), 1005—1013, 1978.
2. **Allard, H. A. and Garner, W. W.,** Further observations of the response of various species of plants to length of day, *U.S. Dep. Agric. Tech. Bull.*, 727, 1940.
3. **Schwabe, W. W.,** *Aster novi-belgii*, in *Handbook of Flowering*, Vol. 5, Halevy, A. H., Ed., CRC Press, Boca Raton, Fla., 1985, 29—41.
4. **Kadman-Zahavi, A. and Yahel, H.,** unpublished results, 1984.

AURINIA SAXATILIS

En. Golden tuft; Fr. Corbielle d'Or; Ge. Echtes Steinkraut; Sp. Alkeli

Thomas C. Weiler and Karen G. Shedron

INTRODUCTION

Aurinia, once described as *Alyssum,* is a genus of about seven biennial or perennial herbaceous species native to central and southern Europe and Turkey. The basal leaves of this species form a tufted rosette, while stem leaves are reduced in size. As a member of the Cruciferae, the fruit is a flattened or slightly inflated silicle.[1]

Aurinia saxatilis (L.) Desv. (Figure 1) is commonly called in English golden tuft, basket-of-gold, madwort, goldentuft alyssum, gold-dust, or rock madwort. Plants are woody at the stem base and form panicles of intensely yellow flowers. The panicle expands in size during seed set. *A. saxatilis* blooms in early spring and can grow to 45 cm tall in rock gardens and perennial borders. The species is recommended for full-sun on well-drained sites. Flower stems are clipped after blooming to encourage new growth.[2,7,8]

COLD INDUCTION OF FLOWERING

Cold treatment promotes flowering in the many edible *Brassica* crucifers (broccoli, cabbage, cauliflower, etc.[5]), *Mathiola incana,* florists' stock,[4] and *Alyssum alyssoides.*[3] Cold treatment also appears to be the primary regulant of flowering in *A. saxatilis.*[9]

Since the plants remained green throughout treatment, low levels of cool white fluorescent lighting (7.2 to 18 μmol^{-2} sec^{-1}) for 10 hr each day were given during cold treatment. After 12 weeks of cold treatment at 4.5°C plants flowered in about 25 days.

The transition from juvenile to mature plants was studied in seed-propagated plants. Plant age was characterized by the number of axillary crowns formed per plant. Plants at the 4- to 5-crown age could not be successfully induced to flower even with 16 weeks of cold treatment, while all plants with 10 or more crowns each flowered after 12 weeks of cold treatment.

Plants induced to flower and brought to the greenhouse for forcing formed macroscopically visible flower buds within 7 to 16 days of growth at 18°C night temperature and 8 to 16 days at 13°C. Plants reached anthesis 17 to 27 days from visible buds at 18°C and 22 to 30 days at 13°C. This rapid flowering correlates well with the ability of the species to flower early in spring in the garden or nature, but the apparent ability of the species to initiate inflorescences at 4.5°C clouds measurements of the precise duration required for cold induction.

PHOTOPERIOD AFTER COLD TREATMENT

Flowering in some crucifers, including *Alyssum alyssoides,*[6] is promoted by long photoperiods. Similarly, *A. saxatilis*[9] growing at 18°C night temperatures, flowered about 2 weeks earlier when grown in long photoperiods after cold treatment (43 days of greenhouse forcing in 10-hr photoperiods compared to 29 days in 18-hr photoperiods). This was a 33% improvement in speed of flowering; however, lighted plants contained fewer inflorescences — e.g., eight inflorescences in a 10-hr photoperiod compared to two inflorescences in an 18-hr photoperiod.

FIGURE 1. *Aurinia saxatilis* (L.) Desv.

REFERENCES

1. **Bailey, L. H. (Staff of Hortorium),** *Hortus Third,* Macmillan, New York, 1976.
2. **Bailey, L. H.,** *Alyssum,* in *Standard Cyclopedia of Horticulture,* Bailey, L. H., Ed., Macmillan, New York, 1942, 268—269.
3. **Baskin, J. M. and Baskin, C. C.,** Effect of vernalization on flowering of the winter annual *Alyssum alyssoides, Bull. Torrey Bot. Club,* 101, 210—213, 1974.
4. **Cockshull, K. E.,** *Matthiola incana,* in *CRC Handbook of Flowering,* Vol. 3, Halevy, A. H., Ed., CRC Press, Boca Raton, Fla., 1985, 363—367.
5. **Friend, D. J. C.,** *Brassica,* in *CRC Handbook of Flowering,* Vol. 2, Halevy, A. H., Ed., CRC Press, Boca Raton, Fla., 1985, 48—77.
6. **Funke, G. L.,** The photoperiodicity of flowering under short days with supplemental light of different wave lengths, in *Vernalization and Photoperiodism,* Murneek, A. E. and Whyte, R. O., Eds., Chronica Bot., Waltham, Mass., 1948, 79—82.
7. **Hebb, R. S.,** *Low Maintenance Perennials,* Macmillan, New York, 1976.
8. **Nehrling, A. and Nehrling, I.,** *The Picture Book of Perennials,* Hearthside, New York, 1964.
9. **Shedron, K. G. and Weiler, T. C.,** Regulation of growth and flowering in basket of gold, *Aurinia saxatilis* (L.) Desv., *HortScience,* 17, 338—340, 1982.

BOUVARDIA

C. Vonk Noordegraaf

INTRODUCTION

Bouvardia is an ornamental crop with attractive flowers in a great variety of colors. In temperate climates production takes place mainly in glasshouses, as the plants are not winter hardy. Only one cv of *B. longiflora,* usually called *B. humboldtii,* is grown outdoors during summer. This plant blooms with large, creamy white flowers which have a strong scent. The other cvs are hybrids with white, pink, and red flowers. Better planning and control of production, and improvement of the vase life (by using preservatives) have renewed commercial interest in this old crop.

TAXONOMY

Bouvardia Salisb., a genus of mainly small shrubs, belongs to the family Rubiaceae. The leaves are simple, opposite or whorled, with narrow acute stipules adnate to the stalk. Flowers are situated on terminal, 3-flowered peduncles or in corymbs with 3-forked branches (Figure 1). The flowers contain a 4-lobed calyx and a long tubular, funnel-shaped corolla with 4 spreading lobes. There are 4 stamens, alternate with the corolla lobes and attached to the corolla tube. There are 2 carpels and 2 stigmas and the fruit is a capsule.

ORIGIN AND HISTORY OF CULTURE

The genus is native to Mexico and some other countries of Central America, and is named after the French medical doctor Charles Bouvard (1572 to 1657). A few species were introduced into England in 1845 and were later used for breeding. Parsons at Brighton, England, and Lemoine at Nancy, France, produced a number of hybrids. Undoubtedly, *B. leiantha* and *B. triphylla* have been used for these crosses. Some of the old hybrids are 'Hogarth', scarlet; 'President Cleveland', brilliant scarlet with crimson tube; 'President Garfield', double pink; 'The Bride', white; 'Triomphe de Nancy', double, orange red.[1]

Bouvardia was an important cut-flower crop around 1920 in England and Germany. Cultivars at present are: 'Red King', red; 'Sappho', dark pink; 'Aretusa', bright pink; 'Daphne', salmon pink; and 'Arteny', white.[3] Selected clones of these cvs are now used in cultivation, as there are great differences in growth, color, and health among the various genotypes.

PROPAGATION

Bouvardia is propagated by soft tip cuttings. In January to February, pruned mother plants are forced (20 to 22°C) under LD conditions to prevent flower bud formation. Young shoots with at least two well-developed leaf pairs are rooted in a mixture of peat and sand.

The temperature in the propagating house must be 23 to 25°C, with soil temperature of 22°C. Air humidity should be high to prevent wilting. Rooting can be done under a plastic tent or under mist. On sunny days shading is necessary.

Rooting can be promoted by using auxin powder (IAA 1% or IBA 2%).[3] Propagation can be also done by root cuttings although this is not a common procedure. Pieces (3 to 4 cm) of the main roots are placed at a soil temperature of 22°C and air temperature of 15 to 20°C. After 3 to 4 weeks, buds will develop from the inner parts of the roots and form young shoots. This method cannot be used with double-flowered cvs, as these are periclinal chimaeras.

FIGURE 1. Flowering shoot of a *Bouvardia* hybrid.

CULTIVATION

Bouvardia likes a humus soil. Rooted cuttings are planted in spring (March through May, 16 to m^2 of bench area). Growth starts slowly. After 3 to 4 weeks the plants are pinched at 2 to 3 leaf pairs. Later in the season and in the second year, bottom breaks will arise, which are important for good production. These breaks are also pinched at 2 to 3 leaf pairs. It is important to grow a strong, well-branched plant, since a *Bouvardia* crop lives for 5 to 6 years.

Bouvardia needs plenty of light. To obtain flowers of a good quality, the normal temperature in winter should not be lower than 12 to 15°C. In summer, the minimum night temperature can be 15 to 17°C. During the rest period in winter, the temperature can be dropped to 5 to 8°C and plants should be kept dry. However it is not necessary to give a rest period. Plants can be pruned throughout the year, both after a rest and during summer.

FLOWER INITIATION

Under greenhouse conditions in northern Europe, *Bouvardia* can flower the whole year. The length of the shoots and the number of leaf pairs at which flowers are initiated vary. In summer, when growth is strong, only long shoots will initiate flowers. To achieve flowering shoots of good quality, many vegetative lateral shoots must be removed by hand. These lateral shoots have a negative influence on the vase life of the cut flower shoots.[2]

In winter, however, flowers are initiated early, on very short shoots. Therefore, there is little or no vegetative growth in winter and the shrubs become very weak and can easily die.

Bouvardia is a SDP, and by controlling the photoperiod, FI can be induced at any desired time; in winter, vegetative growth can be promoted. The critical daylength for flower initiation is between 10 to 12 hr.[7,11] Two to three weeks of SD (less than 12 hr) are sufficient to initiate flowers.[6] Other growth conditions, such as temperature, can influence the photoperiodic reaction to some extent. High temperatures (above 21 to 25°C) retard flower induction in comparison to lower temperatures.[5,9,11] Vegetative growth can be promoted and FI prevented in winter by LD lighting with incandescent lamps (150 to 300 mW m^{-2}). Cyclic lighting (10 min light and 20 min dark) for several hr in the middle of the night (as used for chrysanthemums) will inhibit flowering.[10] The time to start SD treatment or to discontinue lighting in winter depends on the desired shoot length at harvest time. Above the last-developed leaf pair before induction, 6 to 7 leaf pairs will develop, of which, usually, three bear flowers in the axils.

Continuation of the SD treatment after induction does not influence flower development, but has small influence on leaf size and length of internodes in the inflorescence; both are promoted by LD.[8]

POSTHARVEST TREATMENT

Bouvardia stems are usually harvested when 2 to 3 flowers are open. They are cut at 1 or 2 leaf pairs above the last cut. The flower stems must be placed in water immediately after cutting, to prevent them from wilting. Excessive transpiration of the leaves must be prevented. After sorting and grading, the flowers are placed in storage at 2 to 4°C in water containing a preservative.

REFERENCES

1. **Bailey, L. H.,** *The Standard Encyclopedia of Horticulture,* Vol. 1, Macmillan, New York, 1930, 536—537.
2. **Barendse, L. V. J.,** Houdbaarheidsproeven met geplozen *Bouvardia, Vakbl. Bloemisterij,* 36(12), 44—45, 1981.
3. **Berg, A. J. van den and de Joode, A.,** Teelt van *Bouvardia, Bloementeeltinformatie,* No. 25, Proefstation voor de Bloemisterij, 1983, 59.
4. **Papenhagen, A.,** Bouvardien, *Gärtnerbörse Gartenwelt,* 78.13, 287—288, 1978.
5. **Pol, P. H. van den and Elands, J.,** Kwaliteitsverbetering met energiesbesparing bij *Bouvardia, Vakbl. Bloemisterij,* 21, 38—39, 1983.
6. **Vonk Noordegraaf, C.,** Periode korte dag die nodig is voor bloeiinductie., *Bloemisterijonderzoek in Nederland 1981,* Res. Stn. Floriculture, Aalsmeer 1981, 71—72.
7. **Vonk Noordegraaf, C.,** Invloed van daglengte op de bloei bij *Bouvardia, Bloemisterijonderzoek in Nederland 1981,* Res. Stn. Floriculture, Aalsmeer, 1981, 72—73.

8. **Vink Noordegraaf, C.,** Invloed van daglengte op scheutlengte en bloemontwikkeling bij *Bouvardia,* Aalsmeer, *Bloemisterijonderzoek in Nederland 1982,* Res. Stn. Floriculture, Aalsmeer, 1982, 92—93.
9. **Vonk Noordegraaf, C.,** Invloed van temperatuur en korte dag op de bloei van *Bouvardia, Bloemisterijonderzoek in Nederland 1982,* Res. Stn. Floricutlture, Aalsmeer, 1982, 93—94.
10. **Vonk Noordegraaf, C.,** Bloeispreiding bij *Bouvardia, Vakbl. Bloemisterij,* 38(21), 40—41, 1983.
11. **Vonk Noordegraaf. C.,** The influence of light and temperature on growth and development of *Bouvardia, Acta Hortic.,* 147, 187—191, 1983.

CALENDULA OFFICINALIS

En. Calendula, Pot marigold; Fr. Calendule, Souci; Ge. Kalendel, Ringelblume, Ringelrose

H. F. Wilkins

INTRODUCTION

Calendula officinalis L. is a garden plant, grown also for cut-flower production during the winter in fields under Mediterranean climate conditions or under glass in temperate climates. Flowering naturally in northern gardens in the spring and summer, plants can endure light frosts and light snow in the autumn and winter.[2,10]

This herbaceous ornamental annual has a basal rosette of leaves, but stems are up to 60 cm in height when they elongate after floral initiation and floral development. Leaves are entire to slightly toothed, oblong to oblong-obovate, and slightly clasping the stems.[2]

Flower heads are 4 to 15 cm across, with ray and disc flowers which are pale yellow to a deep orange color. This member of the Asteraceae is a native of southern Europe and occasionally escapes from gardens under favorable climatic conditions.[2]

PHOTOPERIOD

Baszynski[3] reported that *C. officinalis* is a LDP while Van de Pol[13] and Vince-Prue[14] stated it is a day-neutral plant, preferring LD. Han and Yeam[4] stated LD promoted reproductive growth, but SD did not repress vegetative growth since leaf numbers and stem diameters were similar, but under SD plants were shorter.

This controversy was resolved by Kamlesh and Kohli,[6] who demonstrated that this plant is an absolute LDP with a very short critical day of only 6 $^1/_2$ hr. Under a photoperiod of 5 $^1/_2$ hr no floral initiation occurred; at six hr only one plant in ten initiated floral primordia. There were no differences in height or leaf number between plants exposed to photoperiods of 6 $^1/_2$ hr or longer. Plants under only a 4-hr photoperiod were very similar in appearance and height to plants under longer photoperiods, except that they failed to flower. Plants are not sensitive to photoperiod until four leaves unfold.[5]

The SDP *Xanthium strumarium* flowered under LD conditions when shoots of reproductive *C. officinalis* were grafted onto vegetative *Xanthium* stock; but flowering did not occur when reproductive *Calendula* was the stock. Too, reproductive *Calendula* scion shoots increased the propensity for maleness when the SD *Xanthium* plants flowered under LD conditions.[13]

TEMPERATURE

Plants grow well at 10°C night temperature. In northern greenhouses, flowering is confined to the period from October to May. In warmer climates, production is during the winter when temperatures are cool. High temperatures result in poorer quality and capitula possessing only one whorl of ray florets.[10]

GROWTH REGULATORS

Maleic hydrazide (50 ppm) reduced the elongation of the main stem by 50%, and increased the number of laterals, but flowering was delayed by 12 days.[12] GA_3 increased the length of the first five axillary branches; however, depending on the concentration, GA_3 could inhibit or increase axillary elongation of lower-order branches.[8] Chlorflurenol caused flower

FIGURE 1. The effect of vitamin E sprays on flower development of *Calendula officinalis* L. All plants were grown under fluorescent light (5.5 klx). The plant on the far left was grown under a 16-hr (LD) photoperiod; the three plants on the right were grown under an 8-hr (SD) photoperiod. The two center plants, designated SDT_1 or SDT_2, were sprayed, respectively, with 30 or 10 μg of vitamin E. The LD plants flowered within 45 days, the SDT_1 plants within 64 days, and the SDT_2 within 81 days. The SD plants never flowered. (From Baszynski, T., *Naturwissenschaften*, 54, 339, 1967. With permission.)

malformation and delayed early winter flowering, but growth and production was enhanced when flowering did occur.[9] Cycloheximide failed to inhibit flowering.[7]

DL-α-Tocopherol (vitamin E) sprays induced flowering under SD.[3] These sprays also resulted in greater multiple flowering and increased vegetative growth under SD when compared to LD-exposed, unsprayed plants. Vitamin E (10 to 80 μg) was sprayed ten times every other day, with flowering under SD within 64 days after seed germination as compared with 45 days for the LD-treated plants. Seeds of *Calendula officinalis* were found to contain α-, γ-, and δ-tocopherol[3] (Figure 1).

CULTURE

Plants grow best under full light intensities in the fall, winter, and spring. In the summer, a light shade will increase flower quality.[10]

This plant is seed propagated at 21°C. Seeds are sown 1 month prior to the date they are to be transplanted.[1] From seed to flower for August and September seed dates requires 3 to 4 months but only 2 months are required for an April seeding to flower.[10] Reger[11] demonstrated that when *Calendula* plants were grown dry, flowering was delayed.

REFERENCES

1. **Auman, C. W. and Larson, R. L.,** Minor cut crops, in *Introduction to Floriculture,* Academic Press, New York, 1980, 190.
2. **Bailey, L. H. (Staff of Hortorium),** *Hortus Third, A Concise Dictionary of Plants Cultivated in the United States and Canada,* Macmillan, New York, 1979, 200.

3. **Baszynski, T.,** The effect of Vitamin E on flower initiation in *Calendula officinalis* L., grown in short-day, *Naturwissenschaften,* 54, 339—340, 1967.
4. **Han, I.-S. and Yeam, D.-Y.,** The effect of photoperiod on the growth and flowering of marigold, salvia, calendula, petunia, and zinnia plants, *J. Korean Soc. Hortic. Sci.,* 19, 117—128, 1978.
5. **Kamlesh, S. S.,** Effect of Some Regulatory Substances on Growth and Development of *Calendula officinalis* in Relation to Photoperiod, M.Sc. thesis, Guru Nanak Dev University, Amritsar, India, 1977.
6. **Kamlesh, S. S. and Kohli, R. K.,** *Calendula officinalis* L., a long day plant with an exceptionally low photoperiodic requirement for flowering, *Indian J. Plant Physiol.,* 24, 299—303, 1981.
7. **Kamlesh, S. S. and Sawhney, N.,** Failure of cycloheximide to inhibit flowering in *Impatiens balsamina* and *Calendula officinalis, Acta Bot. Indica,* 10, 104—106, 1982.
8. **Mohan Ram, H. Y. and Mehta, U.,** Effect of gibberellic acid on the growth of main shoot and axillary branches in *Calendula officinalis, Proc. Indian Acad. Sci.,* 87, 255—270, 1978.
9. **Mehta, U. and Mohan Ram, H. Y.,** Modification of the capitulum in *Calendula officinalis* by chlorflorenol, *Phytomorphology,* 28, 351—369, 1978.
10. **Post, K.,** *Florist Crop Production,* Orange Judd, New York, 1952, 172, 358—360.
11. **Reger, M. W.,** The relation of soil moisture to the time of bloom of calendulas, larkspurs and geraniums, *Proc. Am. Soc. Hortic. Sci.,* 39, 381—383, 1941.
12. **Srivastava, V. K. and Bajpai, P. N.,** A note on maleic-hydrazide treated calendula, *Madras Agric. J.,* 51, 515—516, 1964.
13. **Van de Pol, P. A.,** The floral hormones of *Helianthus annuus* L., *Calendula officinalis* L. and *Xanthium strumarium* L., *Proc. K. Ned. Akad. Wet.,* 74, 449—454, 1971.
14. **Vince-Prue, D.,** *Photoperiodism in Plants,* McGraw-Hill, London, 1975, 16.

CELOSIA

En. Cockscomb; Fr. Celosie; Gr. Brandschopf, Hahnenkamm*

Allan M. Armitage

INTRODUCTION

The genus *Celosia* L. is a member of the Amaranthaceae family and consists of approximately 60 species. They are annual or perenial herbs native to warm regions of America and Africa,[4] commonly used as a leafy vegetable and pot herb in many West African countries. As an ornamental flower, they are used as potted plants, bedding plants, or cut flowers. The common flowering annual *Celosia cristata* L. consists of two distinct flower forms. The crested group has fasciated convoluted inflorescences, and the plumosa group has feather-like flowers. These forms are tetraploid cultigens and range from white, and yellow, to various shades of red.[4] Both flowering forms contain a large number of hybrids and are well established in the floricultural trade.[14]

MORPHOLOGY

The leaves of *C. cristata* are alternate and are entire or rarely lobed. The individual flowers are small and inconspicuous; however, the numerous densely spaced flowers comprising the inflorescence are showy. The terminal inflorescence of the crested forms has been bred into widely fasciated convoluted combs, whereas the terminal flowers of the plumosa form are held above the leaves in a dense panicle. Lateral flowers are much more prevalent on plumosa forms compared with crested forms.[5]

The flowers are bisexual and consist of a 5-parted perianth and 5 stamens, and the stamen filaments are united at the base. The perianth is very persistent and may last many months under garden culture. The fruit is a dehiscent utricle with 2 or more ovaries.[4,5]

ENVIRONMENTAL EFFECTS ON FLOWERING

Photoperiod

C. cristata was originally thought to be day neutral with respect to flowering;[1] however, further studies have since determined that flowering in *C. cristata* is a quantitative SD response.[11,19] Piringer and Borthwick,[19] working with one plumose and three cristata cvs, determined that a photoperiod or 14 hr or less resulted in early flower initiation compared with 16 hr or continuous photoperiod. With an 8-hr photoperiod, night interruptions with 3 hr of incandescent light resulted in significant delay of flowering in two cultivars compared with plants that received an 8-hr day extension. On the other hand, little or no difference in flowering time occurred between day extension and night interruption in two other cvs.[19] There were no differences in flowering time when the daylength was extended to 12 hr by 4 hr of incandescent or fluorescent lamps at equal illuminance. When 8-hr light extensions were imposed to provide a 16-hr photoperiod, there were no significant differences in flowering time between light sources for all but one cv.[19]

Driss-Ecole[11] noted that when seedlings were placed under photoperiods of 8 to 12 hr, flower initiation occurred by 40 days or when 7 to 12 leaves had initiated. Under a 16-hr photoperiod 60 to 70 days were required for floral initiation to occur. Transferring plants from an 8- to a 16-hr photoperiod after 30 days resulted in significantly higher percentages

**Celosia cristata*

of flowering plants compared with transferring them after 20 days.[11] However, all plants transferred from a 16- to an 8-hr photoperiod either after 20 or 30 LD reached anthesis by 70 days from sowing.[12] These data support the hypothesis that activation of the meristem (meristem corpus) began under LD and was followed by activation of the axial tunica and formation of prefloral meristem under inductive SD. This supports Lance's[16] findings of an intermediate state of the meristem (*etat intermediare*) under noninductive photoperiods. Driss-Ecole[13] also showed that some shift of cells in the meristem of cristata types from G_2 to G_1 phase of the mitotic cycle had occurred at the time of irreversible commitment to flower, similar to other species.[6,15] By looking at the distribution of nuclear DNA, she determined that the critical period for ontogenesis of the shoot apex was between 18 and 24 days.[13]

Light Intensity and Supplemental Light

Little work has been conducted on the effects of light intensity and high-intensity supplemental light on *Celosia* flowering. It is likely that reduced light would delay induction of the prefloral stage and possibly organogenesis.[3,20] Boodley[7] grew many annuals under fluorescent light at 15 to 21°C and obtained salable plants of *Celosia* in 5 weeks. Other annuals have shown reduced flowering time with fluorescent and HID lamps,[8] and it is likely *Celosia* would react similarly.

FASCIATION

Fasciation is most pronounced in the cristata form compared with the plumosa form of *Celosia*. The development of fasciation involves a broadening and reorganization of the meristem (Figures 1 to 3). Fasciation of the flowering stem is most prominent under photoperiods of 16 hr or longer.[11,19] Sixty-nine percent of apical meristems of plants in constant 16-hr days became fasciated and the vegetative phase was prolonged, while constant 8-hr photoperiod resulted in only 3% fasciation and plants reached the prefloral stage much more rapidly.[11,12] Plants given 12-hr photoperiods were intermediate in their development of fasciation.[11] Plants transferred to different photoperiods after 20 LD or 30 SD had little fasciation. After 20 SD or 30 LD, however, 45 and 50% of the plants showed fasciation, respectively.[12] This indicates that development of fasciation in *Celosia* occurs early in the development of the plant (20 to 30 days under LD). Fasciation was also much more frequent in plants grown at the widest spacing compared with those grown at very close spacing, and apparently adequate nutrition and water resulted in intensified fasciation.[17]

GROWTH REGULATORS

Ancymidol is the accepted commercial recommendation for height reduction in *Celosia*.[9] Nijs and Boesman[18] showed that 4000 ppm chlormequat not only reduced height, but also delayed flowering by 22 days. GA_3 at 100 ppm reduced internode elongation without delaying flowering, while IBA had no effect on growth or flowering.[18]

SCHEDULING

The advent of increasing light intensity and temperature from winter to spring resulted in earlier flowering of *Celosia cristata* in recent studies (Figure 4).[2] At 33°N latitude, photoperiod is never greater than 16 hr and a LD delay in flower initiation does not occur. Total time to flower varied from 14 weeks during winter months to 6 weeks during summer months. Dietz[10] recommended a total time of 9 weeks at 17°C and 10 weeks at 12°C night temperature.

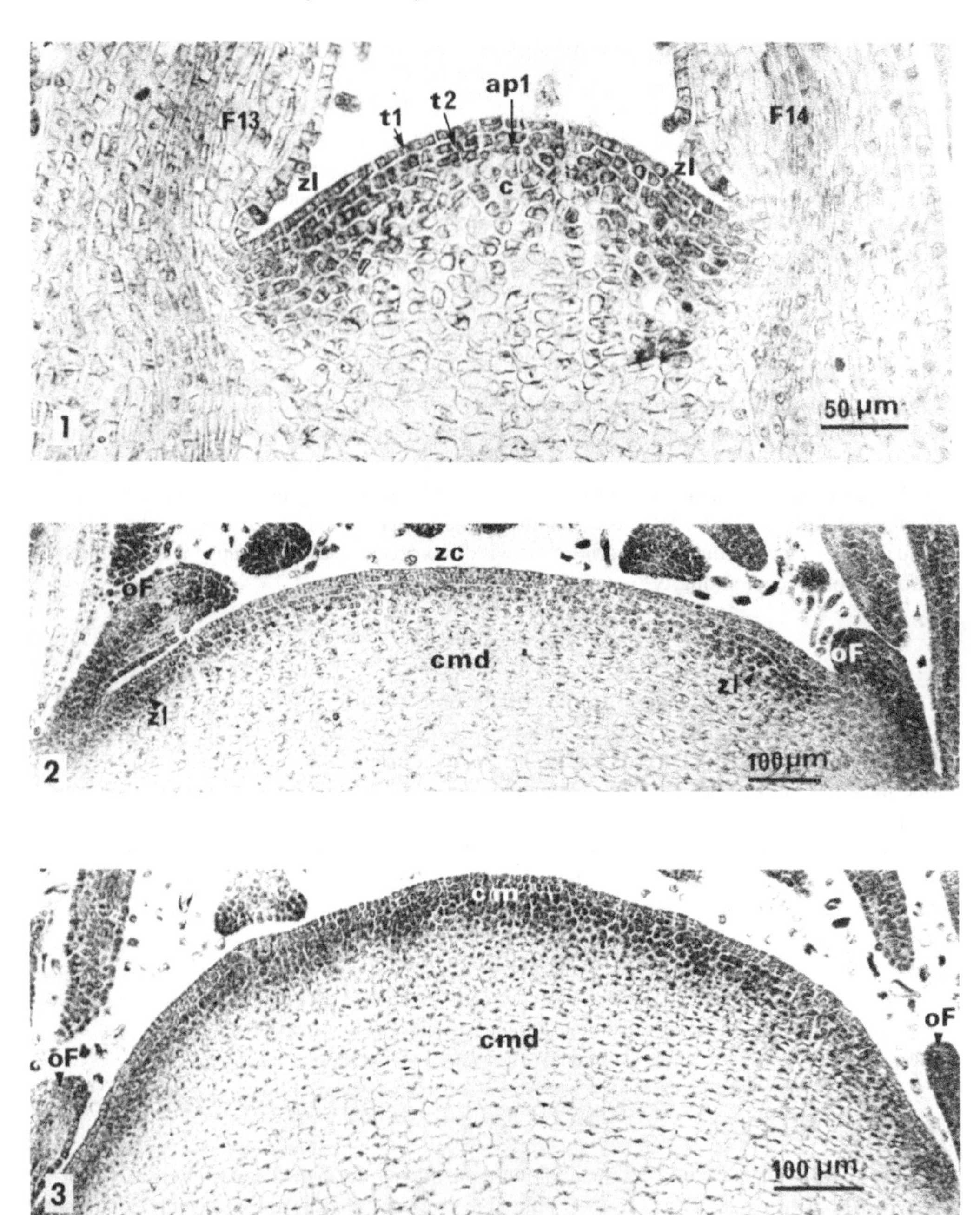

Figures 1 to 3. Development of fasciation in *Celosia*. The apical meristems of plants subjected to 16-hr days undergo fasciation. Figure 1. Beginning of the fasciation of the meristem. A new layer of cells (ap1) has differentiated, beginning with the corpus cells. Figure 2. End of the vegetative phase of the fasciated meristem. The meristem has become very large and remains zoned. The old axial layer connecting the *tunica* and *corpus* has given way to a central zone (zc) forming 4 regularly arranged layers of cells. The lateral zones (zl) are initiating leaf organelles (oF). Figure 3. The prefloral phase of the fasciated meristem ends with the homogenation of the 4 superficial layers and with the rising of the prefloral crest above the last initiating leaves. Abbreviations: c, corpus; cm, meristematic cells; cmd, differentiated medullary cells; F_6 to F_{14}, leaf numbers 6 to 14; t_1 and t_2, tunica layers; 21, initiating lateral zone from leaves or from leaf organs that have leaves. (From Driss-Ecole, D., *Can. J. Bot.*, 55, 1488—1500, 1977. With permission.)

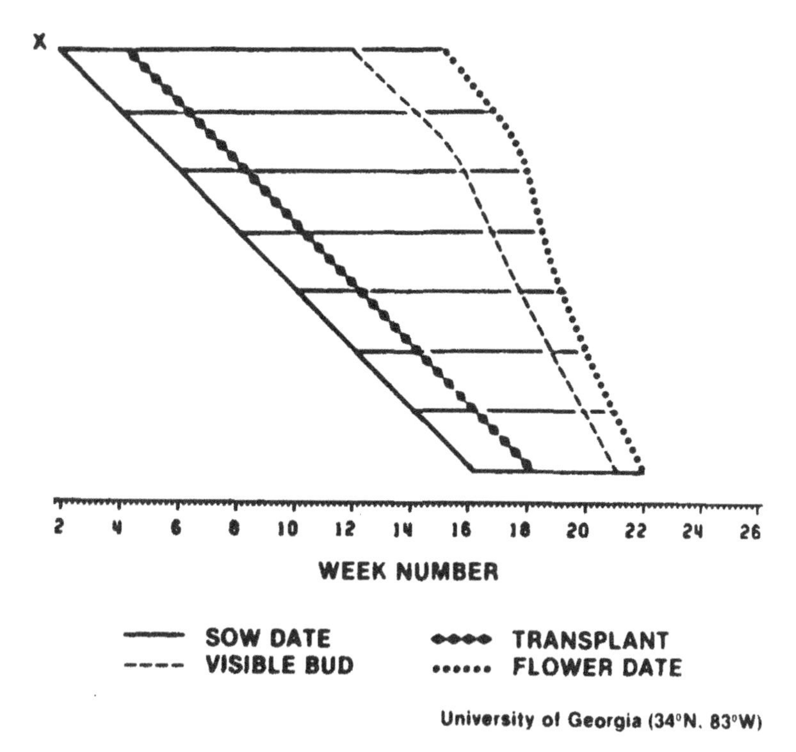

Figure 4. Greenhouse scheduling of *C. argentea* as influenced by sowing date.[2] Instructions: if seed were sown on week 2, follow the horizontal line starting at week 2 (X) until it intersects the transplant line (week 4), the visible bud line (week 12), and the flower date (week 15). This indicates that for seed sown on week 2, transplanting should be done week 4, flower buds will be visible week 12, and anthesis will occur around week 15. Total crop time: 13 weeks; if seed were sown on week 16, flowering would occur on week 22. Total crop time — 6 weeks. Cultivars tested: 'Brushfire' and 'Geisha Mix'.

REFERENCES

1. **Allard, H. A. and Garner, W. W.,** Further observations on the response of various species of plants to length of day, *U.S. Dep. Agric. Tech. Bull.* No. 727, 1940.
2. **Armitage, A. M.,** Determining optimum sowing time of bedding plants for extended marketing periods, *Acta Hortic.*, 147, 1983.
3. **Armitage, A. M. and Wetzstein, H. Y.,** Influence of light intensity on flower initiation and differentiation in hybrid geranium, *HortScience*, 19(1), 114—116, 1984.
4. **Bailey, L. H. (Staff of Hortorium),** *Hortus Third, A Concise Dictionary of Plants Cultivated in the United States and Canada*, Macmillan, New York, 1976.
5. **Bailey, L. H.,** *Manual of Cultivated Plants*, Macmillan, New York, 1951.
6. **Bernier, G., Kinet, J. M., and Bronchart, R.,** Cellular events at the meristem during floral induction in *Sinapsis alba* L., *Physiol. Veg.*, 5, 311—324, 1967.
7. **Boodley, J. W.,** Starting annuals under fluorescent lights, *Bull. N.Y. State Flor. Grow.*, 189, 1, 3—4, 1961.
8. **Carpenter, W. S. and Beck, J. R.,** High intensity supplementary lighting of bedding plants after transplanting, *HortScience*, 8, 482—483, 1973.
9. **Cathey, H. M.,** Growth regulators, in *Bedding Plants, A Manual on the Culture of Bedding Plants as a Greenhouse Crop*, Mastalez, J. W., Ed., Pennsylvania Flower Growers Association, University Park, 1976, 177—189.
10. **Dietz, C.,** Sowing schedules, in *Bedding Plants, A Manual on the Culture of Bedding Plants as a Greenhouse Crop*, Mastalez, J. W., Ed., Pennsylvania Flower Growers Association, University Park, 1976, 55—65.

11. **Driss-Ecole, D.,** Influence de la photoperiode sur de comportement du meristeme caulinaire du *Celosia cristata, Can. J. Bot.,* 55, 1488—1500, 1977.
12. **Driss-Ecole, D.,** Influence de la photoperiode sur la fasciation et la phase reproductrice du *Celosia cristata* (Amarantacees), *Can. J. Bot.,* 56, 166—169, 1978.
13. **Driss-Ecole, D.,** Etude de la teneur en DNA nucleaire dans le meristeme apical des tiges du *Celosia cristata* (Amarantacees) en fonction de la photoperiode, *Can. J. Bot.,* 57, 2760—2765, 1979.
14. **Harbaugh, B. K. and Waters, W. E.,** Evaluation of flowering potted plants under simulated home conditions, *HortScience,* 14, 743—745, 1979.
15. **Jacqmand, A. and Miksche, J. P.,** Cell population and quantitative changes of DNA in the shoot apex of *Sinapsis alba* during floral induction, *Bot. Gaz. (Chicago),* 132, 364—366, 1971.
16. **Lance, A.,** Recherches cytologiques sur l'evocation de quelques meristems apecaux et sur ses variations provoquees par des traitements photoperiodiques, *Ann. Sci. Nat. Bot. Sci.,* 11, 18, 91—421, 1967.
17. **Lebedeva, T. I.,** Conditions for the manifestation of fasciation, *Bot. Z.,* 51 (Abstr.), 1316—1318, 1966.
18. **Nijs, L. P. de and Boesman, G.,** Growth and flowering of *Celosia cristata* L. 'Tokyo Market' after treatment with different growth regulators, *Meded. Fac. Landbouwwet. Rijksuniv. Gent,* 46, 223—226, 1981.
19. **Piringer, A. A. and Borthwick, H. A.,** Effects of photoperiod and kind of supplemental light on growth, flowering, and stem fasciation of *Celosia, Am. J. Bot.,* 48, 588—592, 1961.
20. **Wetzstein, H. Y. and Armitage, A. M.,** Inflorescence and floral development, in *Pelargonium* × *hortorum, J. Am. Soc. Hortic. Sci.,* 108, 595—600, 1983.

COLA

En. Kola tree; Fr. Kolatier; Ge. Kolabaum; Sp. Kolatero

C. L. M. van Eijnatten

INTRODUCTION

Kola, of which the embryos form the "kola nuts" of the trade, may belong to several species of evergreen and sizeable trees, growing up to 25 m in height. These species belong to the genus *Cola* (Schott and Endlicher) within the family Sterculeaceae. The most important species is *Cola nitida* (Ventenat) Schott and Endlicher, the major contributor to the trade in kola nuts and originating from the west African forest zone west of the coastal savanna in Benin, Ghana, and Togo. Second in importance is *Cola acuminata* (Beauvoir) Schott and Endlicher, originating from the rain forest areas east of the coastal savanna in west Africa, mainly beyond the river Niger, including coastal areas of Cameroon, Gabon, and Zaire. The kola tree has been cultivated for its seeds, of which the fleshy cotyledons contain some alkaloids (caffein, kolanin, theobromin) which dispel sleep, thirst, and hunger. In west Africa the kola nuts still play an important role and they are interwoven with many traditional, social customs. The embryos of the seed, i.e., the kola nuts, are chewed fresh in west Africa. In Central and South America they may be dried and utilized in the preparation of an infusion.

Kola trees are typically cultivated in humid tropical lowland areas, characterized by only short dry seasons. The crop is propagated by seed, although attempts are being made to achieve vegetative propagation through rooted cuttings and budding/grafting procedures.[1] The need for vegetative propagation is pressing because of great variability in productivity of individual trees and the cross-pollinating nature of the crop.[2] Most of the crop is cultivated or "gathered" in the original areas of distribution, chiefly in the forest zone of west Africa and to some extent in central Africa. Some kola is produced in the Caribbean islands and around Bahia in Brazil.

THE MATURE PHASE OF THE KOLA TREE

Upon germination, the kola tree passes through a juvenile stage for 2 to 3 years. The seedling has an indeterminate type of growth, regularly increasing in length from the apical bud. If this bud remains undamaged, it produces a stem along which leaves are distributed more or less regularly. If any damage to the apical bud occurs, the seedling may form a multiple fork of which the branches, however, retain the monopodial type of growth.[3] The tree never produces flowers as long as this monopodial growth pattern is maintained.

During the 3rd or 4th year the pattern of growth becomes periodic and occurs in bursts. The number of leaves on any one shoot is, from that moment onwards, determined in the bud from which newly developed branches emanate. The emergence of young shoots occurs in regular "flushes", causing the tree to grow discontinuously. However, the terminal buds remain dormant, abort, or develop into one or a few leaves at the time of the next flushing period. One or few of the lateral buds below the apex may give rise to the next flush, thus taking over dominance from the apical bud. The growth rhythm, therefore, has become sympodial.

Flowering may occur as soon as the kola tree changes from a monopodial to a sympodial growth type, i.e., in the 3rd or 4th year of growth. Usually, however, flowering starts in the 6th or 7th year. Observations show that only well-developed seedlings initiate flowering in the 3rd year, indicating that in other trees environmental factors probably prevent anthesis.[3] Deviating tree types occur, which have steeply ascending "orthotropic" branches, which never produce flowers.[4] Such trees presumably failed to develop the sympodial type of growth, a requisite for the occurrence of flowers.

THE FLUSH

Shoots developing during the mature phase (sympodial growth rhythm) have their leaves clustered towards the apex. Leaves are arranged increasingly close to each other, thus giving the impression that foliage is placed almost in whorls at the tips of the flushes. Further down along the stem of the flush, a few widely spaced caducous bracts occur. The upper lateral buds on any one flush may give rise to new flushes. The lower lateral buds may develop into inflorescences or remain dormant, thus constituting reserves for flushing or flowering in the next 3 or 4 years.

The flushes develop more or less simultaneously after the onset of the main rainy season. Subsequent, often partial, flushing occurs at later intervals of some 10 to 12 weeks, but less uniformly than the first flushing period. A second general flush occurs after fruits have matured in November/December at the onset of the long dry season. Many of the shoots developed during the first flush at the beginning of the main rainy season bear flowers during the flowering season from August to October. Some 10 to 20% of the flowers are produced in other times of the year.

FLORAL MORPHOLOGY AND BIOLOGY

Flowers are borne in inflorescences of which the terminal flower develops first. The inflorescence is a paniculate cyme. The flowers are of two types, either male with a rudimentary gynoecium, or hermaphrodite with an (apparently) developed androecium and a functionable gynoecium (see Figure 1). The two types of flowers occur in separate inflorescences or are mixed within the same inflorescence. This is strongly related to the position of the inflorescence on the shoot. The lowest axillary buds of a flush or shoot may give rise to inflorescences with wholy hermaphrodite flowers or with both male and hermaphrodite flowers. The relative number of hermaphrodite flowers decreases from the base of the flush upwards. It often happens, however, that no hermaphrodite flowers are formed at all in inflorescences arising from one flush. Furthermore, during the normal flowering season the percentage of hermaphrodite flowers is 10 to 30%,[4,5] but inflorescences appearing outside the normal flowering season may contain much higher percentages of hermaphrodite flowers. These proportions also vary from tree to tree. It is suggested that the proportion of hermaphrodite flowers in the inflorescences is likely to be determined partly by the balance of various plant hormones.

Flowering occurs on young shoots 4 to 5 months old. Buds developing into inflorescences have a thickly rounded appearance as distinct from vegetative buds, which have a thin conical shape. The inflorescence bud develops its thickly rounded shape some weeks prior to the shedding of the first bract (see Figure 2).

Shedding of the first or outer bract of this bud marks the first day of development of the inflorescence. Until that moment, the bud has a globular primordium, developing into an inflorescence with partly differentiated flowers in the period between shedding of the first, brown and coriaceous bract and the second, white bract. This period lasts about 3 days. When the second bract is shed, the inflorescence becomes visible and the flowers continue their differentiation. Sepals are formed first, followed by the androecium and later the gynoecium.

The reduction division in the pollen mother cells of the most advanced terminal flower

FIGURE 1. Inflorescence of kola. (From Eijnatten, C. L. M. van, Commun. No. 59, Department of Agricultural Research, Royal Tropical Institute, Amsterdam, 1969. With permission.)

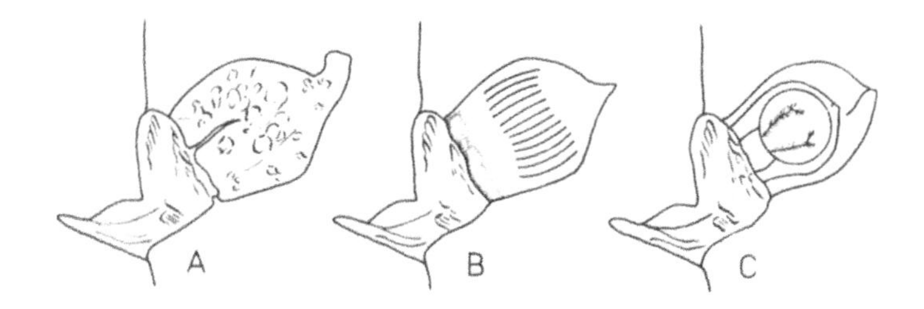

FIGURE 2. Three initial stages of development of an inflorescence bud: (A) shedding of the first bract; (B) the second bract enveloping the inflorescence; and (C) the inflorescence after shedding of the second bract. (From Eijnatten, C. L. M. van, Commun. No. 59, Department of Agricultural Research, Royal Tropical Institute, Amsterdam, 1969. With permission.)

coincides with shedding of the second bract surrounding the inflorescence. During the first metaphase of the reduction division, 20 bivalents become visible. The chromosome number ($2n$) is, therefore, 40. Within 4 to 5 days after the reduction divisions of the pollen mother cells the pollen nuclei divide into two. It is only at this time that the gynoecium starts to develop up to the formation of the ovules. In the case of hermaphrodite flowers, ovules are formed subsequently. In male flowers, however, development halts at this stage and, in that case, the gynoecium consists of sterile carpidia.

When the flowers are ready to open, the male flowers have a subspherical shape and measure 12 to 20 mm in diameter. Hermaphrodite flowers have at this stage a more oval shape and are larger: 30 to 40 mm in diamter.[3,6] The flowers have 5 connate creamy-white sepals each with 3 parallel red lines which vary in width and length. Occasionally this red

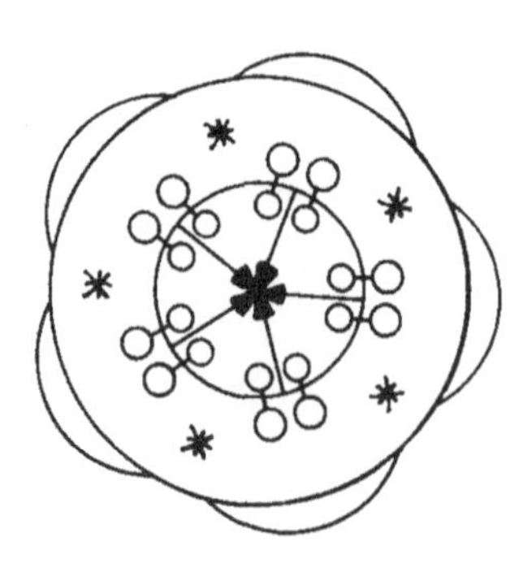

FIGURE 3. Floral diagram of a male kola flower, showing connate sepals, absence of petals, staminal groups, and carpidia. (From Eijnatten, C. L. M. van, Commun. No. 59, Department of Agricultural Research, Royal Tropical Institute, Amsterdam, 1969. With permission.)

coloration may be absent altogether, while also the other extreme, complete red coloration of the adaxial side of the sepals, has been observed. Petals are absent. The androecium consists of 5 connate staminal groups of 4 pollen sacs, which sporulate freely. The structure of the male flower is schematically presented in a floral diagram (Figure 3). In hermaphrodite flowers the pollen sacs are indehiscent and do not sporulate. Although pollen from pollen sacs of hermaphrodite flowers is viable, it has proven to be nonfunctional when used in pollination. The 5 carpels of the hermaphrodite flowers contain 10 to 12 anatropous ovules implanted along the ventral, adaxial side of the carpels.

INFLUENCING THE FLOWERING PHENOMENON

Early attempts to study the influence of hormonal effects indicated that IAA, IBA, or NAA, when applied to inflorescence buds at the moment of shedding of the first bract, induced a more rapid development of the inflorescence. However, the inflorescence remained compact and soon turned brown. Application of these plant hormones to the foliage in low concentration dusts or sprays resulted in much longer retention of the flowers than would normally be the case. The growth hormones 2,4-D and 2,4,5 TP also had such effects;[3] and the application of bimonthly sprays of TIBA was shown to positively influence the number of inflorescences emerging.[8]

Interruption of normal growth may have a positive effect on productivity of kola trees. This may be achieved by ringing the tree or by otherwise interrupting the normal transfer of photosynthates through the phloem tissues of the bark. Part of the effects of such actions appears an induction of dormant "floral" buds to develop, thus increasing the intensity of flowering of the trees.[9]

REFERENCES

1. **Veen, H. A. G., Awonusi, E. O., Furste, L. J., Odusolu, E. O., Eijnatten, C. L. M. van, and Olaniran, Y. A. O.,** Higher productivity of our kola farms, final rep., Kola Pilot Project Ifo-Ogun State; Ministry of Economic Development and Reconstruction, and Ministry of Agriculture Ogun State, Nigeria, and Ministry of Foreign Affairs, Netherlands, 1977.
2. **Eijnatten, C. L. M. van,** Kolanut; in Ferwerda, F. P. and Wit, F., Outlines of perennial crop breeding in the tropics, *Misc. Pap.*, No. 4, Agricultural University of Wageningen, Netherlands, 1969.

3. **Bodard, M.,** Contribution à l'étude systématique du genre *Cola* en Afrique Occidentale, Annales Faculté de Sciences, University of Dakar, Dakar, Senegal, 1962.
4. **Dublin, P.,** Colatier, *Rapp. Ann. 1961,* Centre Recherches Agronomiques, Boukoko, Rép. Centrafricaine, 1961.
5. **Eijnatten, C. L. M. van,** Kola, its botany and cultivation, Commun. No. 59, Department of Agricultural Research, Royal Tropical Institute, Amsterdam, 1969.
6. **Russell, T. A.,** The kola of Nigeria and the Cameroons, *Trop. Agric. Trinidad,* 32, 210—240, 1955.
7. **Bodard, M.,** Contribution à l'étude de *Cola nitida.* Croissance et biologie florale, Bull. No. 11, Centre Recherches Agronomiques, Bingerville, Ivory Coast, 1955, 3—28.
8. **Ashiru, G. A.,** The influence of TIBA on the growth and flowering of kola, Annu. Rep. Cocoa Research Institute, Ibadan, Nigeria, 1970/71, 1972, 81—82.
9. **Eijnatten, C. L. M. van,** Kola, a review of the literature, *Trop. Abstr.,* 28(8), 541—550, 1973.

COLLOMIA GRANDIFLORA

En. Collomia; Fr. Collomie; Ge. Heimsaat, Knaulblume

Teresa C. Minter-Proctor*

INTRODUCTION

Collomia grandiflora Dougl. ex Lindl. is a herbaceous annual species in the Polemoniaceae, or phlox family. A single plant typically produces both closed or cleistogamous (CL), self-pollinating flowers, and open or chasmogamous (CH), often cross-pollinating flowers.[1-5] This phenomenon occurs throughout the geographic range of the species, which is the Pacific slope of North America.[5] The CL flowers show reduction in both the androecium and corolla.[3] In the cymose inflorescence of the primary axis, CL flowers are typically produced first, followed by a progression of intermediates, and then CH flowers.[2,3] There is then a basipetal progression of lateral inflorescences which are predominantly CH, then finally entirely CL at the base of the plant.[2,5] Entirely CL plants do occur in natural populations.[5,6]

COMPARATIVE FLORAL DEVELOPMENT

Anther Development

The CL and CH flowers of *C. grandiflora* are indistinguishable as meristems and remain so until all organs have been initiated (Figures 1 through 4). Divergence in development is first detected with light microscope observation of serial sections in the anthers at archesporial cell stage. Archesporial cells are observed in the four locules of CH anthers and only in the two dorsal locules of CL anthers (Figures 5 and 6). Enlarged archesporial cells were not observed in the much-reduced ventral locules of CL anthers, although they do eventually develop an endothecium.[7]

Archesporial cells in the four locules of the CH anther and in the two dorsal locules of the CL anther divide periclinally to give rise to an inner layer of primary sporogenous cells and an outer layer of primary parietal cells. The primary parietal cells divide periclinally to give rise to an inner and outer layer of secondary parietal cells. The inner layer develops into the tapetum, and the outer layer divides once again to produce the endothecium to the outside and the middle layer to the inside. The primary sporogenous cells give rise to the pollen mother cells (PMC) which are surrounded by a single glandular tapetal layer and a single middle layer, forming the two locules of the CL anther and the four locules of the CH anther. Immediately outside the middle layer is the endothecium surround by the epidermis.[7]

Meiosis occurs earlier in CL anthers (Figure 7) than in CH anthers (Figure 8), and with subsequent development, the anther size difference becomes accentuated (Figures 9 and 10). Fewer PMCs develop in each locule of CL anthers (Figure 11) than in each locule of CH anthers (Figure 12). In both floral types, cytokinesis is simultaneous and the tetrads of microspores formed are tetrahedral. Anther development from meiosis to dehiscence is the same in both floral types. In the CL anther, an absence of ventral locule differentiation, together with a shorter period of time between archesporial cell differentiation and meiosis in the two dorsal locules, results in accelerated anther dehiscence and reduced mature anther size and pollen number (Figures 13 and 14).[7]

* Currently Assistant Professor of Biology, Department of Biology, Porterville College, Porterville, California.

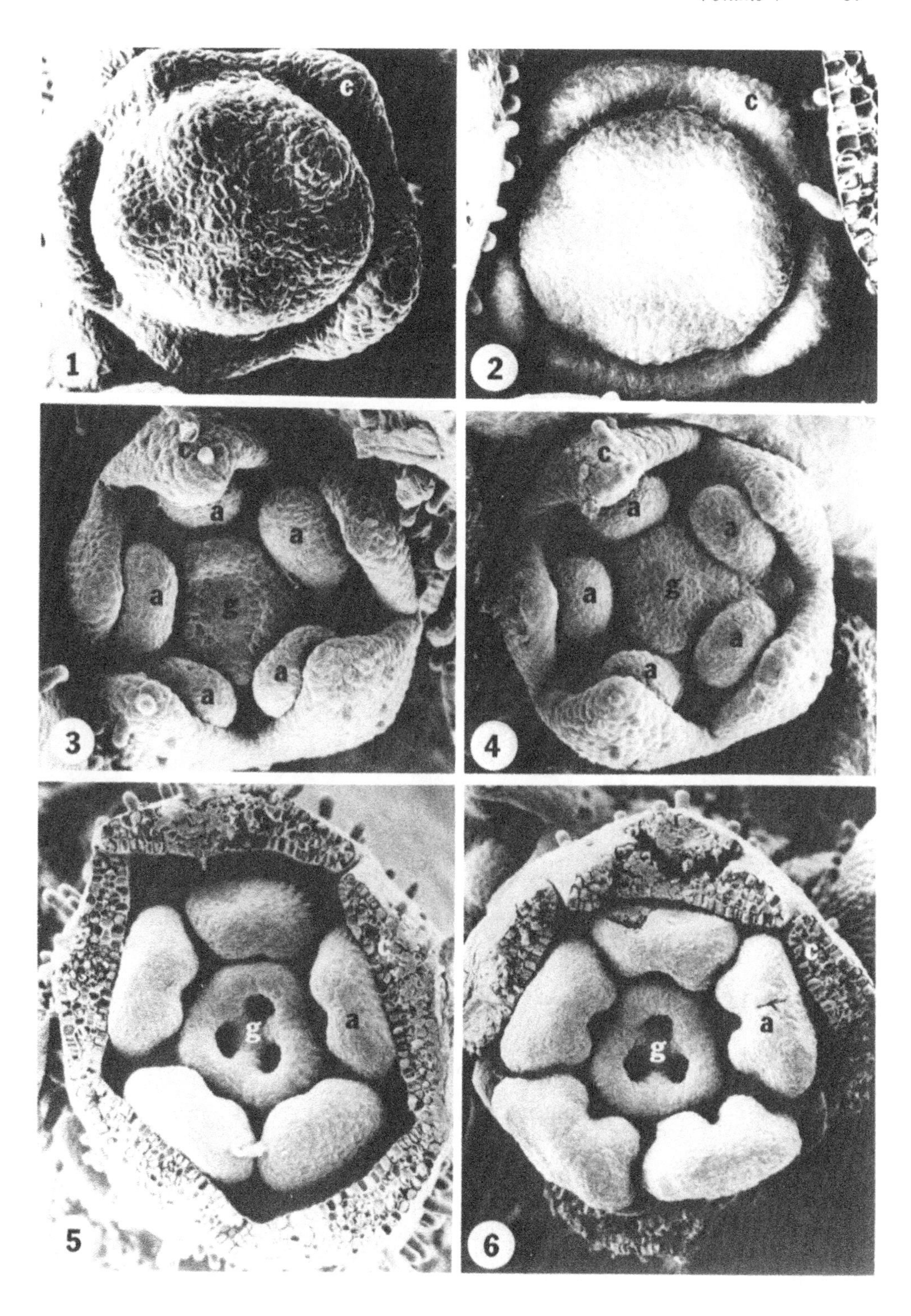

FIGURES 1 to 6. SEM views of *C. grandiflora* floral buds (CL on left, CH on right). Figures 1 and 2 show the CL and CH floral primordia. (Magnification × 270.) In Figures 3 and 4, 0.2—mm CL and CH floral buds with undifferentiated anthers are shown. (Magnification × 180.) In Figures 5 and 6 0.33-mm CL and CH floral buds with calyx partially removed may be seen. Note difference in shape of CL and CH anthers which are at the archesporial cell stage of development. (Magnification × 180.) C = calyx, a = anther, g = gynoecium. (From Minter, T. C. and Lord, E. M., *Am. J. Bot.*, 70, 1499—1508, 1983. With permission.)

FIGURES 7 to 10. SEM views of *C. grandiflora* floral buds (CL on left, CH on right). Figures 7 and 8 show 1.5-mm CL and CH buds with calyx removed. Meiosis has occurred in the anthers of the CL bud, while the anthers of the CH bud are still in the PMC stage. Note reduction in CL anthers. (Magnification × 90.) In Figures 9 and 10, 2-mm CL and CH buds are shown dissected open. Anthers of both buds contain microspores. Note similarity in calyces, gynoecia, and corollas, but reduction in CL anthers. (Magnification × 36.) C = calyx, a = anther, co = corolla, g = gynoecium. (From Minter, T. C. and Lord, E. M., *Am. J. Bot.*, 70, 1499—1508, 1983. With permission.)

Corolla Development

Divergence in corolla development follows the initial alteration in anther form.[7] In CH flowers, microspore mitosis is coincident with an increase in growth in corolla length relative to growth in calyx length which does not occur in CL flowers. Most of the difference in mature corolla length, however, is due to a nonallometric increase in length in the CH corolla which occurs at anthesis.

Both a greater epidermal cell number and size contribute to the increased corolla length and basal width of corolla lobes of CH flowers compared to CL flowers (Table 1). However, the distance and cell count between stamen traces at the base of CH and CL corollas do not vary.[2]

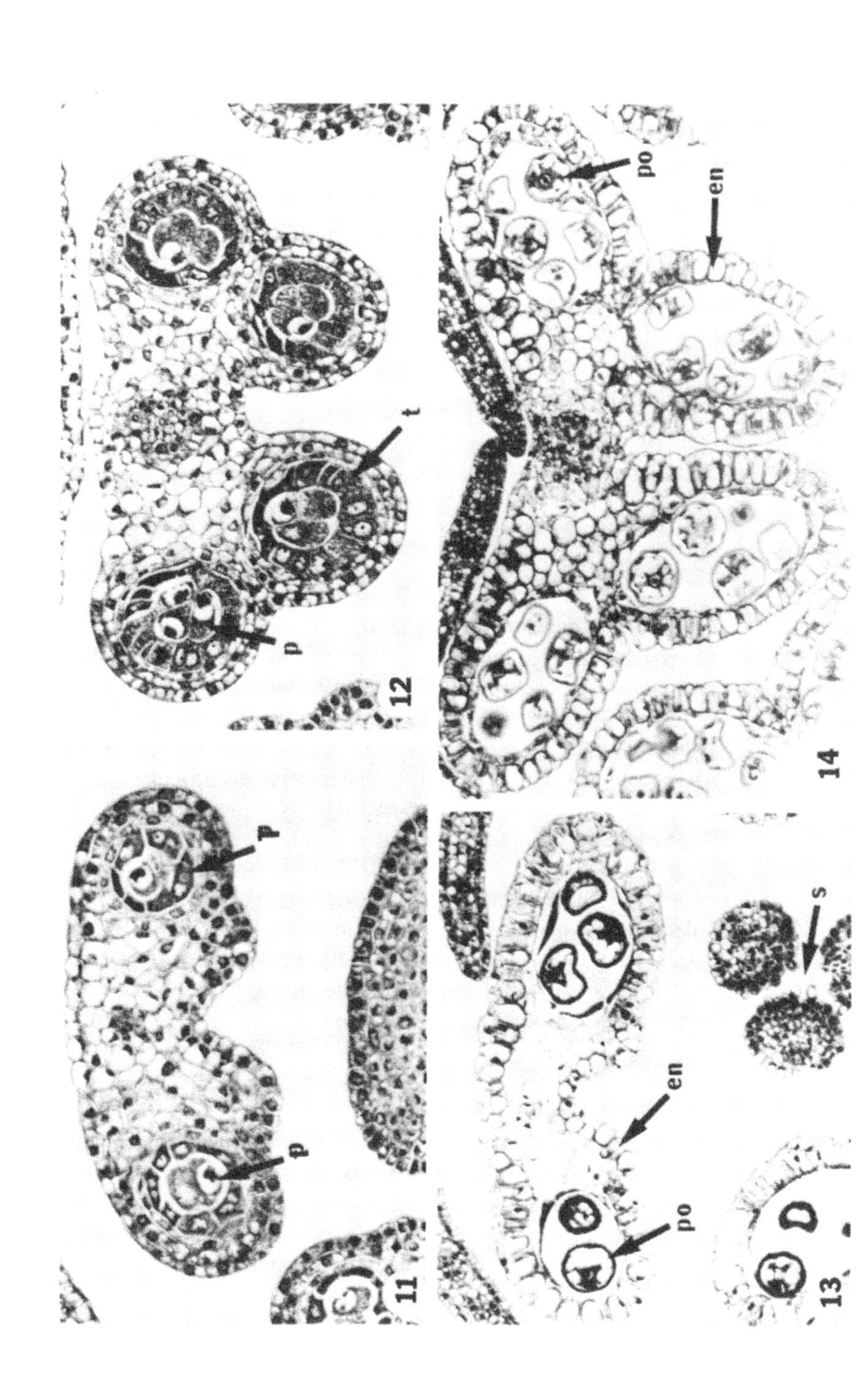
11
p
p
12
p
t
13
po
en
s
14
po
en

Table 1
DIMENSIONS AND EPIDERMAL CELL COUNTS OF COROLLAS FROM CL FLOWER POSITIONS 1—5 AND CH FLOWER POSITIONS 16—20 IN THE MAIN AXIS INFLORESCENCE OF WELL-WATERED *C. GRANDIFLORA* PLANTS

"Measurement"	CL corolla $\bar{X} \pm SD$	CH corolla $\bar{X} \pm SD$
Length, base to sinus of lobes (mm)	2.8 ± 0.2[a]	22.0 ± 1.7[a]
Cell count	85 ± 7[a]	143 ± 9[a]
Width at base of corolla lobe (mm)	0.6 ± 0.1[a]	2.2 ± 0.2[a]
Cell count	82 ± 7[a]	118 ± 9[a]
Distance between corolla stamen traces 0.5 mm from the base (mm)	0.8 ± 0.1	0.9 ± 0.1
Cell count	50 ± 4	45 ± 5
Cell length (μm)[b]	39 ± 6[a]	180 ± 29[a]
Cell width (μm)[b]	13 ± 2[a]	16 ± 2[a]

[a] Significant difference between CL and CH corolla values at 1% level, N = 8.
[b] Length and width of a typical cell located in the area indicated by the arrows in Figure 1.

Calyx and Gynoecial Development

CL and CH calyx growth is similar, and mature calyx size is the same.[7] Gynoecial development appears similar in the two floral forms; however, ovary expansion due to fertilization occurs 2 to 3 days earlier in the CL flower. If ovules in the CH flower are capable of fertilization at the same time as those in the CL flower, then precocious anther differentiation resulting in earlier anther dehiscence, along with accelerated stigma receptivity, may be the sole factors responsible for earlier fertilization in the CL flower. The time from meiosis in the anthers to stigma receptivity and pollination is the same in both flowers (11 to 12 days).[7]

The ovary of each CL and CH flower contains three ovules. The megasporocyte of each ovule undergoes meiosis and wall formation to form a linear tetrad of megaspores. The chalazal megaspore of each tetrad undergoes *Polygonum*-type embryo sac development. CL and CH ovaries are of the same age at the time of the meiotic divisions of the megasporocytes and are indistinguishable from one another.[7]

EFFECTS OF WATER STRESS

The proportion of CL flowers of *C. grandiflora* grown in the field and experimental garden plots increases when plants are subjected to variations of several environmental factors. Presumably, the major cause of this altered floral ratio is a decrease in water availability.[5] Greenhouse-grown, water-stressed plants are shorter and produce fewer total flowers and a lower percentage of CH flowers than well-watered plants (Table 2; Figures 15 through 18).[2] Water-stressed plants produce CL flowers in positions in the inflorescence where CH flowers typically occur (Figures 17 and 18). These water-stress-induced CL flowers are indistinguishable from normal CL flowers.[2] Measurements and epidermal cell counts of their corollas do not differ from those of typical CL flowers of well-watered plants (Table 3). In addition, these induced CL flowers have reduced anthers, similar in size to those of CL flowers of well-watered plants.

Table 2
COMPARISONS OF THE EFFECTS OF WATER STRESS, GA_3, AND ABA ON THE NUMBER OF FLOWERS, AND % CH FLOWERS PRODUCED BY *C. GRANDIFLORA*

Treatment group	Xylem pressure potential (bars)[a]		Main axis inflorescence					Entire inflorescence of plant				
	$\bar{X} \pm SD$ well-watered	$\bar{X} \pm SD$ maximally stressed	No. of flowers			% CH flowers		No. of flowers			% CH flowers	
			N	$\bar{X}$	DMRT[b]	$\bar{X}$	DMRT	N	$\bar{X}$	DMRT	$\bar{X}$	DMRT
Well-watered (control)	−5.2 ± 1.9		19	43	B	60.2	B	3	401	B	14.8	B
Water-stressed	−9.8 ± 3.2	−16.6 ± 1.7	20	26	CD	6.9	D	3	269	C	0.4	C
Well-watered + Triton	−5.2 ± 1.1		20	37	B	50.3	BC	3	431	B	16.0	B
Water-stressed + Triton	−7.5 ± 1.6	−16.0 ± 2.6	20	27	CD	4.5	D	3	149	D	0.0	C
Well-watered + GA_3	−5.7 ± 1.5		20	110	A	84.6	A	3	765	A	80.0	A
Water-stressed + Ga_3	−7.0 ± 2.3	−14.8 ± 1.4	20	27	CD	86.4	A	3	147	D	77.8	A
Well-watered (control)[c]	—		5	28	C	42.9	C	5	282	C	7.1	C
Well-watered + ABA[c]	−3.1 ± 2.6		19	18	DE	9.0	D	3	127	D	0.7	C
Water-stressed + ABA[c]	−6.5 ± 2.9	−16.4 ± 2.2	20	16	E	2.4	D	3	65	D	0.5	C

[a] The xylem pressure potentials of five leaves from five plants per treatment group were measured and the mean values for each group were determined each day measurements were made. All mean potentials determined throughout the life span of the plants in a group were averaged to give the value shown for each well-watered group. Mean potentials determined just before watering when plants were maximally stressed and the mean potentials determined the day after watering were averaged separately to give the two values shown for each water-stressed group.

[b] Duncan's multiple range test, 5% level.

[c] Groups from population grown from April to August; all other groups from population grown from November to May.

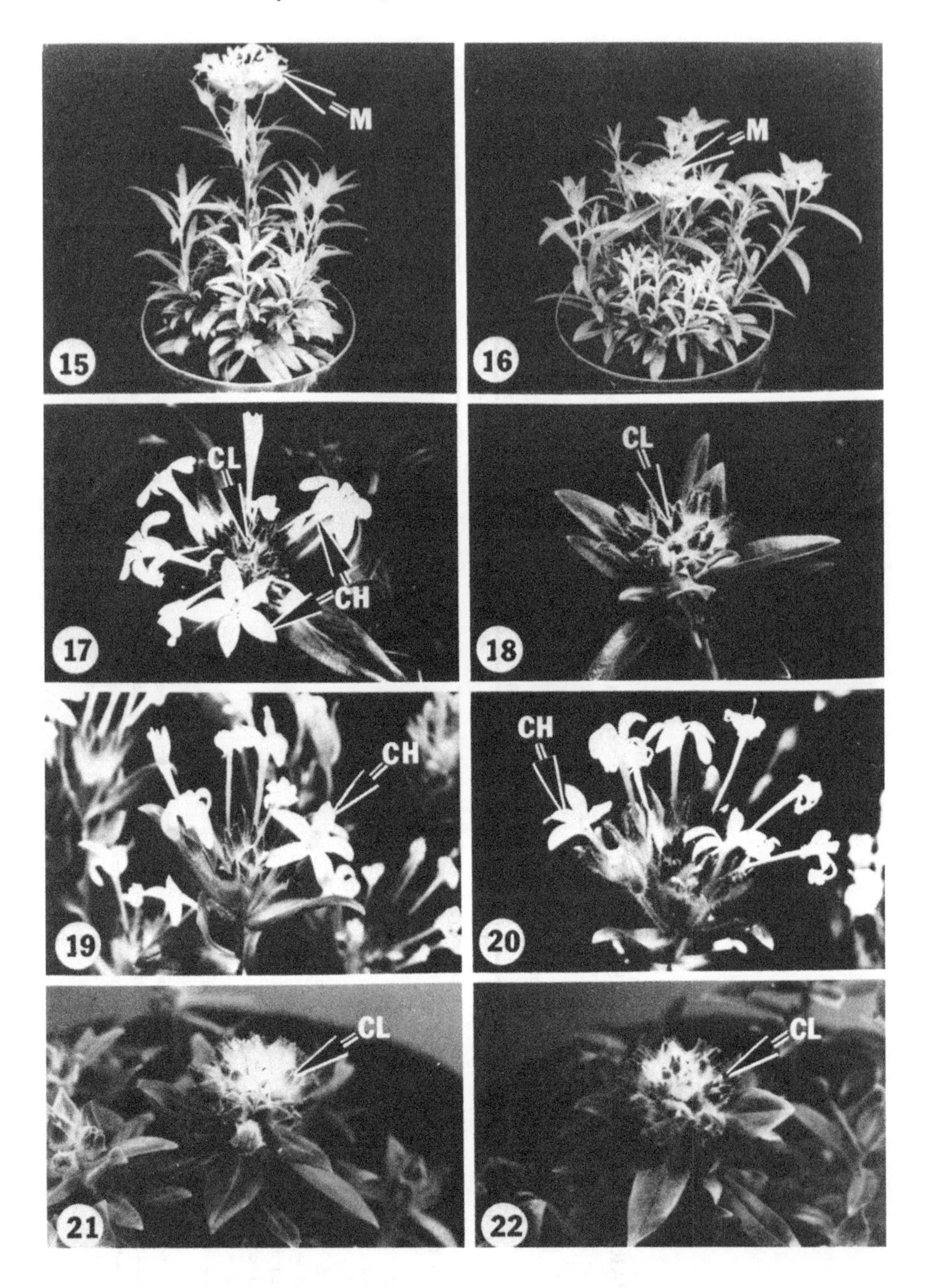

FIGURES 15 to 22. *Collomia grandiflora*. Figure 15 illustrates a well-watered control plant. Figure 16 shows a water-stressed plant. (Magnification × 0.2.) Figures 17 through 22 show examples of main axis inflorescences. (Magnification × 0.75.) Figure 17 shows a well-watered control plant; Figure 18 a water-stressed plant; Figure 19 a well-watered, GA_3-sprayed plant; Figure 20 a water-stressed, GA_3-sprayed plant; Figure 21 a well-watered, ABA-sprayed plant; and Figure 22 a water-stressed, ABA-sprayed plant. M = main axis inflorescence, CH = chasmogamous flower, CL = cleistogamous flower. (From Minter, T. C. and Lord, E. M., *Am. J. Bot.*, 70, 618—624, 1983. With permission.)

Table 3
COMPARISONS OF THE EFFECTS OF WATER STRESS, ABA, AND GA_3 ON COROLLA DIMENSIONS AND EPIDERMAL CELL COUNTS IN *C. GRANDIFLORA*

Treatment group	Flower position	Anthesis[a]	Corolla length, base to sinus of lobes: mm ($\bar{X} \pm SD$)	Cell count $\bar{X} \pm SD$)	Cell length (μm)[b] $\bar{X} \pm SD$	Width at base of corolla lobe (mm; $\bar{X} \pm SD$)
Well-watered (control)	1—5	−	3.1 ± 0.4[c]	86 ± 6[c]	41 ± 5[c]	0.68 ± 0.06[c]
	16—20	+	21.6 ± 1.8[d]	153 ± 6[d]	163 ± 5[d]	2.22 ± 0.24[d]
Water-stressed	1—5	−	2.6 ± 0.1[c]	76 ± 5[c]	36 ± 3[c]	0.63 ± 0.06[c]
	16—20	−	3.0 ± 0.5[c]	82 ± 10[c]	42 ± 5[c]	0.68 ± 0.07[c]
Well-watered + ABA	1—5	−	2.4 ± 0.2[c]	76 ± 5[c]	34 ± 2[c]	0.52 ± 0.03[c]
	16—20	−	2.1 ± 0.1[c]	75 ± 5[c]	35 ± 8[c]	0.51 ± 0.02[c]
Water-stressed + ABA	1—5	−	2.1 ± 0.2[c]	80 ± 7[c]	33 ± 4[c]	0.55 ± 0.02[c]
	16—20	−	2.5 ± 0.1[c]	89 ± 4[c]	33 ± 2[c]	0.56 ± 0.06[c]
Well-watered + GA_3	1—5	+	18.4 ± 0.6	120 ± 4	195 ± 13	1.33 ± 0.08
	16—20	+	21.6 ± 1.3[d]	139 ± 6	180 ± 15[d]	1.42 ± 0.06
Water-stressed + GA_3	1—5	+	20.0 ± 0.9[d]	124 ± 4	191 ± 13	1.35 ± 0.11
	16—20	+	19.7 ± 0.9	134 ± 4	183 ± 17[d]	1.42 ± 0.12

[a] +, Corolla open at maturity; −, corolla closed at maturity.
[b] Length of a typical epidermal cell located in the area indicated by the arrows in Figure 1.
[c] No significant difference at 1% level from value obtained for control treatment group, flower positions 1—5, representing normal CL flower value, N = 3.
[d] No significant difference at 1% level from value obtained for control treatment group, flower positions 16—20, representing normal CH flower value, N = 3.

EFFECTS OF GA APPLICATION

Application of GA_3 (100 ppm solution sprayed on plant three times per week during inflorescence development) induces corolla opening and cell expansion, as well as some cell division, of flowers borne at typically CL positions in well-watered plants, and throughout water-stressed plants, which normally produce very few CH flowers (Figures 19 and 20; Table 2). The anatomy of corollas of these GA_3-induced flowers is intermediate between normal CL and CH flowers in form and epidermal cell number (Table 3); however, the anthers of these flowers are similar in size to those of CL flowers.[2]

EFFECTS OF ABSCISIC ACID APPLICATION

Well-watered *C. grandiflora* sprayed with ABA (100 ppm solution sprayed on plant three times per week during inflorescence development) are shorter and produce significantly fewer total flowers and a lower percentage of CH flowers than nonsprayed, well-watered plants. CL flowers are produced in positions in the inflorescence where CH flowers typically occur (Figures 21 and 22). ABA-induced CL flowers, like water stress-induced CL flowers, are indistinguishable from normal CL flowers (Table 3).[2]

The effect of ABA treatment on flower production is similar to that of water stress. Well-watered plants, sprayed with ABA, produce an average number of flowers and percentage of CH flowers in their main axis inflorescences which are not significantly different from that produced by water-stressed, nonsprayed plants (Table 2).[2]

HYPOTHESIS FOR HORMONAL CONTROL OF FLOWER PRODUCTION

Since divergence in the development of CL and CH flowers in *C. grandiflora* is first detectable in the anthers, it seems likely that they may be the target tissue for hormonal control.[2] A decrease in water availability results in increased production of CL flowers by *C. grandiflora* and other CL species.[2,5,8,9] The concentration of ABA increases in the leaves of water-stressed plants,[10] and injected ABA and spray applications of ABA mimic the morphological effects of water stress.[2,11] During water stress or other conditions that result in CL flower production by *C. grandiflora,* elevated ABA in the floral meristem may cause the development of smaller anthers. Conversely, a low level of ABA relative to that of GA may permit the development of larger CH anthers.[2] Applied GA induces CH flower production in *C. grandiflora* as well as in two other CL species.[2,12,13]

Studies of other species indicate that GA in another tissue may induce filament and corolla expansion.[12,14] It is postulated that after CH flowers are initiated in *C. grandiflora,* a greater accumulation of GA in their larger anthers may act secondarily to cause corolla expansion and opening.[2]

Low levels of GA relative to ABA levels prior to bolting, and then increased GA levels during bolting, might explain the initial transition from CL to CH flower production that occurs in *C. grandiflora* in the absence of water stress. An increased level of ABA with age may be responsible for the final period of CL flower production.[2] It has been observed in plants other than *C. grandiflora* that stomatal conductance decreases with age.[15-17] Since ABA levels or rate of ABA synthesis are inversely correlated with stomatal aperture,[10] this could be evidence for an increase in ABA with age.[2]

REFERENCES

1. **Ludwig, F.,** Über die Kleistogamie von *Collomia grandiflora* Dougl., *Bot. Z.,* 35, 777—780, 1877.
2. **Minter, T. C. and Lord, E. M.,** Effects of water stress, abscisic acid, and gibberellic acid on flower production and differentiation in the cleistogamous species *Collomia grandiflora* Dougl. ex Lindl. (Polemoniaceae), *Am. J. Bot.,* 70, 618—624, 1983.
3. **Ritzerow, H.,** Über Bau und Befruchtung kleistogamer Blüten, *Flora,* 98, 163—212, 1908.
4. **Scharlok, T.,** Über die Blüten der *Collomien, Bot. Z.,* 36, 640—645, 1878.
5. **Wilken, D. H.,** The balance between chasmogamy and cleistogamy in *Collomia grandiflora* (Polemoniaceae), *Am. J. Bot.,* 69, 1326—1333, 1982.
6. **Gillett, G. W., Howell, J. T., and Leschke, H.,** A flora of Lassen Volcanic National Park, California, *Wasmann J. Biol.,* 19, 1—185, 1961.
7. **Minter, T. C. and Lord, E. M.,** A comparison of cleistogamous and chasmogamous floral development, in *Collomia grandiflora* Dougl. ex Lindl. (Polemoniaceae), *Am. J. Bot.,* 70, 1499—1508, 1983.
8. **Brown, W. V.,** The relation of soil moisture to cleistogamy in *Stipa leucotricha, Bot. Gaz.,* 113, 438—444, 1952.
9. **Langer, R. H. M. and Wilson, D.,** Environmental control of cleistogamy in prairie grass *(Bromus unioloides* HBK), *New Phytol.,* 65, 80—85, 1965.
10. **Walton, D. C.,** Biochemistry and physiology of abscisic acid, *Annu. Rev. Plant. Physiol.,* 31, 453—489, 1980.
11. **Quarrie, S. A. and Jones, H. G.,** Effects of abscisic acid and water stress on development and morphology of wheat, *J. Exp. Bot.,* 28, 192—203, 1977.

12. **Lord, E. M.,** Physiological controls on the production of cleistogamous and chasmogamous flowers, in *Lamium amplexicaule* L., *Ann. Bot.*, 44, 757—766, 1980.
13. **Raghuvanshi, S. S., Pathak, C. S., and Singh, R. R.,** Gibberellic acid response and induced chasmogamous variant in cleistogamous *Ruellia* hybrid (*R. tweediana* × *R. tuberosa*), *Bot. Gaz.*, 142, 40—42, 1981.
14. **Greyson, R. I. and Tepfer, S. S.,** Emasculation effects on the stamen filament of *Nigella hispanica* and their partial reversal by gibberellic acid, *Am. J. Bot.*, 54, 971—976, 1967.
15. **Davis, S. D., van Bavel, C. H. M., and McCree, K. J.,** Effect of leaf aging upon stomatal resistance in bean plants, *Crop Sci.*, 17, 640—645, 1977.
16. **Minter, T. C.,** Effect of leaf age on stomatal conductance in a fluctuating environment, *Wasmann J. Biol.*, 38, 30—38, 1980.
17. **Slatyer, R. O. and Bierhuizen, J. F.,** The influence of several transpiration suppressants on transpiration, photosynthesis and water-use efficiency of cotton leaves, *Aust. J. Biol. Sci.*, 17, 131—146, 1964.

CORDIA

Rolf Borchert

INTRODUCTION

The pantropic genus *Cordia* (Boraginaceae) comprises more than 250 species of trees and shrubs, the majority native to the Neotropics.[13] Most species are limited to a rather small range, but a few, such as *C. alliodora*, are widely distributed. While flowers in most species are small and inconspicuous, white or greenish, there are a few species with relatively large (~5 cm), vermilion (*C. dodecandra, C. sebastana*), yellow (*C. dentata*), or white flowers (*C. boissieri*); several Mexican species of the section *Gerascanthus* produce masses of white flowers during the dry season when nearly leafless and are thus quite conspicuous and decorative (*C. gerascanthus, C. globulifera, C. morelana,* and *C. sonorae*).[8] Only *C. dodecandra, C. sebestana,* and *C. boissieri* are commonly cultivated in the southern U.S. and Mexico.[12] Color photographs of cultivated species can be found in Reference 12.

Most species of *Cordia* appear to be fast-growing pioneer trees, common in secondary successions, but rare in tropical climax forests.[14] Trees of the section *Gerascanthus* are locally esteemed timber trees,[8] and *C. alliodora* (laurel, salmwood) is widely planted in the tropics for timber.[7]

MORPHOLOGY OF FLOWERS AND INFLORESCENCES; POLLINATION BIOLOGY

The structure of flowers and inflorescences is highly variable within the genus *Cordia*. The irregular inflorescences range from wide-open panicles (Figure 1) to spike-like (Figure 2), densely glomerate, or capitulate.[13] Inflorescences may be terminal on main or lateral shoots, or they may arise directly from axillary buds (Figures 1 and 2). In *C. furcans*, an irregular dichotomy of the branch system results from one branch at a fork bearing a lateral inflorescence and a few leaves, while the other continues as the main axis of the shoot; the inflorescences appear to arise at any point in an internode along the main shoot, and only by chance do they arise near the leaf axil.[9] While in most other tree genera inflorescences are always terminal (hapaxanthy, e.g., *Tabebuia*[2a,4,6]) or lateral (pleonanthy, *Erythrina* and the majority of temperate trees[2a,4,6]), in *Cordia* the transition from vegetative to reproductive mode may thus occur in any meristem irrespective of its position (Figure 2). Because of the regular termination of main and lateral shoots in inflorescences, new vegetative shoots arise from lateral buds subtending the terminal inflorescences and the shoot system in *Cordia* is commonly modular.[2a,6]

The sessile actinomorphic flowers vary widely in size (Figures 2 and 3). They may be perfect and homomorphic or functionally unisexual and dimorphic;[13,14] some species are functionally dioecious.[13,14] The calyx consists of 5 connate sepals and the corolla has usually 5 connate petals, but flowers with 6 or 7 petals are not uncommon. Corolla size and shape vary widely with the principal pollinators (see below). Stamens, equal in number to the corolla lobes, are inserted at the corolla throat (Figures 1 through 3). The 4-loculed ovary bears a slender, elongate, twice bifid style with 4 stigmas (Figures 2 and 3). Flowers in most species are heterostylous and self-incompatible, showing considerable variation in the length of styles and stamens (Figures 1 through 3).[14] While fruits in small trees and shrubs are mostly fleshy, animal-dispersed drupes, fruits in tall trees may be dry and wind dispersed.[14]

Flowers of most *Cordia* species are visited by a wide array of Lepidoptera, Hymenoptera, and other insects.[14] Floral features determine the type of visitor: species with short, open

FIGURE 1. Flowering shoot and flower organs of *Cordia excelsa*. (From Martius, C. F. P. and Eichler, A. W., *Flora Brasiliensis*, Vol 8/1, Fleischer, Munich, 1856—1872. With permission.)

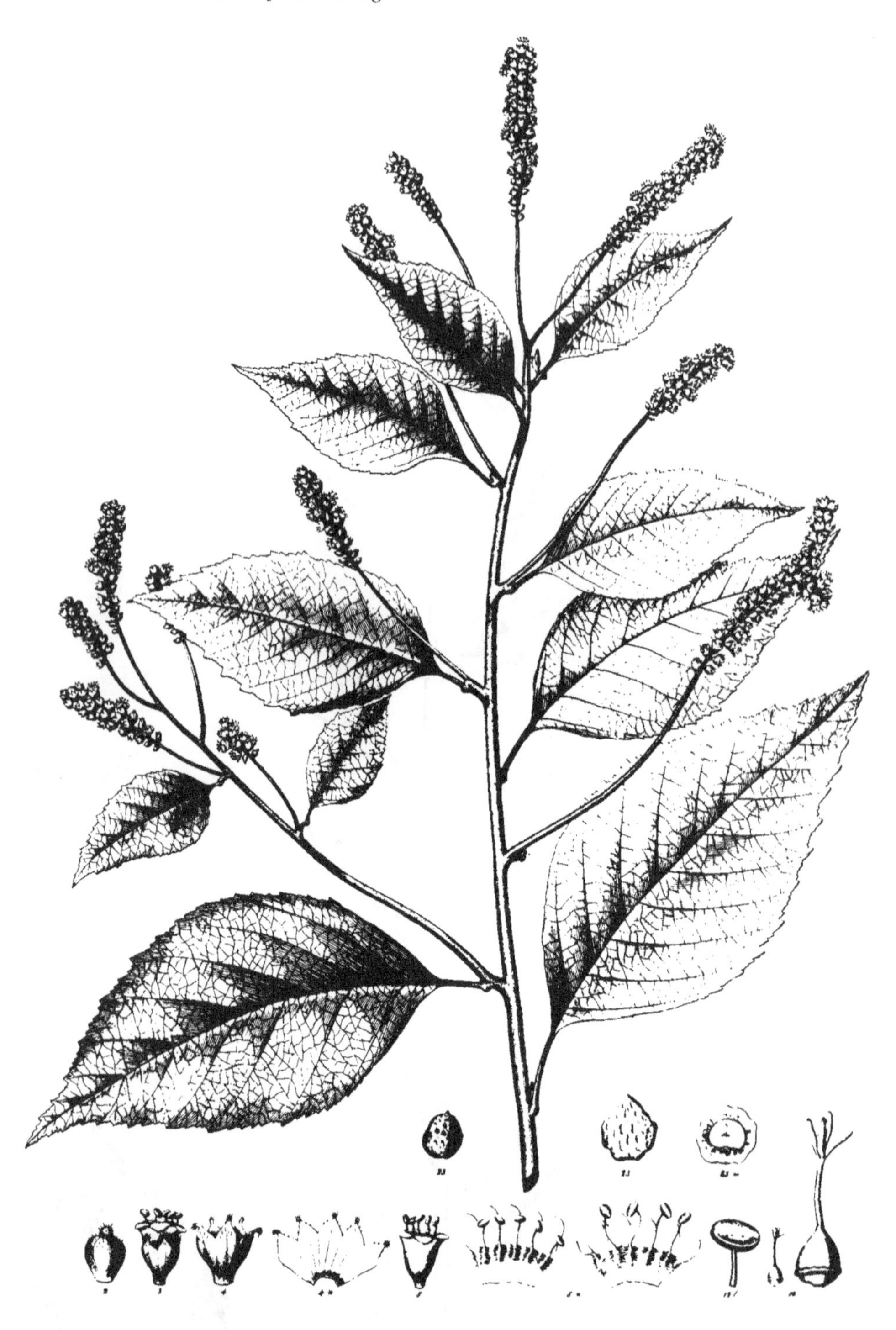

FIGURE 2. Flowering shoot and flower organs of *Cordia multispicata*. (From Martius, C. F. P. and Eichler, A. W., *Flora Brasiliensis*, Vol 8/1, Fleischer, Munich, 1856—1872. With permission.)

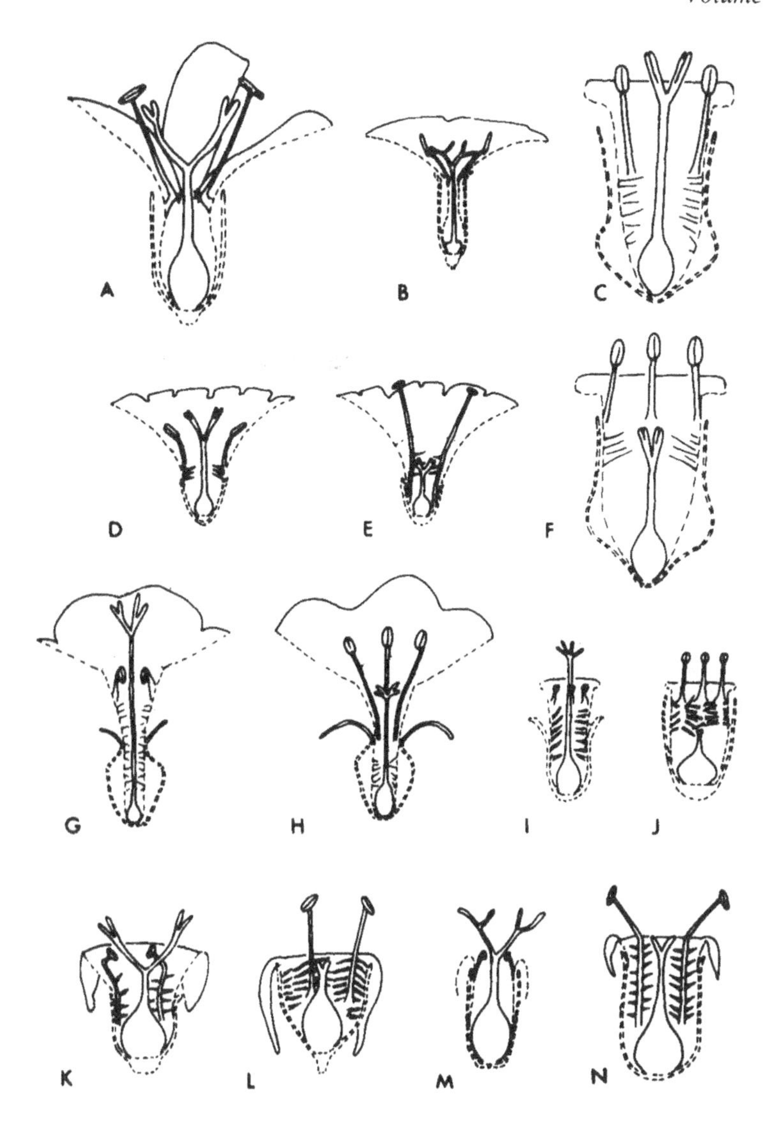

FIGURE 3. Variation in flower structure of *Cordia*. (A) *alliodora*, × 2.8; (B) *gerascanthus*, × 1; (C) *curassavica*, long style, × 3.5; (D) *dentata*, long style, × 3; (E) *dentata*, short style, × 3; (F) *curassavica*, short style, × 3.5; (G) *pringlei*, long style, × 1.7; (H) *pringlei*, short style, × 1.7; (I) *inermis*, long style, × 8; (J) *inermis*, short style, × 8; (K) *collococca*, long style, × 6; (L) *collococca*, short style, × 6; (M) *panamensis*, long style, × 2; and (N) *panamensis*, short style, × 2. (From Opler, P. A., Baker, H. G., and Frankie, G. W., *Biotropica*, 7, 234—247, 1975. With permission.)

flowers have the most numerous flower visitor complements, while species with restricted corolla tubes are visited mainly by long-tongued butterflies, hawkmoths, large bees, and, exceptionally, hummingbirds (e.g., *C. gerascanthus* and *C. sebestena*). Visitors are further restricted by the presence or absence of "landing platforms" and, very specifically, by the time of anthesis, which may occur early in the morning, in late afternoon, or at night; corollas wither or fall within a few hours.[14]

PHENOLOGY AND CONTROL OF FLOWERING

Vegetative and reproductive development are strongly interrelated in all plants, but in trees these relationships are considerably more complex than in herbaceous plants because of the structural complexity of the shoot system (Figures 1 and 2), the life-long alternation between periods of shoot growth and periods of rest, and repeated episodic flowering (Table 1).[2a-4,6,10]

In most tree species, particularly those adapted to markedly seasonal tropical or temperate climates, early differentiation of flower buds is separated from final development of the inflorescence and anthesis by a prolonged rest period (proleptic flowering).[2a-4,10] In contrast, in most *Cordia* species, the transition of apical meristems from the vegetative to the reproductive mode immediately follows the seasonal flush of vegetative growth, and development of the inflorescence proceeds continuously from early differentiation of the floral meristem to anthesis. The differentiation of floral organs may thus occur rather asynchronously in the various branches of a tree throughout the period of vegetative growth (Figure 2). Like the cessation of indeterminate shoot growth during the growing season, flower induction and anthesis must be therefore determined exclusively by internal, correlative mechanisms, and not by environmental cues.[2]

This pattern of flowering is represented in Table 1 by all *Cordia* species which flower during or towards the end of seasonal shoot growth. In *C. glabra* observed in San José, Costa Rica, young shoots emerge soon after leaf shedding during the early dry season, which is identical in duration to that in Guanacaste, Costa Rica (Table 1:17), and the rather synchronous, short flush of shoot growth ends with the formation of terminal inflorescences and anthesis (Table 1:1). In the tropical lowland deciduous forest of Guanacaste, most *Cordia* species flower for prolonged periods after shoot growth has begun with the onset of rains (Table 1:2—6,17). Flowering peaks early in the wet season, indicating cessation of seasonal shoot growth in most branches, and shoot growth and flowering cease entirely with the onset of drought.[14] In the lowland tropical rain forest at La Selva, Costa Rica, most *Cordia* species, like the majority of other trees in the forest,[5,14] shed many leaves during the relatively dry months of January and February (Table 1:18); new shoots emerge soon after leaf shedding and flowering occurs on new shoots (Table 1:7—10) as described above for *C. glabra* and observed at La Selva in *Cordia* species (Table 1:11). Intensive flowering of most *Cordia* species from March to May indicates the extension of the major period of vegetative growth; in *C. bifurcata* and *C. lasiocalyx* a few individuals are in flower throughout the year, reflecting the common asynchrony of vegetative growth in wet, tropical forests (Table 1:9,10). Similarly, many species of the Dipterocarpaceae (*Dipterocarpus* and *Shorea*) and other forest trees flower after rainfalls following the dry season in Malaysian rain forests.[10] Several species of *Cordia* flower throughout the year in Puerto Rico (*C. alba, C. obliqua,* and *C. rickseckeri*), while *C. colococca* does so only in spring and summer (compare Table 1:16).[9a]

In *C. alliodora* and other species of the subgenus *Gerascanthus,* anthesis commonly occurs during the early dry season, before leaf fall (Table 1:12—15); some Mexican species of this subgenus are known for conspicuous mass-flowering of bare trees.[8] This flowering pattern implies that putative flower buds are laid down during shoot growth in the preceding wet season, remain dormant for several months, and then develop into inflorescences during the early dry season. This pattern of flowering is common in many tropical trees native to seasonally dry forests and has been analyzed in some detail for the genera *Erythrina* and *Tabebuia*.[2a-4] In these genera, flowering during the early dry season appears to be triggered by rehydration resulting from leaf fall, but this explanation seems questionable for species like *C. alliodora* which flower before shedding their leaves. *C. alliodora* is unusual also in that adult, but not young, trees regularly remain leafless for several months even when growing in wet lowland forests.[5,7] In the tropical lowland dry forest of Guanacaste, *C. gerascanthus* and *C. collococca* flower synchronously for rather brief periods, early or late in the dry season (Table 1:13b,16); comparison with other tropical trees flowering during

Table 1
PERIODICITY OF FLOWERING AND SEASONAL DEVELOPMENT IN CENTRAL AMERICAN AND MEXICAN SPECIES OF *CORDIA*

No.	Species	Observed at	J	F	M	A	M	J	J	A	S	O	N	D	Ref.[b]
			Month[a]												
1.	*C. glabra*	San José, Costa Rica	\\			[\\\\\\							\\\\\		1
			000000				000	00	0]						
					XXXX	XXX									
2.	*C. curassavica*	Guanacaste, Costa Rica					XX	XX	XXX	XXX	XXX	XX			13
3.	*C. dentata*	Guanacaste					XXXX	XX	XXX	XXX	XXX	XX			13
4.	*C. inermis*	Guanacaste					XX	XX	XXX	XXX	XXX	XX			13
5.	*C. panamensis*	Guanacaste					X	XX	XXX						13
6.	*C. pringlei*	Guanacaste					XXX	XX	XXX	XXX	XXX	XX			13
7.	*C. lucidula*	La Selva, Costa Rica			XXX	XXXXXX									13
8.	*C. protracta*	La Selva			XXX	XXXXXX									13
9.	*C. bifurcata*	La Selva	X X	X	XXX	XXXXXX		X	X X	X	X X	X	X X	X	13
10.	*C. lasiocalyx*	La Selva	X X	X	XXX	XXXXXX		X	X X	X	X X	X	X X	X	13
11.	*C. sp.*	La Selva	\\\000			XXXXXXX		XX							4
12.a	*C. alliodora*	La Selva	XXXXXX		\\\	\\\			000	000					13
b		Guanacaste	XXXXXX												
13.a	*C. gerascanthus*	Mexico	XXXXXX		\\\	\\\									7
b		Guanacaste		XX											13
14.	*C. tinifolia*	Mexico	XXXXXXXXX			X									7
15.	*C. eleagnoides*	Mexico			XXX	XXXX									7
16.	*C. colococca*	Guanacaste			XX										13
17.	Rainfall (cm), 1969	Guanacaste	0	0	8	8	13	45	11	27	25	70	14	1	4

the dry season suggests that these flowering episodes might be triggered by isolated rainfalls during the dry season.[2,4,15]

As might be expected for fast-growing pioneer trees, many small-tree species of *Cordia* start flowering when only 2 or 3 years old; saplings of *C. alliodora* commonly grow 1 to 2 m/year and flower when 5 to 10 years old.[6] Unlike many other tropical trees, most *Cordia* species seem to flower profusely every year. However, in Guanacaste, *C. gerascanthus* flowered only in alternate years,[14] and in a population of *C. glabra*, observed in San José, no flowering occurred in individual trees which changed leaf during the wet, instead of the dry, season (Table 1:1; data in parenthesis).

CONCLUSIONS

In other large genera of tropical flowering trees discussed in this volume, morphology, development of inflorescences and flowers, and the temporal relations between vegetative and reproductive development are remarkably similar throughout the genus: virtually all species of *Tabebuia* bear large, sympetalous flowers on terminal panicles or racemes which develop after the terminal bud has undergone a prolonged rest period[4] (proleptic, hapaxanthic flowering[2a,6]); with few exceptions, inflorescences in *Erythrina* are lateral and usually develop some time after the cessation of shoot growth from axillary buds near the tip of the arrested shoot[3] (proleptic, pleonanthic flowering[2a,6]). In contrast to these genera, in *Cordia* there is great variety in morphology and development of vegetative and reproductive organs (Figures 1 through 3), flowering biology, and reproductive strategies,[14] both within[8] and between[13] species, making the taxonomic treatment of the genus rather difficult.[13]

In an excellent analysis of the adaptive syndromes manifest in the reproductive biology of *Cordia*,[14] two relatively independent sets of adaptive features were recognized: (1) flower size, time of anthesis, amount of nectar, breeding system, and size, type, and diversity of pollinators; and (2) plant stature, successional stage, and seed dispersal type. The convergent evolution of only a limited number of these combinations was considered evidence for the importance of harmonious development and functioning of various aspects of reproductive biology.

Regrettably, as in most evolutionary studies of the reproductive biology of tropical trees,[1,2a] variations in tree architecture and the timing of vegetative and reproductive growth were largely ignored in this analysis. However, key elements of reproductive tree biology, such as size and number of flowers, timing and duration of flowering, seed production, tree stature, and successional status of a species are intimately linked with tree architecture and seasonal growth periodicity, which in turn are partly determined by the prevailing climatic conditions and the habitat of the tree.[2-4,6,16] For instance, available evidence suggests that anthesis on inflorescences bearing large flowers, as prevalent in *Erythrina, Tabebuia*, and subgenus *Gerascanthus* of *Cordia*, is competitively inhibited by the tree foliage and can occur only after partial or complete leaf shedding[2a-4,16] as induced by incipient drought; but drought-induced water stress also inhibits anthesis, and trees must therefore rehydrate before anthesis can occur. Thus, only if two environmentally controlled preconditions, leaf shedding and subsequent rehydration, are met, will mass flowering occur in such trees. Inversely, *Cordia* species flowering when in full leaf usually bear small, inconspicuous flowers (Figures 1 through 3). Implicitly, the above constraints limit synchronous opening of many large flowers to deciduous trees of seasonally dry tropical forests and to deciduous canopy trees of rain forests; it is unlikely to occur in evergreen subcanopy species.

It thus appears that without adequate consideration of the multiple interrelations between reproductive biology, tree architecture, and the environmental controls of periodic tree development the adaptive pressures involved in the evolution of the reproductive biology of tropical trees, such as *Cordia*, cannot be fully understood.[2a]

REFERENCES

1. **Bawa, K. S.,** Patterns of flowering in tropical plants, in *Handbook of Experimental Pollination Biology,* Jones, C. E. and Little, R. J., Eds., Van Nostrand Reinhold, New York, 1982, 394—410.
2. **Borchert, R.,** Feedback control and age-related changes of shoot growth in seasonal and nonseasonal climates, in *Tropical Trees as Living Systems,* Tomlinson, P. B. and Zimmermann, M. H., Eds., Cambridge University Press, Cambridge, 1978, 497—515.

2a. **Borchert, R.,** Phenology and control of flowering in tropical trees, *Biotropica,* 15, 81—89, 1983.

3. **Borchert, R.,** *Erythrina,* in *Handbook of Flowering,* Vol. 5, Halevy, A. H., Ed., CRC Press, Boca Raton, Fla., in press.
4. **Borchert, R.,** *Tabebuia,* in *Handbook of Flowering,* Vol. 5, Halevy, A. H., Ed., CRC Press, Boca Raton, Fla., in press.
5. **Frankie, G. W., Baker, H. G., and Opler, P. A.,** Comparative phenological studies of trees in tropical wet and dry forests in the lowlands of Costa Rica, *J. Ecol.,* 62, 881—919, 1974.
6. **Hallé, F., Oldeman, R. A. A., and Tomlinson, P. B.,** *Tropical Trees and Forests. An Architectural Analysis,* Springer, Berlin, 1978.
7. **Johnson, P. and Morales, R.,** A review of *Cordia alliodora* (Ruiz & Pav.) Oken., *Turrialba,* 22, 210—220, 1972.
8. **Johnston, I. M.,** Studies in Boraginaceae. XIX. *Cordia* section *Gerascanthus* in Mexico and Central America, *J. Arnold Arbor. Harv. Univ.,* 31, 179—187, 1950.
9. **Johnston, I. M.,** Studies in Boraginaceae. XXII. Noteworthy species, chiefly Asian and South American, *J. Arnold Arbor. Harv. Univ.,* 33, 62—78, 1952.

9a. **Little, E. L., Woodbury, R. O., and Wadsworth, F. H.,** *Common trees of Puerto Rico and the Virgin Islands, U.S. Dep. Agric. Handbook No. 449,* Washington, D.C., 1974.

10. **Longman, K. A.,** Tropical forest trees, in *Handbook on Flowering,* Vol. 1, Halevy, A. H., Ed., CRC Press, Boca Raton, Fla., 1985, 23—39.
11. **Martius, C. F. P. and Eichler, A. W.,** *Flora Brasiliensis,* Vol 8/1, Fleischer, Munich, 1856 to 1872.
12. **Menninger, E. A.,** *Flowering Trees of the World for Tropics and Warm Climates,* Hearthside Press, N.Y., 1962.
13. **Nowicke, J. W.,** Flora of Panama: Boraginaceae, *Ann. Mo. Bot. Gard.,* 56, 33—69, 1969.
14. **Opler, P. A., Baker, H. G., and Frankie, G. W.,** Reproductive biology of some Costa Rican *Cordia* species (Boraginaceae), *Biotropica,* 7, 234—247, 1975.
15. **Opler, P. A., Frankie, G. W., and Baker, H. G.,** Rainfall as a factor in the release, timing and synchronization of anthesis by tropical trees and shrubs, *J. Biogeogr.,* 3, 231—236, 1976.
16. **Reich, P. B. and Borchert, R.,** Water stress and the phenology of trees in a tropical lowland dry forest of Costa Rica, *J. Ecol.,* 72, 61—74, 1984.

CORNUS

En. Cornel, Cornelian Cherry, Corneltree, Dog-Berry Tree, Dogwood; Fr. Cornouiller; Ge. Cornelbaum, Cornelle Horstrauch

Dennis P. Stimart

INTRODUCTION

Cornus is a genus of some 45 species of usually deciduous trees or shrubs. They are native to North America, Europe, and Asia, and rare but present in South America and Africa. *Cornus* is a member of the family Cornaceae. As ornamentals in horticulture they are grown as shrub borders and as specimen plants for the color of their flowers, leaves, and stems, and for unusual growth habit. The recommended species for flower color are *C. mas* L., *C. rugosa* Lam., *C. racemosa* Lam., *C. kousa* Hance., *C. florida,* L., *C. nuttallii* Audub., and *C. capitata* Wallich. Those recommended for leaf color are *C. alba* L. cv Spaethii, *C. alba* cv Elegantissima, and *C. mas* cv Aurea Elegantissima. Species grown for colorful stems include *C. alba, C. alba* cv Sibirica, and *C. sericea* L. cv Flaviramea. Those recommended for unusual growth habit include *C. controversa* Hemsl., *C. macrophylla* Wallich, *C. rugosa,* and *C. hessei.*[3]

MORPHOLOGY

The flowers are usually white, sometimes greenish or yellowish, and always small. They are produced in terminal corymbs or cymes or are clustered densely in heads. The parts of each flower are in fours. Flowers are bisexual. The fruit is a drupe which contains a two-celled stone. The leaves are strongly parallel veined, centrally attached with appressed hairs and are opposite except for *C. alternifolia* L.F. and *C. controversa* which have alternately attached leaves. Leaves are usually deciduous except for *C. capitata* which tends to be evergreen in mild climates.[1,2,3,7]

The species of *Cornus* can be classified into three groups.[3] Of these, Group 1 includes those that have flowers borne in corymbose inflorescences which lack both bracts and bracteoles. This includes the largest group and contains all the species except those listed below. In Group 2, the flowers are borne in dense umbels with a yellowish involucre which falls as the flowers open. Included here are *C. mas, C. officinalis* Siebold and Zucc., *C. chinensis,* and *C. sessilis.* In Group 3, the flowers are in dense clusters surrounded by large conspicuous bracts. This group may be subdivided into two subgroupings. In the first, the fruits are clustered but free from each other. Here belong *C. florida* and *C. nuttallii.* In the second subgrouping the fruits are completely united into a fleshy compound fruit. Included here are *C. capitata* and *C. kousa.*

VEGETATIVE AND REPRODUCTIVE DEVELOPMENT

Daylength affects several aspects of growth and development of *Cornus.* A critical daylength exists below which vegetative growth is arrested and above which it continues. This lies between 12 and 15 hr of light for *C. florida.* Actively growing plants stopped growing on 9- to 12-hr days but continued growing under 15-hr days. Also, SD cause the leaves to turn dark green and the stems to become woody and hard. Similar patterns have been observed in *C. nuttallii* where SD stopped growth completely and LD produced uninterrupted growth.[18] The duration of the dark period is actually more critical than the duration of the day. A light

Table 1
FLOWERING TIME OF *CORNUS* SPECIES[2,3,7]

Cornus sp.	Season	*Cornus* sp.	Season
alba	Late spring	*monbeigii*	Late spring
alternifolia	Late spring	*nuttallii*	Early spring to summer; again in fall
amomum	Late spring		
asperifolia	Late spring	*oblonga*	Late fall
australis	Late spring	*occidentalis*	Spring to summer
bretschneideri	Late spring	*officinalis*	Early spring
canadensis	Late spring to early summer	*paucinervis*	Summer
capitata	Early summer	*purpusii*	Spring to summer
controversa	Late spring	*racemosa*	Spring to summer
drummondii	Late spring	*rugosa*	Late spring to early summer
florida	Spring	*sanguinea*	Late spring
glabrata	Late spring	*sericea*	Spring to summer
hemsleyi	Summer	*sessilis*	Spring
hessei	Summer	*stricta*	Late spring
kousa	Late spring to early summer	*suecica*	Summer
macrophylla	Mid to late summer	*walteri*	Late spring
mas	Early spring		

break in the middle of the night successfully overcame the inhibitory effect of long nights in *C. florida* and *C. kousa*.[25,26]

In transfer studies between LD or SD photoperiods, dormant plants in leaf were forced into renewed vegetative growth by long photoperiods and, conversely, actively growing plants were forced into dormancy by short photoperiods.[26] It takes about 2 weeks for *C. florida* growing under a given photoperiod to exhibit the growth pattern typical of the new photoperiodic regime following a transfer.[26]

A single leaf of *C. florida* having just reached its full size was found to inhibit growth when subjected to SD, even though the remaining leaves were kept under LD. If a shoot of *C. florida* growing under LD was decapitated above the pair of leaves which had just reached full size, the two top axillary buds grew. If one of these leaves was given SD, axillary bud growth was inhibited. If the whole plant was subjected to SD no axillary buds grew.[26]

The source of maintaining growth inhibition is not centered in the leaf, since defoliation of inactive plants does not resume bud break.[8] However, chromatographed extracts of *C. florida* var. *rubra* leaves grown under SD contained a preponderance of growth-inhibitory substances, whereas extracts from leaves of plants grown under LD contained growth-stimulating substances.[26]

Most species of *Cornus* flower once a year in the spring to early summer with the exception of *C. oblonga* which blooms in late fall and *C. nuttallii* which will flower in the spring and a second time in the fall (Table 1).[2,3,7]

Relatively few studies have been conducted on the floral initiation and development process in *Cornus*. Flower buds of *C. florida* are normally initiated in the summer months but do not open until the next spring.[17] Daylength seems to have little effect upon floral initiation in *C. florida*.[26] However, LD altered profoundly the development of the various parts of the inflorescence such as by increasing the number and size of the bracts, the reversal of the flower meristem to three-branched vegetative shoots, and the elongation of the peduncle of the inflorescence. Also, some upper leaves retained their leaf morphology, but became colored like bracts.[26]

Cornus mas, which normally flowers in early spring, was found to initiate flowers in early summer.[5]

DORMANCY

The cold acclimation process in dogwood plants appears to be regulated by photoperiod, temperature, and light quality. Plants of *C. sericea* exposed to 8-hr SD for 4 weeks at 20°C day/15°C night became more cold resistant than those kept under SD at 15°C day/5°C night.[10,24] Acclimating *C. sericea* on 8-hr SD and low temperature (15/5°C) lowered stomatal resistance and root conductivity, and increased transpiration rate initially. By the end of acclimation, stomata closed, leaf water potential increased, and stem water content decreased 40 to 50%.[20]

Although the leaves do not themselves become cold hardy, plants of *C. sericea* developed cold resistance when only one leaf pair was exposed to SD.[14] A hardiness promoter from SD leaves appears to be translocated through the phloem.[10]

Daily foliar sprays of ABA (up to 1000 mg l^{-1}) induced the formation of resting buds in *C. florida*, but did not entirely mimic the effect of SD since bud growth resumed once the daily treatment of ABA was discontinued.[6] ABA did not induce cold hardiness in *C. sericea*.[9]

R treatment given as a nightbreak facilitated axillary bud growth and decreased cold acclimation in *C. sericea*. This response was fully reversed by FR given immediately afterwards.[27] SD and end-of-day FR light exposure after LD promoted growth cessation, cold acclimation, and subsequent cold hardening of *C. sericea* stems in response to low temperature.[15,16]

The inheritance of photoperiodically induced cold acclimation in *C. sericea* in crosses between northern and southern ecotypes indicated no maternal influence on cold acclimation but a nuclear effect.[12,13]

The breaking of dormancy in *Cornus* appears to be under the control of temperature and photoperiod. In *C. florida* var. *rubra*, 1000 hr of temperature between 0 and 7°C are needed to break winter dormancy.[11] However, this requirement can apparently be replaced by continuous lighting.[8] The rate of bud break in *C. sericea* was not affected by the length of the photoperiod after cold storage, but shoot elongation was continuous under a 16-hr photoperiod and a cessation of growth occurred under 8 hr.[4]

The flowering date of *C. florida* has been observed to vary both geographically[28] and from year to year at a single location.[22] In a study involving *C. sericea* collected from 21 locations between 40 and 65° N latitude and grown at St. Paul, Minn., flowering occurred within a 10-day period.[23] It was suggested that flowering time was environmentally rather than genetically controlled and was regulated primarily by temperature. In a related study with *C. florida*, the summation of air temperatures above 2.2°C for a 6-week period before flowering explained most of the annual variation in flowering date.[21] Therefore, it appears that spring air temperatures control the rate at which flower buds develop.

In *C. florida*, stored over the winter, there is a carryover effect of photoperiod treatments from the previous season. Plants previously grown under 9-hr days opened buds faster than plants grown under LD (12, 15, or 18 hr). Buds on lower side branches of LD plants opened before buds on top branches which remained actively growing later. It would appear that buds which have been dormant for a longer period of time open before those which went dormant later.[19]

CULTURE AND PROPAGATION

Cornus is easily grown provided plants are cultivated in a good soil. They do poorly in excessive heat and require adequate moisture. Very few pests attack these plants.

Propagation is by seeds, cuttings, and grafting. Seeds can be sown as soon as ripe or they can be stored dry in airtight containers in a cool place for up to 1 year. Seeds of *C. florida*,

C. kousa, and *C. sericea* are usually stratified for 3 months at 4.4°C, then sown. Seeds of others have double dormancy and may have to be stratified at 25°C for 3 to 4 months and then 4.4°C for another 3 to 4 months. Most species root from softwood and hardwood cuttings. In grafting, a crown graft is used; in budding, a shield bud is used.[28]

REFERENCES

1. **Bailey, L. H.,** *Cyclopedia of American Horticulture,* Vol. 2, Doubleday, Page and Company, New York, 1906, 595.
2. **Bailey, L. H.,** *Hortus Third,* Macmillan, New York, 1976, 1290.
3. **Bean, W. J.,** *Trees and Shrubs,* Vol. 1, John Murray, London, 1970, 845.
4. **Benjamin, L. P.,** Responses of woody plants to photoperiod and chilling, *Greenhouse Garden Grass,* 3, 32—33, 1963.
5. **Carpenteur, E. D.,** A Monograph: Morphology of Flower Primordia of Selected Woody Plants, Ph.D. thesis, Michigan State University, East Lansing, 1964.
6. **Cathey, H. M.,** Response of some ornamental plants to synthetic absicisic acid, *J. Am. Soc. Hortic. Sci.,* 93, 693—698, 1968.
7. **Chittenden, F. J. and Synge, P. M.,** *Dictionary of Gardening,* Vol. II, Clarendon Press, Oxford, 1977, 1088.
8. **Downs, R. J. and Borthwick, H. A.,** Effects of photoperiod on growth of trees, *Bot. Gaz.,* 117, 310—326, 1956.
9. **Fuchigami, L. H., Evert, D. R., and Weiser, C. J.,** A translocatable cold hardiness promoter, *Plant Physiol.,* 47, 164—167, 1971.
10. **Fuchigami, L. H., Weiser, C. J., and Evert, D. R.,** Induction of cold acclimation in *Cornus stolonifera, Plant Physiol.,* 47, 98—103, 1971.
11. **Hess, C. E.,** Propagating and over wintering, *Cornus florida* var. *rubra* cuttings, *Proc. Plant Prop. Soc.,* 5, 43—44, 1955.
12. **Hummel, R. L.,** Inheritance of the Photoperiodically Induced Cold Acclimation Response and Self-Incompatibility in *Cornus sericea* L. (*Cornus stolonifera* Michx.), Red-Osier Dogwood, Ph.D. thesis, University of Minnesota, St. Paul, 1981.
13. **Hummel, R. L., Ascher, P. D., and Pellett, H. M.,** Inheritance of the photoperiodically induced cold acclimation response in *Cornus sericea* L., red-osier dogwood, *Theor. Appl. Genet.,* 62, 385—394, 1982.
14. **Hurst, C., Hall, T. C., and Weiser, C. J.,** Reception of the light stimulus for cold acclimation in *Cornus stolonifera* Michx., *HortScience,* 2, 164—166, 1967.
15. **McKenzie, J. S., Weiser, C. J., and Burke, M. J.,** Effects of red and far-red light on the initiation of cold acclimation in *Cornus stolonifera* Michx., *Plant Physiol.,* 53, 783—789, 1974.
16. **McKensie, J. S., Weiser, C. J., and Li, P. H.,** Changes in water relations of *Cornus stolonifera* during cold acclimation, *J. Am. Soc. Hortic. Sci.,* 99, 223—228, 1974.
17. **Morse, W. C.,** Contribution to the life history of *Cornus florida, Ohio Nat.,* 8, 197—204, 1907.
18. **Nitsch, J. P.,** Growth responses of woody plants to photoperiodic stimuli, *Proc. Am. Soc. Hortic. Sci.,* 70, 512—525, 1957.
19. **Nitsch, J. P.,** Photoperiodism in woody plants, *Proc. Am. Soc. Hortic. Sci.,* 70, 526—544, 1957.
20. **Parsons, L. R.,** Water relations, stomatal behavior and root conductivity of red-osier dogwood during acclimation to freezing temperatures, *Plant Physiol.,* 62, 64—70, 1978.
21. **Reader, R. J.,** Effect of air temperature on the flowering of dogwood (*cornus florida*), *Can. J. Bot.,* 53, 1523—1534, 1975.
22. **Smith, J. W.,** Phenological dates and meteorological data recorded by Thomas Mikesell between 1873 and 1912 at Wauseon, Ohio, *Mon. Weather Rev., Suppl.,* 2, 23—93, 1915.
23. **Smithberg, M. H. and Weiser, C. J.,** Patterns of variation among climatic races of red-osier dogwood, *Ecology,* 49, 495—505, 1968.
24. **Van Huystee, R. B., Weiser, C. J., and Li, P. H.,** Cold acclimation in *Cornus stolonifera* under natural and controlled photoperiod and temperature, *Bot. Gaz.,* 28, 200—205, 1967.
25. **Waxman, S.,** The effect of the length of day on the growth of woody plants, *Proc. Plant Prop. Soc.,* 5, 47—49, 1955.
26. **Waxman, S.,** The Development of Woody Plants as Affected by Photoperiodic Treatments, Ph.D. thesis, Cornell University, Ithaca, N.Y., 1957, 193.
27. **Williams, B. J., Pellett, N. E., and Klein, R. M.,** Phytochrome control of growth cessation and initiation of cold acclimation in selected woody plants, *Plant Physiol.,* 50, 262—265, 1972.
28. **Wyman, D.,** Order of bloom, *Arnoldia,* 10, 41—56, 1950.

DELPHINIUM

En. Delphinium, Larkspur; Fr. Dauphinelle; Ge. Ritterspurn

H. F. Wilkins

INTRODUCTION

Delphinium (Ranunculaceae) is a genus with over 300 herbaceous species which are annual, biennial, or perennial plants native to north temperate areas.[1]

The most common delphinium and larkspur species seen in gardens or used in commerce are[1,12]

1. *Consolida ambigua* (L.) P. W. Ball and Heyw. and *Consolida orientalis* (J. Gay) Schröding, the annual larkspurs, were formerly *Delphinium ajacis* and are natives of the Mediterranean region, north Africa, southern Europe, and western Asia.
2. *Delphinium grandiflorum* L., a perennial with spike-like racemes over 1 m long, has many hybrids and is one of the most commonly seen delphiniums, originating from Siberia and China.
3. *Delphinium* × *cultorum* Voss is a name used for many of the various hybrid perennial delphiniums, usually *D. elatum* derivatives.
4. *Delphinium elatum* L., a stately perennial species, has a flowering spicate spike up to 1.8 m in length. It also has many hybrids; native species are found in southern and central Europe and into Siberia.
5. *D. belladonna* Hort. includes branching triploid and hexaploid types probably derived from *D. elatum* and diploid species cultivars of the 19th century.

Flowers are borne on a long, erect, spike-like raceme. Single and double flower forms exist. The calyx is composed of 5 sepals, one of which is spurred. The 2 or 4 petals are smaller and are spurred as well; all are compacted at the throat with many stamens and have 2 to 5 pistils. Flowers are mainly in light- to dark-blue colors; white and pink colors are also common. A few species have red, scarlet, and yellow flowers. Leaves are variously divided or cut to form a palmate outline.[1]

Cultivars of this genus are grown in gardens for spring and early-summer flowering or used in commerce as cut flowers. After flowers of the perennial species fade or are harvested, new secondary flowering stems may arise, but frequently are of lesser quality. Old perennial root clumps are divided in late summer or autumn after flower. There are many beautiful and striking named horticultural perennial forms that are asexually propagated and are hardy even in severe northern climates. They require little special attention, and the annual species reseed themselves with ease. Annual species are propagated from seed. All species are most vigorous in full light and air circulation if grown in well-drained soils.[1,12]

MORPHOLOGY

Ignat'eva[7] has described the morphological and developmental aspects of perennial plants from the seedling stage to those 4 years of age. The number of shoots capable of flowering increases to the 3rd year and then decreases (*D.* × *cultorum*). The average life span of an undivided perennial clump is 4 years; when divided, it can be 5 to 7 years. A dormant bud is vegetative and, when it becomes active in the spring, the balance of leaves are initiated and the inflorescence forms.[7]

TEMPERATURE AND PHOTOPERIOD

It is apparent that the various annual and perennial species require a vernalization treatment for both shoot elongation and flower initiation and development.[1,2,10,12,13]

For the annual delphinium the effective stage of vernalization is the seed at 2°C for around 6 weeks. Further development is best under LD.[13] The annual larkspur seedling will rosette at temperatures below 13°C. The elongation of the stem axis and flowering occurs at higher temperatures. However, long photoperiod (incandescent NB) enhances stem elongation even at 10°C.[12] Hall[6] reported that a 2-hr NB from a 40-W warm white fluorescent lamp also induced larkspur to flower rapidly when plants were grown at a minimum temperature of 15.5°C. However, if very young seedlings are exposed to high temperatures for several days, flower initiation, development, and shoot elongation occur rapidly with very poor flower quality.[12]

Perennial species flower in the year following seeding and respond to similar temperature/photoperiod regimes as with the annual delphiniums.[1,12] Established perennial plants should be kept at low temperatures until February and then forced at 10 to 13°C minimum temperatures.[12]

CULTURE

Propagation is by seed for both the annual and perennial species or by division of the perennial roots and shoot clumps.[1,12] In vitro propagation has not been totally successful.[5]

Seeds of the annual species do not germinate well above 13°C.[12] Culturally, it is best to allow the seedlings to develop for 60 or more days at greenhouse temperatures below 13°C and then expose the rosetted plant to higher night temperatures. In this manner, at temperatures over 13 or 16°C high-quality stems and flowers are produced when stem elongation occurs. LD hasten flowering at 10, 13, or 16°C. LD also increase quality and the number of flowering stems.[12] Seeds can also be sown in the field in the autumn for flowering in the spring.[2]

Seeds of the perennial species *D. grandiflorum* are also sown in the autumn as well and transplanted in the field for the winter with some protection, if needed. These plants will flower in the spring as temperature and daylengths increase.[12] However, with *D.* × *cultorum*, seeds which are sown very early in the spring and transplanted in beds in unheated greenhouses or in the field after danger of frost will flower in late summer. High-quality stems will be harvested the following spring and late summer.[12] Plömacher[10,11] found that the perennial species, *D.* × *cultorum* and *D. belladonna*, can be forced in heated greenhouses for flowering in April and later, provided temperatures do not exceed 12°C day/6°C nights. Greenhouses (plastic) were moved over the crop in the field in late February. Flowering was in April, some 30 to 45 days sooner than in nature. Flowers grown in these environments had a greater postharvest life than the controls.

GROWTH REGULATORS

Chlormequat at 4000 ppm or daminozide at 5000 ppm reduced *Delphinium* height.[3] GA has been reported to have significantly accelerated flower formation and date of flowering of annual larkspur.[4,9]

POST HARVEST

Silver thiosulfate (STS) has prevented normal and ethylene-induced floret shattering in *Delphinium*.[14]

REFERENCES

1. **Bailey, L. H. (Staff of Hortorium),** *Hortus Third, A Concise Dictionary of Plants Cultivated in the United States and Canada*, Macmillan, New York, 1979, 367—369.
2. **Bajpai, P. N. and Nerikar, V. N.,** Effect of sowing at different dates on the growth and flowering of *Cosmos, Caryopsis* and larkspur, *Sci. Cult.*, 25, 140—142, 1959.
3. **Bhattacharjee, S. K.,** Influence of photoperiod on the action of growth retardants in some herbaceous ornamental plants, *Punjab Hortic. J.*, 19, 198—204, 1979.
4. **Bose, T. K.,** Effect of growth substances on growth and flowering of ornamental annuals, *Sci. Cult.*, 31, 34—36, 1965.
5. **Bott, J. C.,** Tissue culture of delphiniums. Preliminary experiments with *D. elatum* and University Hybrids, *Plantsman*, 2(3), 169—171, 1980.
6. **Hall, O. G.,** Lamps for daylength control, *Shinfield Progr.*, 7, 40—50, 1965.
7. **Ignat'eva, I. P.,** Vegetative regeneration of delphiniums (in Russian), *Dokl. Mosk. Skh. Akad.*, 25, 102—129, 1956.
8. **Ignat'eva, I. P.,** Some characteristics in the development of the delphinium (in Russian), *Izv. Timiryazevsk. Skh. Akad.*, 4, 43—55, 1960.
9. **Lindstrom, R. S., Wittwer, S. H., and Bukovac, M. J.,** Gibberellin and higher plants. IV. Flowering responses of some flower crops, *Q. Bull. Mich. Agric. Exp. Stn.*, 39, 673—681, 1957.
10. **Plömacher, H.,** Vervroegen delphinium mogelijk (Advancing flowering of delphiniums is possible), *Vakb. Bloemisterij*, 34(50—51), 89, 1979.
11. **Plömacher, H.,** Schnittstauden unter Folie. Erste Kulturerfahrungen mit Delphinium (Plants grown under plastic for cutting, preliminary results for cultivation of delphiniums), *Zierpflanzenbau*, 20(1), 15—16, 1980.
12. **Post, K.,** *Flower Crop Production*, Orange, Judd, New York, 1952, 446—451.
13. **Roger, D. and Séchet, J.,** Sur la printanisation des Végétaux cultivés et indigenés, *C.R. Acad. Sci.*, 225, 763—765, 1947.
14. **Shillo, R., Mor, Y., and Halevy, A. H.,** Prevention of flower drop in cut sweet peas and delphinium (in Hebrew), *Hassadeh*, 61, 274—276, 1980.

DICENTRA SPECTABILIS

En. Bleeding heart; Fr. Dicentre; Ge. Flammendes Herz; Sp. Alkeli doble

Thomas C. Weiler and Luiz Carlos Lopes

INTRODUCTION

The plant is native to Japan and is a member of the Fumariaceae or Papaveraceae, depending on the authority.[2,4] *D. spectabilis* (L.) Lem. flowers are zygomorphic and in a simple raceme. The sap of the plant is watery and the species grows sympodially (i.e., produces axillary inflorescences). The genus name *Dicentra* is derived from the Greek meaning two-spurred and refers to the corolla. The plants are used in cultivated or naturalistic gardens, and occasionally are grown as flowering potted plants for Valentine's Day or spring displays in botanical gardens.

ONSET AND REMISSION OF DORMANCY

Flowering of *D. spectabilis* is the usual result of a flush of growth. Lopes and Weiler[1] found that low-light, high-temperature conditions inhibited the flowering of growing plants, but that onset and remission of dormancy were the primary regulants of when plants flower. Dormancy is caused by short photoperiodic conditions.[6] When days consisting of 8 hr of sunlight were extended by various durations of 7 to 18 μmol m^{-2} sec^{-1} incandescent lighting, plants which received 12 or fewer hours of light each day ceased to grow and died back to the perennating crown. Plants in photoperiods of 14 or more hours grew and flowered.

While short photoperiods caused dormancy, crowns were brought out of dormancy by cold treatment.[1,3] Such crowns grew and flowered in short photoperiods. Long photoperiods were ineffective for breaking dormancy.[1] In one experiment, 16 weeks of 5° storage were required for all plants to flower. Partially cold-treated plants grew but did not flower.[1]

Once growing, long photoperiods indefinitely promote continued growth and flowering, especially when combined with periodic cutting back of plants to promote new flushes of growth.

CONTROL OF PLANT GROWTH

Based on this research, two approaches to plant culture have been proposed: spring growth and year-around growth. Growing plants for spring flowering is simplest by planting cold-stored crowns. Plants flower 3.5 to 4.5 weeks after planting at 11°C night temperature and 2.5 to 3.5 weeks after planting at 15°C night temperature.[5] Plants are stocky when grown in short photoperiods, although a 1250 mg/ℓ spray of daminozide when the leaves begin to unfurl promotes compact upper stem internodes (Figure 1).

Year-around culture of *D. spectabilis* is feasible through production of stock plants, and propagation and growth of plants in long photoperiods (an 18-hr/day exposure to continuous light).[6] Rooted cuttings usually are allowed to establish and then are pinched. Since stem elongation is promoted by the long photoperiods, plants sprayed with a 5000 to 7500 mg/ℓ daminozide spray on the date of pinch achieve an acceptable form.

FIGURE 1. *D. spectabilis* forced in early spring.

OTHER SPECIES

D. eximia, an acaulescent species with a scapose inflorescence and perennating scaley rhizome,[4] sometimes is cultivated for spring flowers. Little research has been reported, but the species requires cold treatment to flower and blooms in spring after about 6 weeks of growth at a 12°C night temperature.

REFERENCES

1. **Lopes, L. C. and Weiler, T. C.,** Light and temperature effects on the growth and flowering of *Dicentra spectabilis* (L.) Lem., *J. Am. Soc. Hortic. Sci.,* 102, 388—390, 1977.
2. **Miller, W. and Bailey, L. H.,** *Dicentra,* in *Cyclopedia of Horticulture,* Bailey, L. H., Ed., Macmillan, New York, 1001—1003, 1942.
3. **Risser, P. and Cottam, G.,** Influence of temperature on the dormancy of some spring ephemerals, *Ecology,* 48, 500—503, 1967.
4. **Stern, K. R.,** Revision of *Dicentra* (Fumariaceae), *Brittonia,* 13, 1—57, 1961.
5. **Weiler, T. C. and Kirk, T. I.,** unpublished data, 1982.
6. **Weiler, T. C. and Lopes, L. C.,** Photoregulated *Dicentra spectabilis* (L.) Lem. as a potential potted plant, *Acta Hortic.,* 64, 191—195, 1976.

ERYTHRINA

En. Coral tree, Coral bean

Rolf Borchert

INTRODUCTION

Erythrina (Fabaceae) is a genus of 108 trees and shrubs of wide distribution in the tropical and subtropical portions of the world. Because of their brilliant red or orange flowers, usually displayed when trees are bare, coral trees are widely planted in gardens and parks of regions with a warm climate.[14,19,20,22,24] Species cultivated in California and Florida have been described,[20,22] and excellent color photographs of many species of *Erythrina* can be found.[17,19a,22]

Species such as *E. fusca, E. poeppigiana,* and others have been used extensively as shade trees in coffee and cacao plantations. Many species are planted as living fences simply by thrusting large branches into the ground where they take root; easy vegetative reproduction is common in the genus. *E. berteroana* and *E. americana* are planted because their young leaves and inflorescences are edible. Many species of *Erythrina* are thus found well outside their natural ranges as a result of human cultivation.[19,20,24]

Erythrina is one of the best studied genera of tropical trees. Its taxonomy, biogeography, floral biology, biochemical systematics, and biosystematics have been discussed in four recent symposia.[24-27] The large genus is unusually close knit. Characteristics of leaves and inflorescences are generally monotonous, and classification of the species is based mainly on characters of the flowers and fruits (Figures 1 and 2).[13,14] With two exceptions, all species examined so far have a chromosome number of 2n = 42, and natural hybridization appears to be frequent where more than one species of *Erythrina* occur together. Several natural and artificial hybrids have been described.[20,24] Extensive hybridization is currently done with species cultivated at the Pacific Botanical Garden at Lawai, Kauai, Hawaii; this work will greatly enhance our understanding of the control of patterns of vegetative and reproductive growth as discussed below.[22b]

Available evidence does not allow a conclusion about whether *Erythrina* originated in South America, Africa, or tropical Asia. It is likely that there have been always species like *E. fusca* and *E. variegata* which are widely dispersed, because their seeds may float for months in seawater retaining their viability.[24]

FLORAL STRUCTURE AND BIOLOGY

The red or orange papilionaceous flowers are inserted on racemose, lateral inflorescences singly or in clusters of two to five (Figures 1 and 2). Petals differ markedly in size; the standard (banner) is always the longest, and wings and keel are usually less than half as long as the standard. The androecium comprises 10 stamens, with 9 or 10 being united basally. The single stigma is scarcely enlarged.[13]

In general, the style is held away from the anthers, and pollen is not deposited directly on the stigma of the same flower. There may be also differences in the relative timing of maturation in the stigma and the anthers. Although transfer of pollen from one flower to another on the same individual is possible, pollination may not occur in the absence of the appropriate pollen vectors. Outcrossing is therefore the norm for the genus, but seed set in self-compatible species has been reported.[24] All but 12 New World species are pollinated by hummingbirds, while the remaining Old and New World species are pollinated by a

FIGURE 1. *Erythrina caffra.* (From Kunkel, G., *Flowering Trees in Subtropical Gardens,* W. Junk, The Hague, 1978. With permission.)

variety of relatively large passerine birds.[10,11,26,29] Species pollinated by passerine birds possess somewhat "gaping" flowers; the standard tends to be ovate or obovate and the keel and the wings are more or less conspicuously exerted from the calyx (Figure 1). Such flowers produce large quantities of nectar of a low sucrose/hexose ratio and a high amino acid content. Flowers face inwards and are arranged on horizontal, sometimes even inverted, racemes, providing a good landing perch and access to flowers for perching birds.[11,26] In hummingbird-pollinated New World species the corolla is long, narrow, and tubular, and

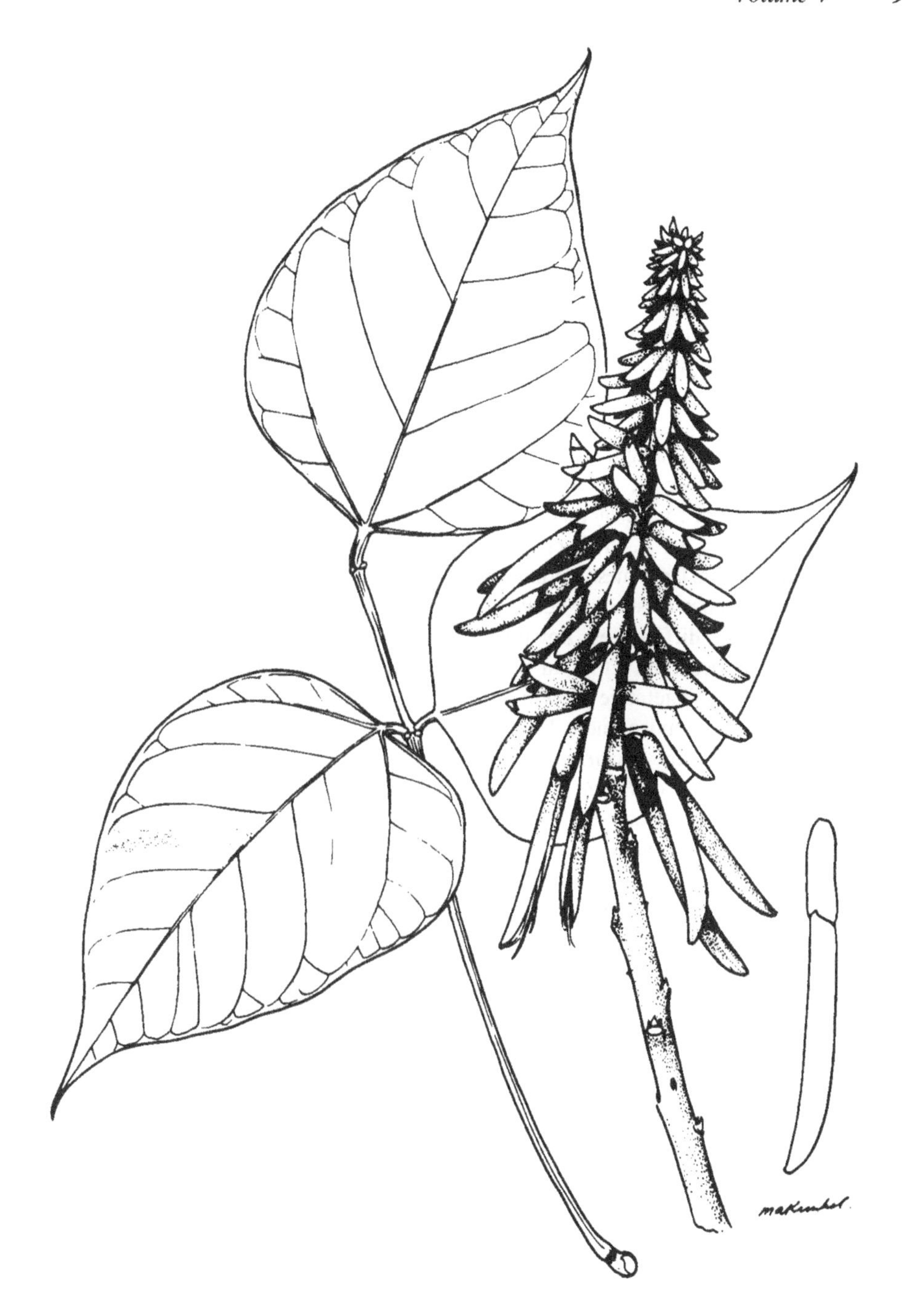

FIGURE 2. *Erythrina variegata.* (From Kunkel, G., *Flowering Trees in Subtropical Gardens,* W. Junk, The Hague, 1978. With permission.)

the reduced wings and keel are enclosed by the elongated standard (Figure 2); flowers are oriented outward on a vertical inflorescence, a landing perch is absent, and the nectar is high in sugars and low in amino acids.[10,11,26,29]

PERIODICITY OF FLOWERING

Flowering in all species of *Erythrina* is episodic. The time of flowering has been mentioned for many species in field studies,[8,10,29] taxonomic descriptions,[13,14] and descriptions of cultivated species.[14,19,20,22] However, with two exceptions, all available information is anecdotal. Flowering periods were reported for more than 30 species cultivated at the Pacific Botanical Garden at Lawai, Kauai, Hawaii.[18,19] Relations between flowering, vegetative growth and climatic conditions, and variability of flowering within tree populations and individual trees have been analyzed in *E. poeppigiana* cultivated in Costa Rica.[2]

Information illustrating the variability of flowering episodes in *Erythrina* over much of its range was selected from various sources and summarized in Table 1. In view of the anecdotal nature of this information and the complete lack of careful phenological studies in the natural range of any species, the following evaluation of the data in Table 1 and the subsequent discussion of the control of flowering should be considered highly tentative. It is hoped that the questions raised will encourage future, more detailed, and comprehensive observation.

Erythrina species have adapted to a variety of habitats ranging from tropical rain forests to tropical deciduous forests, savannahs, and deserts, and from tropical lowlands to high altitudes (e.g., 2700 m for *E. leptorhiza* and *E. montana* in Mexico[10]) and to the northern limits of the tropics (e.g., *E. herbacea* and *E. flabelliformis*[6,10,30]). Morphologically, *Erythrina* species range from large mesomorphic trees to highly xeromorphic shrubs and geophytes.

Even where a number of species occur in the same geographic region, they appear to be separated by ecologic differences. For instance, in species-rich northern Central America, Andean South America, and Ethiopia various species grow at different elevations and in different habitats.[10,24] Vegetative growth and, implicitly, flowering thus appear to be finely tuned to the prevailing environmental conditions.

Most areas inhabited by *Erythrina* are characterized by a more or less well-pronounced dry season. The majority of *Erythrina* species flower during the dry season when leafless, i.e., after drought-induced leaf shedding; vegetative growth occurs mostly during the early rainy season. These observations suggest that the periodicity of vegetative growth and flowering in *Erythrina* is mostly determined by seasonal variation in water availability, well known to be the principal determinant of plant growth in the tropics.[12,23] In spite of the lack of detailed information on the origin of specimens and prevailing climate at the time of the observed flowering episodes, the following discussion of the data compiled in Table 1 will show that the majority of species appear to flower during or immediately after the dry season.

In theory, the following annual distribution of rainfall exists in the tropics:[12,23]

1. Near the equator, there are two maxima and two periods of less rain, the maxima occurring about 1 month after the equinoxes.
2. With increasing distance from the equator (~10°) the two seasons of maximum rainfall move together, reducing one dry season in length, while the other becomes longer and more intense. The long dry period always occurs during the "winter", i.e., when daily photoperiods are shortest. This coincidence between dry season and short photoperiods is probably the reason for the widespread, but erroneous, notion that flowering in *Erythrina* and other tropical trees is induced by short photoperiods.[3] Inversely, during the period of heavy rains the sun is overhead and the days are longest.
3. Further away from the equator (~20°), the period of maximum rainfall becomes weaker and shorter.

Over the tropical continents, and specifically over the South American continent, seasonal rainfall distribution is much more complicated. Furthermore, moisture availability generally increases with altitude, and the numerous *Erythrina* species living at high altitudes are thus less likely to be exposed to seasonal drought than those living in the lowlands.

Throughout Central America and Mexico the rainy season lasts from April/May through

Table 1
PERIODICITY OF FLOWERING IN VARIOUS SPECIES OF *ERYTHRINA*

No.	Species	Origin	Observed at	Site[a]	Month[b]: O N D	J F M	A M J	J A S	Ref.[c]
					New World Species				
1. a	*E. americana*	Central America, Mexico	Mexico	N	XXXX	XXXXXXX			13
b			California	N		XXXXX	XXXXXX		20
2. a	*E. chiapasana*	C. America	Mexico	N		XXX	XXX		13
b			Hawaii	N	0000	00000	000		18
3. a	*E. coralloides*	C. America	Mexico	N		XX	XXXXXXX		13
b			Hawaii	N		000	0000		18
c			California	N		X XX	XXXXXXXXX	X	21
4. a	*E. guatemalensis*	C. America	Guatemala	N	XXXX	XXXXX			13
b			Hawaii	N	0000	00000	0000		18
c			California	N			XXX	X	21
5.	*E. leptorhiza*	C. America	Mexico	N			******	***	10
6.	*E. olivae*	C. America	Mexico	N			*********		29
7.	*E. flabelliformis*	S.W. U.S.	S.W. U.S.	N			******		6, 30
8. a	*E. herbacea*	S. U.S., Mexico	Mexico	N			******		30
b			California	N		XXXXXXX	XX		22b
c			Hawaii	N			XXX	XXXXXXXXX	21
d				N	0 0 0 0 0	0 0 0 0	0 0 0 0 0	0 0 0 0	18
9. a	*E. berteroana*	C. America, West Indies, Colombia	West Indies, Colombia	N		XXXXXXXX	XXX		13
b				N	000	000000000	0000		18
c			Hawaii California	N	X X		X X		21
10. a	*E. chiriquensis*	Panama	Panama	N				XXXXXX	13
b			Hawaii	N	00	000000			18

Table 1 (continued)
PERIODICITY OF FLOWERING IN VARIOUS SPECIES OF *ERYTHRINA*

No.	Species	Origin	Observed at	Site[a]	Month[b] O N D	J F M	A M J	J A S	Ref.[c]
					New World Species				
13. a	*E. fusca*	C. America,	W. Indies,	N		XXXXXX			13
b	(= *E. glauca*)	W. Indies,	Colombia	S				XXXXXX	13
		S. America	Brazil, Ecuador,						
			Guayanas						
14. a	*E. amazonica*	S. America	Brazil, Peru	S			XXX	XXXXXXXX	13
b			Hawaii	N	00	00000000			18
15. a	*E. crista-galli*	S. America	Brazil, Argentina	S	XXXXXXXX				13
b			Hawaii	N	0 0 0 0 0	0 0 0 0	0 0 0 0 0	0 0 0 0	19
c			California	N		XXXXX	XXXXXXXXX	XXXXXXXXX	20
16. a	*E. dominguezii*	S. America	Brazil,	S	XXXXX			XXXXX	13
b			Paraguay,	N		000	00000		18
			Argentina						
			Hawaii						
17. a	*E. falcata*	S. America	Argentina,	S	XXX			XXX	13
b			Paraguay	S	XXXXX		XXXXXXXX	XXXXX	13
c			Peru, Bolivia	N		00	0000		18
d			Hawaii	N			XXXXXX		21
			California						
18. a	*E. speciosa*	S. America	S. Brazil	S				XXXXXXXX	13
b			Hawaii	N	0000	000000000	00000		18
c			California	N	X	XXXXXXXXX			21

Old World Species

No.	Species	Native range	Observation site	N/S[a]	Group 1	Group 2	Group 3	Group 4	Ref.[c]
19. a	*E. caffra*	S. Africa	S. Africa	S				*********	8
b			California	N		XXX	XXXXXXXXX		20
20. a	*E. humeana*	S. Africa	S. Africa	S	**	*********	*****		8
b			Hawaii	N	000000		000000000	000000000	18
c			California	N	X X			X X X	21
21. a	*E. latissima*	S. Africa	S. Africa	S	******			*****	8
b			Hawaii	N		00	00		18
22. a	*E. lysistemon*	S. Africa	S. Africa	S	***			******	8
b			Hawaii	N	00	0000000			
23.	*E. tahitensis*	Tahiti	Hawaii	N		00	0000		18
24.	*E. sandwicensis*	Hawaii	Hawaii	N	000			000000	18
25. a	Monthly precipitation at Lawai, Hawaii (cm)				10 4 13	10 7 1	2 0 1	1 5 6	
b	Precipitation per 10 days; X, > 5 cm; x, > 2 cm				X X XX	x X x x	XX X	XXX XX	

[a] N, observation site located north of climatic equator; S, observation site located south of climatic equator.

[b] O, flowering observed and recorded for species cultivated at the Pacific Botanical Garden in Lawai, Kauai, Hawaii; *, field observations of flowering in the natural range of a species; X, flowering times obtained from collection dates of flowering herbarium material.

[c] Indicates references for data sources.

October/November, and practically all *Erythrina* flower when leafless, either early in the dry season (Table 1: 1a,4a), later in the dry season (Table 1: 2a,3a,8a), or, especially in arid habitats, after the first heavy rains (Table 1: 5—7).[10] In the southern U.S., *E. herbacea* flowers during the early growing season (Table 1: 8a). Central American species cultivated in California often flower later than in the wild (Table 1: 1b,4b).

Over the South American continent, the intertropical convergence zone rarely moves to the Southern Hemisphere and the "climatic equator" normally remains at around 5 to 8° N throughout the year.[23] Flowering periodicity of erythrinas native to northern South America shows considerable variability, probably reflecting local variations in rainfall patterns. While two species flower early in the year (Table 1: 9a,12a), one flowers in July to August (Table 1: 10a), and *E. rubrinerva,* like other species native to the permanently moist high altitudes in the Colombian Andes,[13] flowers year around (Table 1: 11a).

Flowering of *Erythrina* in southern South America reflects the absence of a marked dry season in the Amazon Basin (Table 1: 17b) or the occurrence of a dry season beginning around June/July (Table 1: 12b,13b,14—18a).[23]

Variation of flowering with the timing of the dry season is especially pronounced in species found both north and south of the equator (Table 1: 12a,b;13a,b) or in areas with or without a distinct dry season (Table 1: 17a,b). *E. fusca* (synonymous with the neotropical *E. glauca*[22b]) flowers simultaneously during the dry seasons of Southern Hemisphere Brazil and the Northern Hemisphere Guayanas (4° N) (Table 1: 13b), but at different times elsewhere in the Northern Hemisphere, both in the neotropics (Table 1: 13a) and in the paleotropics (India, Malaysia, Thailand, and Vietnam).[22b] Near the equator, flowering is quite variable: in equatorial South America flowering has been observed in almost every month of the year.[22b]

In South Africa, as in southern South America, most *Erythrina* flower when leafless during the later months of the year (Table 1: 19a,21a,22a), but in the shrubby *E. humeana* flowers form on leafy shoots during the rainy season[8] (Table 1: 20a).

Many New and Old World species of *Erythrina* are cultivated at the Pacific Botanical Garden at Lawai, Kauai, Hawaii,[19] located at 22° N, corresponding in latitude to Central Mexico and the natural range of several *Erythrina* species listed in Table 1. Normally winters at Lawai are cool and wet, but in 1980/81, during which observations listed in Table 1 were made, winter months were unusually dry[18] (Table 1: 24a). Irrespective of their origin, most species cultivated at Lawai stood leafless and began flowering either in November/December (Table 1: 2b,4b,9b,10b,11b,13b,18b,22b) or in March (Table 1: 3b,16b,17b); i.e., during the two driest months (Table 1: 24a). Some species flowered a second time after heavy rains in April (Table 1: 2b,4b,24b), which also triggered flowering on leafy shoots in *E. humeana* (Table 1: 20b), known to flower in its native range during the rainy season when bearing leaves.[8,19a] The only two species flowering more or less year around at Lawai, but not in their native habitat (*E. crista-galli* and *E. herbacea;* Table 1: 18,15), differ notably in their growth habit from all other species (see below).

Flowering periodicity of cultivated trees at Lawai thus clearly illustrates the response of many species of *Erythrina* to the same temporal variations in water availability. However, there is considerable intraspecific variation in flowering periodicity, particularly between trees of different geographic origin.[18]

Variability of vegetative growth and flowering within populations of *E. poeppigiana* cultivated in the Central Valley of Costa Rica is illustrated in Table 2. These observations provide the basis for the following analysis of the causal relations between water availability, vegetative growth, and flowering in *Erythrina*.

PHENOLOGY AND CONTROL OF FLOWERING

Vegetative and reproductive development are strongly interrelated in all plants, but in

Table 2
PERIODICITY OF FLOWERING AND VEGETATIVE GROWTH IN *E. POEPPIGIANA* CULTIVATED IN THE CENTRAL VALLEY OF COSTA RICA, C.A., ALONG AN 18-KM GRADIENT RANGING FROM 1030 TO 1340 M ALT[2]

	Month			
	O N D	J F M	A M J	J A S
A. Large trees at dry sites; 1200 m	III	I	III	
	XX	XXXXX		
		000000	0000	0000
B. Medium trees at dry sites; 1200 m	IIIII	IIIII		
		XXXXXX	X	
	00000	0	000000	
C. Trees at wet sites; 1200 m	IIIII	IIIII	I	
		XXXX	XXX	
	00000		0000000	
D. Sprouts on pruned trees; 1200 m	? ?	I	IIIIII	
			XXXXX	
			000000	0
E. Medium trees at cool sites; 1030 m	III	II		
	X	XXXX		
		o o o	o o 0000	000
F. Medium trees at dry sites; 1340 m	? ?	II	IIII	
		XX	XXX	
			0000000	0
Precipitation in San José (cm)[a]	23 13 0	0 1 0	3 23	
Precipitation per 10 days; X, >5 cm; x, >2 cm	XXXXxX		xxXXXXX	

Note: XXXX, flowering; IIII, leaf fall; OOOooo, production of new shoots with fully (OO) or partially (oo) expanded leaves.

[a] Data from Servicio Meteorológico de Costa Rica.

trees these relationships are considerably more complex than in herbaceous plants because of the structural complexity of the shoot system, the life-long alternation between periods of vegetative growth and periods of rest, and repeated episodic flowering.[3]

In most species of *Erythrina,* flower initiation (i.e., the transition of a meristem from the vegetative to the reproductive state) is confined to lateral meristems. While development of the apical meristem of the main shoot is indeterminate and alternates between periods of shoot growth (flush) and periods of rest, some lateral meristems, located usually near the arrested naked, terminal bud (Figures 1 and 2) develop into determinate, lateral inflorescences (pleonanthy[3,9]); most axillary buds remain subject to apical dominance and never develop. If the apical meristem of a terminal branch is damaged, new vegetative shoots may arise from axillary buds near the end of the damaged branch,[7] but inflorescences have been observed to grow from resting axillary buds below decapitated inflorescences.[22b] Site and timing of the development of lateral inflorescences thus appear to be determined by the pattern of vegetative growth and by internal correlations, but not by environmental cues.

In response to often subtle variations in environmental conditions and endogenous factors, the relative timing of seasonal vegetative growth, leaf fall, and flowering may vary widely both within (Table 2) and between species (Table 1). Flowering may occur on old, leafy shoots, on bare trees shortly or long after leaf fall, or simultaneously with the emergence of new vegetative shoots. Timing and control of flowering are thus intricately interwoven with the pattern and control of vegetative growth and cannot be considered separately. Relations between flowering and vegetative growth will be first analyzed for the most common pattern, flowering in leafless trees during the dry season, then for other patterns.

In the majority of *Erythrina* species flowering occurs during the dry season when trees are leafless, and lateral inflorescences develop near the tip of the shoot formed during the previous growing season (Figures 1 and 2). Formation of lateral flower buds near the tip of a resting terminal shoot is common in trees.[16,17] For *Erythrina* it means that only one lateral bud, or in most cases a few lateral buds, laid down shortly before the arrest of shoot growth, are induced to become inflorescences. Cessation of indeterminate shoot growth occurs in *E. poeppigiana*[2] and probably in most other species of *Erythrina* during the wet season, i.e., under environmental conditions favorable for plant growth. Arrest of shoot growth must therefore be caused by internal, correlative inhibition, not adverse environmental conditions.[2]

Cessation of shoot growth does not necessarily cause induction of inflorescences. (1) While some species of *Erythrina* flower when fairly small,[22] others have a prolonged juvenile vegetative period. (2) Trees cultivated outside their native habitat may remain vegetative for many years or never flower (e.g., *E. fusca, E. rubrinerva, or E. latissima* cultivated in California[20,21] and *E. sandwicensis* in Hawaii when growing at riparian instead of dry sites[18]). (3) In *E. poeppigiana* there are two annual flushes of shoot growth, but most trees flower only once a year (Table 2); a second annual flowering episode may occur in the upper, bare branches of tall trees while the lower branches retain their leaves and remain vegetative.[2]

In *E. falcata, E. dominguezii,* and probably most other species there are no visible differences among small, resting axillary buds of terminal shoots that will give rise to inflorescences or vegetative shoots or remain dormant.[18] As in most temperate deciduous trees, induction of flowering and early differentiation of flower buds thus take place in the resting axillary bud.[3,16,22a] Resting lateral buds may grow into fully developed inflorescences in the course of 2 to 3 weeks, just prior to anthesis.[18] In *Erythrina* species flowering during the dry season, the development of flower buds is thus discontinuous, and anthesis is separated from early inflorescence development by a prolonged rest period (proleptic flowering[3,4]).

In proleptic flowering, flower induction and anthesis must be controlled by two different sets of factors. Our understanding of flower induction in all trees, temperate or tropical, is very scanty.[16,17] In *Erythrina,* as in most trees, induction and early development of inflorescences appear to be stimulated by conditions that inhibit vegetative growth near the potential flower bud and favor the establishment of high carbohydrate levels and a favorable hormone balance in the inducible meristem.[3,12a] Flower induction thus appears to be controlled by endogenous, correlative factors, and not environmental cues.[3] The control of anthesis is basically different from that of flower induction. While the latter implies a switch from vegetative to reproductive development (i.e., a basic redirection of development in the lateral meristem), the triggering of inflorescence growth and anthesis merely involves release from an inhibitory endogenous or environmental condition which prevents the inflorescence from completing its development.

Growth of the inflorescence and flower opening involve cell expansion, known to be inhibited by even moderate water stress.[28] Production of large quantities of nectar, as observed in many species of *Erythrina,* should also require a positive water balance. In moderately water-stressed *E. poeppigiana* and other tropical trees, leaf fall during the early dry season has been observed to eliminate tree water deficits by reducing transpirational water loss and to cause anthesis and, sometimes, new vegetative growth in the absence of rainfall.[2,3,28]

Elimination of water stress is thus the most likely trigger of inflorescence development and anthesis in the species of *Erythrina* flowering during the dry season. Similarly, rain following a period of moisture stress greatly stimulated flowering in *Theobroma cacao,*[1] induced a flush of shoot growth and subsequent formation of terminal inflorescences in several species of *Cordia,*[4] and is probably the cause for the observed mass flowering in *Dipterocarpus, Shorea* (Dipterocarpaceae), and other forest trees in Malaysian rain forests.[17]

In *E. poeppigiana* growing at a 1200 m altitude in the Central Valley of Costa Rica under identical macroclimatic conditions, the following patterns of leaf fall, flowering, and vegetative growth were observed (Table 2). (1) Very large trees growing at dry sites shed leaves and flowered in December/January and formed new shoots in February/March (Table 2,A). (2) Medium-sized trees at dry sites shed leaves in January, flowered in February/March, and formed new leaves in April (Table 2,B). (3) Trees growing near creeks began to flower in March before leaf fall was complete, and new shoots emerged during flowering (Table 2,C). (4) In vigorous sprouts on pruned shade trees in a coffee plantation, shoot growth continued through April, when trees began to shed their leaves, flower, and leaf out in rapid sequence (Table 2,D); while mature trees bore only a few inflorescences near the tip of the last shoot growth increment, dozens of lateral buds along the sprouts of a length of several meters developed into inflorescences. At cool sites of low evaporative demand (1340 m altitude), flowers and new leaves appeared late in the dry season before old leaves had fallen (Table 2,F). Conversely, at dry sites in an area of high evaporative demand (1030 m altitude), trees shed leaves and flowered early in the dry season, remained bare, or formed only very small leaves during the rest of the dry season, and then flushed vigorously after the first heavy rains (Table 2,E). Within a single species of *Erythrina* various combinations of factors affecting tree water status, such as climate, soil water availability, tree size, and root/shoot ratio thus caused variations of several months in the timing of flowering and vegetative growth.

Variations in seasonal development of *E. poeppigiana* with increasing drought strikingly illustrate the relative changes in the control of tree development associated with the transition from an almost aseasonal to a strongly seasonal climate. In the absence of appreciable water stress, the phases of seasonal development, leaf fall, flowering, and shoot emergence take place simultaneously on the same tree (Table 2,F). In such evergreen trees the behavior of the individual branches may be quite asynchronous, indicating almost exclusive control of development by internal correlations and negligible timing by environmental inputs. Strictly endogenous control of flower induction is also manifest in *E. crista-galli,* which represents one of the few *Erythrina* species with terminal, not lateral, inflorescences.[19a] At Lawai, Hawaii, the gradual transition from vegetative shoot growth to the formation of small clusters of individual flowers in the axils of subtending leaves to a terminal, loose inflorescence (hapaxanthy[9]) occurs year around in different branches of the evergreen *E. crista-galli*[19] and is thus entirely independent of environmental cues; in California and elsewhere the widely cultivated tree is deciduous and tolerant of frost-induced dieback, and forms terminal inflorescences on new growth throughout the growing season;[13,20] in its native range the tree flowers only during the early rainy season, i.e., on new flushes of shoot growth (Table 1: 15a).

Increasing drought enhances synchronization of leaf fall and tree development within populations and causes the temporal separation of successive developmental phases (Table 2,E). Rehydration and flowering after leaf fall result from changes in the functional relationship between root and shoot and are not caused by rainfall.[2,3,28]

As drought becomes more severe, residual soil moisture available after leaf fall may suffice only for development of a fraction of a tree's flower buds, and rainfalls toward the end of the rainy season may induce a second flowering episode. This pattern has been described for *E. macrophylla*[15] and other *Erythrina* growing in Lawai (Table 1: 2b,4b), where the second flowering episode was probably induced by heavy rainfalls in early April (Table 1: 24b). Similar flowering patterns have been observed in other deciduous tropical trees.[5]

In tropical lowland deciduous forests subject to severe seasonal drought, mass flowering of short duration (less than 1 week) is caused in many tree species by isolated heavy rainfalls during the late dry season.[3,5] Such a pattern of flowering appears to exist in species native to thorn forests or very dry deciduous woodlands of western Mexico (*E. lanata, E. goldmanii,* and *E. pudica*), which have been observed to flower during the height of the dry season.[22b] In other species subject to severe water stress, flowering may occur at the beginning of the rainy season, shortly before or simultaneously with new vegetative growth. This pattern has been described for trees native to Mexico (e.g., *E. olivae;*[29] Table 1:6) and for xeromorphic shrubs and geophytes living in arid habitats (*E. flabelliformis,*[20,30] *E. leptorhiza,*[10] *E. acanthocarpa,* and *E. zeyheri;*[8] Table 1: 5,7).

In the majority of *Erythrina* species the seasonal periodicities of vegetative growth and flowering thus appear to be closely linked and are determined mainly by seasonal variation in tree water status, which in turn depends on seasonal variation in rainfall, local variation in soil moisture availability, root/shoot ratios, leaf morphology, and other factors.[2]

Seasonally low temperatures, which inhibit shoot growth or flowering of fully hydrated trees, may further complicate the control of seasonal development in *Erythrina* inhabiting high altitudes in tropical mountains or cultivated in areas near the northern temperate zone. For instance, in its native habitat, the Himalayas, *E. arborescens* flowers late in the growing season, shortly before leaves are shed;[14] in trees cultivated in California, flowers appear with the new shoots in spring.[20] Such a delay in the opening of axillary flower buds, probably caused by cool winters rather than severe drought, has been also observed in other tropical trees growing near the northern end of their range[3] and may explain the flowering periodicity of other *Erythrina* species cultivated in California as well[20] (Table 1: 1b,4b,17b). Alternatively, late flowering in California may be the result of delayed leaf shedding during the cool, moist California winters.

Eight unrelated species, native to high altitudes in Asia, Africa, and America or to the temperate zone, where they are regularly exposed to frost and snow, have become semi-herbaceous, geophytic perennials with large fleshy or woody rootstocks.[13,24] Their shoots die back at the end of the growing season, and inflorescences either emerge in spring, directly from the rootstock (e.g., *E. resupinata* and *E. herbacea*[13]), or form late in the growing season from axillary buds on current shoots (e.g., *E. zeyheri*[8]). An unusually wide range of growth habits is manifest among populations of *E. herbacea* of different geographic origin. In the southern U.S., at the northern end of its range, the species grows as a geophyte forming several annual stems. Inflorescences may emerge directly from the rootstock[13,30] or form laterally or terminally on annual shoots[22b] (Table 1: 8a). Populations in southern Florida and Mexico are shrubs or small trees; in Hawaii these ecotypes are evergreen and flower almost year around from lateral buds on current growth.[18] Differences in growth habit are retained among different ecotypes cultivated in Hawaii and are thus genetically determined.[22b]

Different species of *Erythrina* have thus adapted to a wide range of tropical and subtropical habitats, and individual species display considerable phenotypic plasticity in vegetative development and flowering periodicity. Because of this phenotypic plasticity, many species can be cultivated in warm and temporarily cool climates under a variety of conditions, but their flowering behavior varies with the pattern of vegetative growth and cannot be understood or predicted unless growing conditions are carefully controlled.

ACKNOWLEDGMENTS

I gratefully acknowledge valuable advice and information received during the preparation of the manuscript: Prof. H. Eckstein, Guatemala City, observed inflorescence development in *E. atitlanensis* and *E. fusca;* Dr. S. A. Lucas, Lawai, shared phenological records and observations on *Erythrina* cultivated in the Pacific Tropical Botanical Garden at Lawai, Hawaii; Dr. E. McClintock, San Francisco, provided flowering dates for *Erythrina* cultivated in California; and Mr. David Neill, Missouri Botanical Garden, St. Louis, helped in the literature search and generously shared information resulting from his extensive biogeographical studies of *Erythrina*.

REFERENCES

1. **Alvim, P. de T.,** *Theobroma cacao,* in *Handbook of Flowering,* Vol. 5, Halevy, A. H., Ed., CRC Press, Boca Raton, Fla., in press.
2. **Borchert, R.,** Phenology and ecophysiology of a tropical tree, *Erythrina poeppigiana* O. F. Cook, *Ecology,* 61, 1065—1074, 1980.
3. **Borchert, R.,** Phenology and control of flowering in tropical trees, *Biotropica,* 15, 81—89, 1983.
4. **Borchert, R.,** *Cordia,* in *Handbook of Flowering,* Vol. 5, Halevy, A. H., Ed., CRC Press, Boca Raton, Fla., in press.
5. **Borchert, R.,** *Tabebuia* — trumpet tree, *Handbook of Flowering,* Vol. 5, Halevy, A. H., Ed., CRC Press, Boca Raton, Fla., in press.
6. **Conn, J. S. and Snyder-Conn, E. K.,** The relationship of the rock outcrop microhabitat to germination, water relations, and phenology of *Erythrina flabelliformis* (Fabaceae) in southern Arizona, Southwest. Nat., 25, 443—451, 1981.
7. **Eckstein, H.,** personal communication, 1984.
8. **Guillarmod, A. J., Judd, R. A., and Skead, C. J.,** Field studies of six Southern African species of *Erythrina, Ann. Mo. Bot. Gard.,* 66, 521—527, 1979.
9. **Hallé, F., Oldeman, R. A. A., and Tomlinson, P. B.,** *Tropical Trees and Forests.* An *Architectural Analysis,* Springer, Berlin, 1978.
10. **Hernandez, H. M. and Toledo, V. M.,** The role of nectar robbers and pollinators (birds) in the reproduction of *Erythrina leptorhiza, Ann. Mo. Bot. Gard.,* 66, 512—520, 1979.
11. **Hernandez, H. M. and Toledo, V. M.,** Floral biology of *Erythrina batolobium* and the evolution of pollination systems in American species of the genus, *Allertonia,* 3, 77—84, 1982.
12. **Jackson, I. J.,** *Climate, Water and Agriculture in the Tropics,* Longman, London, 1977.

12a. **Juntilla, O.,** Flower bud differentiation in *Salix pentandra* as affected by photoperiod, temperature, and growth regulators, *Physiol. Plant.,* 49, 127—134, 1980.

13. **Krukoff, B. A.,** The American species of *Erythrina, Brittonia,* 3, 205—337, 1939.
14. **Krukoff, B. A. and Barneby, R. C.,** Conspectus of species of the genus *Erythrina, Lloydia,* 37, 332—459, 1974.
15. **Kunkel, G.,** *Flowering Trees in Subtropical Gardens,* W. Junk, The Hague, 1978.
16. **Lavender, D. L.,** Angiospermous forest trees of temperate zones, in *Handbook of Flowering,* Vol. 1, Halevy, A. H., Ed., CRC Press, Boca Raton, Fla., 1985, 15—22.
17. **Longman, K. A.,** Tropical forest trees, in *Handbook of Flowering,* Vol. 1, Halevy, A. H., Ed., CRC Press, Boca Raton, Fla., 1985, 23—39.
18. **Lucas, S. A.,** personal communication, 1984.
19. **Lucas, S. A. and Theobald, W. L.,** Observations of flowering behavior in selected species of *Erythrina* in cultivation in Hawaii, *Allertonia,* 3, 85—119, 1982.

19a. **Mathias, M. E., Ed.,** *Flowering Plants in the Landscape,* University of California Press, Berkeley, 1982.

20. **McClintock, E.,** *Erythrinas* cultivated in California, *Allertonia,* 3, 139—154, 1982.
21. **McClintock, E.,** personal communication, 1984.
22. **Menninger, E. A.,** *Flowering Trees of the World for Tropics and Warm Climates,* Heartside Press, New York, 1962.

22a. **Nasr, T. A. A. and Wareing, P. F.,** Studies on flower initiation in black currant; some internal factors affecting flowering, *J. Hortic. Sci.,* 36, 1—10, 11—17, 1961.

22b. **Neill, D.,** personal communication, 1984.

23. **Nieuvolt, S.,** *Tropical Climatology. An Introduction to the Climates of the Low Latitudes,* Wiley, London, 1977.
24. **Raven, P. H.,** *Erythrina* (Fabaceae): achievements and opportunities, *Lloydia,* 37, 321—331, 1974.

25. **Raven, P. H.,** *Erythrina* symposium. II. *Erythrina* (Fabaceae: Faboideae); introduction to symposium II, *Lloydia*, 40, 401—406, 1977.
26. **Raven, P. H.,** *Erythrina* symposium. III. *Erythrina* (Fabaceae: Faboideae): introduction to symposium III (pollination relationships), *Ann. Mo. Bot. Gard.*, 66, 417—421, 1979.
27. **Raven, P. H.,** *Erythrina* (Fabaceae: Faboideae). Introduction to Symposium IV (Systematics, evolution, ecology, and horticulture of the coral trees), *Allertonia*, 3, 1—6, 1982.
28. **Reich, P. B. and Borchert, R.,** Water stress and the phenology of trees in a tropical lowland dry forest of Costa Rica, *J. Ecol.*, 72, 61—74, 1984.
29. **Toledo, V. M.,** Observations on the relationship between hummingbirds and Erythrina species, *Lloydia*, 37, 482—487, 1974.
30. **Vines, R. A.,** *Trees, Shrubs and Woody Vines of the Southwest*, University of Texas Press, Austin, 1960.

ESCHSCHOLTZIA CALIFORNICA

En. California poppy; Fr. Eschscholtzie de Californie; Ge. Kalifornischer Mohn; Sp. Adormidera de California

Robert E. Lyons and Jody Booze-Daniels

INTRODUCTION

Eschscholtzia californica Cham. is one of seven to twelve species within the genus, and is a member of the Papaveraceae. The native distribution of this species includes Oregon, California, and northern Mexico, but it has been introduced and become naturalized in Australia, New Zealand, and India.[1,2,4,7,10,14] This widespread and diverse habitat can be attributed to the ability of the species to tolerate extremes in temperature and moisture availability. *Eschscholtzia* species are most commonly found along coastlines, in valleys, and among foothills.[6,7] Four different ecotypes of California poppy have been identified.[7,10] The "typical" perennial form has yellow flowers and smooth, glaucous leaves which are broadly compact. It is generally found along the Pacific coastline of North America and Mexico. An "inland" ecotype, also perennial, has been classified as the variety *crocea* by Jepson[7] and grows within valleys. The smooth, glaucous leaves are less compact than the "typical" type, and the flower color and size vary uniquely with the season. The first flush of spring flowers is large and deep orange, quite different from the smaller and pale yellow blooms which are subsequently produced during the summer. Discrepancies occur within the literature as to the distinction between the "typical" form and variety *crocea*. Hoover[6] makes no distinction between the two and classifies the group as *E. californica* var. *californica*. The variety *peninsularis* (Greene)Munz is another "inland" ecotype which is morphologically similar to the variety *crocea;* however, *peninsularis* is an annual. The fourth form, *E. californica* var. *maritima* Jepson, is perennial and grows primarily on coastal dunes. The leaves are prostrate, grey, pitted, and pubescent; the flowers are lemon yellow with an orange dot at the base of the petal.

It is not known whether a specific annual or perennial ecotype of *E. californica* is produced by the ornamental industry. However, most locations treat it as a strict or self-sowing annual because it will flower the first year after seeding in cultivation, but will not necessarily survive the winter (U.S. Department of Agriculture hardiness zone 9). Generally, *E. californica* will grow to be 25 to 60 cm tall with a narrow taproot. The erect, but weak stems arise from a rosette of basal leaves and tend to fall over with age. The leaves are glaucous, blue-green, alternate, cauline, and ternately dissected into fine segments.[3,7,9,12]

FLORAL MORPHOLOGY

Flowers are regular, complete, and perigynous (Figure 1). Each single flower terminates a 3- to 15-cm long peduncle.[10] The four petals are attached on the inside rim of a hollow, funnel-shaped receptacle.[5,7,13] Petals range from 2 to 6 cm long and possess a satiny texture.[9] The coherent (nonseparating) sepals and hollow receptacle are unique features of the genus *Eschscholtzia*. A specialized, "miter-shaped" calyx called the calyptra is pushed off in its entirety as the petals expand.[1,7,13] The spreading rim of the receptacle forms a collar around the sessile ovary. The gynoecium consists of a superior, 1-celled, multiseeded ovary possessing a short style with a 4-lobed, spreading stigma.[1,3,7,11,13] The numerous bilocular anthers are usually longer than the filaments which are attached to the base of the petals.[3,5,7,11,13] Capsules are 1-celled, slender, 10-ribbed structures measuring 2.5 to 10 cm

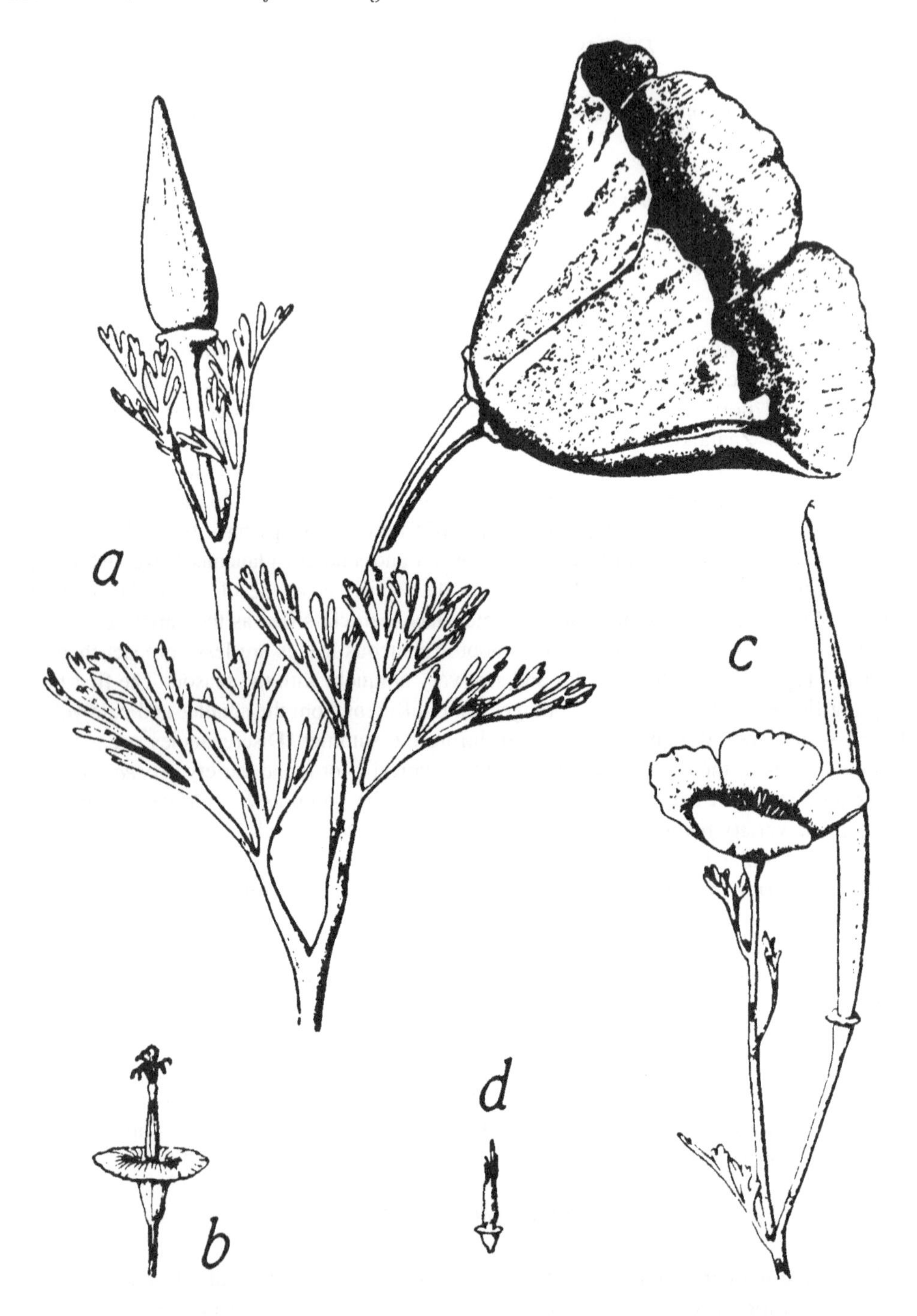

FIGURE 1. The two flowering forms of *Eschscholtzia californica* var. *crocea* Jepson: (a) the vernal stage with (b) a broad receptacle rim, and (c) the summer/autumnal stage with (d) a narrow rim. (From Jepson, W. L., *Manual of the Flowering Plants of California,* University of California Press, Berkeley, 1970, 403. With permission.)

long. Seed dehiscence occurs from 2 valves which open while the capsule is still upright. The globose to slightly elongated, reticulate seeds are 1.2 to 1.5 mm long and usually brown to gray in color.[1,3,10]

PHOTOPERIOD

The California poppy is a qualitative LDP which bolts from a rosette habit under inductive conditions.[8,15] Although stem and peduncle tissue comprise the entire "bolting height", the length of the peduncle is often larger than the contribution of the stem.[9]

When seedlings are maintained under different photoperiods upon germination, those under LD begin stem elongation within 35 days, and show visible flower buds after 50 days. Although the continuous SD prohibited flowering, there was slight stem elongation after 98 days.[15] In an attempt to find out when the plant becomes receptive to the inductive photoperiod, Lyons and Neale[9] germinated seeds under SD. Seedlings were subsequently moved to inductive LD upon germination, and at the 2-, 4-, 6-, or 8-true-leaf stage. A significant negative linear correlation was found between leaf stage and the rapidity of flowering, indicating that as the plant grew older it became more responsive to the inductive photoperiod. Leaf counts made at the time of anthesis showed that, in actuality, a minimum of 18 to 24 leaves (including primordia) were required to perceive the inductive stimulus.

The exact number of LD cycles required for floral initiation and development cannot be stated conclusively. When plants possessing 8 to 10 unfolded leaves were exposed to LD, 5 cycles were necessary for floral initiation and 10 were required to ensure complete development and anthesis.[11] However, obtaining similar results for experimental material which is younger or older than the 8- to 10-unfolded-leaf stage is questionable, given the previously stated dependence of flowering on plant age.

Photoperiod also directs the patterns of plant habit. SD promote rosette flowering with acropetal branching, first evident 105 days after seeding. Branching occurs basipetally under LD, but begins after the main stem has flowered. Only the three uppermost branches form secondary flowers; the remainder elongate but remain vegetative.[15]

TEMPERATURE

Went[16] indicated that the overall response of California poppy to night temperature (nyctotemperature) was closely parabolic from the point of seedling emergence through flowering. While germination is best at approximately 20°C, subsequent growth is optimum if a gradual drop of 10°C is experienced by the end of 60 days. From that point, up to and including flowering, optimal temperatures began a slow incline back towards 20°C, the level which had previously best suited the initial growth of the newly germinated seedling. Yet these optimal temperatures are usually not precisely reproduced in either cultivation or experimentation. This, of course, leads to a discussion on the interaction between photoperiod and temperature which is commonly found in many species. Unfortunately, few data have been published which attempt to clarify this interaction in California poppy, and it is rudimentary. Lewis and Went[8] illustrated several features of California poppy (*E. c.* var. *crocea*) growth under variable temperature and photoperiod, not all of which pertained to flowering. First, there was a "probable positive correlation" between night temperature and the rate of leaf production, whereby 19°C was optimal for lamina development. Photoperiod (8, 12, 14, 18, or 24 hr) had no effect on either of these observations. Under inductive LD, floral initiation was unaffected by those day/night temperature combinations tested; 19/26.5°, 19/19°, 19/13°, 19/7°, 26.5/26.5°, and 26.5/13°C. However, first flower anthesis was accelerated by 42 days when the temperature was held constant at 19°C and the photoperiod was extended from 18 to 24 hr. Flowering was not observed under SD (8, 10, 12, or 14 hr) at night temperatures of 7, 13, 26.5, or the optimal 19°C.

GROWTH REGULATORS

The effects of exogenously applied GA were shown to depend upon the frequency of application and prevailing photoperiod. Under SD a single treatment with 0.25 mg $GA_{4/7}$ per plant significantly increased stem elongation, decreased the rate of leaf production, increased the rate of lateral branch emergence, and did not induce flowering. The same treatment applied during LD significantly decreased the rate of lateral branch emergence, did not accelerate flowering, and significantly increased stem, but not peduncle, elongation.[9]

Although repeated applications of $GA_{4/7}$ have caused flower initiation under SD, most buds usually abort prior to anthesis.[11] Under LD, $GA_{4/7}$ stimulated flowering earlier than either the controls or those treated under SD. The effects of exogenously applied cyclic 3′,5′-adenosine monophosphate (cAMP) were also examined along with $GA_{4/7}$. Under SD, all cAMP-induced floral buds aborted, while under LD an acceleratory effect, comparable to the $GA_{4/7}$, was observed. It was concluded that endogenous GA was essential to California poppy flowering, as is commonly the case in many LDP. However, cAMP may, in some way, be a prerequisite to endogenous GA production, and either substance must persist if flower buds are to develop completely.[11]

The GA antagonist ancymidol does not interfere with flower initiation, but does reduce plant height at flowering. This effect is primarily due to an inhibition of stem rather than peduncle tissue.[9]

REFERENCES

1. **Bailey, L. H.,** *Manual of Cultivated Plants,* Macmillan, New York, 1949, 300—301.
2. **Benson, L.,** *Plant Classification,* Heath, Lexington, Mass., 1979, 155.
3. **Gilkey, H. M. and Dennis, L. J.,** *Handbook of Northwestern Plants,* Oregon State University Bookstore Publishers, Corvallis, 1967, 150.
4. **Good, R.,** *Geography of the Flowering Plants,* John Wiley & Sons, New York, 1953, 182.
5. **Hitchcock, C. L. and Cronquist, A.,** *Flora of the Pacific Northwest,* University of Washington Press, Seattle, 1973, 143.
6. **Hoover, R. F.,** *The Vascular Plants of San Luis Obispo County, California,* University of California Press, Berkeley, 1970, 139.
7. **Jepson, W. L.,** *A Manual of the Flowering Plants of California,* University of California Press, Berkeley, 1970, 402—408.
8. **Lewis, H. and Went, F. W.,** Plant growth under controlled conditions. IV. Response of California annuals to photoperiod and temperature, *Am. J. Bot.,* 32, 1—12, 1945.
9. **Lyons, R. E. and Neale, L.,** Effect of photoperiod, gibberellin and ancymidol on flowering and vegetative development of California poppy (*Eschscholtzia californica* Cham.), *HortScience,* 18(4) (Abstr.), 573 1983.
10. **Munz, P. A.,** *A Flora of Southern California,* University of California Press, Berkeley, 1974, 626—627.
11. **Nanda, K. K. and Sharma, R.,** Effects of gibberellic acid and cyclic 3′,5′-adenosine monophosphate on the flowering of *Eschscholtzia californica* Cham., a qualitative long day plant, *Plant Cell Physiol.,* 17, 1093—1095, 1976.
12. **Rendle, A. B.,** *The Classification of Flowering Plants,* Vol. 2, Cambridge University Press, Cambridge, 1963, 168.
13. **Rickett, H. W.,** *Wild Flowers of the United States,* Vol. 5, McGraw Hill, New York, 1971, 132.
14. **Schmidt, M. G.,** *Growing California Native Plants,* University of California Press, Berkeley, 1980, 57—59.
15. **Sharma, R. and Nanda, K. K.,** Effect of photoperiod on growth and development of *Eschscholtzia californica* Cham., *Indian J. Plant Physiol.,* 19, 202—206, 1976.
16. **Went, F. W.,** *The Experimental Control of Plant Growth,* Vol. 17, Chronica Botanica, Waltham, Mass., 1957, 204—260.

FORSYTHIA × *INTERMEDIA*

En. Forsythia, Golden-bells; Fr. Forsythie; Ge. Forsythie, Goldflöckchen

H. F. Wilkins

INTRODUCTION

Forsythia (*Forsythia* × *intermedia* Zab.: *F. suspensa* × *F. viridissia),* which belongs to the Oleaceae family, is a deciduous ornamental woody landscape shrub with arching or spreading branches that can be grown up to 3 m in height. There are also dwarfed selections which can be used as a ground cover or pot plants. This shrub is grown in gardens for its profuse and early spring flowering habit as the flowers are showy, yellow, and appear prior to leaf development. China and Korea are the sites of origin.[2]

PLANT AND FLOWER MORPHOLOGY

The opposite leaves are simple or three-parted and can be up to 1.5 cm long. The corolla is deeply parted and composed of four oblong lobes.[2] Flower buds are axillary and are on shoots which develop during the current season. The terminal bud is always vegetative.[4,6]

FLORAL INITIATION AND DEVELOPMENT

Axillary flower bud initiation and development along the vegetative shoot are controlled by shoot maturity and temperature; flower bud opening is prevented by dormancy.[6,7,11,12]

Floral initiation commenced in late June or early July and development was completed by late September or early October.[4,11-13] These observations for *F. suspensa* and *F. viridissima* were similar to data for *F. ovata*.[5,6]

Any treatment which resulted in earlier vegetative shoot elongation also resulted in earlier flower formation. Floral initiation was similar between 15 to 25°C treatments. Development was, however, inhibited at 25°C, with development occurring between 5 and 20°C. The optimum temperature, as measured by the number of florets per leaf axil, was 10°C. Photoperiodic treatments during the middle of May to the middle of June did not influence floral cultivation.[12]

A daminozide and ethephon combination spray in early spring hastened flower bud formation, resulted in shorter lateral flower bud growth, and improved ornamental value when branches were forced.[20]

FLOWER BUD DORMANCY

Axillary flower bud dormancy in *Forsythia* can be divided into three categories: (1) summer dormancy resulting from an inhibition system(s) inside the plant, but not inside the flower bud; (2) winter dormancy resulting from an inhibitor system(s) within the bud; or (3) imposed dormancy resulting from environmental conditions unfavorable for growth.[8,9]

"Summer dormancy" is not an apical-dominance phenomenon, nor is it related to inhibitors found in leaves; when leafy plants with roots were ringed (stem cambium contact broken), axillary flower buds opened. "Summer dormancy" was broken only for those buds above the "ring" or when branches were harvested and flowers opened in solution (see section on "Postharvest Handling."[9]

Furthermore, in August, when leafy shoots with dormant axillary flower buds were har-

FIGURE 1. The effect of a hot-water bath on breaking dormancy of cut forsythia branches when winter dormancy was at its deepest point (October 27). From left to right: controls at 3, 6, and 12 hr in water of 30°C. (From Doorenbos, J., *Meded. Dir. Tuinbouw (Neth.)*, 533—543, 1953. With permission.)

vested, shoots with leaves rooted and flowered; shoots which were defoliated did not root or flower, but the dormancy of the terminal vegetative bud broke and started growing. It was suggested that during this time period flowers do not open under natural conditions because there is a lack of photosynthetic products.[9]

During summer dormancy in August, cut *Forsythia* branches can be forced into flower by warm temperatures (immersed in a 30°C water bath for 12 hr).[10] The response to this treatment varies according to the degree that winter dormancy has developed or is present[8,9] (Figure 1). The flowering response to a warm-water treatment alone was 50% in late October when winter dormancy was at its deepest. At this time, a cold treatment (5°C) for 4 to 8 weeks would be adequate to break dormancy and cause the branches begin flowering[9,10] (Figure 2). Cold treatments (6 weeks at −2°C) are not effective until after ovule formation has occurred in late October or early November.[11] Forcing of branches is best at 15°C.[11]

During deep-summer dormancy, when August-harvested branches were warm-water treated and placed under continuous lighting, flowering was reduced when compared to warm-water treatments alone[9] (Figure 1). As summer dormancy decreased and winter dormancy began (September to early October) full flowering of branches was possible when heat treated and placed under lights. However, when winter dormancy was fully developed (early October to mid November) only 50% flowering was possible with warm-water and/or light treatment.[7,9]

Full flowering capacities return with time. When branches were warm-water treated and lighted, acceptable flowering was again possible by November 21. By December 5, lights were not needed and heat treatment was adequate.[7,9] Heat treatments can continue to be used with January- or February-harvested branches as Meeus[15] reported; a 10-sec dip in 40°C water was effective for rapid flower opening at this date.

During October and early November, 2 weeks of cold- plus warm-temperature treatment is no better than warm-temperature treatments plus lighting. Cathey[7] reported that when early-November-harvested branches were cold stored (0.5°C) for 20 or 30 days and then heat treated, 41 to 60% flowering of the axillary buds resulted. Full flowering, however, can be accomplished for branches harvested on November 14 for the cold- plus warm-temperature treatments vs. November 21, for the above-mentioned warm-temperature-plus-

FIGURE 2. The effect of a cold treatment on breaking dormancy of cut forsythia branches. From left to right: 5°C for 0, 2, 4, and 6 weeks, respectively. Branches cut on October 10 and photographed December 9. (From Doorenbos, J., *Meded. Dir. Tuinbouw (Neth.)*, 533—543, 1953. With permission.)

lighting treatments, or on December 5 for the above-mentioned warm-water treatments only.[9]

When "imposed dormancy" is present, forcing occurs with ease during mid winter;[11,16] however, flowering can be hastened if additional lighting with mercury lamps or incandescent lamps is used at night to supplement normal winter-day intensities.[9]

POSTHARVEST HANDLING

Harvested stems and branches of *Forsythia* are easily forced into flower in water or preservative solutions in summer, autumn, or early winter after certain treatments (immersed in warm water and/or lighted) are given, or without treatments after December or January 15[3,4,8,9,11] (see the preceding section on "Flower Bud Dormancy").

Branches of *Forsythia* with leaves harvested in August and September or during summer dormancy will flower more rapidly in appropriate nutrient solutions than in tap water.[4,9,17]

Rupprecht[17] reported that forcing was more rapid, flower color better, and postharvest life increased when sucrose solutions were 1.5 to 3%. Cathey[7] reported that a solution of sodium azide (10^{-2} *M*) promoted flowering on late-November-harvested branches, but no promotion was seen on later harvest dates.

Branches should be stored at 95% relative humidity[18] or in plastic bags.[4] Soaking branches in water after long-term storage is a common practice.[4]

It is possible to store branches at below 0°C from February to September and flower them after removal from storage.[19] Branches should not be stored with ethylene-emitting fruit.[1,4,14]

REFERENCES

1. **Anon.,** Houdbaarheid van forsythia, *Vakbl. Bloemist.*, 16, 24, 1961.
2. **Bailey, L. H. (Staff of Hortorium),** *Hortus Third, A Concise Dictionary of Plants Cultivated in the United States and Canada,* Macmillan, New York, 1979, 482—483.
3. **Broertjes, C.,** Trekproeven met *Forsythia, Jaarb. Proefstn. Boomkwekerij. Boskoop,* 72—75, 1953.
4. **Broertjes, C.,** The forcing of *Forsythia intermedia spectabilis* Khne, *Rep. 14th Int. Hortic. Congr. 1955,* Vol. 2, H. Veenman & Zonen, Wageningen, Netherlands, 1955, 1065—1071.
5. **Carpenter, E. D.,** A Monograph: Morphology of Flower Primordia of Selected Woody Plants, Ph.D. thesis, Michigan State University, East Lansing, 1964.
6. **Carpenter, E. D. and Watson, D. P.,** The initiation of flower buds in *Forsythia ovata* and *Indigofera potanini, HortScience,* 2, 25, 1967.
7. **Cathey, H. M.,** Het verbreken van de winterrust bij *Forsythia* door warm water en chemicaliën, *Meded. Dir. Tuinbouw (Neth.).*, 19, 768—773, 1956.
8. **Doorenbos, J.,** A literature review on bud dormancy in woody plants, *Meded. Landbouwhogesh. Wageningen,* 53, 1—24, 1953.
9. **Doorenbos, J.,** Oriënterend onderzoek over het forceren van *Forsythia* en *Rhodendendron, Overdruk uit, Meded. Dir. Tuinbouw (Neth.).*, 16, 533—543, 1953.
10. **Marchal, J.,** *Forsythia, Jaarversl. Proefstn. Bloem. Aalsmeer,* 1959, 60—61.
11. **Goi, M.,** Studies on the acceleration of flowering in woody ornamentals by low temperature treatments. III. Forcing of *Forsythia viridissima* Lindl., *Tech. Bull. Fac. Agric. Kagawa Univ.*, 25(2), 207—217, 1974.
12. **Goi, M.,** Studies on the acceleration of flowering in woody ornamentals by low temperature treatments. VI. The factors involved in flower formation in *Forsythia viridissima* Lindl., *Tech. Bull. Faculty Agric. Kagawa Univ.*, 27(59), 63—75, 1976.
13. **Hanaoka, Y., Yokoi, M., Kosugu, K., and Chihiru, T.,** Studies on flower bud differentiation and development in ornamental trees and shrubs, *Tech. Bull. Fac. Hortic. Chiba Univ.*, 17, 7—12, 1969.
14. **Marchal, J.,** *Forsythia,* Vervroegde bloei van struiken door koudebehandeling, *Jaarversl. Proefstn. Bloem. Aalsmeer,* 1953, 41.
15. **Meeus, M.,** Keeping quality of cut flowers in water, *Tuinbouwber.*, 21, 93, 1957.
16. **Palanchan, A. I.,** Forcing cut branches of flowering shrubs, *Izv. Akad. Nauk Mold. S.S.R. Biol. Khim. Nauk,* 5, 5—8, 1978.
17. **Rupprecht, H.,** Neue Wege in der Treiberei von Blülengehölzen, *Dtsch Gartenbau,* 8, 341—342, 1961.
18. **Rupprecht, H.,** Further results on forcing branches of our woody ornamentals, *Dtsch Gartenbau,* 9, 330—332, 1962.
19. **Sytsema, W.,** Forcing, *16th Int. Hortic. Congr.-1962,* 38—45, 1962.
20. **Sytsema, W. and Glas, A.,** The influence of daminozide and ethephon on growth and flower bud formation on *Forsythia intermedia* 'Lynwood Gold', *Acta Hortic.*, 137, 251—253, 1983.

GAILLARDIA

En. Indian blanket flower, Gaillardia; Fr. Gaillarde; Ge. Kokardenblume, Gaillardia; Sp. Gaillardia

Ross E. Koning*

INTRODUCTION

Gaillardia is a highly ornamental genus of Asteraceae. Its common names include gaillardia and Indian blanket flower. The latter apparently refers to its bright red and yellow colors and its natural distribution in the American southwest.

Gaillardia was first described in 1783 from the American flora.[8] Since then, many natural species of *Gaillardia* have been found (Table 1), and new ones are continually discovered in the floras of the southwestern U.S. and northern Mexico.[3,44-47] The natural species are self-incompatible,[41] hybridize frequently, and produce allotetraploids spontaneously.[36-38,41] The latter have larger flowers, larger inflorescences, and more robust vegetative parts than natural species.[36,38] Some 41 showy common cultivars have been assigned to the hybrid *G. grandiflora* and the natural *G. pulchella* species.[38] Other important cvs include *G. pulchella* 'Lorenziana', which is an annual plant with ray flowers distributed among the disc flowers giving an appearance of "double" form.[38] A natural white-flowered variety of *G. aestivalis* may have potential horticultural value.[47]

Although *G. pulchella* varieties are annuals, the other cvs of *Gaillardia* are misleadingly listed as biennials in seed catalogs. While the plants do usually form a rosette of leaves the first year from seed and bolt to flower the second year, new rosettes are produced at the base of the plant each fall and will bolt and flower the next spring. Thus, the *G. grandiflora* cvs are usually grown and sold as perennials. Interestingly, these plants are thought to be derived from a spontaneous hybrid between *G. aristata*, a perennial, and *G. pulchella*, an annual.[38] Nevertheless, most of the *G. grandiflora* cvs are hardy perennials and survive severe winters in Michigan.

The plants, especially established perennial plants, form a mound of foliage covered with flowers from early summer until frost. They tolerate very poor, dry, hot, alkaline soils, and prefer full sunlight. Thus, they are particularly suited to use in perennial borders around foundations, sidewalks, and limestone driveways. *Gaillardia* has been successfully used with other wildflowers to form sods for roadside beautification and erosion-control projects.[1] The plants have also been used to control erosion on sandy coastal dunes.[5] The inflorescences, usually produced singly on a long stem, are suitable for cutting, and last for more than 2 weeks in a vase if the inflorescences are cut before any disc flowers are open. The number and size of inflorescences for cutting are increased if the senescent ones are promptly removed from the plant.

Gaillardia is also used in pharmaceutical research. The plant produces the sesquiterpene lactones spathulins,[9,11,12,16,18] pulchellins,[9,11,12,19,29] and gaillardins,[20] among several other pseudoguaianolides.[9-12,14,15,19] Spathulin and pulchellin are antibiotic to *Staphylococcus* and *Streptococcus*.[9] Spathulin and other pseudoguaianolides are fungistatic.[9] Spathulin and gaillardin inhibit human nasopharynx carcinomas in cell cultures.[9,21,27] Helenalin, from *G. megapotamica*, *G. multiceps*, and *G. pinnatifida*, is allergenic and causes contact dermatitis in some persons.[6] The distribution of these sesquiterpenes and flavonoids helps determine the biosystematic relationships within the genus *Gaillardia*.[4,39,40,49]

Table 1
THE NATURAL SPECIES AND CULTIVARS OF THE GENUS *GAILLARDIA*

Cultivated types

***Gaillardia pulchella* — annual**

Fiesta	Red Giant
Fireball	Sunshine
Gaiety	Tetra Fiesta
Indian Chief	Tetra Giant
Lorenziana	

***Gaillardia aristata* — perennial**

No cultivars, but genetic source for many below

***Gaillardia grandiflora* — hybrid perennial**

Aurea	Golden Goddess	Sun God
Aurea Pura	Goldkobold	Sunset
Baby Cole	Grandiflora	Superba
Bremen	Kobold	Tangerine
Burgundy	Maxima	The King
Compacta	Monarch	The Prince
Dazzler	Mr. Sherbrooke	The Warrior
General Patton	Nana Nieske	Tokaier
Goblin	Portola	Tokaj
Golden Goblin	Ruby	Tokay
	Sun	Yellow Queen

Natural species from common soil types

G. aestivalis
G. amblyodon
G. coahuilensis
G. parryi
G. pinnatifida
G. suavis — fragrant flowers

Saline soil types

G. comosa

Gypsum soil types

G. gypsophila
G. henricksonii
G. multiceps
G. powellii
G. turneri

Note: Updated but based largely upon Stoutamire[38] and Turner.[44]

FIGURE 1. An inflorescence of *Gaillardia grandiflora*. The small fertile disc flowers are surrounded by showy sterile ray flowers (Magnification × 2.)

PLANT DESCRIPTION

Gaillardia are herbaceous plants which vary in height from 30 to 100 cm, depending upon the cv. Seedlings have a rosette of alternate, clasping, hairy, and variously lobed leaves during the first season of growth. In late summer for annuals and annual-perennial hybrids,[33] and in the following spring for perennials, the stem elongates and produces a terminal inflorescence (Figure 1). During bolting, the new leaves are progressively smaller and less deeply and less numerously lobed. The inflorescence is composed of both ray and disc flowers. The ray flowers are ligulate and showy (Figure 2); they have 2 to 5 lobes and are usually red at the base and yellow at the tip of the corolla. A few cvs have completely yellow ray flowers.[33] The ray flowers are sterile and have only a vestigial intersexual or female organ; these flowers serve primarily as attraction devices and landing platforms for pollinators. The fertile disc flowers are tubular and smaller than the ray flowers (Figure 3); nevertheless they are similar in size to disc flowers of *Helianthus*. The disc flowers have 5 red to purple corolla lobes, 5 stamens fused at the yellow anthers to form a tube, and a central single but branched deep-red style-stigma unit attached to the inferior ovary. The fruit is a cypsela with a number of awl-shaped calyx parts remaining from the flower.

PROPAGATION AND CULTURE

The annual varieties are usually propagated from seed. The seed of some species has little dormancy, but *G. pulchella* requires a heat treatment (50°C for 2 weeks under dry conditions) and *G. pinnatifida* and *G. multiceps* require stratification (3°C for several weeks under moist conditions) for reasonable germination.[34]

FIGURE 2. Five stages in the growth and opening of the *Gaillardia grandiflora* ray flower. A flower passes through these stages in roughly 48 hr. (Magnification × 3.)

Gaillardia perennials can be propagated by stem cuttings, but an auxin rooting compound should be applied to enhance rooting. Alternatively, stem cuttings with a developing apical inflorescence will root without the hormone; this is presumably due to the high levels of natural auxin produced during flower opening.[23-25] For the perennial varieties, division of robust plants is the most common and reliable method of vegetative propagation.

The native soils for some *Gaillardia* species are highly saline (*G. comosa*) or contain gypsum ($CaSO_4$) (*G. multiceps, G. gypsophila, G. powellii, G. turneri,* and *G. henricksonii*).[3,44,45] Other species do not grow near the gypsum (*G. pulchella* and *G. coahuilensis*).[34,46,48] Seed germination for gypsum-tolerant species (*G. pinnatifida*) is slightly promoted by low concentrations (10 ppm) of $CaSO_4$ and inhibited by higher ones (1000 ppm). For gypsophilous *G. multiceps,* these same concentrations greatly promote seed germination.[34] The sulfate may be sequestered in the plants; K_2SO_4 may be a major cell wall component in *Gaillardia.*[35]

Gaillardia is not particularly susceptible to many plant diseases[30,31,41] or insect pests[7] when cultivated in the field. In fact, *Gaillardia* plants can inhibit population expansion of *Pratylenchus penetrans* and *Ditylenchus dipsaci* nematodes in garden soil as do *Tagetes* plants.[13] Ether extracts contain two active materials: pentaynene and 2,3-dihydro-2-hydroxy-3-methylene-6-methylbenzofuran. The latter compound is similar to the systemic nematicide carbofuran.[13] *G. grandiflora* 'Burgundy' was the second most potent nematicidal plant of 35 Asteraceae species tested.[17] This variety could be beneficial as part of the field crop rotation in perennial crop nurseries.[17] *Gaillardia pulchella* cvs 'Picta' and 'Lorenziana' have no significant effect upon nematode population dynamics.[17]

FLOWERING

Besides the fact that it is a LDP,[2,28,32] very little is known about the induction of flowering

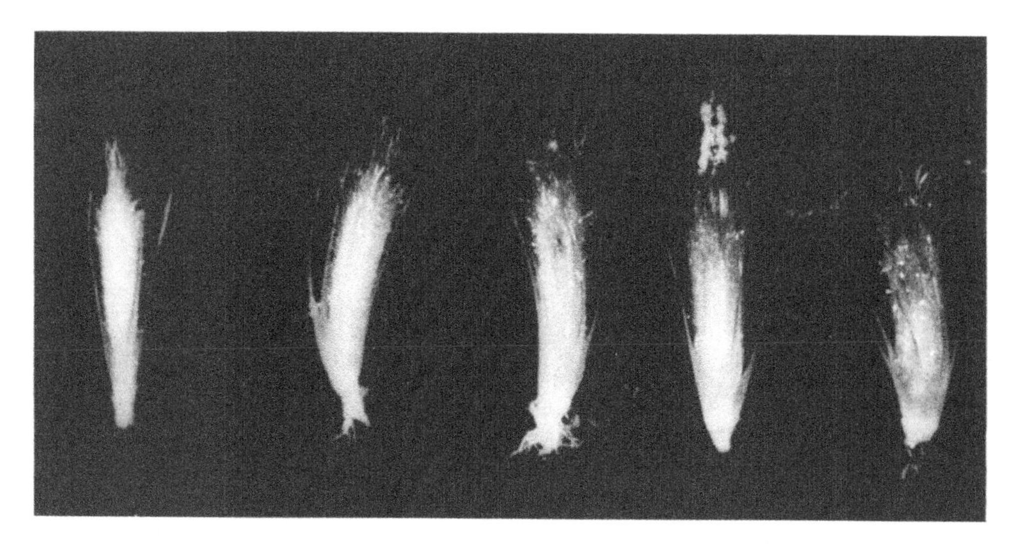

FIGURE 3. Five stages in the growth and opening of the *Gaillardia grandiflora* disc flower. A flower passes through these stages in roughly 48 hr. (Magnification × 4.)

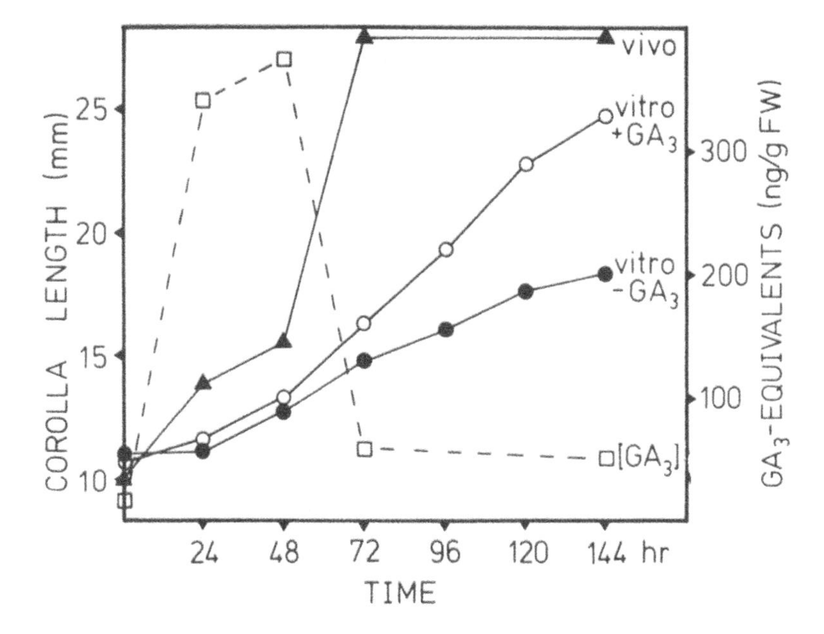

FIGURE 4. The natural growth of the *Gaillardia grandiflora* ray flower corolla (▲) is preceded by an increase in the natural gibberellin activity (□). Applied GA_3 promotes growth of corollas isolated in vitro (○) as compared with controls (●). (Adapted from Koning.[26])

in *Gaillardia*. The critical photoperiod and the required number of photoinductive cycles is not known. Under SD (10 hr) the plants form a rosette of leaves and only bolt to flower under LD conditions or when the temperature exceeds 27°C.[43] During bolting, a change in leaf form occurs; the usually lobate leaves of the rosette intergrade to smaller ovate leaves higher on the inflorescence stem. Continuous light temporarily promotes flowering but the subsequent rapid production of fruits induces senescence.[43] Only after these fruits mature do lateral stems elongate and new inflorescences appear above the older ones.[43] If senescing inflorescences are continually removed, the inflorescence size can be maintained throughout

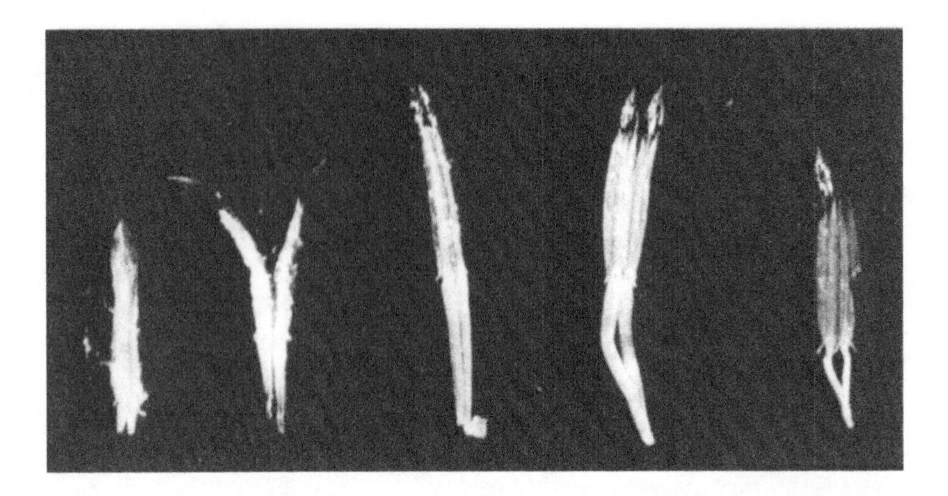

FIGURE 5. Five stages in the growth of stamens in *Gaillardia grandiflora* disc flowers. Most of the rapid filament elongation occurs in the few hours preceding anthesis. (Magnification × 5.)

the season;[22] otherwise the presence of infructescences brings on a decline in inflorescence size.[43] With proper inflorescence removal, the plant becomes a mound of stems and leaves and is covered with inflorescences from early summer until frost.[22]

Weekly spray applications of GA_3 at 10 mg/ℓ to rosette-form plants promote bolting of the stem, accelerate the conversion of leaf form from lobate to ovate, and cause earlier flower production. Flower opening is delayed by the GA treatments, but the open flowers are much larger than those of untreated plants.[42]

The phenomenon of flowering also includes the rapid growth of flower parts prior to pollination. *Gaillardia* is one of the very few plants in which these processes have been studied.

The sterile ray flowers elongate over a 2-day period (Figure 2) by means of fresh- and dry-weight increase. This rapid corolla elongation is accomplished by cell elongation and not by cell division. Corollas isolated from young ray flowers elongate in vitro in response to GAs (Figure 4) but do not respond to auxins, cytokinins, ABA, ethylene, or inhibitors of ethylene biosynthesis. Sequential and simultaneous hormone applications show no additive or synergistic relationships between the hormones. A 20-fold increase in the endogenous GA activity occurs just prior to rapid corolla elongation. There are no significant changes in endogenous auxin levels or ethylene production associated with corolla growth. Thus it appears that corolla expansion (and therefore the inflorescence diameter) is controlled, at least in part, by an increase in GA activity.[26] This also explains why application of GA to young inflorescences can increase the size of cut inflorescences.[42]

In the fertile disc flowers, the corolla does not grow (Figure 3); instead auxin-induced growth of the filament and style accomplish the final positioning of flower parts for pollination.

Filament elongation occurs in the few hours just prior to anthesis (Figure 5) and is accomplished by cell elongation and fresh-weight increase. Isolated filaments exposed to IAA elongate in vitro with kinetics similar to those of filaments in vivo (Figure 6). Ethylene applications only slightly promote and GA, cytokinin, and ABA applications have no effect on filament growth. The endogenous IAA levels increase just prior to the rapid filament elongation and apparently induce a slight increase in ethylene production. Endogenous GAs do not change significantly. Thus, filament elongation is apparently controlled, at least in part, by auxin production.[23,24]

The elongation of the style is an extremely important part of flowering in the Asteraceae. The pollen is shed inside the flower directly upon the outer surfaces of the stigma. The

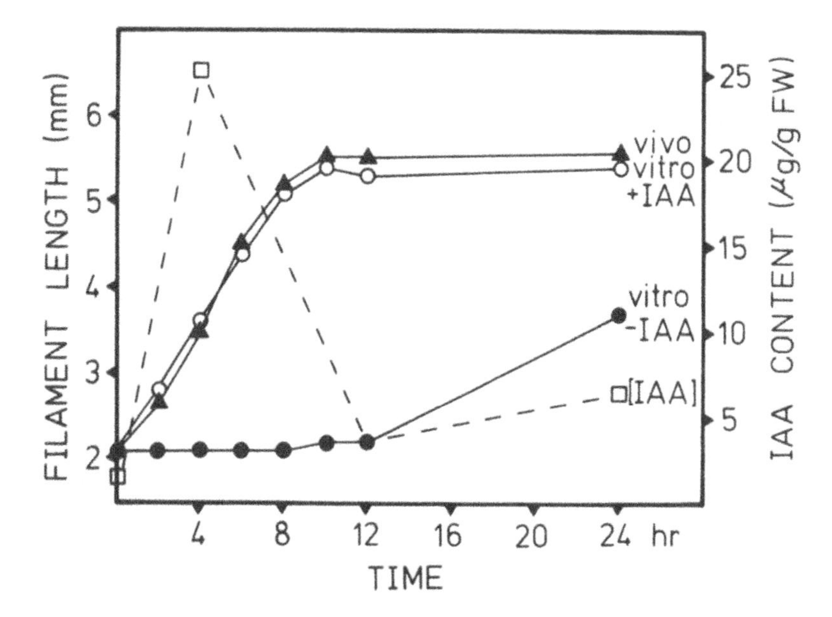

FIGURE 6. The natural growth of the *Gaillardia grandiflora* filament (▲) is accompanied by an increase in the natural auxin concentration (□) of disc flowers. Applied auxin promotes growth of filaments isolated in vitro (○) as compared to controls (●). (From Koning, R. E., *Bioscience*, 33, 458—459, 1983. Copyright 1983 by the American Institute of Biological Science. With permission.)

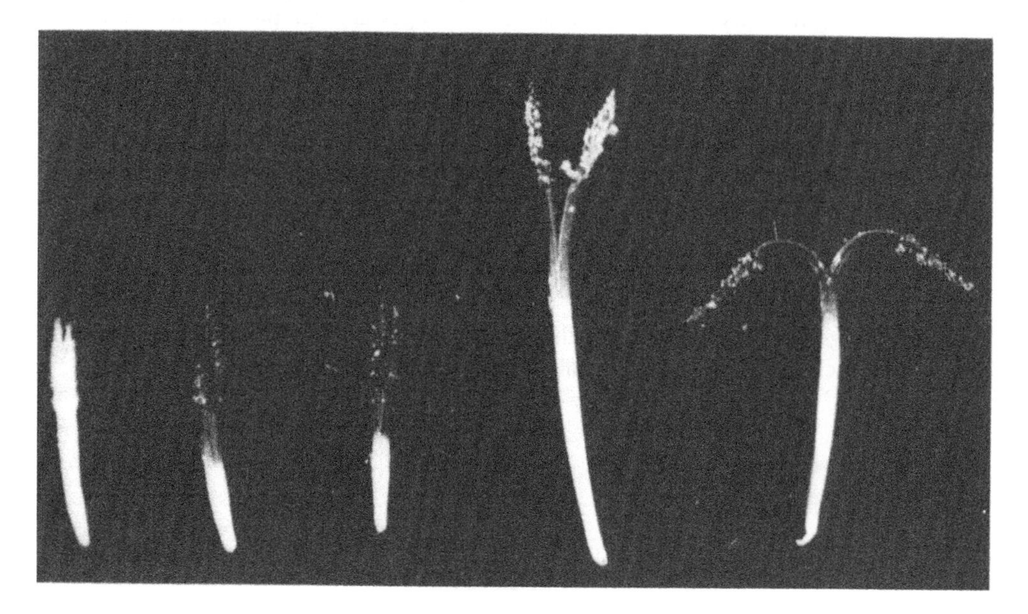

FIGURE 7. Five stages in the growth of style-stigma units in *Gaillardia grandiflora* disc flowers. Style elongation takes a few hours and occurs after anthesis. (Magnification × 5.)

subsequent growth of the style pushes the pollen out of the tubular flower and the pollen is removed by pollinators. Style elongation takes a few hours (Figure 7) and is accompanied by cell elongation and fresh-weight increase. Style elongation in vitro is promoted by applied auxin (Figure 8), inhibited by applied GA and ethylene, and is unaffected by other hormones. The endogenous GA levels and ethylene production do not vary in a parallel or inverse manner with style elongation, but the endogenous IAA level increases prior to style elon-

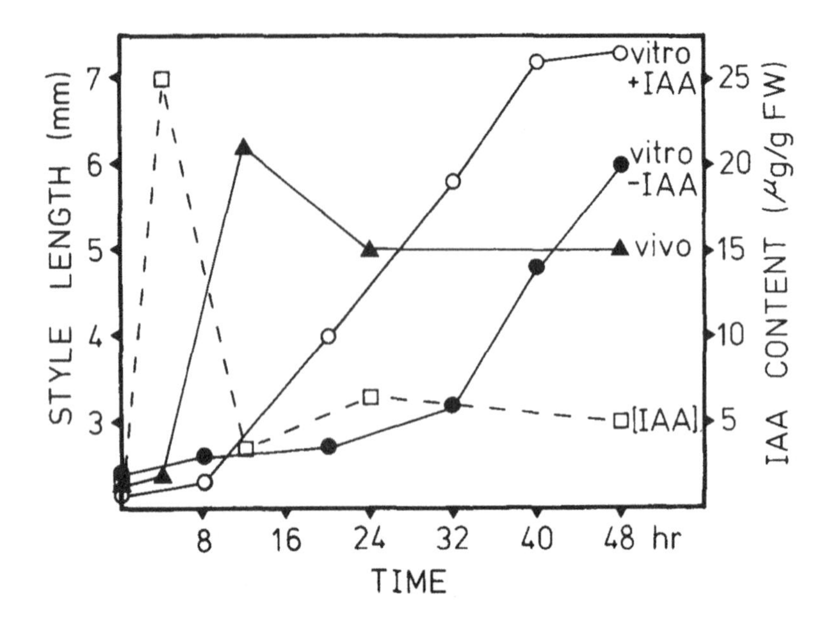

FIGURE 8. The natural growth of the *Gaillardia grandiflora* style (▲) is preceded by an increase in the natural auxin concentration (□) of disc flowers. Applied auxin promotes growth of styles isolated in vitro (○) as compared to controls (●). (Adapted from Koning.[25])

gation, indicating that this growth is controlled by changes in auxin production.[25] Since the emergence of the stigmas causes release of copious sticky pollen and ruins the appearance of cut inflorescences, auxin biosynthesis and action inhibitors might be used to delay or prevent this development.

In summary, while *Gaillardia* has provided much of our knowledge as to how flowers open and position their parts for pollination, many questions about induction of its flowering remain unanswered.

ACKNOWLEDGMENT

The author gratefully acknowledges the support of grants from the Rutgers University Research Council and DNA Plant Technology Corp., Cinnaminson, N.J.

REFERENCES

1. **Airhart, D. L., Falls, K. M., and Hosmer, T.,** Developing wildflower sods, *HortScience*, 18, 89—91, 1983.
2. **Allard, H. A. and Garner, W. W.,** Further observations of the response of various species of plants to length of day, *U.S. Dep. Agric. Tech. Bull.*, 727, 1—64, 1940.
3. **Averett, J. E.,** A new gypsophilous species of *Gaillardia* (Asteraceae) from Chihuahua, Mexico, *Ann. Mo. Bot. Gard.*, 63, 375—377, 1976.
4. **Carman, N. J., Watson, T., Bierner, M. W., Averett, J., Sanderson, S., Seaman, F. C., and Mabry, T. J.,** 6-Methoxy-apigenin from thirty-four species of compositae, *Phytochemistry*, 11, 3271—3272, 1972.
5. **Craig, R. M.,** Herbaceous plants for coastal dune areas, *Proc. Fla. State Hortic. Soc.*, 90, 108—110, 1977.

6. **Evans, F. J. and Schmidt, R. J.,** Plants and plant products that induce contact dermatitis, *Planta Med.*, 38, 289—316, 1980.
7. **Ferner, J. W. and Rosenthal, M.,** A cryptic moth, *Schinia Masoni* (Noctuidae), on *Gaillardia aristata* (Compositae) in Colorado, *Southwest. Nat.*, 26, 88—90, 1981.
8. **Fougeroux de Bondaroy,** Description d'un nouveau genre de plante, *Mem. Acad. Sci. Paris*, 1786, 1—8, 1786.
9. **Gill, S., Barteczko, I., Dembińska-Migas, W., and Zielińska-Stasiek, M.,** Investigation of chemistry and biological activity of lactones of *Gaillardia aristata* Pursh. Herb. II. Antibiotic and anticancer properties of isolated sesquiterpene lactones, *Herba Pol.*, 27, 341—346, 1981.
10. **Gill, S., Dembińska-Migas, W., Sliwińska, E., Daniewski, W., and Bohlmann, F.,** Pseudoguaianolides from *Gaillardia grandiflora, Phytochemistry,* 19, 2049—2051, 1980.
11. **Gill, S., Dembińska-Migas, W., and Zielińska-Stasiek, M.,** Investigation of chemistry and biological activity of lactones of *Gaillardia aristata* Pursh. Herb. I. Isolation of sesquiterpene lactones, *Herba Pol.*, 27, 213—219, 1981.
12. **Gill, S., Dembińska-Migas, W., Zielińska-Stasiek, M., Daniewski, W. M., and Wawrzuń, A.,** Pseudoguaianolides from *Gaillardia aristata, Phytochemistry,* 22, 599—600, 1983.
13. **Gommers, F. J.,** Nematicidal principles from roots of some Compositae, *Acta Bot. Neerl.*, 21, 111—112, 1972.
14. **Herz, W. and Srinivasan, A.,** Pseudoguaianolides of *Gaillardia amblyodon, Phytochemistry,* 11, 2093—2095, 1972.
15. **Herz, W. and Srinivasan, A.,** Reexamination of *Gaillardia amblyodon:* isolation of new pseudoguaianolides, *Phytochemistry,* 13, 1187—1192, 1974.
16. **Herz, W. and Srinivasan, A.,** Stereochemistry of spathulin, *Phytochemistry,* 13, 1171—1173, 1974.
17. **Hijink, M. J. and Suatmadji, R. W.,** Influence of different Compositae on population density of *Pratylenchus penetrans* and some other root-infesting nematodes, *Neth. J. Plant Pathol.*, 73, 71—82, 1967.
18. **Inayama, S., Ohkura, T., and Iitaka, Y.,** The complete structure of spathulin, a crystallographic study of diacetylspathulin, *Chem. Pharm. Bull.*, 25, 1928—1933, 1977.
19. **Inayama, S., Kawamata, T., Ohkura, T., Itai, A., and Iitaka, Y.,** A novel pseudotwistane, pulchellon from *Gaillardia pulchella, Chem. Pharm. Bull.*, 23, 2998—3007, 1975.
20. **Inayama, S., Kawamata, T., and Yanagita, M.,** Sesquiterpene lactones of *Gaillardia pulchella, Phytochemistry,* 12, 1741—1743, 1973.
21. **Kawamata, T. and Inayama, S.,** A Michael addition of amines to α-methylene-γ-butyrolactones, *Chem. Pharm. Bull.*, 19, 643—644, 1971.
22. **Koning, R. E.,** Control of Flower Opening by Plant Hormones in *Gaillardia grandiflora,* Ph.D thesis, University of Michigan, Ann Arbor, 1981.
23. **Koning, R. E.,** The roles of auxin, ethylene, and acid growth in filament elongation in *Gaillardia grandiflora* (Asteraceae), *Am. J. Bot.*, 70, 602—610, 1983.
24. **Koning, R. E.,** Plant hormones control filament growth in *Gaillardia, Bioscience,* 33, 458—459, 1983.
25. **Koning, R. E.,** The roles of plant hormones in style and stigma growth in *Gaillardia grandiflora* (Asteraceae), *Am. J. Bot.*, 70, 978—986, 1983.
26. **Koning, R. E.,** The roles of plant hormones in the growth of the corolla of *Gaillardia grandiflora* (Asteraceae) ray flowers, *Am. J. Bot.*, 71, 1—8, 1984.
27. **Kupchan, S. M., Cassady, J. M., Bailey, J., and Knox, J. R.,** Tumor inhibitors. XII. Gaillardin, a new cytotoxic sesquiterpene lactone from *Gaillardia pulchella, J. Pharm. Sci.*, 54, 1703—1705, 1965.
28. **Laurie, A. and Poesch, G. H.,** Photoperiodism: the value of supplementary illumination and reduction of light on flowering plants in the greenhouse, *Ohio Agric. Exp. Stn. Bull.*, 512, 1—42, 1932.
29. **Mitchell, D. R. and Asplund, R. O.,** Pulchellins-C and -E in *Gaillardia aristata, Phytochemistry,* 12, 2541, 1973.
30. **Olsen, C. M.,** A fusarium wilt of *Tagetes* marigolds, *Phytopathology,* 55, 711—714, 1965.
31. **Rai, R. A. and Agnihotri, J. P.,** Foot rot of *Gaillardia pulchella* Foug, *Sci. Cult.*, 36, 284—286, 1970.
32. **Ramaley, F.,** Influence of supplemental light on blooming, *Bot. Gaz.*, 96, 165—174, 1934.
33. **Schneider, C.,** From a German hybridizer, three new annuals, *Horticulture,* 29, 17, 1951.
34. **Secor, J. B. and Farhadnejad, D. O.,** The seed germination ecology of three species of *Gaillardia* that occur in the gypsumland areas of eastern New Mexico, *Southwest. Nat.*, 23, 181—186, 1978.
35. **Sen, J.,** A preliminary note on the X-ray diffraction pattern of pith cell walls, *Bot. Not.*, 111, 474—475, 1958.
36. **Stoutamire, W. P.,** Cytological differentiation in *Gaillardia pulchella, Am. J. Bot.*, 42, 912—916, 1955.
37. **Stoutamire, W. P.,** Cytological variation in Texas *Gaillardias, Brittonia,* 10, 97—103, 1958.
38. **Stoutamire, W. P.,** The history of cultivated *Gaillardias, Baileya,* 8, 12—17, 1960.
39. **Stoutamire, W. P.,** The relation of anthocyanins to *Gaillardia* taxonomy, *Pap. Mich. Acad. Sci. Arts Lett.*, 45, 35—39, 1960.

40. **Stoutamire, W. P.,** Distribution of sesquiterpenes in the genus *Gaillardia* (Compositae), *Am. J. Bot.*, 53, 638, 1966.
41. **Stoutamire, W. P.,** Chromosome races of *Gaillardia pulchella* (Asteraceae), *Brittonia*, 29, 297—309, 1977.
42. **Trippi, V. S.,** Studies on ontogeny and senility in plants. VI. Reversion in *Acacia melanoxylon* and morphogenetic changes in *Gaillardia pulchella*, *Phyton*, 20, 172—174, 1963.
43. **Trippi, V. S.,** Studies on ontogeny and senility in plants. XI. Leaf shape and longevity in relation to photoperiodism in *Gaillardia pulchella*, *Phyton*, 22, 113—117, 1965.
44. **Turner, B. L.,** Two new gypsophilous species of *Gaillardia* (Compositae) from northcentral Mexico, *Southwest. Nat.*, 17, 181—190, 1972.
45. **Turner, B. L.,** A new gypsophilous species of *Gaillardia* (Compositae) from Coahuila, Mexico, *Wrightia*, 5, 305—307, 1976.
46. **Turner, B. L.,** A new species of *Gaillardia* (Asteraceae-Heliantheae) from northcentral Mexico and adjacent Texas, *Southwest Nat.*, 21, 539—541, 1977.
47. **Turner, B. L.,** *Gaillardia aestivalis* var. *winkleri* (Asteraceae), a white-flowered tetraploid taxon endemic to southeastern Texas, *Southwest. Nat.*, 24, 621—624, 1979.
48. **Turner, B. L. and Whalen, M.,** Taxonomic study of *Gaillardia pulchella* (Asteraceae-Heliantheae), *Wrightia*, 5, 189—192, 1975.
49. **Wagner, H., Iyengar, M. A., and Herz, W.,** Flavone-C-glycosides in a coastal race of *Gaillardia pulchella*, *Phytochemistry*, 11, 851, 1972.

GARDENIA JASMINOIDES

En. Gardenia, Common gardenia, Cape jasmine; Fr. Gardenie; Ge. Gardenie, Jasminglanz

H. F. Wilkins

INTRODUCTION

The genus *Gardenia* Ellis consists of some 200 tropical or subtropical shrubs or small trees which belong to the Rubiaceae family. They are native to South Africa, western tropical Africa, India, the Malay Peninsula, the Philippines, Indonesia, and China. *Gardenia jasminoides* Ellis (*G. florida* L.; *G. grandiflora* Lour.; *G. radicans* Thunb.), a native of China, is the most common and popular species seen in southern gardens or grown commercially under protection for greenhouse forcing.[2,18]

Gardenia jasminoides is a shrub which can grow to be 2 m in height. It has opposite, thick, dark green, shiny, evergreen leaves which are lanceolate to obovate in form and some 10 cm long. The calyx has five long teeth. The multipetaled flowers are white and can be some 8 to 10 cm across. Flowers are very fragrant and borne on a short stem. Plants can be used as a flowering potted plant, or the individual flowers may be used as cut flowers for corsages and wedding bouquets.[2,18] Popularity has greatly decreased since the 1940s and 1950s, but the author notes that there has been a renewed interest in the gardenia as a potted plant in recent years.

MORPHOLOGY

Flower buds are terminal on new growth. Leaves are mostly opposite with a rare whorl of three leaves. Axillary shoots arise from one or from both of the two opposite buds at the leaf axils of the upper node when apical dominance is broken. An increment of vegetative growth, following apical dominance removal, most frequently consists of one set of opposite leaves and then the shoot terminates in a reproductive bud. However, two or three sets of leaves may develop.[7,15]

Kosugi et al.[15] report that in nature the primary apical flower is initiated in late July. Afterwards, two to three vegetative shoots appear from the axillary buds and these in turn initiate secondary flowers in early September. Floral development continues and the stamens and pistils can be macroscopically observed 40 days after the respective meristems become reproductive. These dates and time span may vary with cultivars and the environmental conditions. Flowering is in late June and early July the following year. Two to three vegetative shoots will appear at the base of the secondary flower buds, and the cycle is repeated.[15]

FLORAL INITIATION AND DEVELOPMENT

There is no distinct apical cellular organization which can be termed tunica or corpus zones.[7] However, two indistinct tissue zones can be observed with a transition area between them. The outer zone has anticlinal division and the inner zone divides at all planes. The apical meristems become flattened when floral differentiation commences. When apical dominance is lost as a response to terminal buds becoming reproductive, lateral bud cell division commences and lateral buds in turn become reproductive.[7]

After floral initiation occurs, stage 1 of flower bud development continues until the buds are some 0.75 to 1.50 cm long. From micro- to macroscopically visible flowers, some 40 days are required.[1,15] Floral initiation and early development can occur at 20°C or higher temperatures.[1] Afterwards, low night temperatures are required for further differentiation and development (stage 2).[1] If environmental conditions are unfavorable for further development past stage 1, buds will persist for several months, then abort. Stage 2 is low-night-temperature dependent (15.5 to 16.5°C) for flower bud development from 1.50 to 3.0 cm in length. Stage 3, from buds some 3.0 cm in length to anthesis, requires only a few days under LD in the winter. However, some 2 months are required for stage 2 to be completed.

Under natural daylight (ND) greenhouse conditions in Ohio, Davis[7] found that when floral initiation occurred in September, October, November, or December, 158, 165, 152, or 144 days, respectively, were required to reach anthesis. When initiation occurred in March, April, May, or June, 116, 114, 103, or 101 days were required for anthesis. Higher light levels and temperatures in the spring were attributed for the more rapid development of the floral buds.[7] When plants were sheared back on May 24, SD treatments commenced on July 23 and ended on August 13, floral initiation was observed in two out of six shoots 13 days after SD commenced. Some 19 days later, four out of five shoots had initiated flower primordia; 27 days later all meristems were reproductive. With the ND controls only 5 out of 70 meristems had floral primordia by August 31.[7] Flower buds develop and open rapidly when given NB of light under winter greenhouse conditions.[1,2,4,11,16-18]

LIGHT INTENSITY

Reducing light intensities during the summer is a common practice in greenhouse production.[18] However, with greenhouse-grown plants in Florida a 50% reduction of natural light intensity reduced flower bud numbers when compared to plants grown in full light from early October to early May.[19] Similar negative responses to light-intensity reductions were observed in Ohio.[10]

During the winter under greenhouse forcing conditions in the north, 500-W incandescent light treatments during the night for 4- to 6-hr NB were found to favor continuous flower bud development, reduced abortions, and hastened flower bud development.[1] This supplementary high-intensity light treatment may have an effect on flower development in addition to its photoperiodic effect (see the section below on "Photoperiod").

TEMPERATURE

Flower buds initiate and develop through stage 1 (see earlier section titled "Floral Initiation and Development") under many conditions of temperature and photoperiod.[1] Kosugi et al.[15] found that flower initiation in nature occurred in a natural temperature range of 20 to 28.5°C. Flower initiation will occur in greenhouses at 21°C, but development will not continue.[1,7,9,10] Uematsu and Tomita[20] state that a night/day temperature of 18/28°C promoted flower initiation and rapid flowering regardless of photoperiod (ND, 8 or 16 hr). These treatments (temperature and photoperiod) commenced in early February.

Arthur and Harvill[1] have shown that for continued flower bud development (stage 2), low night temperatures are required. Night temperatures below 18°C are required and day temperatures as high as 35 to 38°C did not interfere with development. Keyes[13] noted that flower bud development could be accelerated by 20 to 21°C day, and 15.5 to 16.5°C night temperatures. Night temperatures of 13 to 15.5°C reduced flower bud abscission under unfavorable light conditions, but development was slow.[4-6] Night temperatures above 14°C were required for commercial forcing. Flower bud abscission occurs when night temperatures are above 18°C, particularly during low-light periods.[18]

The greenhouse temperature data presented by Davis[7] (see earlier section titled "Floral Initiation and Development") indicate that the average day temperatures from early May to early September fell below 27°C only twice and that the average night temperature fell below

21°C only three times from early May to early September. Also in Florida, SD treatments were given in periods of high temperatures.[11,16] Thus, the question still remains as to what is the upper limit of temperature for floral initiation in concert with photoperiod.

PHOTOPERIOD

The gardenia is classified as a DNP.[21] However, floral initiation and early development are hastened by SD treatments.[3,4,7,10,11,16] Afterwards LD treatments also contribute to the continued development of these flower buds.[1,3,4,11]

With plants growing out of doors under ND in Chiba, Japan, which is at the 35.5° north parellel, Kosugi et al.[15] found terminal flowers initiated in late July, and secondary lateral buds in early September. The length of day at this parallel is 13 hr, 43 min for the former date and 12 hr, 36 min for the latter date. Further, temperature data for these time spans were 28.5°C high, 21°C low for the former date and 28.5°C high, 20°C low for the latter date.[15]

With greenhouse plants growing under ND in Ohio, which is at the 40° north parallel, Davis[7] found that terminal shoots from 1-year-old nonsheared plants commenced initiating flowers in September. The natural photoperiod at that time is 13 to 12 hr at this parallel. Further, all new developing shoots were reproductive from October to February. Floral initiation numbers slowly declined from early March onward until only vegetative meristems were found during the months of July and August.[7]

When Ohio greenhouse-grown plants were placed under SD, Baird[3] reported that visible flower buds were present 30 days after the end of an 8- or 10-week SD treatment. SD were best given from July 1 to September 9 or from July 15 to September 15 for flowering pot-plant production for Christmas and early January. Under ND, plants never prolifically flowered until April. In Florida during the summer under high temperatures, similar SD floral initiation responses have been observed, with only 2 weeks of an 8-hr photoperiod being required for acceleration of initiation.[11,16]

LD treatments have also been recognized to contribute to the continued development of initiated flower buds.[1,3,4] Baird[3] found that a 150-W incandescent light treatment as a day continuation from 1700 to 2200 hr hastened late stage 3 flower development by 3 to 4 weeks when commenced in early December. Similar positive responses to additive low levels of light in reducing flower bud abortions have been reported.[4,5,11,13,17]

Joiner and Poole[11] found NB (2 hr) reduced abortions and increased total flower production. They also noted that NB were an additive influence to the previous 2-week SD floral induction treatment. Once floral initiation occurs, SD slowed and LD hastened development, and SD increased and LD decreased flower bud abortion. Uematsu and Tomita[20] found that under ND light conditions in February, a photoperiod totaling 16 hr resulted in the lowest number of bud abortions. The critical fact for further flower bud development is that night temperatures are below 18°.[1] (See preceding section, "Temperature.")

GROWTH REGULATORS

Gardenia plants sprayed with daminozide or with chlormequat during production will have reduced heights.[11,16,20] A treatment of chlormequat hastened flower initiation and consequently the date of flower as well as the number of flowers by reducing the number of flower bud abortions.[16] Chlormequat treatments were more effective when applied before the start of SD treatments or prior to floral initiation.[11]

GA used as a foliar spray also increased the number of flowers by reducing the number of buds which abort.[16] To a degree, flower bud abscission appears a natural occurrence. GA is reported to hasten vegetative growth rates but reduced the resistance of the gardenia to cold.[14]

Joiner and Poole[11] and Poole[16] indicate that the gardenia can initiate and develop more flower buds under high temperatures with the use of chlormequat and SD and then LD. Alpha-naphthalene acetamine reduced flower bud abscission.[3]

CULTURE

Propagation is by terminal cuttings from November to February under high humidity at temperatures near 24°C under reduced light intensities.[18] Vegetative growth should be under full light at 18°C night temperatures and no water stress. Soil pH is held below 6.0.[18,19]

Burkhard and Biekart[5] observed that after floral initiation, high nutritional regimes are conducive for vegetative growth. Abrupt shifts in nutrient levels resulted in increased flower bud abscission. Flower quality is better at low nutrient levels. Water stress or any abrupt interruption of bud development will also increase abscission. High humidities are also thought to be needed.[18] At temperatures below 15.5°C, foliar chlorosis may occur and is caused by low soil temperatures.[6,12] Optimal conditions are a minimum soil temperature of 19°C and night air temperatures of 16.5° or higher and day temperatures of 21°C. Soil temperatures of 26°C or higher increased flower bud abscission and decreased foliar chlorosis.[3,4,8,9]

DISCUSSION

In northern Europe, greenhouse-grown plants under ND sporatically flower the year round.[22] In California, in the cool coastal areas near San Francisco, plants are under commercial production and also flower the year round.[18] However, SD hastens floral initiation under the constraints of high summer temperatures in Ohio and Florida, and consequently the gardenia could be classed as a quantitative SD plant.[3,4,7,10,11,16] As of this time no one has apparently determined the critical photoperiod and temperature interaction for floral initiation of this plant.[13,15,20] Data illustrate that the gardenia will rapidly initiate flowers under SD at high temperatures and is a DNP under cool temperatures. Further, flower development is most rapid under LD as long as night temperatures are 15.5 to 16.5°C.[1,3,4,7,11] Apparently, this plant should be considered as a quantitative SL-LD plant, as suggested by Poole.[16]

REFERENCES

1. **Arthur, J. M. and Harvill, E. K.,** Forcing flower buds in gardenia with low temperature and light, *Boyce Thompson Inst. Contrib., 1936—1937,* 8, 405—412, 1937.
2. **Bailey, L. H. (Staff of Hortorium),** *Hortus Third, A Concise Dictionary of Plants Cultivated in the United States and Canada,* Macmillan, New York, 1976, 495.
3. **Baird, E.,** Effect of environmental factors on gardenia production, *Flor. Rev.,* 89(2291), 24—25, 1941.
4. **Baird, E. and Laurie, A.,** Studies of the effect of environmental factors and cultural practices on bud initiation, bud abscission and bud development of the gardenia, *Proc. Am. Soc. Hortic. Sci.,* 40, 585—588, 1942.

Burkhart, L. and Biekart, H. M., Gardenia nutrition in relation to flower bud development, *Proc. Am. ˇoc. Hortic. Sci.,* 35, 768—769, 1937.

vidson, O. W., Effects of temperature on growth and flower production of gardenia, *Proc. Am. Soc. ˙ic. Sci.,* 39, 387—390, 1941.

W. E., Relation of Time of Year and Short Photoperiod to Floral Initiation and Development in *ˈa grandiflora,* Ph.D. thesis, Ohio State University, Columbus, 1952, 44.

ʾ., Gardenias 1946—1947, *Bull. Ohio Flor. Assoc.,* 214, 5—8, 1947.

Observation on gardenia flower production at high air and soil temperatures, *Proc. Am. Soc.* 51, 610—612, 1948.

10. **Hasek, R.**, Gardenia flower production and bud drop, *Bull. Ohio Flor. Assoc.*, 243, 3—5, 1949.
11. **Joiner, J. N. and Poole, R. T.**, Variable photoperiod and CCC effects on growth and flowering of *Garnedia jasminoides* 'Veitchii', *Proc. Fla. State Hortic. Sci.*, 75, 449—451, 1963.
12. **Jones, L. H.**, Relation of soil temperature to chlorosis of gardenia, *J. Agric. Res.*, 57, 611—621, 1938.
13. **Keyes, C. G.**, Bud formation, abscission and flower production of gardenia as affected by light and temperature, *Proc. Am. Soc. Hortic. Sci.*, 37, 1034—1036, 1940.
14. **Kitamura, F.**, Effects of gibberellin on the growth and the flowering of several evergreen trees and shrubs, *J. Jpn. Inst. Landsc. Archit.*, 23, 5—8, 1960.
15. **Kosugi, K., Oyoshi, K., Sumitomo, A., and Kaneko, M.**, Studies on flower bud differentiation and development in some ornamental trees and shrubs. VI. On the time of flower bud differentiation and flower bud development in *Gardenia jasminoides* Ellis, *G. jasminoides* Ellis var. *ovalifolia* Nakai and *Magnolia denudata* Desr., *J. Hortic. Assoc. Jpn.*, 26, 37—42, 1957.
16. **Poole, R. T.**, Effects of gibberellic acid and 2-chloroethyl trimethylammonium chloride on growth and flowering of *Gardenia jasminoides* "Veitchii", *Proc. Fla. State Hortic. Soc.*, 76, 474—477, 1964.
17. **Post, K.**, Effects of daylength and temperature on growth and flowering of some florist crops, *N.Y. Agric. Exp. Stn. Cornell Univ. Bull.*, 787, 1—70, 1942.
18. **Post, K.**, *Florist Crop Production*, Orange, Judd, New York, 1952, 519—529.
19. **Rose, S. A. and Dickey, R. D.**, The effect of light, plunging medium and fertilization on bud set of *Gardenia jasminoides* 'Veitchii', *Proc. Fla. State Hortic. Soc.*, 73, 362—363, 1961.
20. **Uematsu, L. and Tomita, H.**, Studies on gardenia pot plant production, *Bull. Saitama Hortic. Exp. Stn.*, 10, 17—29, 1981.
21. **Vince-Prue, D.**, *Photoperiodism in Plants*, McGraw Hill, London, 1975, 16.
22. **Wilkins, H. F.**, personal observation.

HEVEA

En. Rubber; Fr. Hévéa; Ge. Federharzbaum, Parakautschütbaum

K. Paranjothy

INTRODUCTION

The genus *Hevea,* a member of the Euphorbiaceae, is endemic to the Amazon basin and parts of Matto Grosso, the Upper Orinoco, and the Guianas in South America.[28] Nine species are now recognized in this genus. Of these, *H. brasiliensis* is the most important economically, accounting for virtually all the natural rubber produced today.

Cultivated rubber is found mostly within 15°N and 10°S of the equator in high rainfall areas. Rubber-growing countries, in decreasing order of importance, include Malaysia, Indonesia, Thailand, Sri Lanka, and South India. Rubber is also cultivated to a lesser extent in Africa, principally in Liberia and Nigeria. Cultivation in the South American tropics has been severely limited by South American Leaf Blight caused by *Microcyclus ulei.* Purseglove[20] gives an interesting historical account of the early introduction of *Hevea* to the Old World.

The total natural rubber production in 1982 amounted to 3.7M tonnes. Over 41% of this production came from peninsular Malaysia, the average yield being over 1100 kg/ha/year. About 45% of natural rubber produced in 1982 was consumed by the industrialized countries in the EEC and by the U.S. and Japan.

Chemically, the pure rubber obtained from *H. brasiliensis* is *cis*-1,4-polyisoprene $(C_5H_8)_n$. The molecular weight of the polymer ranges from 0.7 to 1.4 million daltons. The rubber is found in the form of spherical, ovoid, or eccentrically pear-shaped particles, ranging in size from 5 to 6 μm in diameter or length, in laticifers in various tissues of the tree. The bark is the only tissue that is commercially exploited. The bark is tapped at periodic intervals by removal of a thin shaving. The latex that flows out of the severed laticifers is a complex exudate comprising various cellular organelles and rubber particles. The latter comprise about 90% of the dry matter of the exudate.

The trees reach a height of 16 to 18 m within 11 years under plantation conditions and, at the time of replanting, after about 25 to 30 years, may reach heights of 25 m. Wild trees over 40 m have been recorded. Propagation in commerical holdings is largely by bud-grafting.

Chromosome counts of somatic cells of all members of the genus studied to date indicate $2n$ to be 36. Allotetraploidy, with $x = 9$, has been suggested.[16]

FLORAL BIOLOGY

H. brasiliensis is monoecious, with small short-stalked unisexual flowers. The inflorescence is a panicle (Figure 1), with the male flowers arranged spirally along the main and branch axes. The female flowers are found at the tips of the axes. The ratio of male to female flowers varies considerably between clones, but on average may be around 60:1.

Petals are absent in *Hevea* flowers. The calyx is yellow and divided into 5 narrow lobes. The male flower, approximately 5 mm long, has a staminal column on which 10 sessile anthers are arranged in 2 whorls of 5 each. The female flower, usually larger than the male, and approximately 8 mm long, has a superior tricarpic ovary with 3 sessile stigmas.

Anthesis of male flowers is a little earlier than female flowers within a single inflorescence, but there is considerable overlap between inflorescences within a tree. The male flowers remain on the inflorescence for 1 or 2 days. Female flowers remain receptive for 3 or 4

FIGURE 1. An inflorescence of *Hevea brasiliensis* (magnification × 1). By courtesy of Rubber Research Institute of Malaysia, Kuala Lumpur, Malaysia.

days after opening. Detailed observations on the floral biology of *H. brasiliensis* are found in Heuseur.[12] Pollination in nature is carried out by insects, most of them weak-flying thrips and midges.

GROWTH AND DEVELOPMENT

The normal growth and development of *H. brasiliensis* during the first 3 to 5 years is purely vegetative. Growth during this phase is accompanied by rapid physiological changes. Thus, cuttings from young juvenile plants root readily, while this ability is rapidly reduced as the plants age.[8,33] Leaf cuttings, carrying one leaf and one bud on a short piece of stem approximately 1 to 2 cm, have been found to root only when taken from seedlings less than

6 months old (Table 1). Wiersum[33] noted that cuttings taken from young seedlings produced pseudo-taproots, while Yoon and Leong[37] found that cuttings taken from source bushes generally lacked taproots.

Seedlings have a characteristic conical stem. Buddings, unlike seedlings, generally have cylindrical boles, uniform in diameter from the collar upwards. However, buds taken progressively lower down the bole of a seedling have been noted to produce buddings with correspondingly greater seedling-like stem conicity.[7,17,22]

The physiological basis of these phenomena is not understood, though it is clear that they resemble the juvenility phenomenon well recognized in many woody plants (see, for example, Doorenbos[6]). Muzik and Cruzado,[18] on the basis of experiments which involved grafting of buds from mature clones onto young seedling scions, believed that the young seedling form may transmit a substance to the adult capable of inducing the latter to assume juvenile characteristics. The chemical transmission of juvenility is also implicit in experiments of Gregory[8] in which rooting was more readily induced after grafting juvenile type buds onto mature shoot cuttings.

WINTERING AND FLOWERING

H. brasiliensis sheds its leaves at least once a year. This phenomenon, commonly referred to as wintering, has been observed in most countries where the species is cultivated. Wintering usually follows a dry spell and therefore occurs at different times in different localities.

Within 2 weeks following leaf-fall, the inflorescences start sprouting from buds at the ends of bare shoots or buds at the axils of scale leaves of young leaf flushes. While the sprouting of floral inflorescences in relation to leaf-fall is well known, there does not appear to be any information, based on anatomical studies, on floral bud initiation in relation to shoot growth and environmental factors. Thus the dormancy period, if any, before floral buds sprout has not been experimentally established. From the work of Halle and Martin[10] it might be inferred that floral buds sprout within 40 days of their initiation. Schweizer,[29] on the basis of artificial defoliation experiments, argued that floral buds were initiated shortly before wintering, unlike those of trees in temperate zones.

In peninsular Malaysia, mature trees of *H. brasiliensis* shed their leaves twice a year, usually between March and April and between August and September. Defoliation is usually more severe (and flowering more intense) during the first season, reflecting the severity of the corresponding dry season that precedes leaf-fall.

A survey of wintering carried out by Wycherley[36] in 1954 reflects the sensitivity of the phenomenon to both environmental and genotypic influences. Two clones, PB 86 and Tjir 1, were studied in this survey. In northwest peninsular Malaysia, where prolonged dry spells are the rule, wintering was found to be early and complete. In other areas, wintering was late, incomplete, or irregular. Of the two clones, PB 86 showed a lower proportion of incomplete or irregular winterings than Tjir 1 and, in general, wintered somewhat earlier and recovered sooner. In Sri Lanka, Liyanage[14] found that older trees defoliated and refoliated much earlier than younger trees (Table 1). Liyanage[14] also observed that defoliation and refoliation occurred earlier at lower elevations than at higher elevations (Table 2). Planting density did not appear to influence the pattern of wintering.

Water stress is believed to be the most important factor affecting leaf-fall and flowering in deciduous primary forest trees in Malaysia,[32] but this suspected relationship is yet to be proven experimentally. Changes of photoperiod and temperature in peninsular Malaysia, though slight, are sufficient to trigger flowering in some species in the region.[32] Though systematic studies on photoperiodic and thermoperiodic responses in *Hevea* have not been carried out, it would be difficult to explain the second midyear wintering and associated flowering on the basis of such factors in *Hevea*.

Table 1
THE INFLUENCE OF AGE ON WINTERING PATTERN

Age (year)	Onset of defoliation	Completion defoliation	Defoliation phase (days)	Onset refoliation	Completion refoliation	Refoliation phase (days)	Overlap, refoliation and defoliation phases (days)	Gap between onset of defoliation and refoliation phases (days)
Less than 10	Jan. 21	Feb. 12	22	Feb. 2	Feb. 23	21	10	12
10—11	Jan. 17	Feb. 11	25	Jan. 29	Feb. 25	27	13	12
12—13	Jan. 16	Feb. 13	28	Jan. 28	Mar. 1	32	16	12
14—15	Jan. 11	Feb. 9	29	Jan. 24	Feb. 28	35	16	13
16 and above	Jan. 8	Feb. 18	41	Jan. 20	Mar. 8	47	29	12
$F_{4,48}$	*	*	*	*	*	*	3.16	<1.00

* Denotes significance at 0.05%.

From Liyanage, A. de S., *J. Rubb. Res. Inst. Sri Lanka*, 53, 31—38, 1976. With permission.

Table 2
THE WINTERING PATTERN AT DIFFERENT ELEVATIONS

Elevation (m)	Onset of defoliation	Completion defoliation	Defoliation phase (days)	Onset refoliation	Completion refoliation	Refoliation phase (days)	Overlap, refoliation and defoliation phases (days)	Gap between onset of defoliation and refoliation phases (days)
Less than 30	Jan. 14	Feb. 12	29	Jan. 26	Feb. 17	22	17	12
30—90	Jan. 11	Feb. 14	34	Jan. 24	Feb. 20	27	21	13
90—150	Jan. 21	Mar. 9	47	Feb. 5	Mar. 23	46	32	15
150 and above	Feb. 2	Mar. 23	49	Feb. 17	Apr. 15	57	34	15
$F_{3,144}$	*	*	*	*	*	*	*	*

* Denotes significance at 0.05%.

From Liyanage, A. de S., *J. Rubb. Res. Inst. Sri Lanka*, 53, 31—38, 1976. With permission.

Schweizer[29] argued that wintering periodicity in *H. brasiliensis* reflected an inherent genetically determined internal rhythm of the tree. In support of this contention he pointed out that wintering severity and commencement were clonal features that could be transmitted genetically in breeding experiments. He conducted experiments which showed that wintering periodicity was maintained even in trees that were defoliated at different times of the year, thus ruling out leaf age as a primary determinant of wintering. External influences were only seen to modify this rhythm. Thus fertilizer treatments were found to delay (but not prevent) wintering while climatological factors were seen in Schweizer's studies to shift wintering periodicity by 1 or 2 months.

Halle and Martin,[10] and Halle and Oldeman[11] also argued that flowering is controlled by an endogenous rhythm. Their studies revealed that under continuous favorable conditions of growth the apical meristem of *Hevea* maintains a rhythmic 40-day cycle of growth: 30 days of mitotic activity and 10 days of inactivity on average. This leads to the production of short units of morphogenesis (equivalent to caulomeres). Their experimental work led them to believe that the rhythm resulted from a competition for water between leaves and the apical meristem.

Wintering periodicity is reflected clearly in growth ring formation in the wood.[27] Soong[30] demonstrated significant changes in feeder root development in relation to wintering. Maximum root development occurred in February/March, corresponding to the period of active refoliation and peak uptake of moisture and nutrients by the tree. Minimum root development was seen around August to December.

Chua[4] found that a critical balance between growth promoter and inhibitor activity in the leaves determined the onset of their senescence. Water stress is well known to modulate hormone levels[35] and it is possible that in *Hevea*, too, drought effects on severity of wintering, for example, are manifested through hormone levels.

A great deal of information is available on seasonal variations in composition of latices and leaves, mostly collated with practical considerations in mind. Early work concerning seasonal variation in composition of latices has been reviewed by Wiltshire.[34] Much of this early work lacks information concerning exact dates of defoliation and refoliation and for this reason it is difficult to correlate adequately changes in composition of latices, which were studied intensively, with wintering. Furthermore, data were usually presented as monthly averages.

More recently, variation in nutrient content of leaves with age has been studied in detail by Guha and Narayanan,[9] the main aim of this study being to utilize leaf analysis as a guide to the nutritional requirements of trees. Nitrogen, phosphorus, potassium, and magnesium content in the leaf showed sharp drops in concentration during the first 3 weeks after leaf emergence, probably due to the dilution caused by rapid increase in dry weight of leaves at this stage. During the following 8 months, a more gradual change in concentration of all elements was observed. Nitrogen, phosphorus, and potassium content showed negative linear relationships with leaf age, while calcium and manganese content showed positive linear relationships. Magnesium content, however, did not show any such relationship with leaf age. Chua[4] reported somewhat similar changes and suggested that losses in senescing leaves may partly be attributable to leaching by rain.

In a study covering changes in major mineral nutrients in leaf, bark, shoot, and root with respect to wintering, Tan[31] reported that maximum concentrations of mineral nutrients were found in the green shoot, bark, and root tissues at the end of leaf-fall, when the concentration of most of the major nutrients in the leaf was lowest, with the exception of calcium. A mobilization of the nutrients from shoot, bark, and root tissues to the newly developing leaves following wintering was evident, continuing during leaf maturation and reaching a minimum only after 90 to 120 days of refoliation. It was thus evident in this study that green shoot, bark, and roots serve as storage tissues for nutrients which are utilized during refoliation and subsequent leaf growth.

Table 3
COMPARISON OF YIELDS DURING REFOLIATION

Mean yield during refoliation (as % of annual mean)	31—40	41—50	51—60	61—70	71—80
Number of clones	4	4	6	4	3
Example	Tjir 1	RRIM 610	RRIM 501, 600	RRIM 605	RRIM 603

From Liyanage, A. de S., *J. Rubb. Res. Inst. Sri Lanka*, 53, 31—38, 1976.

Yield depression during refoliation can be severe in some clones[36] (Table 3). Dry rubber content has also been reported to be lower during wintering,[13] especially during refoliation.[1] In addition, it is likely that an accentuation of plugging during the wintering period is also responsible for lower yields.[20] The reasons for the increase in plugging index during wintering are not clear. Latices of several clones are known to be prone to pre-coagulation on the tapping cut during wintering.[23]

FLOWER INDUCTION

The main impetus for attempts at inducing flowering in juvenile plants derives from various difficulties encountered by breeders. Flowering in mature trees is brief and periodic and nonsynchrony of flowering between parents further accentuates the problem.

Expensive scaffolding is necessary for hand pollinators and, in addition to this, control of fungal pathogens and insects is an essential requisite for successful hand-pollinations. Thus the anticipated convenience of year-round pollination and control measures from the ground has been a compelling force in work aimed at inducing young plants to flower.

Young root-diseased and water-logged trees are well known to flower in *Hevea* but have obviously been of little or no value for breeding purposes. Various workers have reported that gravimorphic and ring-barking treatments induce young buddings to flower.[2,3,5,19,21] Flowering has been hastened by combining such treatments with foliar applications of TIBA and coumarin.[2] A cyclical pattern of flowering, in synchrony with that in mature trees, has been reported in young girdled trees.[26] Ring-barking has also been effective in inducing off-season flowering in mature trees.[15,21,24]

There are evidently clonal differences in amenability to induction of flowering. Girdling methods, while effective in inducing flowering in young buddings, are ineffective in young seedlings, even with the aid of growth substances.[19]

CONCLUDING REMARKS

There can be little doubt that there is much scope for experimental studies on flowering in *Hevea*. This crop offers the advantage of widespread cultivation of a large number of clones in several parts in the tropics. The induction of flowering in young buddings by simple horticultural methods would no doubt be a useful tool in these studies. Two views that have emerged would seem to merit experimental attention, first, the relationship of drought to flowering and second, the nature and role of endogenous rhythms in flowering. Answers to these questions may have important practical implications, particularly in breeding practices, not only in *Hevea* but in other tropical dicotyledonous perennials with similar flowering patterns.

REFERENCES

1. **Adjuwana, H. and Soerianegara, I.,** Fotosintesa dan produksi lateks pada tiga klon karet (*Hevea brasiliensis*) (Photosynthesis and latex production in three clones of rubber), *Menara Perkebunan,* 39 (5/6), 77—86, 1970.
2. **Camacho, E. V. and Jimenez, E. S.,** Resultados preliminares de una prueba de induction de floracion premature an arboles jovenes de *Hevea, Turrialba,* 13, 186—188, 1963.
3. **Campaignolie, J. and Bouthillion, J.,** Pollination artificielle sur journes leveas conduits on espalier *Rapp. Ann. Inst. Caoutch, Indoch.,* 1954, 64—65.
4. **Chua, S. E.,** The physiology of foliar senescence and abscission in *Hevea brasiliensis* Muell. Arg., Rubber Research Institute Research Archives Doc. 65, Rubber Research Institute of Malaysia, Kuala Lumpur, Malaysia, 1970.
5. **De Silva, C. A. and Chandrasekera, L. B.,** A method of inducing floral stimulus for early flowering of *Hevea brasiliensis, Q. J. Rubb. Res. Inst. Ceylon,* 35(3), 50—55, 1959.
6. **Doorenbos, J.,** Juvenile and adult phases in woody plants, *Encyclopedia of Plant Physiology,* Ruhland, W., Ed., 15(1), 1965, 1222—1235.
7. **Ferweda, F. P.,** A possible explanation of the divergence between juvenile type budgrafts and their seedling mother trees in *Hevea, Euphytica,* 2, 15—24, 1953.
8. **Gregory, L. E.,** A note on the rooting of *Hevea* clones, *Turrialba,* 1, 201—203, 1951.
9. **Guha, M. M. and Narayanan, R.,** Variation in leaf nutrient content of *Hevea* with clone and age of leaf, *J. Rubb. Res. Inst. Malaya,* 22, 225—239, 1969.
10. **Halle, F. and Martin, R.,** Etude de al croissance rythmique chez l' Hevea (*Hevea brasiliensis* Mull.-Arg., Euphorbiacees-Crotonoidees), *Adansonia,* 2(8), 475—503, 1968.
11. **Halle, F. and Oldeman, R. A. A.,** An essay on the architecture and dynamics of growth of tropical trees, University of Malaya, Kuala Lumpur, 1975, 1—156.
12. **Heuseur, C.,** The sexual organs of *Hevea brasiliensis, Arch. Rubb. Cult.,* 3(11), 455—514, 1919.
13. **Lee, C. K.,** Variation in Yield and Dry Rubber Content in Modern *Hevea* Clones, M. Sc. thesis, University of Sydney, Sydney, Australia, 1974.
14. **Liyanage, A. De S.,** Influence of some factors on the pattern of wintering and on the incidence of *Oidium* leaf fall in clone PB 86, *J. Rubb. Res. Inst. Sri Lanka,* 53, 31—38, 1976.
15. **Masjid, A., Maslichah, D., Saleh, and Bunjamin,** Percobaan menginduksi pembungaan pada *Hevea* (Flower induction trials in *Hevea), Menara Perkebunan,* 45(1), 23—29, 1977.
16. **Majumder, S. K.,** Chromosome studies of some species of *Hevea, J. Rubb. Res. Inst. Malaya,* 18, 269—275, 1964.
17. **McIndoe, K. G.,** The development of clonal rootstocks in *Hevea, Rubb. Res. Inst. Ceylon Q. J.,* 34, 39—57, 1958.
18. **Muzik, T. J. and Cruzado, H. J.,** Transmission of juvenile rooting ability from seedlings to adults of *Hevea brasiliensis, Nature (London),* 181, 1288, 1958.
19. **Najib, L. and Paranjothy, K.,** Control and induction of flowering in *Hevea, J. Rubb. Res. Inst. Malaysia,* 26, 123—134, 1978.
20. **Narayanan, R. and Abraham, P. D.,** Latex vessel plugging indices and their variations for clones, sites and tapping systems, *J. Rubb. Inst. Malaysia,* 24, 248—260, 1977.
21. **Ong, S. H.,** Flower induction in *Hevea, 1st Symp. Int. Crop Hevea Breeding Kuala Lumpur,* Preprint No. 14, Rubber Research Institute of Malaya, Kuala Lumpur, 1972.
22. **Paardekooper, E. C.,** The occurrence, properties and possible application of juvenile type buddings, *Arch. Rubb. Cult.,* 33, 141—157, 1956.
23. **Paardekooper, E. C.,** *Clones of Hevea brasiliensis of Commercial Interest in Malaya,* RRIM Planting Manual No. 11, Rubber Research Institute of Malaya, Kuala Lumpur, Malaysia, 1965.
24. **Premakumari, D. and Bhaskaran Nair, V. K.,** Induction of off-season flowering in *Hevea brasiliensis, Rubb. Board Bull.,* 13(3), 43—44, 1976.
25. **Purseglove, J. W.,** *Tropical Crops,* Longman, Green & Co., London, 1982, 146—171.
26. **Rohani Othman and Paranjothy, K.,** Induced flowering in young *Hevea* budding, *J. Rubb. Res. Inst. Malaysia,* 28(3), 149—156, 1980.
27. Rubber Research Institute of Malaya, *Annu. Rep. Rubb. Res. Inst. Malaya,* p. 37, 1953.
28. **Schultes, R. E.,** The history of taxonomic studies in *Hevea, Bot. Rev.,* 36, 197—276, 1970.
29. **Schweizer, J.,** Leaf-fall phenomena in *Hevea brasiliensis* — a contribution to the problem of periodicity in tropical trees, Biological and Agricultural Work in the N. Indies No. 16, RRIM Trans. No. 102, Rubber Research Institute of Malaysia, Kuala Lumpur, Malaysia, 1933, 22 pp.
30. **Soong, N. K.,** Feeder root development of *Hevea brasiliensis* in relation to clones and environment, *J. Rubb. Res. Inst. Malaya,* 24, 283—298, 1976.
31. **Tan, K. T.,** Seasonal changes in the concentration of nutrients in mature *Hevea, Proc. Int. Rubb. Conf. 1975, Kuala Lumpur,* 111, 73—83, 1975.

32. **Whitmore, T. C.,** *Tropical Rain Forests of the Far East,* Oxford University Press, London, 1975.
33. **Wiersum, L. K.,** Observations on the rooting of *Hevea* cuttings, *Arch. Rubb. Cult.,* 32, 213—243, 1955.
34. **Wiltshire, J. L.,** Variations in the composition of latex from clone and seedling rubber, *Rubb. Res. Inst. Malaya, Bull. No. 5,* Rubber Research Institute of Malaya, Kuala Lumpur, Malaysia, 1934, 61 pp.
35. **Wright, S. T. C. and Hiron, R. W. P.,** The accumulation of abscissic acid in plants during wilting and under stress conditions, in *Plant Growth Substances 1970,* Carr, D. J., Ed., Springer-Verlag, Berlin, 1972, 291—298.
36. **Wycherley, P. R.,** Variation in the performance of *Hevea* in Malaya, *J. Trop. Geogr.,* 17, 143—171, 1963.
37. **Yoon, P. K. and Leong, S. K.,** Induction of pseudotaproots of cuttings and production of clonal root-stocks in *Hevea, Proc. Int. Rubb. Conf. 1975 Kuala Lumpur,* Vol. 2, Rubber Research Institute of Malaysia, Kuala Lumpur, Malaysia, 1976, 85—108.

HIBISCUS ROSA-SINENSIS

En. Chinese hibiscus, Hawaiian hibiscus, Rose-of-China, China rose, Blacking plant; Fr. Hibiscus, Ketmie; Ge. Eibisch

H. F. Wilkins

INTRODUCTION

Hibiscus rosa-sinensis L. (*H. chinensis* Hort., *H. sinensis* Hort.) is one of some 250 species of herbs, shrubs, or trees found growing from the tropics to temperate zones. Belonging to the Malvaceae, *Hibiscus rosa-sinensis* L. is the most highly prized ornamental species and has been so commonly grown in tropical and subtropical gardens that its exact site of origin is lost. No doubt it originated in China or in Vietnam, and it is commonly found in the East Indies as well. Illustrations are found in ancient art and recorded in ancient writings.[1,6]

We do not know for certain how many cultivars have been developed over the centuries, or how many have been lost since being introduced to Europe in 1731.[6] However, certain cultivars are commonly used for greenhouse production.[18]

PLANT AND FLOWER MORPHOLOGY

This woody shrub can be tree-like (7 to 13.5 m). The terminal bud is always vegetative. At each node there is a vegetative and reproductive bud. The reason that *H. rosa-sinensis* is so prized is because of the large individual single five-petalled or double flowers which can be some 5 to 12.5 cm across. Flower colors are white to brilliant reds, oranges, yellows, or with various shading combinations. The stamens are exserted and are united to form a tubular column; they are usually longer than the petals. The style is 5-branched at the apex and can be quite proliferated and ornamentally modified. The ovary is a 5-celled structure with 3 or more seeds per cell. The calyx is 5-lobed and bell-shaped. The glossy green leaves may be up to 6 in. long, ovate, with serrated margins.[1,6]

CULTURE

This plant has become popular in recent years in northern Europe and North America as the plant acclimatizes to the home environment surprisingly well. This plant has become a speciality commercial production item for northern homes as well as for southern gardens and landscapes where it thrives under frost-free conditions. It is tolerant of high temperatures and full sun in the south as well as being an excellent patio plant in the north.[1,4,6,18,19]

Propagation of hibiscus can be achieved by seed, cutting, grafting, or layering. Cutting is most common. There is an abundance of information on cultural techniques and greenhouse production.[1,4,6,8,11-13,17-19]

GROWTH AND FLOWERING

H. rosa-sinensis is an autonomously inductive plant and flowering is not photoperiodically controlled. The number of flowers depends on the rate of growth and lateral shoot production. Vegetative growth rates and flower production numbers are dependent upon total light energy.[15,17] Flower production peaks in April and May in Minnesota, 45° north parallel.[15] Plants flower year round, even in the winter, at 60° north latitudes.[18]

GROWTH REGULATORS

In commerce, for house-plant and patio specimens growth regulators are commonly used. Sprays of chloromequat not only induce shorter internodes and darker green leaves, but also hasten date of flowering under greenhouse conditions and increase branching and consequently flower numbers.[2-5,7,9,10,14,18,19]

FLOWER OPENING AND SENESCENCE

Flowers are ephemeral, lasting only 1 day on the plant from opening to wilting; however, selections do exist whose flowers last somewhat longer. The author has observed in Hawaii that a detached flower harvested in the morning will remain turgid and open for 20-plus hours without water.

Sprays of silver thiosulfate (STS) have not been able to prevent the rapid 1-day flower opening to senescence. However, STS sprays have been able to reduce flower bud abscission during transporting or low-light stress periods.[6a]

Swanson and Wilkins[16] studied development of flowers and found a parallel between ethylene production and abscisic acid levels during development and senescence to abscission of the flower, calyx, and peduncle segments.

REFERENCES

1. **Bailey, L. H. (Staff of Hortorium),** *Hortus Third,* A Concise Dictionary of Plants Cultivated in the United States and Canada, Macmillan, New York, 1977, 560—562.
2. **Bhattacharjee, S. K., Mukherjee, T. P., and Bose, T. K.,** Dwarfing of *Hibiscus rosa-sinensis* L. using CCC as a soil drench, *Punjab Hortic. J.,* 17, 164—167, 1977.
3. **Bose, T. K., Hore, B. K., and Mukherjee, D.,** Dwarfing of some malvaceous ornamental plants as a nursery practice, *HortScience,* 3, 171—180, 1968.
4. **Criley, R. A.,** Potted flowering hibiscus, *Flor. Rev.,* 165 (4290), 48—49, 64, 1980.
5. **Criley, R. A.,** Vegetative growth control of *Hibiscus rosa-sinensis* hedges with chlormequat, *HortScience,* 16, 343—344, 1981.
6. **DeWolf, G.,** Hibiscus, *Horticulture,* 60(6), 26—27, 1980.

6a. **Heins, R. D.,** personal communication, 1984.

7. **Hore, B. K. and Bose, T. K.,** Effect of growth retarding chemicals on tropic ornamental shrubs, *Indian Agric.,* 15, 115—125, 1971.
8. **Jensen, H. E. K. and Rasmussen, P. M.,** Hypobaric storage of cuttings. I. Principles and preliminary experiments, *Tidsskr. Planteavl,* 82, 623—632, 1978.
9. **Lamper, J.,** Investigations with growth-retardants on *Hibiscus* and *Crassula, Gartenwelt,* 65, 79—80, 1965.
10. **Morioka, K., Yonemura, K., Sakai, K., Fukuda, M., and Higuchi, H.,** Controlling the height of some pot plants by the use of growth retardants, *Res. Bull. Aichi-Ken Agric. Res. Centre Hortic.,* 10, 36—41, 1978.
11. **Mukherjee, T. P., Roy, T., and Bose, T. K.,** Standardization of propagation from cuttings under mist, *Punjab Hortic. J.,* 4, 153—155, 1964.
12. **Olson, C. H.,** Some new 4-inch crops and how to grow them, *Flor. Rev.,* 170(4397), 9, 42—43, 1982.
13. **Pizzette, I. and Crocker, H.,** *Flowers: A Guide For Your Garden,* H. N. Abrams, New York, 1975, 590—594.
14. **Shanks, J. B.,** Chemical control of growth and flowering and hibiscus, *HortScience,* 7, 574, 1971.
15. **Swanson, B. T.,** Ethylene in Relation to Phytogerentology, Ph.D. thesis, University of Minnesota, St. Paul, 1972, 326.
16. **Swanson, B. T. and Wilkins, H. F.,** Endogenous ethylene and abscisic acid relative to phytogerentology, *Plant Physiol.,* 55, 370—376, 1975.
17. **Von Hentig, W.-V. and Heiman, M.,** *Hibiscus rosa-sinensis, Kulturkartei Ziorpflanzenbau,* Paul Parey, Hamburg, 1977.
18. **Wikesjö, K.,** Odling ai *Hibiscus rosa-sinensis, Konsulentavdelningens Rapporta, Tradgard 205,* Swedish University of Agricultural Science, Alnarp, 1981.
19. **Wilkins, H. F. and Kotecki, D.,** *Hibiscus rosa-sinensis* L., the Chinese hibiscus, Hawaiian hibiscus, rose-of-China, China rose, blackening plant, *Minn. State Flor. Bull.,* 31(4), 3—7, 1982.

HYOSCYAMUS NIGER

En. Black henbane; Fr. Jusquiame noire; Ge. Bilsenkraut; Rus. Belena

Anton Lang

INTRODUCTION; HISTORY OF USE

Upon my secure hour thy uncle stole,
With juice of cursed hebona in a vial,
And in the porches of mine ear did pour
The leperous distilment; whose effect
Holds such an enmity with blood of man
That swift as quicksilver it courses through
The natural gates and alleys of the body,
And curd, like eager droppings into milk,
The thin and wholesome blood. So did it mine;
And a most instant tetter bark'd about
Most lazar-like, with vile and loathsome crust,
All my smooth body.[65]

Mode of administration, mode of action, and clinical symptoms may be described with poetic license, but the above citation illustrates what undoubtedly has been the oldest use of hebona = henbane = *Hyoscyamus niger* L. Because of its content in alkaloids (hyoscyamine and scopolamine = hyoscine) it can be and has been used to expedite the demise of people unloved for one reason or the other. The widest use of this sort, which the plant shared with another member of the Solanaceae, *Scopolia carniolica,* was in parts of central and southeastern Europe. In those regions, farmers did not die; they faded away. When getting old they retired, turning their farms over to their offspring who, in return, were supposed to care for their parents' needs for the rest of the latters' lives. The offspring, however, did not always care for this part of the bargain: and a dose of henbane or *Scopolia,* judiciously administered in a tasty dish, resolved the problem and earned the plants the somewhat grisly name of ''old peasants' herb'' (Altsitzerkraut).

Henbane has, however, had also more salutary uses, in olden times and until quite recently. The same alkaloids that will kill a man at high doses have beneficial effects at lower ones, and the plant has served as a source for these alkaloids and has been grown commercially, although on a relatively small scale. Today, it has been displaced in this role by other plants, particularly *Duboisia,* a shrub of Australian origin which has several times the alkaloid content of *H. niger*. But *H. niger*, although apparently no longer cultured as a crop plant, is still collected from wild stands and used either in the form of powdered leaves, or as an extract from this powder, about four times as strong as the latter, mainly as a relaxant to relieve tensions in the gastrointestinal tract and perhaps to ward off gastritis and ulcers. However, man having developed much more effective means for killing man, and his tensions nowadays requiring more potent remedies, it seems this remaining use for *H. niger* will continue to decline and the day will come when we shall know of the uses of hebona only from William Shakespeare, from ancient pharmacopoeias, and from books on the physiology of flowering.

Use of *H. niger* as a guinea pig — a more appropriate term would perhaps be a warhorse — by students of flowering dates back to 1904 when C. E. Correns, one of the rediscoverers of Mendel's laws, reported that the biennial and the annual habits were inherited as a simple Mendelian character, the former according to Correns being dominant and the latter recessive.[12] This was among the earliest demonstrations, if not the very earliest one, that

FIGURE 1. *H. niger*, plants of one of the annual (left) and one of the biennial (right) strains of Melchers. Both plants on LD; the biennial one nonthermoinduced. (Original photographs courtesy of Professor G. Melchers, Max-Planck-Institut für Biologie, Tübingen, FRG.)

Mendel's laws held also for purely physiological (Correns said "biological") characters. Today this may seem a trivial matter, but in the early era of genetics it was an important discovery, proving that the newly (re)discovered laws of heredity determined not only the shape of the organism but also its function and its adjustment to environment.

Correns[12] also showed that the purple flower color of most *H. niger* strains and the pale yellow of some others were controlled by one gene (locus), heterozygotes in this case, although visually barely distinguishable from dominant homozygotes, having an intermediate pigment content. However, Correns' work with *H. niger* remained an isolated effort and his stocks were not maintained. Thus, when about 30 years later Melchers, working at the Kaiser-Wilhelm (later, Max-Planck) Institute of Biology in Berlin-Dahlem (later, Tübingen) was searching for a handy plant with both annual and biennial strains and decided, from Correns' accounts, that *H. niger* should meet the needs, he had to obtain seed samples from a number of sources and in only two of these, from the Hungarian Drug Research Institute of Budapest and the Botanical Institute of the U.S.S.R. Academy of Sciences in Leningrad, did he find biennial plants. He[47] confirmed Correns' results on the genetic basis of biennial vs. annual habit and purple vs. yellow flower color, and obtained evidence, although perhaps less unequivocal, that another physiological character, the need for freezing in seed germination, was also determined as a simple dominant Mendelian character. He gave the genes the following Latin names: annuus (*ann*) for that determining the annual vs. the biennial (ann^+) habit; pallidus = pale (*pal*) for that determining yellow vs. purple (pal^+) flowers; and calidus = warm (*cal*) for that responsible for absence vs. presence (cal^+) of the freezing requirement.

It is the Melchers strains, specifically those derived from the Budapest material, which have been used in almost all studies in the physiology of flowering that were carried out with *H. niger* (Figure 1). These strains have been maintained by inbreeding for numerous generations, at least in part (in Melchers' laboratory) using progenies from individual plants, and possess unquestionably as high a degree of genetic homogeneity as it is possible to obtain. It should perhaps be pointed out, however, that these strains are not isogenic (although without doubt closely related in genetic terms), as a breeding program which would have been needed to reach this condition has never been carried out.

The two strains which have been most extensively used are a purple-flowered biennial $\left(\frac{ann^+\ pall^+}{ann^+\ pall^+}\right)$ and a yellow-flowered annual one $\left(\frac{ann\ pall}{ann\ pall}\right)$. This preference arose more or less by chance, but at least with the former it proved to be a serendipitous one, as the yellow-flowered biennial strain has a tendency to flower with no cold treatment, although only after prolonged periods of time.[49]

CULTURE METHODS; GRAFTING

General Culture

The culture of *H. niger* is quite simple if one is careful to observe a few rules and precautions, the main ones being ample mineral nutrition, avoidance of excessively high temperatures, control of insect pests, and prevention of virus infections. The following procedures have been used by Melchers and myself with very satisfactory results but it is not claimed that others may not be as good.

The plant can be grown in liquid culture, in soil, and in an inert medium (e.g., equal parts of crushed rock and medium-mesh vermiculite). Seeds are best germinated on coarse sand and covered with medium-mesh (not with fine) vermiculite, and watered, preferably by soaking. A good germination temperature is 27°C for 8 hr a day and 19°C for the remaining 16 hr, but a constant temperature of 25 to 30°C is also satisfactory. The biennials usually take a little longer to germinate than the annuals and if freshly harvested the seed may display some dormancy. In this case the containers should be placed for about 4 weeks in the cold (ca. 5°C) or daily frozen and thawed for about 5 to 10 days. (The marked freezing requirement that had been present in the original material — coded by the cal^+ allele — has, however, in all probability been eliminated by continued selection of fast-germinating seedlings.)

When the seedlings are established (that is, when the cotyledons are fully grown but before the first leaf begins to emerge), they are ready for further handling. For liquid culture, they can be directly transferred to the containers selected. In fastening the plants on the containers it should be remembered that the hypocotyl will develop into a tuber which can reach 1 cm and more in diameter. An excellent nutrient solution of *H. niger,* developed by Melchers,[49] is given in Table 1, Section A. The recipe is the original one; however, the "soluble ferric phosphate" can be replaced by one of the chelated forms of iron now readily available.

If culture in a solid medium is desired, the plants should be transplanted twice, each time selecting for uniformity; in this way, one obtains very uniform and sturdy material. The seedlings can be conveniently planted into trays (wood or plastic) about 7.5 to 10 cm deep. They should be spaced approximately 5 cm apart. The second transplanting, into 10-cm pots, should be done when the leaves of neighboring plants begin to overlap, usually around 4 weeks after the first. If gravel-vermiculite is used, the plants must be watered with nutrient solution (Table 1-A); if soil is to be used, it should be of high nutrient content (see Table 1-C) and the plants should receive supplementary mineral fertilization at regular intervals, using a nutrient solution like the one listed in Table 1-B.

Plants of the annual strains, being LDP, must be grown on SD (8 to 9 hr); biennial plants should, except for some special experimental purpose requiring SD culture, be grown on LD, as these are favorable for rapid growth. If artificial light is used, the light flux (illu-

Table 1
CULTURE METHODS FOR *H. NIGER*[a]

A. Nutrient Solution for Liquid and Gravel-Vermiculite Culture

Compound	Amount
NH_4NO_3	0.25 g
$CaCl_2 \cdot 6H_2O$	0.5 g
$MgSO_4 \cdot 7H_2O$	0.25 g
KH_2PO_4	0.25 g
Microelement solution	1 mℓ
Soluble ferric phosphate solution, diluted	5 mℓ
Water	To 1 ℓ
Microelement solution:	
H_3BO_3	1 g
$ZnSO_4 \cdot 7H_2O$	0.25 g
$MnSO_4 \cdot 4H_2O$	0.5 g
$Cu_2(SO_4)_3$	0.1 g
Water	1 ℓ
Soluble ferric phosphate solution	
Stock solution:	
Ferric citrate	50 g
$NaH_2PO_4 \cdot 12H_2O$	55 mg
Water	100 ℓ

Dissolve by shaking on shaking apparatus (may take several days). Solution usable as long as dark reddish-brown. For use, dilute 1:99, use 5 mℓ/1 ℓ nutrient solution.

B. Nutrient Solution for Potted Plants

Compound	Stock solution (g/ℓ)
NH_4NO_3	100
$Ca(NO_3)_2 \cdot 4H_2O$	200
$MgSO_4 \cdot 7H_2O$	50
KH_2PO_4	50
KNO_3 (optional)	25

For use, 5 mℓ stock solution in 1 ℓ water

C. Soil Mixture[b]

Component	Parts by volume
Compost or high-grade garden (top) soil	3
Leaf mold	1
Manure mold (from hot frames)	1
Peat moss	0.5—2[b]

[a] As practiced in Professor G. Melchers' laboratories since approximately 1933, with modifications.

[b] All should be sieved. When used for established plants, add 1 part bone meal per 100 parts of soil.

[c] Depending on compactness of soil.

minance) should be at least 10 klx and preferably more. Light from fluorescent tubes (warm-white and daylight types seem to be somewhat superior to cool-whites) with an admixture of about (in wattage) 5% of light from incandescent lamps is very satisfactory. When grown in the greenhouse at relatively high latitudes, the plants may do rather poorly in winter unless given supplementary light.

A good temperature regime for the first 4 weeks after transplanting the seedlings is 27/19°C (8/16 hr) for the biennials and 23/19°C for the annuals. Thereafter, the biennials may be kept at 23/19°C; the annuals can be continued at this temperature, too, but they do even better in 19/19°C. Temperatures in excess of approximately 27°C, particularly if prevailing for prolonged periods of time, are distinctly less favorable, and it is high night temperatures that should be avoided in particular.[41] Lower temperatures, although quite good for growth, must be used with care as they may cause thermoinduction in biennial strains and lower the critical photoperiod in annual ones (see "Kinetics" in the section titled "Thermoinduction" and "Temperature Effects" in the section titled "Photoperiodic Induction", respectively) and thus result in premature flower initiation.

For many vernalization and photoperiodic studies the plants can be used when 4 to 6 weeks old; by this time they have reached maximal sensitivity to both thermo- and photoinduction. For defoliation and grafting experiments they should be preferably transplanted one more time, into 12- to 15-cm pots or 1-ℓ plastic containers, and grown until the tuber has reached a diameter of at least 1 cm. For easy handling in such kinds of experiments it is advisable to plant the plants so that most of the tuber is above the medium, with its base at the level of the edge of the container.

Quite in contrast to Hamlet's father and humans in general, aphids, red spider mites, and white flies (*Sitotroga*) are not sensitive to the juice of hebona and can cause abundant infestations, but if this is done early they are easy to control with the common insecticides and miticides. More serious are attacks by molds and by larvae of *Sciara;* they are particularly a problem with defoliated plants and with grafts, as these have to be kept at least temporarily at high humidities which are enjoyable for these pests. Fungal infestation can be combatted with the usual fungicides. The only means against *Sciara* larvae is hot water. The most serious problem in work with *Hyoscyamus* is the great susceptibility of the plant to tobacco mosaic and a variety of other viruses. If a virus infection is discovered (the first symptoms appear on the young, growing leaves in the center of the rosette and consist of a dull, somewhat grayish color, stunting, and a crumpled appearance) the only chance, if any, for salvaging the population is ruthless removal not only of the diseased plants but also of their neighbors. If there is a chance of contamination (e.g., because other virus-susceptible plants such as tobacco are grown nearby) it is advisable before handling *Hyoscyamus* cultures to wash one's hands with a detergent. Smokers should do this as a routine since cigarettes and other smoking material can be a serious source of contamination. This author, who doesn't smoke, performed a simple but highly successful experiment: he asked a smoker, after finishing a cigarette but before washing his hands to grasp a door handle; then, having washed his own hands, he grasped the same handle and rubbed the leaves of some *H. niger* plants. The infection was 100%.

Culture on Nutrient Agar

Hsu and Hamner[24] have cultured *H. niger* in test tubes on White's nutrient medium with 5% sucrose and 1% agar. This method was particularly designed for experiments in which the plants were exposed to long periods of darkness, under which conditions they do not survive without supplementary organic nutrition.

Grafting

As *H. niger* has been extensively used in grafting experiments, some words are in order on this technique.

When *Hyoscyamus* is to serve as stock, make a cut into the tuber, about two thirds or three quarters of its length and as close as possible (but not right through!) the growing point. When *Hyoscyamus* rosettes are serving as scions, the tuber is trimmed to a wedge; if an elongating shoot or a single leaf is to serve this function, the same is done with the stem or petiole. (It should be noted that rosette leaves of annual strains graft very readily but that rosette leaves of unvernalized biennial plants and any stem leaves do not.) Insert the scion, trimmed to a wedge, into the cut so that the cut portion of the scion is slightly below the upper edge of the cut in the tuber. Tie the two together quite firmly but without crushing the tissue, first at the top of the graft region, to hold the scion in place. The best material for the tie seems still to be old-fashioned raffia which apparently has the optimal combination of strength and resilience. Two or three ties are usually enough.

If biennial *H. niger* is to serve as receptor, the procedure of Melchers is to be highly recommended. The rosette is almost completely defoliated, leaving only the youngest leaf initials which surround and barely cover the apical meristem, and this degree of defoliation is maintained throughout the experiment, permitting direct, continuous observation of the response. It is important, particularly in the case of the young leaves, to remove all of each leaf, including the base, as the latter, if left on the rosette, will grow and may obscure the growing point.* Axillary shoots on the rosette axis which may grow out after defoliation of the rosette should also be removed regularly as they occasionally give erratic flowering responses. If the defoliated plant is used as stock, the cut for the scion is made quite close to the growing tip. Annual *H. niger*, if defoliated, forms flowers independent of photoperiod (see "Defoliating Experiments" in the later section titled "Inhibitory Processes and Factors") and if it is to serve as receptor, complete defoliation is obviously self-defeating. However, it is still advisable to remove most of the older leaves and any outgrowing axillary shoots.

In the course of the first 8 to 10 days the grafts should be kept in high humidity, either by maintaining such humidity in the entire facility in which the grafts are made, or by placing transparent plastic bags of an appropriate size over the scions and tying them around the stocks. They should also be protected from direct sunlight; this can be achieved by two or three layers of cheesecloth (heavier shading is disadvantageous).

For certain purposes, particularly combinations between two *Hyoscyamus* plants or a *Hyoscyamus* and another rosette plant, approach grafts may be preferable. For this, the plants are removed from their containers and half of the tuber of each partner is cut off, again as close to the growing points as practical and, if a biennial plant is to serve as receptor, after defoliation. Half of the entire root system is also cut off; the remaining halves are placed together in one container, bringing the half-tubers to the same level, and the latter are tied together. In this case, there is no need for high humidity and other precautions.

It is the experience of this author that perhaps with the exception of approach grafts, grafts are more successful in the greenhouse than under artificial light. In higher latitudes, however, grafts should not be made in the winter months.

GROWTH HABIT AND INFLORESCENCE MORPHOLOGY

Hyoscyamus niger is a typical pseudorosette plant. It passes the vegetative part of its life cycle in the rosette habit, making growth in thickness but none in length, but upon transition to reproductive development forms a stem with elongate internodes and terminating in an inflorescence. The plants possess a tuber which is derived from the hypocotyl and the upper portion of the root. They are usually monocarpic (hapaxanthic) but on rare occasions a new

* The leaf bases, except of the youngest leaf initials, contain a purple pigment. This tissue, and homologous tissue of the youngest initials, should be removed (e.g., by gentle scraping with a spear-headed dissecting needle), as it is capable of proliferation.

rosette may be formed from an axillary bud on the tuber and carry the plant into another growth cycle.

The inflorescence of *Hyoscyamus*, although often described as a spike, is a scorpioid cyme or cincinnus, i.e., a monochasium in which the successive branches (flowers) are arranged tridimensionally on alternating sides. The first branches can develop into secondary inflorescences, especially on vigorous plants which then carry several inflorescences attached very near each other. Each flower is supported by a well-developed, sessile bract. The flower is slightly zygomorphic. The fruit is a capsule which remains enclosed in the calyx and has a circumscissile dehiscence, as is typical for the whole genus.

CRITERIA FOR FLOWERING

For measuring the floral response, the following criteria have been used: percent of flowering individuals; days to incipient stem elongation which usually, although not always, is soon followed by the initiation of the first flower primordia; days to visible flower buds or to the first open flower; and number of leaves produced from the beginning of the experiment up to the first flower ("leaf increment"; the leaf number at the start of the experiment is determined on a representative sample of plants). Some investigators[20,25,63] have also used the number and/or developmental stage of flowers formed, but a carefully designed, highly reproducible quantitative method based on flower number or flower development, comparable, for example, to the stage systems in *Xanthium* or the score system in rye,[22a,60a] does not seem to have been elaborated. However, combined use of one of the temporal criteria and the leaf increment results generally in accurate and reproducible data. Since flower primordia may be initiated but develop only into quite small, incomplete buds it is important to examine the shoot apex of individuals with no visible flowers or flower buds, under the dissecting microscope.

One of the remarkable features of *H. niger*, first recognized by Melchers,[47] is that flower initiation can be studied directly at the shoot meristem with all leaves and most leaf primordia removed. The only precaution is a relatively humid (but by no means saturated) atmosphere and protection from direct sunlight. The drawings in Figure 2, showing the typical sequence of flower initiation and early development, in this case induced by a grafting partner (donor), were made from a single, defoliated plant at intervals of 24 hr. There are few plants that will stand a similar insult. The advantage of using defoliated receptors in grafts with biennial *H. niger* has already been mentioned. However, nonthermoinduced defoliated biennial plants with no grafting partner make only limited growth; for some unknown reason they seem unable to mobilize the storage materials in the tuber as efficiently and continuously as defoliated annual and thermoinduced biennial individuals.

EFFECTS OF PLANT AGE

Thermoinduction in biennial *H. niger* is not possible before an age of about 10 days. Thereafter, the responsiveness to low temperature increases rather sharply up to an age of 30 days and then remains unchanged for at least another 50 to 60 days[61] and very probably for indefinite periods of time. (As mentioned in the introduction to this chapter, the yellow-flowered biennial strain of Melchers initiates flowers without any cold treatment, although quite late. It is not clear whether its cold requirement has a genuine quantitative character or whether its thermoinductive temperatures range higher than those of the purple-flowered strain and induction is consummated, even though slowly, at temperatures already strictly noninductive for this latter strain.)

Systematic studies on the age dependence of photoinductivity in the annual strains do not

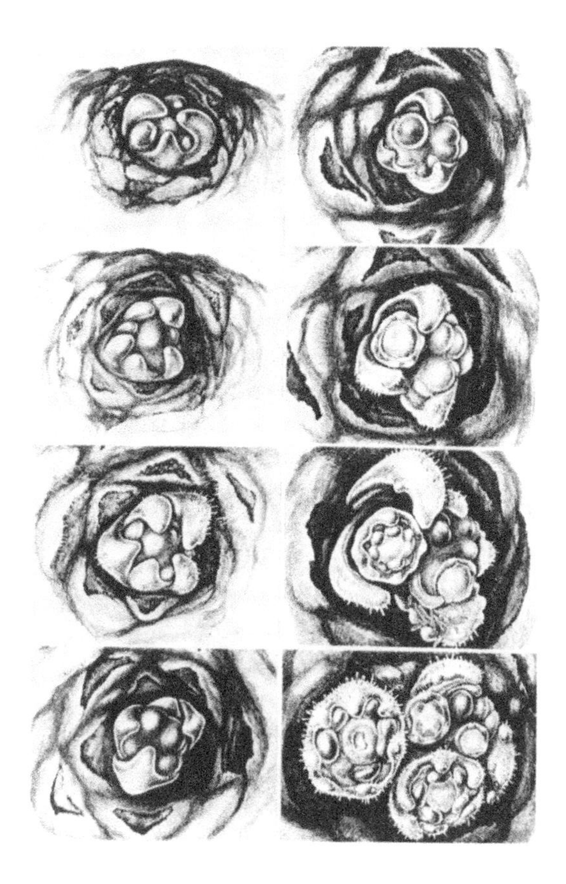

FIGURE 2. Transition from vegetative growth to reproductive development of the shoot meristem of a nonthermoinduced, defoliated biennial plant under the influence of a *Nicotiana tabacum* cv. 'Cavalla' donor scion. Drawings made 24 hr apart. The top left-hand panel illustrates the vegetative growing point. Continuing downward, the second panel on the left side shows that the growing point has enlarged. This is the first sign of transition to flower initiation, but at this stage the plant can still revert to vegetative growth. In the third panel, the growing point has divided, producing the first flower primordium (positioned at ca. 10 o'clock). The lower left-hand panel shows that the growing point has divided again, producing the second flower primordium (ca. 5 o'clock). Panels on the right illustrate the subsequent development: the growing point continues dividing, each time producing a flower primordium and a growing point which then undergoes the same type of division; the primordia develop into complete but still microscopic flower buds. (After Melchers, G., *Biol. Zentralbl.*, 57, 568—614, 1937. With permission.)

seem to have been made. Plants raised from germination in 15- to 16-hr LD needed 17 to 19 days for floral induction.[64] Since older plants can be induced by three similar LD[42] and, at least partially, by a single day of continuous light,[72] photoinductivity in annual plants seems to be reached about as early as thermoinductivity in biennial ones, at least in the Melchers strains.

Biennial *H. niger* is completely unresponsive to photoinduction until thermoinduced (Figure 3). When maximally thermoinduced, the photoperiodic behavior, although it has not been studied in much detail, seems to be quite similar to that of the annual strains.[41,43,49]

THERMOINDUCTION

Kinetics

Correns[12] had established the genetic basis of the biennial habit in *H. niger*. Melchers[47]

FIGURE 3. Biennial *H. niger*, nonthermoinduced (above) and thermoinduced (below) on LD (left) and SD (right). Drawings from photographs. (After Melchers, G. and Lang, A., *Biol. Zentralbl.*, 67, 105—174, 1948. With permission.)

Table 2
FLOWER FORMATION IN SHOOT APEX/TUBER GRAFTS IN BIENNIAL *H. NIGER*

Plants which supplied the . . .			Grafts in which apex formed flowers	
Apex	Tuber	No. surviving grafts	No.	%
Chilled	Chilled	35	30	57
Chilled	Unchilled	38	28	73
Unchilled	Chilled	37	0	0
Unchilled	Unchilled	30	0	0

Data from Melchers.[47]

showed that the physiological basis was cold requirement; i.e., that this was a case of vernalization. He also showed that defoliated plants do respond to thermoinduction (cold treatment) and proved, by grafting shoot tips of thermoinduced, defoliated plants onto the tubers of noninduced ones, and reciprocally, that flower formation occurred only if the shoot apex had been directly exposed to cold (see Table 2). Thus, the shoot tip (meristem and perhaps the youngest leaf primordia) is the main site of thermoinduction in *H. niger*. It should however be noted that defoliated plants respond to thermoinduction less well than intact ones. Also, if plants are given a suboptimal cold treatment and thereupon defoliated, they do not proceed to flower formation while intact ones do.[61]

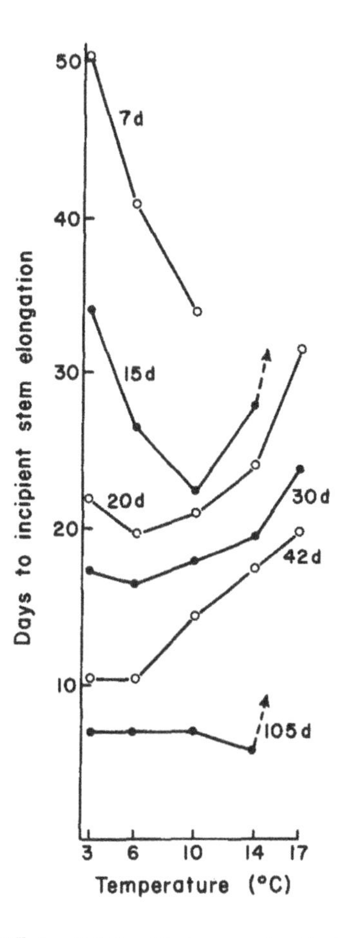

FIGURE 4. Relationship between length of thermoinduction, temperature, and flower formation in biennial *H. niger*. (From Lang, A., in *Encyclopedia of Plant Physiology*, Vol. XV (Part 1), Ruhland, W., Ed., Springer-Verlag, Berlin, 1965, 1380—1536. With permission.)

A quantitative study of thermoinduction in biennial *H. niger* has given the following results (Figure 4):[36]

1. Temperatures between 0 and approximately 17°C are thermoinductive. (*H. niger*, at least the strain used, does not stand freezing temperatures; this feature determines the lower limit of the inductive temperature range.)
2. The effect of temperature has a distinctly quantitative character, with the response (time to incipient stem elongation = flower bud initiation) decreasing with increasing length of the inductive treatment.
3. All inductive temperatures, except perhaps those very close to the upper limit of the inductive range, if applied long enough result in the same, maximal rsponse (FI within 6 to 7 days after the end of the treatment).
4. There is no constant optimum for the inductive temperature. The optimum shifts with extended periods of induction from relatively higher temperatures (10°C) to relatively lower ones (3 to 6°C).

Devernalization

The effect of thermoinduction can be nullified by exposing the plants to a relatively high

Table 3
DEVERNALIZATION IN BIENNIAL *H. NIGER*

Warm treatment	No. plants with flower buds	No. plants with no flower primordia[a]	Score[b]
None	11	0	4.5
30°C, 10 days	2	10	2.7
36—37°C, 10 days	0	11	1.5
36—37°C, 20 days	0	8	1.1

Note: Plants were thermoinduced for 50 days at 1 to 3°C; the warm treatment was given in the dark.

[a] Part of the plants exhibited some stem elongation and increase in size of the apical meristem; these are signs of transition to flower initiation.

[b] 1 = vegetative; 2 = incipient stem elongation, meristem slightly enlarged; 3 = distinct stem elongation, meristem enlarged; 4 = elongated stem, microscopic flower buds; 5 = elongated stem, visible flower buds.

Data from Lang and Melchers.[43]

Table 4
STABILIZATION OF THE VERNALIZED CONDITION IN BIENNIAL *H. NIGER*

1st period: no. short days at 22—24°	2nd period: no. short days at 38—39°	3rd period: no. short days at 22—24°	No. plants forming flowers	No. plants vegetative
0	10	15	0	5
1	10	14	2	4
2	10	13	1	5
4	10	11	7	0
8	10	7	6	0
15	10	—	2	0
25	—	—	4	0

Note: All plants were thermoinduced for 60 days at 3 to 5°C, then subjected to 25 SD at the temperature schedules shown; thereafter all were moved to continuous light.

Data from Lang and Melchers.[43]

temperature (30°C and above, see Table 3). However, this ''devernalization'' is complete only if the warm treatment is given immediately after the cold. If a period at an intermediate temperature (22 to 24°C) is intercalated, the devernalizing effect decreases rapidly and is soon lost; the vernalized state becomes ''stabilized'' (Table 4).[43]

The available information on vernalization in biennial *H. niger* can be interpreted, in a general manner, by a scheme suggested by Lang and Melchers:[43]

$$A \rightarrow A' \begin{array}{l} \nearrow B \\ \searrow B' \end{array}$$

where A is a ''precursor'', A′ is a ''labile intermediary'', and B is the ''stable end product''

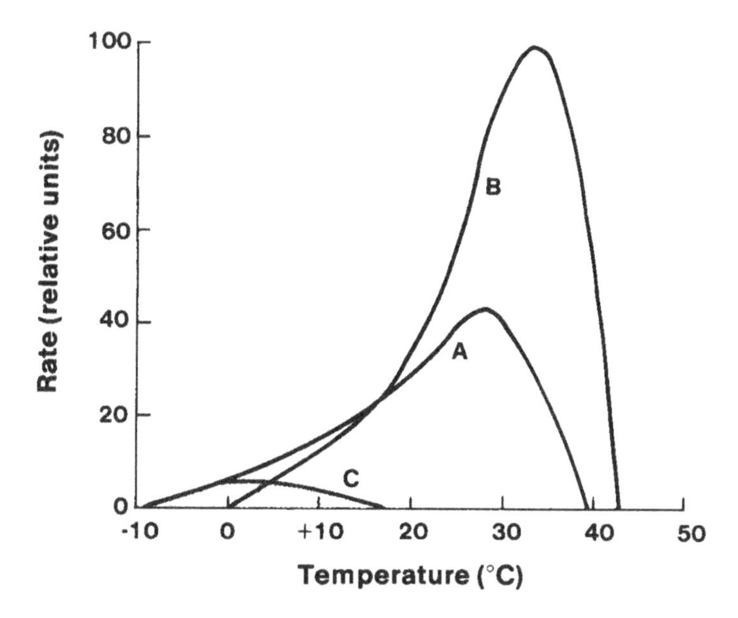

FIGURE 5. Diagrammatic illustration of thermoinduction as the balance between promotive and inhibitory processes differing in temperature minima, optima, and quotients. Only two processes are assumed:

Process	Relationship to floral induction	Minimum (°C)	Optimum (°C)	Q_{10}
A	Promotive	−10	28	2
B	Inhibitory	0	33	3

C is the difference between the two processes, = the temperature region where flower initiation will occur. (From Lang, A., in *Encyclopedia of Plant Physiology*, Vol. XV (Part 1), Ruhland, W., Ed., Springer-Verlag, Berlin, 1965, 1380—1536. With permission.)

of vernalization essential for flower initiation, while B′ is some inactive material. The individual processes in this scheme are assumed to differ somewhat in their temperature relationships, so that at lower temperatures some B is formed, while at higher ones the competing process A′ → B′ predominates, preventing any production of B (Figure 5). The scheme permits understanding thermoinduction on the basis of regular physiological or biochemical processes, rather than, for example, postulating a process with a negative temperature quotient. However, it should be stressed that A, A′, B, and B′ are physiological entities; nothing is known about the biochemical bases.

Once the thermoinduced state in *H. niger* has become stabilized it exhibits very substantial persistence. If thermoinduced plants are maintained on SD, in which no flowers are initiated, for as long as 190 days and then exposed to LD, they initiate flowers as rapidly as plants transferred to LD directly after the end of the cold treatment.[43,61] Only after more than 300 SD were there marked signs of a decline in the vernalized state.[61] One hundred and ninety days after the end of cold treatment all leaves and leaf primordia that were present during the treatment are gone.[61] Thus, the thermoinduced state is somehow maintained in the functioning apical meristem without undergoing any decrease for long periods of time, but apparently not indefinitely.

The production of a hormone-like, flower-inducing material (vernalin) as the result of thermoinduction of biennial *H. niger* is discussed in section "Flower Hormones (Florigen, Vernalin)", subsection "Grafts with Biennial *H. niger*".

Histology and Histochemistry

Histological and histochemical examination of the shoot apex of biennial *H. niger* during and after thermoinduction[16] showed that during the cold treatment, although growth was slow, the meristem increased markedly in size and the stem began to elongate, although only slightly. Nuclei and nucleoli in the meristem — both in the apical and the peripheral zone — and the surrounding young leaves became larger and stainability with pyronin increased, indicating an increase in RNA content. After transfer to the greenhouse, stem elongation increased but transition to flower formation (appearance of the first, terminal floral primordium) occurred only after 2 to 4 weeks; nuclear and nucleolar size and pyronin affinity were reduced. Unfortunately, the meaning of these changes with respect to floral initiation is far from clear. As the authors themselves point out the changes may be related to cold hardening since similar changes have been observed in other plants during acquisition of cold resistance. Apart from this possibility, the thermoinductive treatment, which consisted of daily cycles of 11 and 3°C, 12:12 hr and was given in light, was such that it would permit active growth of the plants, and the observed changes may be related to this, rather than to processes leading to floral initiation. Moreover, this regime was clearly suboptimal for floral induction since after optimal thermoinduction flower primordia appear much sooner — as already mentioned, within 6 to 7 days after the end of the cold treatment.

PHOTOPERIODIC INDUCTION

Localization

It is generally agreed that the main site of the perception of the light energy required for the photoperiodic regulation of flowering is the leaf — in most cases the young, fully grown one, in *Xanthium* the leaf having reached about half the mature length (see the chapter on *Xanthium*, in Volume IV[60a]) exhibiting maximal sensitivity. In *Hyoscyamus,* the sensitivity of leaves of different age has not been studied, but the fact that a single mature leaf of annual *H. niger* can serve as an effective donor for receptors of biennial *H. niger*[48] and 'Maryland Mammoth' tobacco[51] indicates that in this plant, too, the leaves are the main sites of photoperiodic induction.

Moreover, this author has been able to obtain induction in detached leaves of annual *H. niger,* by exposing them to LD and then grafting onto nonthermoinduced, defoliated biennial receptors.[41] This was possible only when the detached leaves were maintained at a relatively low temperature (14°C) and natural daylight (extended by light from incandescent lamps); leaves kept at 19°C and higher or in artificial light throughout (ca. 6 klx) could not be grafted. Of the 14°C leaves, graft union with the receptors was obtained in only 19 out of 50 grafts; and of the 19, flower initiation in the receptors was induced in only 6 while in the others the grafted leaf underwent rapid senescence. However, in grafts with leaves treated in the same manner but in SD, in 17 "takes" no case of flower induction was obtained, although the condition of the grafted leaves was equal to that in the grafts with LD leaves. It would be interesting to see whether these results can be improved by treating the detached leaves with cytokinin, which is known to delay senescence in detached leaves of many plants.

Length and Number of Cycles

Figure 6 illustrates the dependence of flower formation of annual *H. niger* on the daily light-dark regime. Flower formation occurs with light periods of 10 hr and 40 min and above; the critical photoperiod is between 10 hr and 10 hr and 40 min. With longer light periods the response is at first greatly accelerated; however, this effect becomes rapidly smaller, and above 16 hr there is but little further acceleration.[42] It should be noted that this response curve, and particularly the position of the critical daylength, is valid only for the

Table 5
DEPENDENCE OF PHOTOINDUCTION IN ANNUAL *H. NIGER* ON DURATION AND NUMBER OF PHOTOPERIODS

	12-hr days		16-hr days		24-hr days	
No. cycles	No. plants induced[a]	Leaf increment	No. plants induced[a]	Leaf increment	No. plants induced[a]	Leaf increment
1	—	—	0	∞	0	∞
2	—	—	—	—	0	∞
3	—	—	1	8.0	5	6.4
4	—	—	4	7.8	5	7.0
6	0	∞	5	8.2	5	6.8
8	2	7.5	—	—	—	—
10	5	10.0	5	7.4	5	7.6
15	5	11.8	5	7.0	—	—

[a] Out of five.

Data from Lang and Melchers.[42]

conditions of the experiment (light from incandescent lamps and a temperature of approximately 22°C). If light predominantly from fluorescent lamps is used, the critical daylength seems to lie in the range of 12 to 13 hr (see data in References 24, 56, and 64).

Table 5 shows the dependence of the response on number of LD and the length of photoperiod.[42] Two important facts emerge:

1. Fewer cycles are needed for floral induction with longer photoperiods. In the experiment of Table 5, the minimum was 3 days of continuous light, but in other experiments some flower initiation in some plants was achieved with a single continuous-light day.[72]
2. Once the minimum inductive number of photoperiods of a given length has been given, the response (measured as leaf increment) does not improve.

The first fact serves to explain the photoperiodic response curve of the plant: it reflects the minimal number of cycles of a given length needed to consummate floral induction (see Figure 7). The second fact points up a marked difference with vernalization. The effect of thermoinduction was clearly quantitative. The effect of photoinduction has — as far as the formation of the first flower primordium is concerned — a qualitative or all-or-none character: photoinduction appears to consist in the accumulation of something which has to attain a threshold level; once this threshold has been attained further accumulation is without effect. The "something" is probably florigen (see "Grafts with Annual *H. niger*" in the section titled "Flower Hormones (Florigen, Vernalin)".

The floral response in light-dark cycles longer than 24 hr will be reviewed in connection with the question of endogenous rhythms (see the subsection "Endogenous Rhythms" below.)

Light Energy and Night Breaks

The importance of the level of light energy for photoperiodic induction in both LDP and SDP has mostly been studied by giving the plants a "main" period, shorter than the critical daylength, of high-energy light and either following it with relatively long periods of light ("supplementary light") or interrupting the dark periods with relatively short periods of

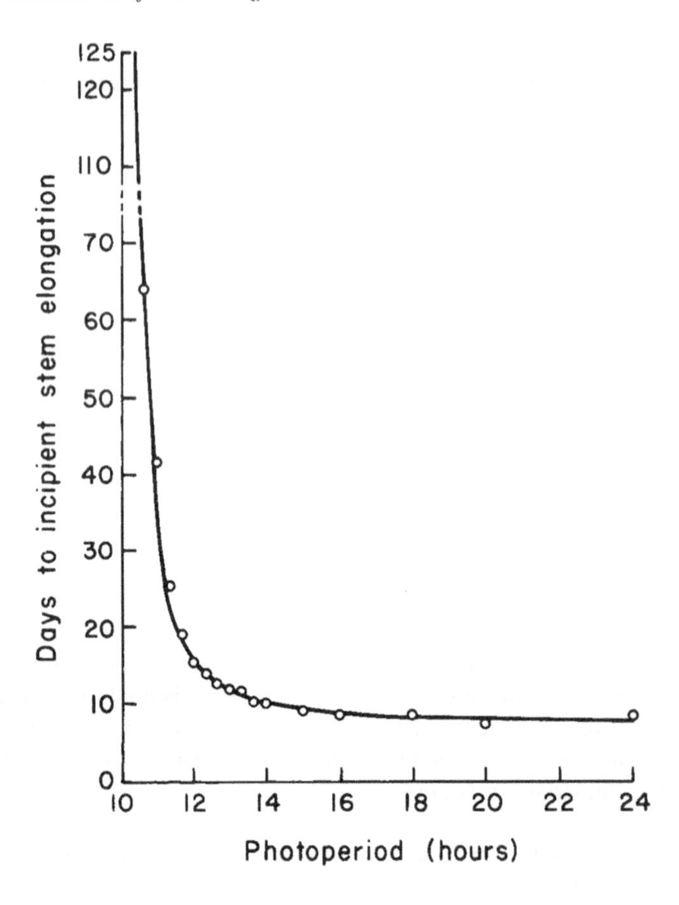

FIGURE 6. Photoperiodic response curve of annual *H. niger*. (From Lang, A., in *Encyclopedia of Plant Physiology*, Vol. XV (Part 1), Ruhland, W., Ed., Springer-Verlag, Berlin, 1965, 1380—1536. With permission.)

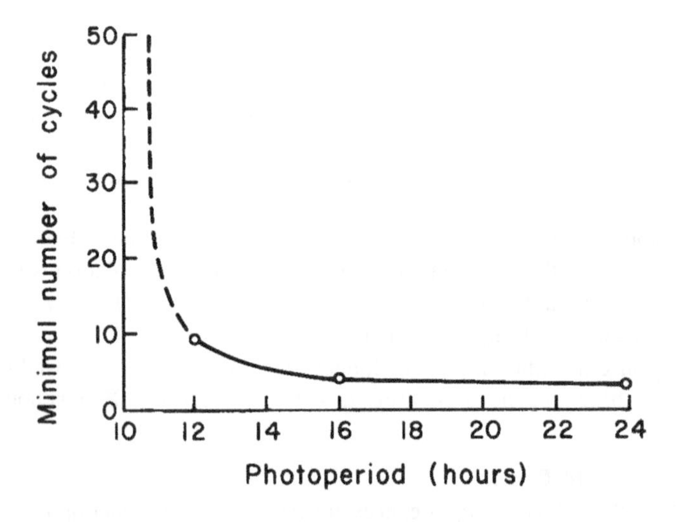

FIGURE 7. Relationship between length and minimal number of inductive photoperiods necessary for floral induction in annual *H. niger*. (From Lang, A., in *Encyclopedia of Plant Physiology*, Vol. XV (Part 1), Ruhland, W., Ed., Springer-Verlag, Berlin, 1965, 1380—1536. With permission.)

light ("night breaks"), in either case usually at lower energy levels. In annual *H. niger*, however, the importance of light energy in induction by 2 to 4 days of continuous (white) light has also been investigated.[72] With energy fluence rates of 0.16 and 0.32 W m^{-2}, the plants remained vegetative; with fluence rates of 1.6 W m^{-2} and above, and also with increasing length of the treatment, the response — measured as percent of plants initiating flowers — increased, reaching 100% after 4 days at 6.4 and 16 W m^{-2}. Thus, not very surprisingly, some light of high energy is necessary for the consummation of photoperiodic induction in *H. niger* and, undoubtedly, in other photoperiodic plants. It is most probably based upon the need of photosynthate and/or reducing power generated by photosynthesis. Photosynthesis, measured as ^{14}C incorporation, at 0.16 and 0.32 W m^{-2} was less than 1% of that at the highest fluence used.[72]

When a SD main light period is extended with light (from incandescent lamps), a light flux as low as 0.6 lux or a fluence of 0.032 W m^{-2} given throughout the night period is sufficient to induce some flower initiation in *H. niger*. A flux of 10 lux or a fluence of 32 W m^{-2}, if given for enough cycles, is saturating or nearly so.[10,72] The reciprocity law does not seem to hold: long exposures at relatively low light energies are more effective than short exposures at relatively high ones.

Night breaks with white light also induce flower initiation, and the latter can then occur with total periods of light per 24 hr that are far below the critical daylength of the plant; e.g., with 5 + 1 = 6 hr of light.[10] The effect increases with the duration of the NB, the duration of the main light period, and the light flux used during the break,[56] and it depends very much on the time of the NB, with breaks given a certain time after the start of the main light period (about 15 to 16 hr) being much more effective than breaks given earlier or later. This apparent variation of the effectiveness of light accounts without doubt for the seeming lack of reciprocity in the effectiveness of supplementary light starting immediately after the main light period: as these periods are extended, light impinges on the plant during the period of increasing light responsiveness. If light is given during the period of its maximal effectiveness, reciprocity does hold,[56] at least above certain energy levels.

Light Quality

Annual *H. niger* was one of the first LDP for which a precise photoperiodic action spectrum was established, by Parker, Borthwick, and Hendricks at the U.S. Department of Agriculture laboratories at Beltsville, Md.[56] In plants receiving 12-hr photoperiods and NB in the period of maximal effectiveness of the latter, flower formation was promoted mainly in the R, with a broad optimum at about 630 nm, and to a lesser extent also in the B. Although because of energy limitations it was not possible to establish a continuous response curve for the blue region, the action spectrum was very similar to the action spectra established somewhat earlier for two SDP (*Xanthium* and 'Biloxi' soybean) and for another LDP, the barley cv 'Wintex', and was characteristic for phytochrome-mediated responses. Somewhat later, Downs[17] showed that the effect of R night breaks could be at least partly reversed by exposing the plants, immediately after the R, to FR light — another very characteristic feature of phytochrome responses. Thus, it is evident that phytochrome participates in the absorption of the light energy involved in the photoperiodic regulation of flower formation in *H. niger*. Phytochrome has been shown to be present in extracts of soluble proteins from leaves of annual *H. niger*.[33a]

From later work it appears that Parker and associates and Downs had been blessed with a good deal of serendipity. Under the conditions they were using, the plants were evidently close to the critical daylength, and in this situation both the phytochrome action spectrum and the R-FR reversibility can be demonstrated in *H. niger* rather unequivocally. Later work, carried out at Beltsville,[33,62] at the Agriculture University of Wageningen in the Netherlands[13-15,20,66] and some other laboratories,[26,72] has yielded quite complicated and in part contradictory results. It appears that (1) the photoperiodic response of *H. niger* depends on the length of the main light period and on the spectral composition both of the light in the main light period and of the supplementary or the NB light; (2) that there are complex interdependencies between the two; (3) that the effect of light of a given quality may change in the course of the daily cycle; and (4) that the sequence of different light treatments may also be important for the response. To give some examples, while short R night breaks promote flower formation with 12- or 12.5-hr main light periods, they are not effective if the latter are reduced to 8 hr/day; with such more extreme short days, NB of several hours are required in order to obtain a floral response, and R as well as FR may be promotive.[33] The effectiveness of R seems to increase, and that of FR to decrease as the night progresses.[20] *H. niger* plants kept on LD in light of different quality flowered in the B and violet but not in the R or green.[20,66] Admixture of either FR or B to the R resulted in a rapid response; the effect of the B admixture did not appear explainable by contamination with FR.[15] Generally, light in the main light period is particularly effective in promoting the floral response when it is high in FR.[26,33] If plants are given daily cycles with 8 hr of white light followed by some FR (30 min or above) they form flowers even though the total duration of light is below the critical daylength,[14,33,66] whereas they remain vegetative in 8 hr of white plus 8 hr of R per day.[14,15] The sequence white light-dark-R is more effective in inducing flower formation than the sequence white-R-dark.[33,62]

Some authors have interpreted these and other results to mean that for photoperiodic induction of flowering in *H. niger* to occur, phytochrome must be cycling between the R- and the FR-absorbing form. They have also suggested that whereas at lower light fluxes the photoperiodic response of this plant is mediated by phytochrome, a high-irradiance response comes into play at higher fluxes, i.e., that photoperiodic regulation of flower formation in *H. niger* may involve two light-absorbing "systems".[62]

The precise meaning of those entire results, however, is difficult to assess for several reasons. The various experiments were done, in part, under different conditions, and with different light sources (including sunlight passed through color filters[26]). The spectral composition of sunlight undergoes only quantitative changes but there are considerable changes in the R/FR ratio, both in the course of the day and depending on the degree of shading (in canopies). Thus, use of "pure" spectral regions — even if rather broad ones — would seem a rather unphysiological approach, and experiments with light sources producing both R and FR but at different ratios (see, for example, References 18, 33, and 62) would appear to be more meaningful. In some of the experiments, stem elongation rather than flower initiation served as the main parameter for floral induction but while the two processes are closely coordinated under natural conditions, they can proceed separately under some experimental ones. Where no elongation or no visible flowers were found it is not always clear whether the plants were examined for the presence of microscopic flower buds or primorida. The most important source of ambiguity, however, may have been that the treatments were applied until the plants, or at least some of them, did form flowers. In a recent very interesting and penetrating study Downs and Thomas[18] have shown that *H. niger* exhibits a marked difference in the optimal light regimes for floral induction; i.e., the inductive processes prior to the actual formation of flower primordia, and the development of the initiated flowers (Table 6). When plants were given 10 LD of white light plus 5 or 20 min of FR at the end of each light period, stem elongation and flower initiation were affected but little. But when the plants were given 10 *more* LD, 5 min of FR at the end of the light periods very markedly promoted either response (Table 7) and this effect was abolished by 5 min of R given immediately after the FR. When used to supplement a period of 8 hr of daylight, light from fluorescent lamps — which is high in R and thus established a high Pfr level in the plants — not only delayed anthesis but reduced fruit set. Thus, whereas a high level of Pfr in the plants (leaves), particularly near the middle of the dark period, is, as shown by the early

Table 6
EFFECT ON FLOWERING OF *H. NIGER* OF FR GIVEN AT THE CLOSE OF EACH 16-HR PERIOD OF LIGHT FROM FLUORESCENT LAMPS

FR (min)	Period of FR treatment (days)	No. plants flowering (out of 7)	Mean stage of flower	Stem length (mm)
0	None	2	0.3	6
5	1—10	4	1.1	9
20	1—10	5	1.4	9
5	10—20	7	3.6	24
20	10—20	7	3.6	24

From Downs, R. J. and Thomas, J. F., *Plant Physiol.*, 70, 898—900, 1982. With permission.

Table 7
FRACTIONAL INDUCTION IN ANNUAL *H. NIGER*

	9.5-hr SD			6-hr SD		
Treatment	No. plants forming flowers	No. plants vegetative	Leaf increment	No. plants forming flowers	No. plants vegetative	Leaf increment
6 16-hr days, uninterrupted	6	0	—			
2 16-hr days, uninterrupted	0	6	—			
3 × 2 16-hr days, interrupted by						
2 SD	6	0	—[a]	6	0	—[a]
5 SD	6	0	8.2	6	0	4.8
10 SD	6	0	10.9	6	0	7.0

[a] Not determined.

After Lang, A. and Melchers, G., *Planta*, 33, 653—702, 1943. With permission.

data,[17,56] favorable for induction in *H. niger*, a lower level is more favorable for further flower development up to fruit formation. Obviously, disregard of such differences in the spectral requirements of different stages of the flowering process as a whole in this or any other plant may easily result in a confusing picture.

Temperature Effects

When annual *H. niger* was exposed to photoperiods of different lengths under different temperatures, the critical daylength was found to decrease with increasing temperature, at least in the range tested; it was approximately 8.5 hr at 15.5°C and approximately 11.5 hr at 28.5° (Figure 8).[42] This effect was mainly dependent on the dark-period temperature and may be interpreted as a result of slowing down inhibitory processes taking place in the dark periods (see "Kinetic Considerations" in the section below titled "Inhibitory Processes and Factors"). There was, however, also a temperature effect in the light periods, the floral response being greater at 21 to 23°C as compared to 14 to 18 and 27.5 to 29°C.

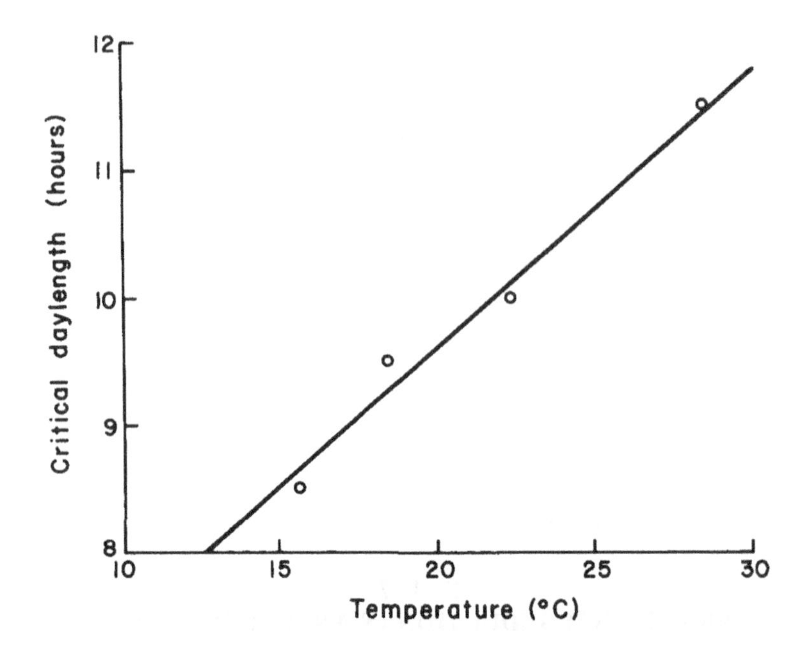

FIGURE 8. Dependence of the critical photoperiod on temperature in annual *H. niger*. (From Lang, A., in *Encyclopedia of Plant Physiology*, Vol. XV (Part 1), Ruhland, W., Ed., Springer-Verlag, Berlin, 1965, 1380—1536. With permission.)

Wagenaar[69] found that when *H. niger* plants were given short light periods extended with low-intensity supplementary light the floral response was reduced by low temperatures during the supplementary-light and the dark periods, although this effect became relatively smaller as the duration of the inductive treatment was increased. When returned from low to higher temperatures the plants exhibited a burst of rapid growth.

Endogenous Rhythms

Search for the participation of an endodiurnal or circadian rhythm in the photoperiodic response of *H. niger* (all this work was done using an annual strain) has followed two approaches which have also been used in analogous work with other plants: the plants were either subjected to cycles consisting of the same light period but dark periods of different length (that is, cycles with a constant light period but of different total length), or they were subjected to long cycles consisting of constant light and dark periods, but were given a light break at different times of the latter.

Either approach has produced one somewhat surprising finding: as the length of the cycle is increased, at least up to a certain limit, the minimal length of the light period needed for floral induction decreases. Thus, *H. niger* plants grown on 33- to 48-hr cycles formed flowers with 8, 9, and 10 hr of light per cycle whereas on 24-hr cycles they required around 11 hr or more.[10,11,21,24]

Either approach has also shown that the floral response of *H. niger* may exhibit cyclic variations depending on the length of the cycle or the time of the light break, with maxima and minima spaced about 24 hr apart.[10,11,21,24] However, this effect is extremely variable. Thus, when cycles of different length (up to 72 hr) with light periods of 6 hr were used, the floral response was relatively high in 18-, 30- to 42-, and 54- to 66-hr cycles, and low or nonexistent in 24- (i.e., ''normal'' short days), 48-, and 72-hr cycles. With light periods of 12 hr, the response was depressed in 24- and 36-hr cycles, but exhibited no cyclic variations; with light periods of 18 hr it was depressed in 36- and again, but only slightly, in 60-hr cycles.[21] With light breaks at different times in the dark period of 72-hr cycles, the middle period of flower promotion was much less conspicuous than the early and the late ones.[11]

When 3-hr periods at 5°C were given to *H. niger* plants kept on 12-hr days, during the 6th to the 9th hour of the dark periods flower formation was promoted while the same cold breaks given from hour 1 to 3 or hour 9 to 12 resulted in a decreased response. Warm breaks (35°) had the opposite effects.[63]

In conclusion, it appears that, under some specialized experimental conditions, circadian rhythms can be shown to affect the photoperiodic response of *H. niger*. These effects cannot, however, be considered as proof that these rhythms are the time-measuring mechanism and thus an integral part of the photoperiodic "machinery" in this plant. It is of greater interest for understanding this machinery that, while some of the "odd" light-dark cycles can induce a flowering response where none occurs in 24-hr cycles, responses clearly and consistently superior to that in continuous light have not been obtained.

Persistence and Reversal

The persistence of the photoinduced "state" in *H. niger* has two different and to some extent contradictory aspects. On the one hand, it has been shown, by "fractional induction" experiments, that the changes which are caused in the plant by exposure to LD are maintained for quite extensive periods of time. Thus, 2 LD are not sufficient to induce flower formation in annual *H. niger* (it goes without saying that such statements refer only to the strain used in, and the conditions of, the particular experiment), but 6 LD are. If 6 LD are given in pairs, separated by SD, the plants initiate flowers and there is no apparent decrease in response, at least up to periods of 10 SD (Table 7; the increase in leaf number with increasing numbers of SD is because the plants continue to form leaves during the SD periods). This remains so even if the SD are made more extreme, with only 6 hr of light per day.[42]*

On the other hand, if a *H. niger* plant is given inductive photoperiods long enough to produce visible flower buds but then transferred to SD, only the first few flowers develop normally, through fruit set and seed production. The following ones are increasingly under-developed and remain sterile (Figure 9). The meristem may, however, continue to initiate flower buds, which macroscopically consist of a small calyx only, for extended periods of time (this author has observed this to happen for up to 3 months) although it may ultimately return to vegetative growth, producing an aerial rosette (Figure 10).[41] It seems that the induced state created in *H. niger* plants by LD, while quite persistent, is not sufficient to maintain an extended normal floral response if the plant is transferred to SD, and may ultimately be lost.

Histological Changes

As in many other plants, the vegetative shoot meristem of *H. niger* has a so-called quiescent central zone, occupying the uppermost portion of the meristem and characterized by low mitotic frequency and relatively vacuolated cells. As a rule, this zone is "activated" (i.e., the cells undergo more frequent mitoses and divisions, become more typically meristematic, and lose their distinction from the cells of the surrounding "peripheral zone") when the plant prepares to undergo flower initiation. In annual *H. niger*, however, the central zone becomes activated if the plants are left on SD long enough, although they of course remain

* Carr,[4] while otherwise confirming and somewhat extending these findings, has found that the floral response in *H. niger* plants given 6 single LD each separated by 12 SD — 6 × (1 + 12) — is considerably less than in plants given three pairs of LD separated by the same number of SD — 3 × (2 + 12). He has placed heavy emphasis on this datum to conclude that the "stimulus" generated in *H. niger* in LD does not persist indefinitely. However, the total length of this treatment (78 days) is a good deal longer than the life span of a *H. niger* leaf. Thus, no leaf of the 6 × (1 + 12) plants received the full 6 LD; the weak response of these plants is because they never reached the full induced state that can be established by this many LD, not because they had lost this state.

FIGURE 9. Reversal of floral induction in annual *H. niger*. Top, inflorescence of a plant on continuous LD; bottom, inflorescence of a plant which was kept on LD until the appearance of the first visible flower buds and then transferred to SD. (A. Lang, unpublished material.)

vegetative.[64] This finding raises doubts about a close connection between activation of the central zone and flower formation, at least in this plant.

FLOWER HORMONES (FLORIGEN, VERNALIN)

In both biennial and annual *H. niger*, flower formation can be obtained in plants maintained on strictly noninductive conditions ("receptors") by graft union with a "donor" either induced to or capable of flower formation.

FIGURE 10. An annual *H. niger* plant that had been transferred from LD to SD as the plant in Figure 9 (bottom), and has continued initiating flowers which, however, failed to develop beyond small, rudimentary buds. The inflorescence branch on the right has now turned into a vegetative aerial rosette. (A. Lang, unpublished material.)

FIGURE 11. Grafts between biennial *H. niger* plants. Graft on the left-hand side: thermoinduced donor (right), nonthermoinduced receptor (left). Graft on the right-hand side: two nonthermoinduced individuals (control). (From Melchers, G., *Biol. Zentralbl.*, 57, 658—614, 1937. With permission.)

Grafts with Biennial *H. niger*

In nonthermoinduced *H. niger* receptors, flower formation has been accomplished with the following donors: vernalized biennial *H. niger* (Figure 11); photoinduced annual *H. niger; H. albus; Petunia hybrida; Nicotiana silvestris;* and *N. tabacum* cvs 'Java', 'Cavalla',

Table 8
LENGTH OF CONNECTION WITH THE RECEPTOR NEEDED FOR FLOWER FORMATION IN THE DONOR IN GRAFTS WITH BIENNIAL *H. NIGER*

No. of days after grafting (donor removed)	No. receptors forming flowers	Days to flower initiation	No. receptors vegetative
∞[a]	5	8—10	1
6	5	8—10	2
5	4	8—10	1
4,3,2	None	—	All

[a] Not removed throughout experiment.

Data from Melchers.[48]

and 'Maryland Mammoth'.[47,48,51] Except for biennial *H. niger* itself, all these plants are without cold requirement; *N. silvestris* is an obligate LDP, 'Java' and 'Cavalla' tobaccos are day-neutral, and 'Maryland Mammoth' is an obligate SDP. In several cases (annual *H. niger, N. silvestris,* and 'Cavalla', 'Java', and 'Maryland Mammoth' tobacco) a single leaf has been sufficient as a donor.[48,51] The response is maximal if the donor is grafted quite close to the growing point of the receptor and flower buds developing on the donor are removed. Only one of the species tried, the tomato, failed to function as donor, despite good graft union as judged by the continued growth of the tomato scion.[47]

In grafts between donors and receptors of biennial *H. niger* it was shown that as little as 5 days of union between the graft partners suffices to obtain a near-maximal floral response (see Table 8).[48] While no detailed histological study has been made, it is quite improbable that this is enough time to establish extensive union of the vascular systems of donor and receptor and to move large quantities of nutrients or other nonspecific metabolites from the former to the latter. On the other hand, some tissue union is necessary. Attempts at obtaining flower formation in nonthermoinduced biennial *H. niger* by donors from an annual strain, while keeping donor and receptor separated by means of filters of different pore size, were entirely negative; whenever an occasional receptor did form flowers, the filter had been ruptured and tissue union had occurred.[53]

Grafts with Annual *H. niger*

In receptors of annual *H. niger,* flower formation has been obtained with annual *H. niger* itself, *Nicotiana silvestris* (Figure 12), 'Maryland Mammoth' tobacco (Figure 13), and the day-neutral tobacco 'Trapezond' as donors.[27,41,44] In turn, flower formation has been obtained with annual *Hyoscyamus* donors in receptors from *N. silvestris* and 'Maryland Mammoth' (in the latter plant with single leaves), and has been hastened in 'Trapezond' receptors.[27,41,44] In grafts between annual *H. niger* and 'Maryland Mammoth' tobacco, in which either partner was maintained on its noninductive photoperiod, no flower formation took place (Figure 13).[27,41,44] These results are reviewed in more detail in the chapter "*Nicotiana*" (Volume 6) where they can be considered in a broader context. They constitute a substantial part of the evidence for the existence of flower hormones. They also show that the flower hormones of LD, SD, and DNP are freely interchangeable between all these photoperiodic response types and between graft partners belonging to the same species, to different species of the same genus, and to species of different genera; i.e., they are most probably identical.

Vernalin and Florigen

While the flower hormones of the different photoperiodic plants can be considered as

FIGURE 12. Grafts of annual *H. niger* receptors (left-hand partners): left, with an annual *H. niger* donor; right, with a *Nicotiana silvestris* (LDP) donor. (A. Lang, unpublished material.)

identical, the hormone first demonstrated in biennial *H. niger* seems to occupy a separate place. In order to test the relationship of this hormone and that demonstrable in annual plants, Melchers made grafts between nonthermoinduced biennial *H. niger* plants and shoots or single leaves of 'Maryland Mammoth' tobacco, and maintained them in either SD or LD conditions.[48] He found that the *Hyoscyamus* partners formed flowers under *either* photoperiod (Figure 14), i.e., even when the "donor" was under noninductive conditions. This finding has been confirmed by Chailakhyan.[6] In reciprocal grafts, in contrast, 'Maryland Mammoth' receptors are induced to form flowers only by thermoinduced biennial *H. niger* donors (Figure 15).[49,54]

These findings, in their entirety, lead to the following ideas:

1. There are two distinct flower hormones;
2. One of these is formed in cold-requiring plants only upon thermoinduction, but is present in non-cold-requiring ones independent of a special temperature regime. The other is formed in photoperiodic plants only under inductive daylengths while in day-neutral plants it is formed independent of photoperiod; this hormone is identical in LD, SD, and day-neutral plants; and
3. The former hormone is, in a *physiological* sense, a precursor of the latter.

FIGURE 13. Response of annual *H. niger* in grafts with the SDP *Nicotiana tabacum* 'Maryland Mammoth'. Left, 'Maryland Mammoth' partner on SD; right, on LD. (A. Lang, unpublished material.) Similar experiments in Khudairi and Lang[27] and Lang and Melchers.[44] (Courtesy of A. Lang.)

Chailakhyan[5] had named the flower hormone demonstrated in grafts with photoperiodic plants *florigen*. Melchers[48] named the one demonstrated in grafts with cold-requiring plants *vernalin*.

In the earlier experiments[48] no difference in the time-to-flower formation in biennial *H. niger* in grafts with 'Maryland Mammoth' had been noted, regardless of whether the grafts were kept on SD or LD. However, in later work, Melchers[49] did find a significant difference; SD-treated (that is, induced) 'Maryland Mammoth' donors caused an earlier response in the receptors than LD-treated (that is, noninduced) ones. This is obviously in agreement with the third item in the list above: if LD 'Maryland Mammoth' donors are used, they supply only vernalin, and florigen has to be produced in the receptors once the latter have received vernalin; SD 'Maryland Mammoth' supply florigen directly. The fact that biennial *H. niger* does not respond to photoinduction before being thermoinduced is, of course, also in full agreement with this idea.

If the idea about the distinct nature and the relationship of florigen and vernalin is valid, one might expect that flower formation in nonvernalized biennial *H. niger* would also be caused by annual *H. niger* donors kept in SD, but experiments undertaken to see whether this was so gave negative results.[51] These experiments were, however, done before it had been recognized that the leaves of SD-treated annual *H. niger* exert an active inhibition on flower induction (see the later section titled "Inhibitory Processes and Factors"). Thus, the negative outcome of the experiments is no longer surprising and does not vitiate the validity of the interpretation based on the biennial *H. niger*/'Maryland Mammoth' grafts. Moreover, when the graft partners were separated 12 days after the graft, 50% of the biennial receptors did initiate flowers,[49] indicating that it may be possible to separate vernalin transfer from and floral inhibition by noninduced annual *H. niger*.

Transport

When annual *H. niger* plants were defoliated except for two leaves, one of which was then given LD and the other SD (the latter is necessary as completely defoliated annual *H. niger* forms flowers also in SD; see "Defoliation Experiments" in the later section "Inhibitory Processes and Factors"), and the petiole of the LD leaf was subjected to low or high temperature by passing water of 5 or about 45°C through copper tubing coiled around part of the petiole, flower initiation was retarded compared to untreated control plants.[41] Thus, transport of the flower-promoting material from the leaf to the shoot apex is reduced by both sub- and supraoptimal temperatures, and it is thus probable that it occurs in the phloem. Analogous extensive experiments have been conducted with several SDP, but to the best of my knowledge with no other LDP.

While most grafts with biennial *H. niger* receptors were done with the latter serving as stocks, the results are the same if they are used as scions (Figure 16). In the grafts between nonthermoinduced biennial *H. niger* and nonphotoinduced 'Maryland Mammoth' tobacco, either partner has been used both as stock[48] and as scion,[6] with about the same success. Annual *H. niger* functions equally well both as donor and receptor whether used as stock or scion. Thus, both vernalin and florigen move readily in both an upward and a downward direction, also supporting the idea of phloem transport which can proceed in either direction, depending on the spatial relationship of source and sink.

Florigenic Extracts

Attempts have been made, using both annual and biennial *H. niger,* to obtain extracts from induced plants that would cause flower formation in noninduced ones. The extracts were prepared by means of various organic solvents, but it was also tried to "flush out" the active material by forcing water through the plants by means of a vacuum. Subsequent purification and fractionation procedures were rather crude by modern standards, and the experiments may therefore be considered less than conclusive, but the fact remains that the results were distressingly negative.[51]

GROWTH SUBSTANCES AND GROWTH REGULATORS

Exogenous Applications: Gibberellin, Growth Retardants

H. niger was the first plant in which it was found that the natural inductive conditions for flower formation can be "substituted" by treatment of the plants with gibberellin (as a rule, gibberellic acid was used, and application was to the shoot apex). This holds for both the cold requirement in the biennial form and the LD requirement in the annual one.[38] In biennial plants (no such experiments were done with the annual ones) the same effect was obtained by treatment with the liquid endosperm of *Echinocystis macrocarpa* (*Marah macrocarpus*), a material very high in native GA, or with an extract thereof.[45] Nonthermoin-

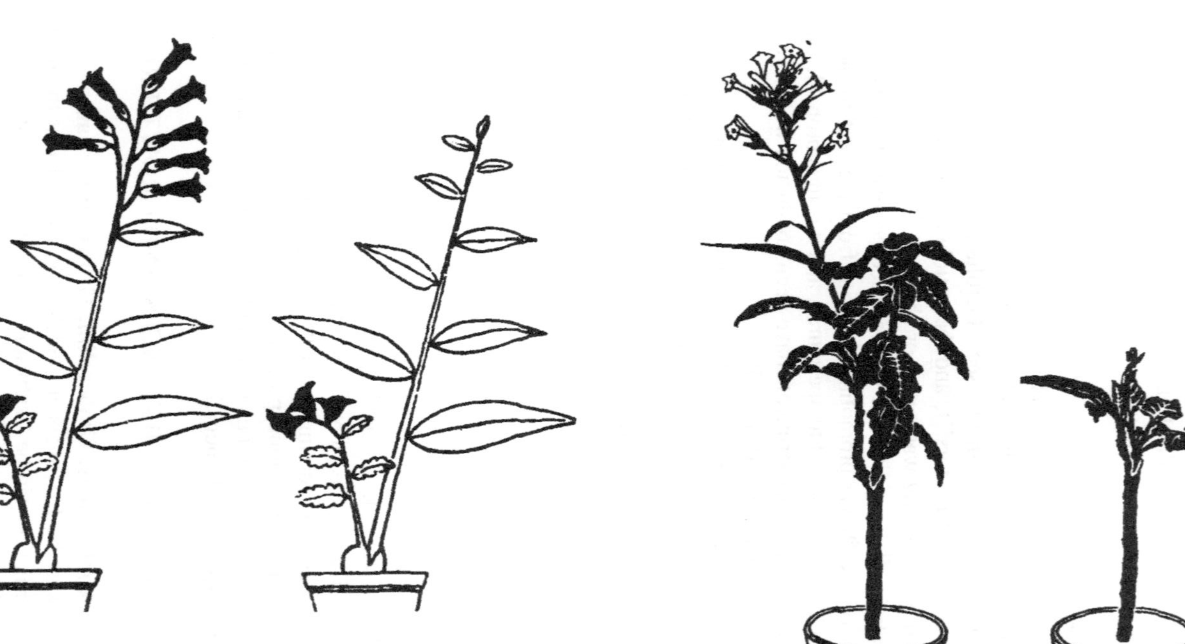

FIGURE 14. Flower formation in nonthermoinduced biennial *H. niger* induced by 'Maryland Mammoth' partners kept in SD (left) and LD (right) conditions. Diagrammatic; in the actual grafts, the *Hyoscyamus* receptors were defoliated and flower formation was observed at the shoot meristem, and in many cases, a single 'Maryland Mammoth' leaf was used as donor. After experiments of Melchers.[48] (From Lang, A., in *Encyclopedia of Plant Physiology*, Vol. XV (Part 1), Ruhland, W., Ed., Springer-Verlag, Berlin, 1965,

FIGURE 15. Response of 'Maryland Mammoth' receptors kept on LD to thermoinduced (left) and nonthermoinduced (right) donors of biennial *H. niger*. Drawings made from photographs. (From Lang, A., in *Encyclopedia of Plant Physiology*, Vol.

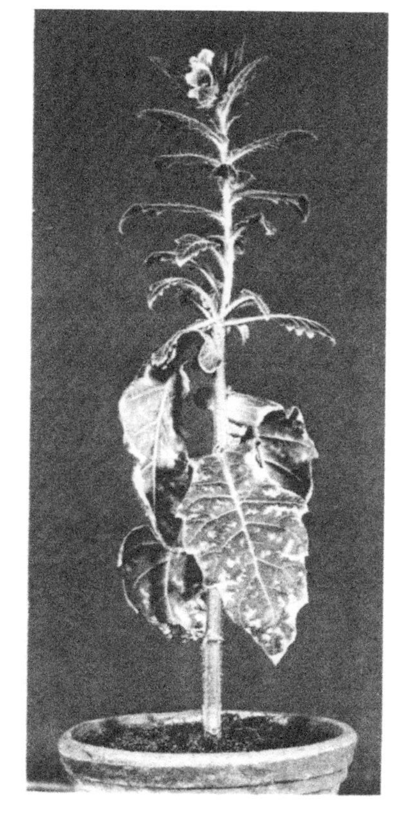

FIGURE 16. Flower formation in a receptor of biennial *H. niger* serving as

duced biennial plants, when kept on LD, responded fairly rapidly to doses of 2 ng per plant per day, applied for 4 weeks (Figure 17). If annual plants were grown in regular short days (greenhouse), the amounts of GA that had to be applied were larger, and the response was slower and incomplete, with only part of the plants forming flowers.[38] However, when annual plants were grown under long photoperiods in green or near-red light, which are relatively ineffective in flower induction (see "Light Quality" in the earlier section "Photoperiodic Induction"), application of GA resulted in as rapid and complete a floral response as that obtained in B.[13]

The first response of *H. niger* (annual plants kept on SD in the greenhouse) to GA application is a great increase in mitotic and cell-division activity in the subapical (rib-meristem) region of the shoot meristem[60] and, as a rule, flower initiation in GA-treated annual as well as biennial plants is preceded by extensive stem elongation — far more so than after thermo- or photoinduction. In *H. niger* too young to flower, and in nonthermoinduced biennial *H. niger* plants kept in SD conditions, GA treatment may result in stem elongation with no flower formation.[2,37,55]* However, Warm[70] has recently shown that when GA is applied to annual *H. niger* plants kept on SD (but by infiltration of the leaves), small doses (a single application of around 1 to 2 ng per plant) induced in part of the plants flower initiation with no stem elongation at all.

When annual *H. niger* was exposed to LD and treated with the growth retardants, chlormequat or daminozide, stem elongation was much reduced, but flower formation little if at all affected; the inhibition of stem elongation was overcome by simultaneous application of GA.[7,41] Chlormequat has been shown to inhibit GA biosynthesis, at least in the mold *Gibberella*.

Although the results of GA application are striking and in fact represent the first case where flower formation in representatives of two important physiological groups of plants, the cold-requiring and the LDPs, has been obtained by chemical means, the physiological meaning of these results is not clear. With regard to stem elongation, the results indicate that GA is indeed *essential,* i.e., normally required for this process. However, essentiality is no proof that the material in question can become limiting and thus perform a *regulatory* function, except when its level in the plant is lowered by experimental means — in this case, by inhibition of its biosynthesis. And those results do not provide evidence that GA plays a role in mediating the action of photoperiod on stem elongation. On the contrary, stem elongation in *H. niger* caused by exogenous GA is not influenced by R or FR light, and elongation promoted by GA and by FR is additive;[18] that is, Pfr does not seem to regulate elongation via GA. With regard to flower formation, the earlier results conveyed the impression that floral induction by GA was somehow a "secondary" response to stem elongation, although the mechanism of such a response would be entirely unclear. The findings of Warm[70] show that GA treatment can induce flower formation in *H. niger* independently of stem elongation. However, they do not prove that GA has an essential or regulatory role in this process and are too limited to serve as basis for conclusions. They are, however, worth pursuing and extending; leaf infiltration would seem a more physiological method of supplying GA to the plant since, if the hormone is indeed involved in flower formation, particularly its photoperiodic induction, it would presumably exert its action in the leaves.

* In some reports of this kind (see, for example, Reference 2), the treated plants, judging by the illustrations provided, had sessile leaves (i.e., bracts) in the upper region of the elongated stem and it is quite likely that these were supporting very poorly developed flowers. Instances like this serve to stress the need for a microscopic examination of the apex of seemingly nonflowering individuals of *H. niger*.

Exogenous Applications: Other Hormones and Regulators

Some promotion of flower formation in annual *H. niger* has also been obtained by spraying the plants with solutions of auxin, with maximal effect (up to 100% of the treated plants forming flowers) at concentrations of about 3 to 10 mg/ℓ, higher concentrations being less effective to ineffective.[46] This response was, however, obtained only when the plants were grown on SD and the light periods extended with supplementary light (from incandescent lamps) at fluxes not quite sufficient for a flowering response. If they were given straightforward light-dark cycles, either slightly below or slightly above the critical daylength, neither auxin (IAA) nor the "antiauxin" TIBA had either a promotive or an inhibitory effect on flower formation.[9] The promotive effect with subinductive supplementary light, on the other hand, was obtained with IAA, IBA, and 1-naphthaleneacetic acid (2-naphthoxyacetic acid and 2,4-D were ineffective, but most probably because of high toxicity). However (and surprisingly) a similar promotive effect was also obtained with TIBA, 2,6-di- and 2,4,6-trichlorophenoxyacetic acid, 4-chloro- and 2,4-dichlorophenoxy*iso*butyric acid, and 2,4-dichloranisol and 2,4-dichlorophenetol — compounds usually considered as antiauxins. Thus, the physiological meaning of these results, if any, is quite difficult to assess. (In most cases, the promotive concentrations were 10^{-5} to 10^{-4} *M*; those of 2,4-dichloroanisol and 2,4-dichlorophenetol were ten times higher, but the effect of these two substances was marginal.)[41]

Auxin (IAA), kinetin, salicylic acid, and acetylsalicylic acid (aspirin), when applied to annual *H. niger* plants by infiltration, did not promote flower formation in SD nor inhibit it in LD.[70] (The two last-named compounds promote flower formation in LD *Lemna;* see chapter "Lemnaceae", in Volume 3.[26a])

Changes in the Endogenous Levels of Hormones and Other Substances

Several investigators have measured the endogenous levels of phytohormones in annual *H. niger,* in order to determine whether photoperiodic induction results in changes which may help us to understand the physiological bases of induction. This has been done for GA,[31,32,39,57] auxin,[3,31,32] cytokinin, and ABA-like materials.[31,32] In most cases it was found that induced plants seemed to have, at one time or another, higher levels of all these hormones than noninduced controls, at least as measured by the activities of extracts in bioassays. In some cases, the increases coincided with the appearance of flower primordia or preceded it, while during later stages they were sometimes followed by declines. But an initial decline of the auxin level after transfer of plants to LD, followed by an increase, has also been reported.[3] The meaning of all these findings is, however, moot. Even if an increase precedes the morphological response it does not follow that it is a causal link between the inducing factor — here, daylength — and the response; it may be an early part of the "floral syndrome" or may be a parallel but causally unrelated "second" response to photoperiod. The latter interpretation is supported by the fact that similar changes (i.e., increased hormone contents under LD) have also been reported in SDP. Last, but not least, all measurements were made with techniques which by now have been superseded by far more sensitive and specific ones; if one feels that such measurements will provide meaningful information about regulation of flower formation they should be repeated with the modern methods.

It has been reported that substances which were considered to be estrogens underwent considerable increases in *H. niger* plants, beginning at flower formation and reaching a maximum in plants with large flower buds.[30] However, a subsequent study casts very serious doubts as to the estrogenic nature of these substances and also on the presence of steroidal estrogens in other plants where they have been reported; the substances in question are monoglycerides of fatty acids and their relation to flowering, if any, is quite unclear.[68]

METABOLIC STUDIES

Sugars; Modified Gas Composition

In Melchers' laboratory, a series of studies have been undertaken, based on a proposition[42] that the inhibitory effect of the daily dark periods in the photoperiodic response of *H. niger* may be based on dissimilatory process(es) resulting in the absence or shortage of material(s) required for flower formation. Infiltrating the leaves of annual *H. niger* plants maintained in SD, but near the critical daylength, with sugar solutions resulted in flower formation. The greatest response (100% of the treated plants initiating flowers) was obtained with glucose, but sucrose, fructose, maltose, and mannose were also effective, even though mannose had rather marked toxic effects. Infiltration with an approximately isotonic salt solution resulted also in occasional flower formation, but the response was considerably less than that to the most effective sugars.[52] Sugar infiltration results in an increase of the sugar level in the leaves, despite an increase in CO_2 evolution.[8] Keeping the plants for part of the dark periods in nitrogen atmosphere also resulted in flower formation; this effect could be obtained in regular SD (9 hr of light daily) and is in this respect more spectacular than the sugar effect.[50] Nitrogen atmosphere reduces CO_2 evolution by the leaves, and so do supplementary light and low temperature during the dark periods.[8] Thus, measures which may be assumed to reduce the relative effectiveness or the rate of dissimilatory processes (increasing the sugar content of the leaves, nitrogen atmosphere, supplementary light, and low temperature) also promote flower induction. The relationship is qualitative but not quantitative, however. In particular, while with respect to flower induction supplementary light reaches saturation at about 10 lux (in the earlier section "Photoperiodic Induction" see the subsection "Light Energy and Night Breaks"), its effect on CO_2 evolution increases linearly to much higher light fluxes.[8] "Preliminary" but quite extensive studies on respiration and starch hydrolysis in detached leaves of annual *H. niger* plants which had been raised under different light regimes[69] have also failed to demonstrate any clearcut relations with flower formation. Thus, any simple, direct connection between dissimilatory processes in the leaves and the photoperiodic regulation of flower formation in *H. niger* remains to be proven.

The results of the sugar infiltration experiments are, at least indirectly, supported by the results of experiments in which CO_2 was withheld from annual *H. niger* plants during different parts of inductive cycles.[22] Withholding CO_2 during the entire light periods proved fatal to the plants, but if the plants were simultaneously fed with sucrose or if they were given, in each light period, first 4 hr in normal air and then 12 hr with no CO_2, part of them survived and the flowering response was reduced as compared with controls having CO_2 throughout the entire light periods. Withdrawal of CO_2 during periods of supplementary light extending a short main light period and during the dark periods was without effect on floral induction. Presence of CO_2 is thus important only in the light periods of inductive cycles. This agrees also quite well with the need for high light fluxes during induction by continuous light (see "Length and Number of Cycles" in the earlier section "Photoperiodic Induction") and may indicate the need for a certain level of photosynthesis.

Protein Changes; Nucleic-Acid Metabolism?

Also in Melchers' laboratory, efforts have been made at determining whether vernalization is associated with changes of the proteins in biennial *H. niger*, and whether it may involve nucleic-acid metabolism. Serological differences between thermoinduced and nonthermoinduced biennial plants were indeed found. However, they were much greater in the purple- than in the yellow-flowered strain, even though their cold requirements do not differ to any comparable extent, and annual *H. niger* plants behaved serologically, both with and without a cold treatment, like the nonthermoinduced biennial ones. Thus, the serological differences found do not seem to be related to the developmental events underlying vernalization.[1] The results show, however, how important it is, in experiments on chemical changes caused by thermoinduction in cold-requiring plants, to include non-cold-requiring strains as controls, and the same holds for analogous work on photoinduction, where day-neutral relatives can serve as controls.

Supplying biennial *Hyoscyamus* plants with guanosine, adenine, guanine, or a RNA hydrolysate during cold treatment had no effect on the flowering response, even though the cold treatment given was suboptimal, to maximize any possible effect. Simultaneous administration of sucrose did not help either.[61]

An attempt at approaching photoperiodic induction in annual *H. niger* by the methods of modern molecular biology has recently been made by Warm.[70a] Polyadenylated RNA (essentially, messenger RNA) was prepared from leaves of plants that had been induced to flower formation by exposure to 58 hr of white light, and control plants maintained on SD, and was then translated in vitro, using a rabbit-reticulocyte protein-synthesis system, in the presence of ^{35}S-methionine; the synthesized polypeptides were separated by two-dimensional electrophoresis and fluorography. To reduce as far as possible unspecific effects of light, the RNA extraction was made after a dark period of 14 hr, since most "light-induced" proteins and their mRNAs are known to disappear in darkness, because of rapid turnover. The induced leaves contained higher amounts of total and polyadenyated RNA than the control leaves, and among the translation products seven were present in higher and three in lower amounts than in the latter leaves. These changes indicate that certain genes in the leaves are under photoperiodic control, and some of the altered polypeptides may be specifically related, e.g., to the formation of a translocatable floral stimulus. The author is commendably conservative in his conclusions and admits the need for a good deal more experimentation. Some "light-induced" RNAs and proteins do persist through the daily dark periods of normal light-dark cycles. The inductive treatment chosen, while it causes, (under the conditions used) flower initiation in 100% of the plants, is far from optimal for *Hyoscyamus;* and even if the specificity of some of the observed changes can be unequivocally demonstrated, it is somewhat difficult to see their meaning as long as no function can be attached to the altered polypeptides.

Cycloheximide, Thiouracil, and Azauracil

Rau and co-workers have found that 2-thiouracil[19] and cycloheximide,[28] when supplied by vacuum infiltration to the leaves of annual *H. niger* plants kept on SD, cause a substantial although not maximal floral response. Application to the apex had no such effect and reduced the response to LD. The two compounds are known to be inhibitors of RNA and protein synthesis, respectively, and infiltrated cycloheximide did reduce incorporation of [^{14}C]leucine into the soluble proteins of *H. niger* leaves.[29] However, the compound was also found to inhibit leucine uptake by the leaf, and seems furthermore to inhibit the export of applied GA from leaf to shoot apex — the latter action being deduced from a reduction by cycloheximide of the shoot elongation response to GA infiltration of the leaves.[71] On this basis the authors propose the following hypothesis: some flower hormone is produced in the leaves of annual *H. niger* even in SD, but not enough to cause flower formation at the shoot meristem. Cycloheximide blocks temporarily the transport of the hormone from the leaves and allows it to accumulate to a level which, once transport to the apex becomes possible again, is sufficient for some floral induction.

6-Azauracil (5×10^{-3} *M*) applied to the apex of annual *H. niger* plants shortly before or at the start of LD treatment reduced or completely suppressed the floral response, also causing damage to the leaves; the inhibition was partly overcome by simultaneous application of the tenfold concentration of uridine.[67]

INHIBITORY PROCESSES AND FACTORS

Kinetic Considerations

The photoperiodic response curve of *H. niger* (Figure 5) exhibits an important feature. The floral response is maximal with 24 hr of light per day (i.e., in continuous light) nor has (as has already been emphasized under "Endogenous Rhythms" in the section "Photoperiodic Induction") a consistently and markedly superior response been obtained with any cycles of other than 24 hr duration and/or with NB. Thus, the dark periods of the daily light-dark cycles have no positive function in the photoperiodic induction of flower formation in this LDP. Do they have any function, or are they mere interruptions of processes occurring only in light and leading to flower initiation? If the latter were true it would be difficult to understand why there is a critical daylength: the plant should be able to flower in any daylength in which it is capable of making growth — and provided light intensity is sufficient, *H. niger* grows quite well in daylengths of 8 hr and even less. Thus, the daily dark periods do have an active function, but in relation to flower formation it is a negative one; dark periods are undoing or countering something that happens or is produced during the light periods. It may be speculated that the processes in the light periods lead ultimately to the formation of florigen, and those in the dark periods to that of an inhibitor of flower formation, antiflorigen (see "Grafting Experiments. Antiflorigen" below).

Some experiments[21] have shown that annual *H. niger*, while capable of flower formation in 12:12-hr light-dark cycles, does not form flowers if the dark periods are lengthened to 18 hr (i.e., cycle length is extended to 30 hr); others[25] have shown that maximal inhibition of flower initiation seems to occur in photoperiods of 5 hr. The latter experiments were done by giving the plants light periods of 0.1 min to 24 hr per day for 6 days and then transferring them to LD, since they do not survive on continuous very short days. On the other hand, whereas *H. niger* plants do not flower on 9:15-hr cycles, they do so if the dark periods are extended to 24 hr, i.e., the cycle length to 33 hr.[21] Thus, the inhibitory effect of a dark period seems to increase with its duration, but then to decline again. At first sight, perhaps surprisingly, the floral response seems to improve (i.e., the inhibitory action of the dark periods seems to decline) if daylength is reduced below 5 hr of light.[25] However, and while this is presently pure speculation, it may be argued that the inhibitory action of dark periods in turn depends on something occurring in the light periods — in the simplest case, photosynthesis — and that 5 hr of light is needed for maximal inhibitory effectiveness while, when the dark periods become longer than 15 hr or somewhat above, the inhibitory effect begins to decline.

Defoliation Experiments

Strong support for the idea of inhibitory effects of the dark periods of nonphotoinductive cycles comes from defoliation experiments.[34,42] If all expanded leaves of annual or of thermoinduced biennial *H. niger* plants are removed and the plants maintained in this condition, they initiate flowers in continuous light,[41] LD, SD, and continuous darkness:* the photoperiodic control of flower formation has disappeared (see Figure 18). Thus, the inhibitory action of dark periods is localized in the leaves. The inhibitory effect seems to become manifest when the leaf begins to unfold (Figure 19). Leaving a single mature leaf on or grafting it back to a defoliated plant restores the inhibitory effect of SD.[41,42]

H. niger is so far the only LDP in which flower initiation upon defoliation has been found to be possible. However, this is because it is the only plant so far found which is capable of maintaining, in the defoliated state, sufficient growth activity to produce flower primordia.

* Continuous darkness was actually not complete. For observation and continued defoliation the plants were handled every few days in dim R. In those days it was not known that R had very great photomorphogenetic activity.

FIGURE 18. Flower formation in a defoliated *H. niger* plant kept on LD. This is a particularly strong response; in most cases, particularly on SD, the flowers remain underdeveloped. (Original photograph courtesy of Professor G. Melchers, Max-Planck Institut für Biologie.)

Other LDP in which such defoliation experiments have been attempted have soon suspended any further growth; that their failure to form flowers was unspecific is evident from the fact that flowers were formed neither in SD nor in LD.

The defoliation experiments might make one think that the inhibitory processes occurring in the leaves during long dark periods are the sole factor controlling flower formation in annual *H. niger*. However, the response in intact plants given LD is markedly faster than in defoliated ones, and if a single mature leaf is left on an otherwise defoliated plant and the plant given LD, flower formation is again more rapid than without that leaf. If up to five leaves of a plant are maintained on SD and only one on LD, flower initiation is as rapid, or almost so, as in a plant left with a single leaf and given LD.[41] Thus, LD do exert a promotive effect which is also localized in the leaves, and which seems to be a good deal greater than the inhibitory effect.

Grafting Experiments. Antiflorigen

The effect of defoliation on flower formation in *H. niger* was first interpreted as a diversion, by the leaves of plants kept on SD from their stems (tubers), of materials needed for flower initiation, e.g., some precursor(s) of florigen.[42] More recent grafting experiments[40,41] have shown, however, that *H. niger* plants produce, while on SD, a translocatable floral inhibitor or "antiflorigen". Like the grafting experiments with *H. niger* which support the florigen concept, these experiments will be reviewed in more detail and in a somewhat broader

FIGURE 19. Stage of defoliation of annual *H. niger* plants necessary to obtain flower formation independently of daylength or presence of light. Plant on left will form flowers, plant on right will not. (From Lang, A. and Melchers, G., *Planta*, 33, 653—702, 1943. With permission.)

FIGURE 20. Grafts of annual *H. niger* on the day-neutral tobacco 'Trapezond'. Left: on LD, both partners flower (flowers on *Hyoscyamus* partner removed); right: on SD, *Hyoscyamus* partner vegetative, flower formation in 'Trapezond' inhibited. (Original photograph courtesy of A. Lang.)

context in the chapter "*Nicotiana*" (Volume 6). Here, let us only point out that *H. niger* has been shown to inhibit flower formation both in a day-neutral and a SD grafting partner (tobacco cvs 'Trapezond' (Figure 20) and 'Maryland Mammoth', respectively). Like florigen, antiflorigen readily moves to, and is active in, plants of a different photoperiodic type and belonging to another genus. Thus, in *H. niger*, as in at least one other LDP, *Nicotiana silvestris*, two graft-transmissible, hormone-like materials participate in the photoperiodic regulation of flower formation: a promoter, florigen, and an inhibitor, antiflorigen, produced (at least predominantly) under LD and SD conditions, respectively.

SHOOT ELONGATION, FLOWER INITIATION, AND FLOWER DEVELOPMENT

Under natural conditons, both thermo- and photoinduction in *H. niger* results first in some elongation of the rosette axis; this is quite rapidly followed by the appearance of the first flower primordia and, if the plants are left in the inductive conditions, the primordia continue to develop into mature flowers which set fruits. However, the three processes can be separated. We have seen that treatment with GA, by application to the shoot apex, of noninduced *H. niger* plants (annual as well as biennial) results in extensive elongation, while flower initiation occurs considerably later and may not occur at all; whereas application by leaf infiltration may result in flower initiation with no stem elongation (see "Exogenous Applications: GA and Growth Retardants" in "Growth Substances and Growth Regulators" above). We have likewise seen that floral induction and the further development of initiated flowers have different spectral requirements (see "Light Quality" in the earlier section titled "Photoperiodic Induction"). Elongation with no flower initiation has also been observed in annual *H. niger* plants kept in darkness, or on marginal photoperiods (10 to 11 hr) at a high temperature (28.5°C)[42] and to a lesser extent and in an abnormal manner after treatment with TIBA.[9] If either thermo- or photoinduction has been marginal, the plants will form flower primordia, but these fail to develop further than into rudimentary buds, with only a small calyx visible. As in annual *H. niger* plants that have been given enough LD for some stem elongation and production of visible buds but then returned to SD (Figures 8 and 9), such buds may continue to be formed for quite extended periods of time, but there is virtually no stem extension and the plants could be mistaken for vegetative rosettes, except that their leaves are sessile, giving them away as enlarged bracts. While quite limited and unsystematic, these observations indicate that, although normally closely coordinated, shoot elongation, flower initiation, and flower development are not controlled by one and the same regulatory mechanism(s).

When biennial *H. niger* plants bearing flower buds were treated with relatively high concentrations of auxin[58] or GA,[59] by injection into the stem, the normal development of the flower parts was inhibited in a progressive manner, beginning with the corolla and stamens, until flowers consisting only of a small calyx and a very rudimentary ovary were formed.

WORK WITH *H. ALBUS*

The only other *Hyoscyamus* species that seems to have been used for studies of the physiology of flower formation is *H. albus* L. It is a plant which, like *H. niger*, will not flower in photoperiods below ca. 11 hr and in which flower formation is greatly accelerated with increasing photoperiods.[35] However, close inspection of the daylength-dependence curve of this species points up a difference with *H. niger* that, although subtle, may be important for understanding the photoperiodic response of different LDP. The curve (drop in the response time) is less steep than that of *H. niger* (Figure 21; compare with Figure

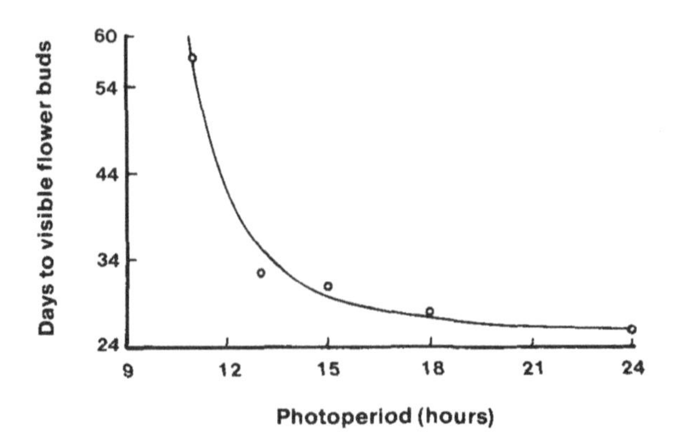

FIGURE 21. Photoperiodic response curve of *H. albus*. (After data in Lang.[35])

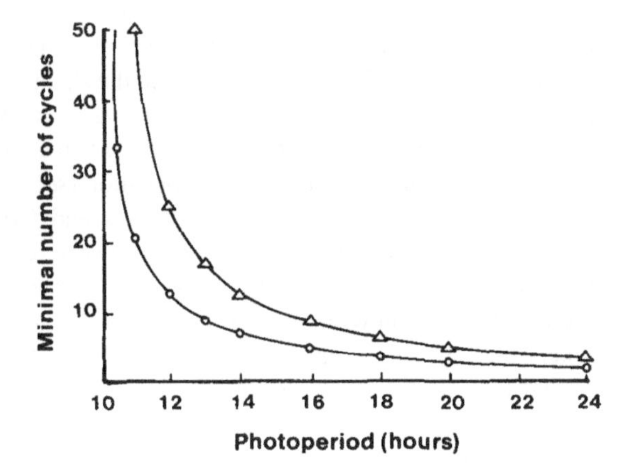

FIGURE 22. Curves showing the minimum number of 24-hr cycles necessary for floral induction, assuming the latter requires 50 hr of light and the effect of the light periods is (lower curve) or is not (upper curve) counteracted by the dark periods. The lower curve resembles the photoperiodic response curve of *H. niger* (compare Figure 6); the upper, that of *H. albus* (compare Figure 21). For further details, see text. (After Melchers, G. and Lang, A., *Biol. Zentralbl.*, 67, 107—174, 1948. With permission.)

6). In *H. niger,* the photoperiod response curve is a reflection of the number of inductive cycles of different length needed for floral induction (see "Localization" in the earlier section "Photoperiodic Induction"). In continuous light, 2 days are sufficient for induction but in light-dark cycles the daily dark periods abolish some of the inductive action occurring in the light periods (see "Kinetic Considerations" in the earlier section titled "Inhibitory Processes and Factors"), until in 10:14-hr light-dark cycles this action of the light periods is completely lost. If we assume that *H. niger* needs 50 hr of light for floral induction, and that the effect of 1 hr of light is nullified by 1.4 hr of darkness, we can construct the lower curve in Figure 22 which is very similar to the photoperiodic response curve of *H. niger*. If we now assume that *H. albus* likewise needs about 50 hr of light for floral induction but that in this plant dark periods have no antagonistic action with regard to the light periods, then we obtain the upper curve in Figure 22, which is quite similar to the photoperiodic response curve of *H. albus*.[54] Thus, *H. albus* may differ from *H. niger* in lacking the flower-inhibitory action of dark periods; the latter would function only as interruptions of light. That under LD conditions *H. albus* donors are effective in inducing flower formation in nonthermoinduced biennial *H. niger* receptors was mentioned earlier in the first subsection of "Flower Hormones (Florigen and Vernalin)".

Treatment with GA enhanced stem elongation and flower formation in *H. albus* under LD and SD conditions; treatment with daminozide reduced or delayed these processes, and this effect was reversed by simultaneous GA application.[23]

GENETIC STUDIES

The genetic basis for the biennial vs. annual habit was elucidated by Correns and confirmed by Melchers (see the opening section in this chapter). Correns considered the biennial habit as dominant, but later it was shown that $\frac{ann^+}{ann^+}$ homozygotes and $\frac{ann^+}{ann}$ heterozygotes are different: the heterozygotes reach thermoinductivity earlier and respond to the same cold treatment with a greater floral response than the homozygotes, and flower ultimately without thermoinduction.[61] The latter behavior may mean, as in the yellow-flowered biennial strain (see "Effects of Plant Age" above), either that the cold requirement in these plants has a quantitative character, or that relatively high temperatures that are no longer thermoinductive in purple-flowered biennial homozygotes are still thermoinductive in the heterozygotes.

Comparisons of the photoperiodic response of di- and autotetraploid annual *H. niger* and of *H. albus*[35] showed that in *H. niger* tetraploids the critical daylength was slightly reduced: they initiated flowers in a daylength (11 hr) in which, in the particular experimental conditions, the diploids all remained vegetative. Flowering (anthesis) in relatively short photoperiods was somewhat later in the tetraploids than the diploids while in longer ones the situation was reversed; stem elongation occurred in the tetraploids throughout the entire daylength range somewhat earlier than in the diploids, however, and leaf increment was smaller. In tetraploid *H. albus* leaf increment was the same as in diploid plants, but flower buds and open flowers appeared somewhat later.

OVERVIEW; ACHIEVEMENTS AND PROBLEMS REMAINING

Work with *Hyoscyamus* has yielded data, much of it "firsts", which have markedly contributed to our insights into the regulation of flower formation by low temperature (vernalization) and daylength (photoperiodism). The two phenomena are undoubtedly of great intrinsic interest. However, their study in *Hyoscyamus* has also led to insights into the physiology of flower formation in general that are now part of our current concepts in this field.

H. niger was among the first plants in which phytochrome has been identified as the pigment absorbing the light energy responsible for the photoperiodic regulation of flower formation. It has also been among the first plants in which the existence of hormone-like, graft-transmissible substances regulating flower formation has been demonstrated. Two of these, vernalin and florigen, are promoters of flower formation, with vernalin being necessary for florigen production; the third, antiflorigen, is an inhibitor of flower formation. Vernalin was discovered using a cold-requiring (biennial) strain of *H. niger;* evidence for the existence of florigen and antiflorigen was obtained in experiments using an annual (non-cold-requiring) strain, a LDP. However, as already pointed out in the subsections "Grafts with Annual *H. niger*" and "Vernalin and Florigen" in "Flower Hormones (Florigen and Vernalin)" and subsection "Grafting Experiments. Antiflorigen" in the section "Inhibitory Factors and Processes" these hormones are freely interchangeable between different response types, different species, and at least two different genera. At least one non-cold-requiring plant, the SD cv 'Maryland Mammoth' of tobacco, clearly can donate vernalin to nonthermoinduced biennial *H. niger,* while annual *H. niger* can donate florigen to LD, SD, and day-neutral species and cultivars of the genus *Nicotiana* and can receive florigen from these species and cultivars; it also can donate antiflorigen to day-neutral and SD plants belonging to a different genus. All three hormones are or can be present and effective (i.e., are most probably identical) in all plants. Cold-requiring and photoperiodic plants are unique only in possessing blocks to the production of vernalin, florigen, and antiflorigen which have to be removed by the appropriate environmental factor.

Last, *H. niger* was the first plant in which treatment with GA was shown to induce stem elongation and flower formation under noninductive conditions, both in biennial and annual strains.

Kinetic studies of thermo- and photoinduction in *H. niger* have pointed up a marked difference between the two. Thermoinduction has a distinct quantitative character, with the floral response becoming more rapid, up to a maximum, with increasing length of low-temperature treatment. Photoinduction has an all-or-none character: the first flower is initiated when the plant has received the minimum number of inductive cycles, but its initiation is not enhanced by further photoinductive treatment. Some factor necessary for flower formation — presumaby, florigen — needs to be accumulated in the plant until a threshold level has been attained.

Thermoinduction can be consummated in the shoot apices of biennial *H. niger* plants. It passes through a reversible stage and can be envisaged as a balancing act between physiological or biochemical processes which have somewhat differing temperature dependencies and some of which are, with regard to flower initiation, promotive while some are inhibitory. At low temperatures, it is the promotive processes that prevail, in relative terms; at higher temperatures, it is the inhibitory ones. Photoinduction takes place in the leaves and also involves promotive and inhibitory processes, at least in *H. niger,* with the former occurring in light and presumably resulting in florigen formation and the latter occurring in the dark periods and perhaps leading to antiflorigen formation. The balance between these processes can be shifted by temperature; presumably as a result of this, the critical daylength also changes, being lower at low temperatures and higher at higher ones. Both thermo- and photoinduction, once consummated, exhibit high degrees of persistence but are not permanent.

Thus, the credit ledger for work on the physiology of flowering in *Hyoscyamus* is quite impressive. However, the debit ledger is quite substantial, too, as yet. One reason is that the various positive accomplishments, while undoubtedly important, are by no means complete. We do not know the exact mode of function of phytochrome in the regulation of flowering in *H. niger*. The chemical nature of vernalin, florigen, and antiflorigen is not known, nor is the nature of the various promotive and inhibitory processes that are participating in both thermo- and photoinduction and ultimately determine whether or not vernalin, florigen, or antiflorigen is produced. We are dealing with physiological concepts rather than chemical entities or definite biochemical reactions.

There are also some findings with *H. niger* which are intriguing, but whose meaning are quite obscure. Thus, flower formation in this plant can be promoted, usually under subinductive photoperiodic conditions, by treating the plants with auxins as well as analogs usually considered as antiauxins, or some compounds mainly known as inhibitors of RNA or protein synthesis (thiouracil or cycloheximide), or by sugar infiltration or nitrogen atmosphere, the latter two being conditions reducing the efficiency of dissimilatory processes. The physiological relevance of these results, if any, is however quite unknown.

Further, we have very little knowledge of the connections that must exist between the various major bodies of facts of which we do know — the involvement of phytochrome, the various hormonal factors, etc. Thus, the photoperiodically active light is absorbed by phytochrome, but how does it ultimately result in the formation of florigen under LD conditions and of antiflorigen under SD conditions? The only information we have is that the inhibitory effect of dark periods — which may ultimately lead to antiflorigen production — depends on a high level of Pfr since short periods of FR given at the end of the daily light periods, permit flower formation to take place even in SD, (that is, they evidently abolish at least part of the inhibitory action of the subsequent dark periods), whereas a high Pfr level later in the daily cycle is, on the contrary, essential for floral induction (that is, presumably for florigen production).

As to GA, the evidence presently available indicates that it is an essential factor in stem elongation in *H. niger* (and quite likely in *H. albus*), but does not indicate that it has a regulatory function; in fact, the evidence argues against the hormone being a link in the chain of events between light absorption by phytochrome and this morphogenetic process. With regard to flower initiation, there is some evidence that GA may induce it with no effect on stem elongation, but only further research can show whether this reflects physiological — essential or even regulatory — role.

As a last point, flower initiation and the processes leading up to it are undoubtedly the most critical stage of the flowering process as a whole, marking as they do the point of no return in the transition from vegetative to reproductive development of the plant, or at least the shoot meristem in question. However, the following stages need also to be understood, for fundamental as well as practical reasons. Although we often use some of these stages as parameters of flower initiation our physiological knowledge about them is next to nothing.

The gaps in our knowledge of the physiology of flowering in *Hyoscyamus* are not unique; in other cold-requiring and photoperiodic plants they are equal and in fact often larger, and point up the fact that our understanding of this important area of developmental plant biology is still very fragmentary.

REFERENCES

1. **Aach, H. G. and Melchers, G.,** Serologische Untersuchungen an kältebedürftigen Pflanzen, *Biol. Zentralbl.*, 76, 466—475, 1957.
2. **Bouillenne-Walrand, M.,** Action des gibberellins sur la croissance et la floraison des végétaux superieurs. Relation avec l'acide β-indoleactique, *Meded. Landbouwhogesch. Gent,* 24, 705—717, 1959.
3. **Bouillenne-Walrand, M., Payne-Gibson, M., and Bouillenne, R.,** Rapport existant entre l'A.I.A. libre et la mise en fleur chez *Hyoscyamus niger* L., *Bull. Soc. R. Sci. Liège,* 29, 228—238, 1960.
4. **Carr, D.,** On the nature of the photoperiodic induction. III. The summation of the effects of inductive photoperiodic cycles, *Physiol. Plant.*, 8, 512—526, 1955.
5. **Chailakhyan, M. Kh.,** New facts in support of the hormonal theory of plant development, *C. R. (Dokl.) Acad. Sci. U.R.S.S.*, n.s., 13 (1936 4), 79—83, 1936.
6. **Chailakhyan, M. Kh.,** Flowering in plants in combination of two vegetating grafting partners (in Russian), *Dokl. Akad. Nauk S.S.S.R.*, 159, 1421—1424, 1964.
7. **Chailakhyan, M. Kh.,** Interrelation of the processes of growth and reproductive development in plants under the action of gibberellins and retardants (in Russian), in *Papers of the Erevan Symposium on Ontogenesis of Higher Plants,* Kazaryan, O. V., Ed., Academy of Science of the Armenian S.S.R., Erevan, 1966, 93—103.
8. **Claes, H.,** Die Beteiligung des dissimilatorischen Stoffwechsels an der photoperiodischen Reaktion von *Hyoscyamus niger, Z. Naturforsch.*, 2b, 45—55, 1947.
9. **Claes, H.,** Die Wirkung von β-Indolylessigsäure und 2,3,5-Trijodbenzoesäure auf die Blütenbildung von *Hyoscyamus niger, Z. Naturforsch.*, 7b, 50—55, 1952.
10. **Claes, H. and Lang, A.,** Die Blütenbildung von *Hyoscyamus niger* in 48-stündigen Licht-Dunkel-Zyklen und in Zyklen mit aufgeteilten Lichtphasen, *Z. Naturforsch.*, 2b, 56—63, 1947.
11. **Clauss, H. and Rau, W.,** Über die Blütenbildung von *Hyoscyamus niger* und *Arabidopsis thaliana* in 72-Stunden-Zyklen, *Z. Bot.*, 44, 437—454, 1956.
12. **Correns, C. E.,** Ein typisch spaltender Bastard zwischen einer einjährigen und einer zweijährigen Sippe des *Hyoscyamus niger, Ber. Dtsch. Bot. Ges.*, 22, 517—524, 1904.
13. **Curry, G. M. and Wassink, E. C.,** Photoperiodic and formative effects of various wavelength regions in *Hyoscyamus niger* as influenced by gibberellic acid, *Meded. Landbouwhogesch. Wageningen,* 56, No. 14, 1956.
14. **de Lint, P. J. A. L.,** Stem formation in *Hyoscyamus niger* under short days including supplementary irradiation with far red, *Meded. Landbouwhogesch. Wageningen,* 58, No. 10, 1958.

15. **de Lint, P. J. A. L.**, An attempt to analysis of the effect of light on stem elongation and flowering in *Hyoscyamus niger* L., *Meded. Landbouwhogesch. Wageningen*, 61, No. 16, 1961.
16. **Diomaiuto-Bonnand, J., Houivet, J.-Y., and Picard, C.**, Vernalisation et mise en fleur de la Jusquiame noire bisannuelle (*Hyoscyamus niger* L.). Recherches ontogéniques, *Bull. Soc. Bot. Fr.*, 127, 427—442, 1980.
17. **Downs, R. J.**, Photoreversibility of flower initiation, *Plant Physiol.*, 31, 279—284, 1956.
18. **Downs, R. J. and Thomas, J. F.**, Phytochrome regulation of flowering in the long-day plant *Hyoscyamus niger*, *Plant Physiol.*, 70, 898—900, 1982.
19. **Eichhoff, E. and Rau, W.**, Auslösung der Blütenbildung bei der Langtagpflanze *Hyoscyamus niger* im Kurztag durch 2-Thiouracil, *Planta*, 97, 290—303, 1969.
20. **El Hattab, A. M.**, Effects of light quality on flowering and morphogenesis in *Hyoscyamus niger* L., *Meded. Landbouwhogesch. Wageningen*, 68, No. 12, 1968.
21. **Finn, J. C., Jr. and Hamner, K. C.**, Investigation of *Hyoscyamus niger* L., a long-day plant, for endodiurnal periodicity in flowering response, *Plant Physiol.*, 35, 982—985, 1960.
22. **Fredericq, H.**, The significance of carbon dioxide of the air for flower bud initiation, *Biol. Jaarb. Naturwet. Genootensch. "Dodonaea" (Gent)*, 25, 53—63, 1958.
22a. **Gregory, F. G. and Purvis, O. N.**, Studies in vernalisation of cereals. II. The vernalisation of excised mature embryos, and of developing ears, *Ann. Bot.*, N.S., 2, 237—241, 1938.
23. **Houivet, J.-Y.**, Action de l'acide gibberellique (GA_3) et de l'acide *N,N*-dimethylsuccinamique (B995) sur l'elongation caulinaire et la floraison de la Jusquiame blanche (*Hyoscyamus albus* L.), *Bull. Soc. Bot. Fr.*, 124, 375—383, 1977.
24. **Hsu, J. C. S. and Hamner, K. C.**, Studies on the involvement of an endogenous rhythm in the photoperiodic response of *Hyoscyamus niger*, *Plant Physiol.*, 42, 725—730, 1967.
25. **Joustra, M. K.**, Daylength dependence of flower initiation in *Hyoscyamus niger* L., *Meded. Landbouwhogesch. Wageningen*, 69 II, No. 13, 1969.
26. **Kadman-Zahavi, A. and Ephrat, E.**, Development of plants in filtered sunlight. III. Interaction of the composition of main light periods with end-of-day red or far-red irradiations and with red night interruptions in bolting and flowering of *Hyoscyamus niger*, *Isr. J. Bot.*, 25, 203—210, 1976.
26a. **Kandeler, R.**, Lemnaceae, in *CRC Handbook of Flowering*, Vol. 3, Halevy, A. H., Ed., CRC Press, Boca Raton, Fla., 1985, 251—279.
27. **Khudairi, A. K. and Lang, A.**, Flowering hormone of short day and long day plants. VIIIe Congr. Internat. Bot. (Paris), Rapp. et Comm. parvenus avant le Congr., Sect. 11 et 12, p. 331, 1954.
28. **Klautke, S. and Rau, W.**, Auslösung der Blütenbildung durch Cycloheximid im Kurztag bei der Langtagpflanze *Hyoscyamus niger*, *Planta*, 112, 25—34, 1973.
29. **Klautke, S. and Rau, W.**, Die Wirkung von Cycloheximid und 2-Thiouracil auf die Proteinsynthese in Blättern von *Hyoscyamus niger* im Zusammenhang mit der Induktion der Blütenbildung, *Ber. Dtsch. Bot. Ges.*, 86, 571—576, 1974.
30. **Kopcewicz, J.**, Oestrogens in the long-day plants *Hyoscyamus niger* and *Salvia splendens* grown under inductive and noninductive conditions, *New Phytol.*, 71, 129—134, 1972.
31. **Kopcewicz, J., Centkowska, G., Kriesel, K., and Zatorska, Z.**, The effect of inductive photoperiod on flower formation and phytohormone level in a long day plant *Hyoscyamus niger* L., *Acta Soc. Bot. Pol.*, 48, 255—265, 1979.
32. **Kopczewicz, J., Centkowska, G., and Kriesel, K.**, Phytohormone levels in the leaves of *Hyoscyamus niger* L. during variable photoperiods at the time of flower initiation and development, *Acta Soc. Bot. Pol.*, 49, 245—258, 1980.
33. **Lane, H. C., Cathey, H. M., and Evans, L. T.**, The dependence of flowering in several long-day plants on the spectral composition of light extending the photoperiod, *Am. J. Bot.*, 52, 1006—1014, 1965.
33a. **Lane, H. C., Siegelman, H. W., Butler, W. L., and Firer, E. M.**, Detection of phytochrome in green plants, *Plant Physiol.*, 38, 414—416, 1963.
34. **Lang, A.**, Über die Bedeutung von Licht und Dunkelheit in der photoperiodischen Reaktion von Langtagpflanzen, *Biol. Zentralbl.*, 61, 427—432, 1941.
35. **Lang, A.**, Beiträge zur Genetik des Photoperiodismus. II. Photoperiodismus und Autopolyploidie, *Z. Naturforsch.*, 2b, 36—44, 1947.
36. **Lang, A.**, Untersuchungen über das Kältebedürfnis von zweijährigem *Hyoscyamus niger*, *Züchter*, 21, 241—243, 1941.
37. **Lang, A.**, Induction of flower formation in biennial *Hyoscyamus* by treatment with gibberellin, *Naturwissenschaften*, 43, 284—285, 1956.
38. **Lang, A.**, The effect of gibberellin upon flower formation, *Proc. Natl. Acad. Sci. U.S.A.*, 43, 709—717, 1957.
39. **Lang, A.**, Gibberellin-like substances in photoinduced and vegetative *Hyoscyamus* plants, *Planta*, 54, 498—504, 1960.

40. **Lang, A.**, Inhibition of flowering in long-day plants, in *Plant Growth Substances 1979*, Skoog, F., Ed., Springer-Verlag, Berlin, 1980, 310—322.
41. **Lang, A.**, unpublished results.
42. **Lang, A. and Melchers, G.**, Die photoperiodische Reaktion von *Hyoscyamus niger*, *Planta*, 33, 653—702, 1943.
43. **Lang, A. and Melchers, G.**, Vernalisation und Devernalisation bei einer zweijährigen Pflanze, *Z. Naturforsch.*, 2b, 444—449, 1947.
44. **Lang, A. and Melchers, G.**, Auslösung der Blütenbildung bei Langtagpflanzen unter Kurztagbedingungen durch Aufpfropfung von Kurztagpflanzen, *Z. Naturforsch.*, 3b, 108—111, 1948.
45. **Lang, A., Sandoval, J. A., and Bedri, A.**, Induction of bolting and flowering in *Hyoscyamus* and *Samolus* by a gibberellin-like material from a seed plant, *Proc. Natl. Acad. Sci. U.S.A.*, 43, 960—963, 1957.
46. **Liverman, J. L. and Lang, A.**, Induction of flowering in long-day plants by applied indoleacetic acid, *Plant Physiol.*, 31, 147—150, 1956.
47. **Melchers, G.**, Die Wirkung von Genen, tiefen Temperaturen und blühenden Pfropfpartnern auf die Blühreife von *Hyoscyamus niger* L., *Biol. Zentralbl.*, 57, 568—614, 1937.
48. **Melchers, G.**, Die Blühhormone, *Ber. Dtsch. Bot. Ges.*, 57, 29—48, 1939.
49. **Melchers, G.**, unpublished results.
50. **Melchers, G. and Claes, H.**, Auslösung der Blütenbildung bei der Langtagpflanze *Hyoscyamus niger* in Kurztagbedingungen durch Hemmung der Atmung in den Dunkelphasen, *Naturwissenschaften*, 31, 249, 1943.
51. **Melchers, G. and Lang, A.**, Weitere Untersuchungen zur Frage der Blühhormone, *Biol. Zentralbl.*, 61, 16—39, 1941.
52. **Melchers, G. and Lang, A.**, Auslösung von Blütenbildung bei *Hyoscyamus niger* durch Infiltration der Blätter mit Zuckerlösungen, *Naturwissenschaften*, 30, 589—590, 1942.
53. **Melchers, G. and Lang, A.**, Versuche zur Auslösung von Blütenbildung an zweijährigen *Hyoscyamus niger* -Pflanzen durch Verbindung mit einjährigen ohne Gewebeverwachsung, *Z. Naturforsch.*, 3b, 105—107, 1948.
54. **Melchers, G. and Lang, A.**, Die Physiologie der Blütenbildung (Ein Übersichtsbericht), *Biol. Zentralbl.*, 67, 105—174, 1948.
55. **Mugnier, C.**, Essai de remplacement de la refrigeration vernalisante par application d'acide gibberellique chez la Jusquiame noire bisannuelle, *Biol. Plant.*, 19, 40—47, 1977.
56. **Parker, M. W., Hendricks, S. B., and Borthwick, H. A.**, Action spectrum for the photoperiodic control of floral initiation of the long-day plant *Hyoscyamus niger*, *Bot. Gaz.*, 111, 242—252, 1950.
57. **Reinhard, E.**, Über Veränderungen des Gibberellingehaltes von *Hyoscyamus niger*. Ein Beitrag zur Analyse von Gibberellinen in höheren Pflanzen, Thesis (Habilitationsschrift), University of Würzburg, W. Germany, 1962.
58. **Resende, F.**, Auxin e sexo em *Hyoscyamus* L., *Bol. Soc. Port. Cien. Nat.*, 2a (Ser. 4), 248—250, 1953.
59. **Resende, F. and Viana, M. J.**, Gibberellin and sex expression, *Port. Acta Biol. Ser. A*, 6, 77—98, 1959.
60. **Sachs, R. M., Bretz, C. F., and Lang, A.**, Shoot histogenesis: the early effects of gibberellin upon stem elongation in two rosette plants, *Am. J. Bot.*, 46, 376—384, 1959.
60a. **Salisbury, F. B.**, *Xanthium*, in *CRC Handbook of Flowering*, Vol. 4, Halevy, A. H., Ed., CRC Press, Boca Raton, Fla., 1985, 473—522.
61. **Sarkar, S.**, Versuche zur Physiologie der Vernalisation, *Biol. Zentralbl.*, 77, 1—49, 1958.
62. **Schneider, M. J., Borthwick, A. H., and Hendricks, S. B.**, Effect of radiation on flowering of *Hyoscyamus niger*, *Am. J. Bot.*, 54, 1241—1249, 1967.
63. **Schwemmle, B.**, Unterschiedliche Schwankungen der Temperaturempfindlichkeit bei Lang- und Kurztagpflanzen (Versuche zur Blütenbildung), *Naturwissenschaften*, 47, 68—69, 1960.
64. **Seidlová, F. and Juraková, J.**, Ein Vergleich der Struktur des Vegotationskegels mit der Länge der photoperiodischen Induktion bei *Hyoscyamus niger*, *Naturwissenschaften*, 51, 442—443, 1964.
65. **Shakespeare, W.**, *The Tragicall Historie of Hamlet, Prince of Denmark*, Newly imprinted and enlarged to almost as much again as it was, according to the true and perfect Coppie (see Act I, Scene 5.), N. L., London, 1604.
66. **Stolwijk, J. A. J. and Zeevaart, J. A. D.**, Wave dependence of different light reactions governing flowering in *Hyoscyamus niger*, *Proc. K. Ned. Akad. Wet. Ser. C*, 58, 386—396, 1955.
67. **Teltscherová, L., Seidlová, F., and Krekule, J.**, Effects of some pyridine analogs on flowering of long-day and short-day plants, *Biol. Plant.*, 9, 234—244, 1967.
68. **van Rompuy, L. L. L. and Zeevaart, J. A. D.**, Are steroidal estrogens natural plant constituents?, *Phytochemistry*, 18, 863—865, 1979.
69. **Wagenaar, S.**, A preliminary study of photoperiodic and formative processes in relation to metabolism, with special reference to the effect of night temperature, *Meded. Landbouwhogesch. Wageningen*, 54, 45—101, 1954.

70. **Warm, E.,** Die Wirkung von Phytohormonen und von Salicylsäure auf die Blütenbildung und Sprossverlängerung bei der Langtagpflanze *Hyoscyamus niger*, *Z. Pflanzenphysiol.*, 99, 325—330, 1980.
70a. **Warm, E.,** Changes in the composition of in vitro translated leaf m-RNA caused by photoperiodic flower induction of *Hyoscyamus niger*, *Physiol. Plant.*, 61, 344—350, 1984.
71. **Warm, E. and Rau, W.,** Investigations on the mechanism of floral induction of the long day plant *Hyoscyamus niger* L. with cycloheximide, *Plant Sci. Lett.*, 16, 273—279, 1979.
72. **Warm, E. and Rau, W.,** A quantitative and cumulative response to photoperiodic induction of *Hyoscyamus niger*, a qualitative long-day plant, *Z. Pflanzenphysiol.*, 105, 111—118, 1982.

LAGERSTROEMIA

En. Crape Myrtle; Fr. Lagerstremie; Ge. Lagerstromie

Dennis P. Stimart

INTRODUCTION

Lagerstroemia L., a member of the Lythraceae, is a genus with some 55 species. Plants are deciduous or evergreen and can be small to large shrubs and trees. All are native to the tropical or subtropical parts of southern and eastern Asia, New Guinea, Australia, and the Pacific islands.[2] These plants are grown in horticulture for their colorful exfoliating bark and flowering habit. Sometimes *Lagerstroemia* are grown as pot plants in the greenhouse,[6,8] but most are grown in the landscape for their ornamental value. The plants prefer full sun in a hot location for optimum culture. One of the most beautiful species and most studied is that of *L. indica* L., the common crape myrtle.

MORPHOLOGY

The flowers are a showy purple, red, or white and are produced in axillary and terminal panicles from the apical meristem on the current year's growth. The pedicels are bracted. The calyx is a funnel-shaped tube containing 6 to 9 lobes. There are usually 6 petals which are crinkled or fringed with long slender claws. Each flower contains 15 to 200 stamens which are long and frequently curved upward. The ovary is 3- to 6-celled with a long bent style and a capitate stigma. The fruit is a capsule containing winged seeds.[2] The leaves are opposite or the uppermost alternate, mostly ovate and entire.

The morphological development of flower buds in dwarf *L. indica* have been outlined by Motoki et al.[9] The nine stages are (1) vegetative, (2) growing point enlarged, (3) sepals formed, (4) petals formed, (5) stamens formed, (6) pistil formed, (7) ovule formed, (8) pollen formed, and (9) flowering. Photomicrographs in Figure 1 illustrate these stages.[7]

FLOWER INITIATION AND DEVELOPMENT

The regulation of flowering in *L. indica* has not been extensively studied, but the available information suggests that temperature and accumulated light intensity elicit some control. Bailey stated that crape myrtle grown indoors will bloom two to three times a year if cut back.[1] This implies that no specific photoperiod requirement exists but that flowering is regulated more by temperature. Furthermore, specimen plants grown outdoors in southern England were found to bloom only during the summer months of years when the temperature was consistently warm.[2]

Under LD (between 14 to 15 hr) in Kagawa, Japan, vigorous vegetative growth and subsequent flower initiation and development of dwarf *L. indica* were enhanced by 25°C, but were inhibited at 20 and 15°C.[5] Similarly, floral bud initiation and subsequent development rate were advanced by 21°C night/28°C days when compared with those grown at 17°C night/24°C days.[7]

Artificial extension of daylength as a day continuation (1700 to 2300 hr) to equal 16 hr of light daily or a NB from 2200 to 0300 hr,[7] or from 2400 to 0200 hr,[5] enhanced flowering in dwarf crape myrtles. Plants grown under SD conditions were inhibited in flowering so that many never initiated flower buds.

Total light energy appears to be a contributing factor in flower production, since supple-

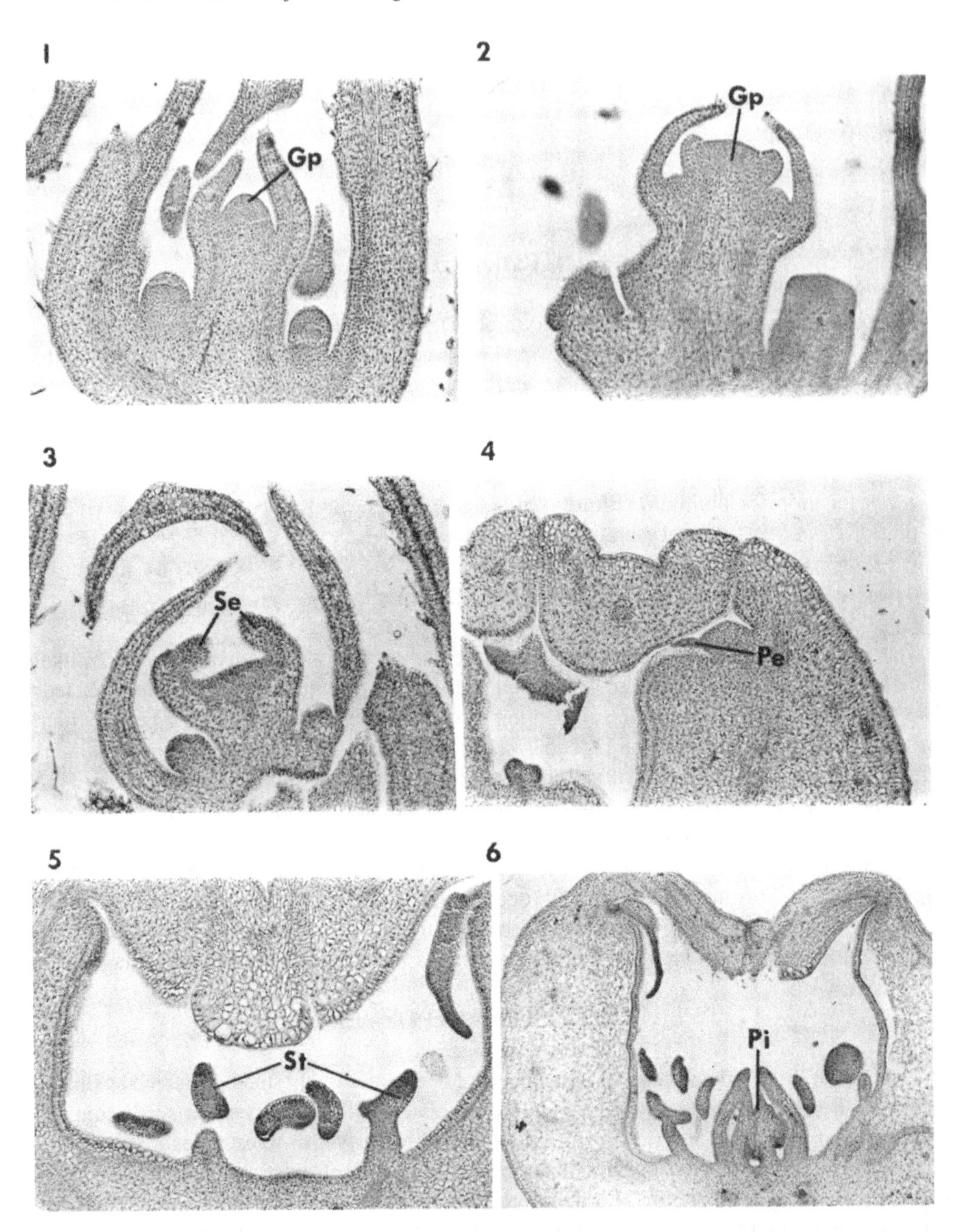

FIGURE 1. Developmental stages of flower buds of *L. indica nana* cv Centennial as described by Motoki, Y., Yokoi, M., and Kosugi, K.[9] (1) Vegetative growing point (Gp); (2) growing point (Gp) enlarged; (3) sepals (Se) formed; (4) petals (Pe) formed; (5) stamens (St) formed; (6) pistils (Pl) formed; (7) ovules (Ov) formed, and (8) pollen (Pol) formed. (Photomicrographs prepared by R. Guidry and A. E. Einert.[8])

mentary irradiation for 24 hr for 8 weeks from high-pressure sodium lamps increased the number of flower buds produced and hastened flowering, as well as increased branching and vegetative growth.[10] Furthermore, during the winter the number of days required to flower crape myrtles was increased when compared to other times of the year.[7]

In Kagawa, Japan, plants growing under natural conditions did not initiate flower buds until mid to late June.[9] At this time the shoot apex increased both in height and in width. From late June to early July flower bud development occurred and stamens and the pistil

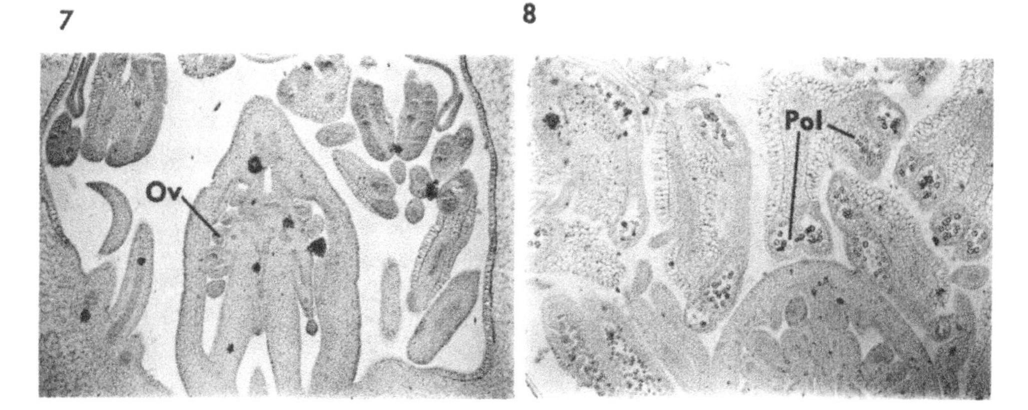

FIGURE 1 (Continued)

differentiated in the floret. Pollen and ovule formation followed in early to mid July. The plants began to flower by mid July with the average date of flowering on July 24th.

CULTURE AND FORCING

The time of flowering affects the time of year in which plants can be pruned. Since flowering is on the current year's growth, plants can be pruned heavily in early spring to make plants more compact and still be expected to produce flowers. Also, they may be pruned during dormant periods (winter). If pruning is done at or just prior to flowering the flowering response is inhibited on the newly formed shoots.[5]

The forcing of dwarf crape myrtles as pot plants can be divided into three phases.[7,8] The first phase is that of propagation. This requires maintaining stock plants in vigorous vegetative growth by LD (16 hr) and a minimum night temperature of 16 to 18°C. The second phase is the growth of the plant which is to be forced. This requires 4 or 5 weeks with at least two pinches. Plants are maintained under LD by day continuation (1700 to 2300 hr) or night interruption (2200 to 0300 hr), and a minimum night temperature of 18°C. The third phase is the forcing phase which begins after the final pinch. High temperatures (21°C night/28°C days), LD (16 hr) and high light intensity favor rapid flower development.

GROWTH REGULATORS

Even though there are genetically dwarf *L. indica* cultivars which have greatly facilitated the production of flowering plants in small containers, growth regulators are needed. An application of ancymidol (0.5 mg per pot) at 1 week after transplanting and again some 5 or 6 weeks following a second pinch appears optimal.[4] Flowering may be delayed by these treatments.

REFERENCES

1. **Bailey, L. H.,** *The Standard Cyclopedia of American Horticulture,* Vol. II, Macmillan, New York, 1906, 1775.
2. **Bean, W. J.,** *Trees and Shrubs Hardy in the British Isles,* Vol. II, Butler and Tanner, Great Britain, 1973, 784.
3. **Chittenden, F. J.,** *Dictionary of Gardening,* Vol. III, 2nd ed., Synge, P. M., Ed., Clarendon Press, Oxford, 1977, 1088.

4. **Einert, A. E.,** The crepemyrtle, *Hortic. Horizons in Oklahoma*, 7, 16—17, 1976.
5. **Goi, M. and Tanaka, Y.,** Effect of photoperiod and temperature on growth and flowering in *Lagerstroemia indica L., Tech. Bull. Fac. Agric. Kagawa Univ.*, 27, 77—83, 1976.
6. **Guidry, R. K. and Einert, A. E.,** Potted dwarf crapemyrtles: a promising new floriculture crop, *Flor. Rev.*, 157(4066), 30, 1975.
7. **Guidry, R. K.,** Forcing Dwarf Crapemyrtles, M.S. thesis, University of Arkansas, Fayetteville, 1977, 45.
8. **Guidry, R. and Einert, A. E.,** Forcing dwarf crapemyrtles, *HortScience*, 12 (Abstr.), 3, 1977.
9. **Motoki, Y., Yokoi, M., and Kosugi, K.,** Floral initiation and development in *Punica granatum* var. *nana* and *Lagerstroemia indica L.* dwarf hybrids, *Tech. Bull. Fac. Hortic. Chiba Univ.*, 20, 31—36, 1972.
10. **Richardson, M.,** The Effects of Continuous High Pressure Sodium Lamp Illumination on the Growth and Flowering of Dwarf Crapemyrtles, Oklahoma State University Research Report, Oklahoma State University, Stillwater, 1975.

LAMIUM AMPLEXICAULE

En. Henbit, Dead nettle; Fr. Lamier; Ge. Ackertaubnessel, Stengelumfassende

Elizabeth M. Lord

INTRODUCTION

Lamium amplexicaule L. (Labiatae), an annual mint, is an agricultural weed introduced into California from Mediterranean areas.[1] The inflorescence contains both cleistogamous (closed, CL) flowers, which self in the bud, as well as chasmogamous (open, CH) flowers, which undergo anthesis and may outcross. This species is the only one of the genus *Lamium* which produces CL flowers, and as a result it is highly inbred.[2-5] The phenomenon of cleistogamy, whereby such dimorphic flowers are produced by an individual, occurs in at least 56 families.[6-8] It provides a means for both inbreeding and outcrossing for the species.

FLORAL FORM AND FUNCTION

Cleistogamous flowers are characterized by a reduction in number and size of floral parts and show precocious pollination and seed set in the bud.[9] The CH flowers open and are visited by bees (*Apis mellifera*) for nectar and may be outcrossed. In *Lamium*, the CL flowers show reduction in cell number and size in the anthers and the corolla; the calyx and gynoecium are unmodified in form, except for a slightly shorter style and abortive nectaries (Figure 1). The inflorescence is a raceme in *Lamium*, and CL flowers are invariably produced first in the axillary cymes at node 4 (Figure 2). It is only in the later nodes that CH flowers appear and then only when the plants are growing under good nutrient, light, and water conditions. In this plant, which is day neutral with respect to floral initiation, long daylengths appear to trigger more CH flower production during spring and summer months in California (Figure 2 and Table 1).

The stigma is of the wet type and begins secreting early in the CL flower when the anthers are pressed close to the stigma lobes in the bud. Anthers dehisce and the pollen germinates in the locules due to the proximity of the highly lipidic, stigmatic secretions nearby.[12] So, precocious and simultaneous maturation of pollen and stigma results in earlier seed set in the CL flower (Figure 3). The CH flower undergoes anthesis before the anthers dehisce with the stigmatic lobes becoming receptive later. Nectar is collected by honey bees, allowing for some outcrossing to occur in the CH flowers. A gene marker for corolla spottedness was used to document that a minimum of 4% outcrossing occurred in a field population in Berkeley, Calif.[13]

COMPARATIVE DEVELOPMENT

The dimorphic flowers of *Lamium* begin as indistinguishable primordia (Figures 4 through 6) and initiate all the floral organs before diverging in their development (Figures 7 through 10).[14] At prophase of meiosis in the anthers, the floral buds are of the same shape (Figures 11 and 12), but already divergence has occurred since the CL anthers are reduced in size. With development, shape differences appear in the corolla which surrounds the gynoecium and androecium (Figures 13 and 14). These differences can be attributed to the differences in anther growth in the two flowers (Figures 15 and 16). At maturity, the CH anthers are larger than those of the CL, and this can be traced back to fewer pollen mother cells being produced in the CL anther (Figures 17 through 19).[15] The development of the corolla is

FIGURE 1. Mature CL and CH corollas of *L. amplexicaule*. (Magnification × 4.0.) (From Lord, E. M. and Mayers, A. M., *Ann. Bot.*, 50, 301—307, 1982. With permission.) FIGURE 2. Main shoot inflorescences of *L. amplexicaule*. Left, a plant grown under LD with CH flowers; right, one grown under SD with only CL flowers. (Magnification × 1.5.) (From Lord, E. M., *Am. J. Bot.*, 67, 529—533, 1980. With permission.)

Table 1
EFFECT OF PHOTOPERIOD ON OPEN (CH) FLOWER PRODUCTION IN *LAMIUM AMPLEXICAULE* L.

Photoperiod	Temperature (°C)	Light intensity[a] (μE m²/sec⁻¹)	CH flowers (%)	Sample no.
Expt 1				
SD 10 hr D[b]	21	140	1 · 0 ± 0 · 83	40
14 hr N[b]	10			
LD 16 hr D	21	140	15 · 2 ± 9 · 53	30
8 hr N	10			
Expt 2				
SD 10 hr D	25	581 · 2	18 · 75 ± 6 · 7	30
14 hr N	13			
LD 16 hr D	25	410 · 25	48 · 65 ± 12 · 9	19
8 hr N	13			

[a] Intensity, as indicated, was maintained throughout LD so the effect of higher total light energy was not eliminated as a variable. There was no significant difference in biomass between LD and SD plants.

[b] D, day; N, night.

From Lord, E. M., *Ann. Bot.*, 49, 261—263, 1982. With permission.

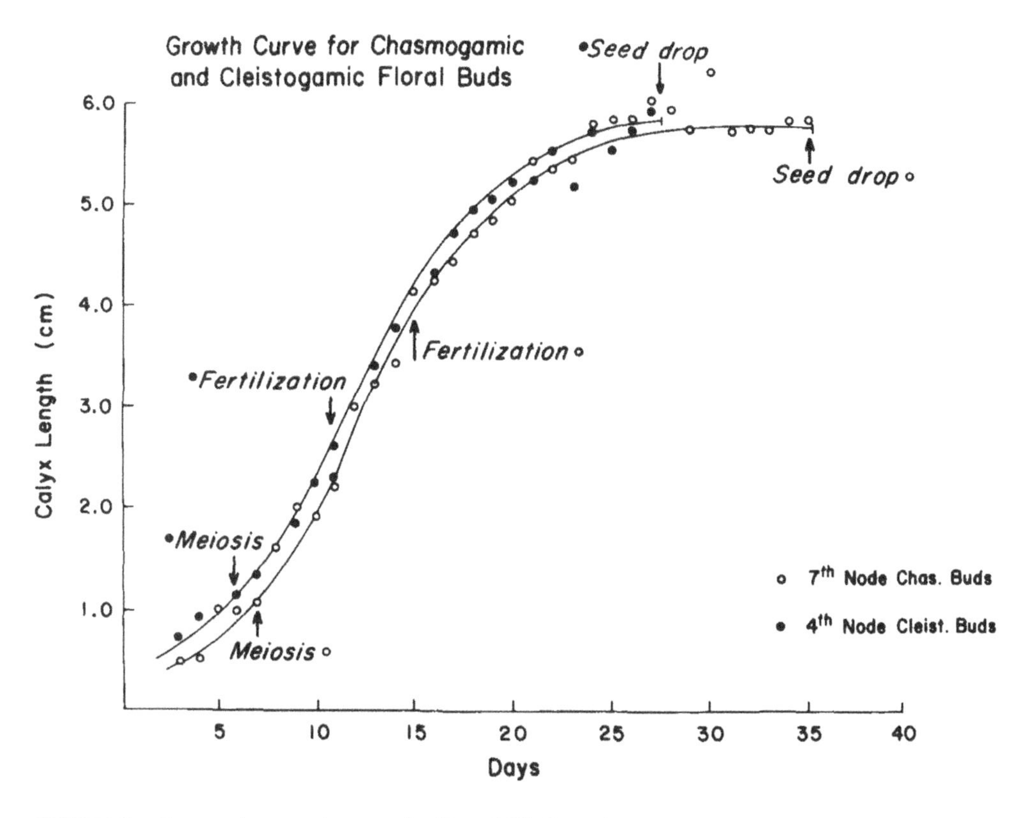

FIGURE 3. Comparative growth curves for CL and CH flower buds using calyx length as bud measurement. The lines are not significantly different. (From Lord, E. M., *Bot. Gaz. (Chicago)*, 140, 39—50, 1979. With permission.)

divergent in the CL form due to slower growth in the upper corolla width zone and lack of cell expansion throughout (Figure 20). The base of the CL corolla, which surrounds the ovary, shows simple cell expansion arrest, but the upper zone shows both cell division and expansion reduction. This dissociation in development of the two corolla zones results in a new corolla shape in the CL flower, one which is not found in the ontogeny of the CH flower (Figure 20). For this reason, the concept of arrest[15] is inappropriate to describe the development and evolution of the CL floral form. Since calyx development in both floral forms is similar, calyx length can be used as a time standard against which the growth of other organs can be measured. The CL corolla shows a simple arrested expansion in corolla length (Figure 21), but an actual decrease in growth rate in corolla width (Figure 22). Since the CL flower looks like a young bud form of the CH flower and is precocious in its sexual function, the model of progenesis can be invoked as a means by which the CL evolved from the CH form (Figure 23).[14] Progenesis is precocious sexual maturation at an early shape with no change in developmental rates. Since the CL is obviously derived from the CH flower and both forms are produced on an individual, cleistogamy provides a good comparative system to test such models. We have proposed a modification of the pure progenesis model to explain the ontogeny and phylogeny of the CL floral form in *Lamium* (Figure 24). The developmental markers used were prophase of meiosis in the anthers and anther dehiscence. Corolla shape development was measured as L/W ratio and calyx length as a time standard. The CL flower is sexually precocious, but the rate of corolla shape change is accelerated and dissociation brings about a new CL shape. The CL flower is not a pure progenetic form, but rather a progenetic dwarf adult form. The defining features are the precocious sexual maturation and scaled-down parts.

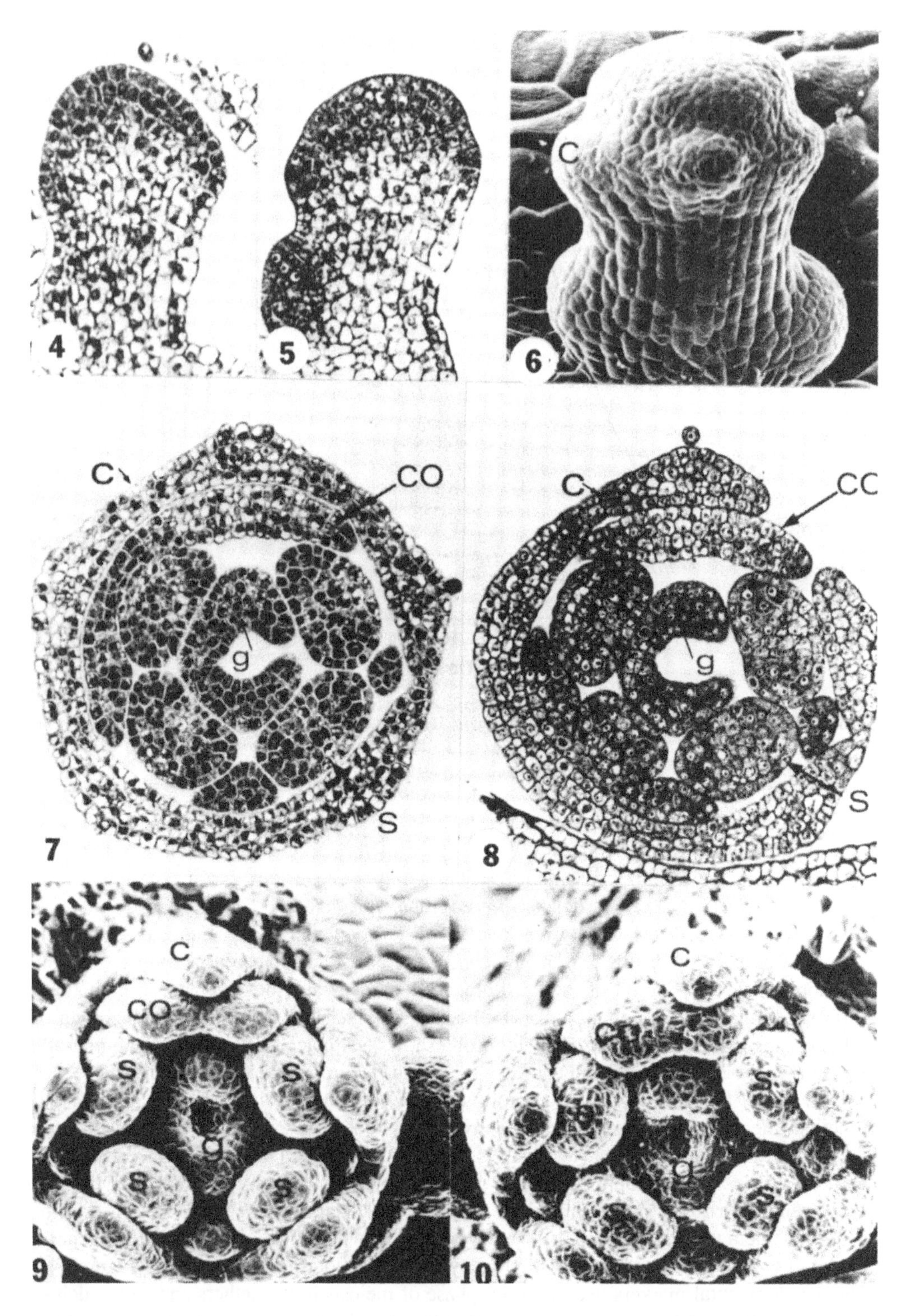

FIGURES 4 to 10. Comparative floral development in *Lamium amplexicaule*. Figure 4. Section of CH floral primordium. (Magnification × 280.) Figure 5. Section of CL floral primordium. (Magnification × 280.) Figure 6. SEM view of CH primordium with calyx being initiated. (Magnification × 360.) Figure 7. Cross-section of CH floral bud. (Magnification × 105.) Figure 8. Cross-section of CL floral bud. (Magnification × 105.) Figure 9. SEM view of CH floral bud. (Magnification × 306.) Figure 10. SEM view of CL floral bud. (Magnification × 306.) C = calyx, co = corolla, s = stamen, g = gynoecium. (From Lord, E. M., *Bot. Gaz. (Chicago)*, 143, 63—72, 1982. With permision.)

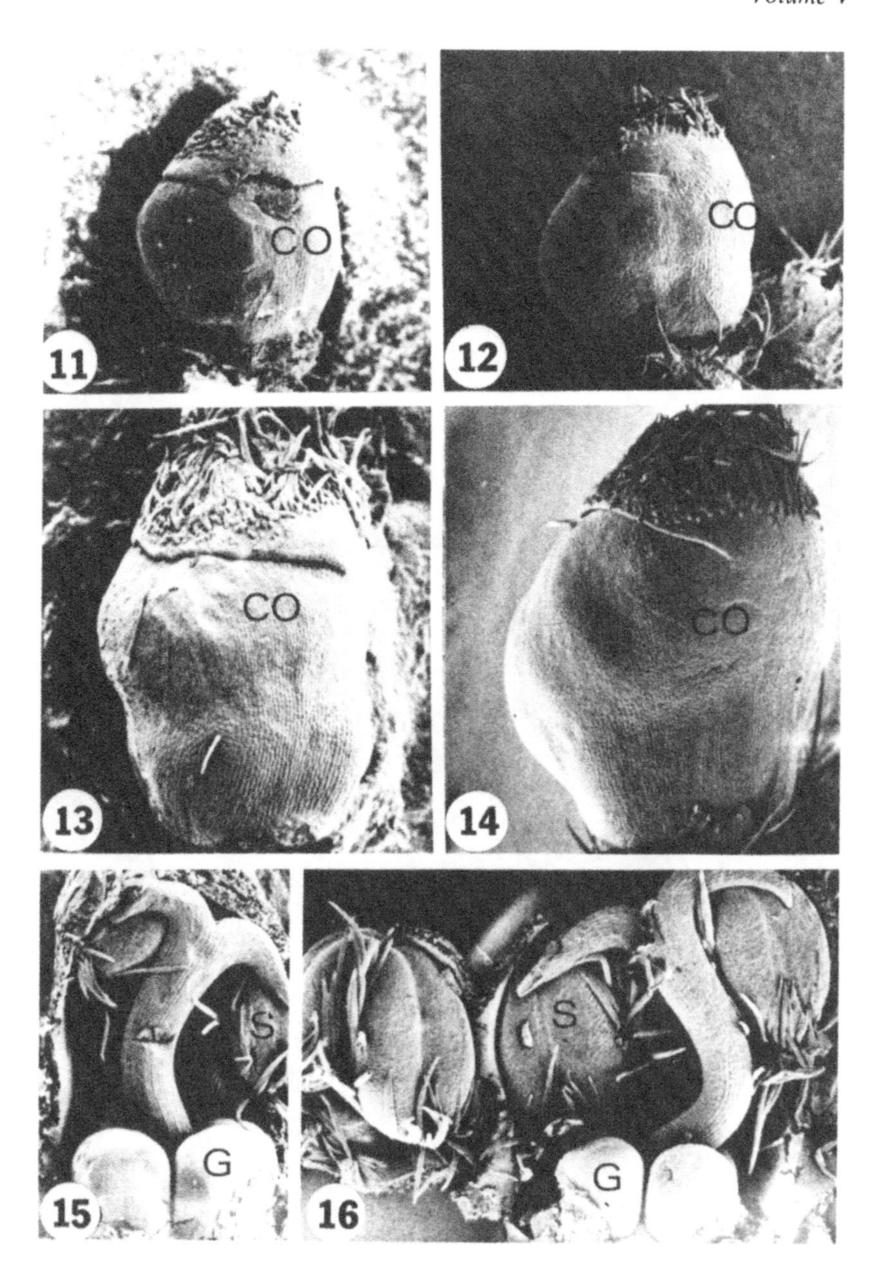

FIGURES 11 to 16. Comparative SEM views of CL (left) and CH (right) floral buds of the same age with calyx removed. (Magnification × 58.) Figures 11 and 12 show CL and CH buds, calyx length 1.0 mm. Figures 13 and 14 show CL and CH buds, calyx length 1.5 mm. Figures 15 and 16 show the CL and CH buds as in Figures 13 and 14, dissected open to reveal anther and gynoecium. Co = corolla, s = stamens, g = gynoecium. (From Lord, E. M., *Bot. Gaz. (Chicago)*, 143, 63—72, 1982. With permission.)

EFFECTS OF GA_3 ON FLORAL FORM IN VIVO AND IN VITRO

Since under short days CL flowers predominate (Figure 25) and under long days CH floral production is enhanced, we proposed that a build-up of gibberellins in plants growing under LD conditions triggers corolla expansion.[17] When GA_3 is applied to plants growing under conditions that normally induce predominantly CL flowers, entirely CH flowers are produced

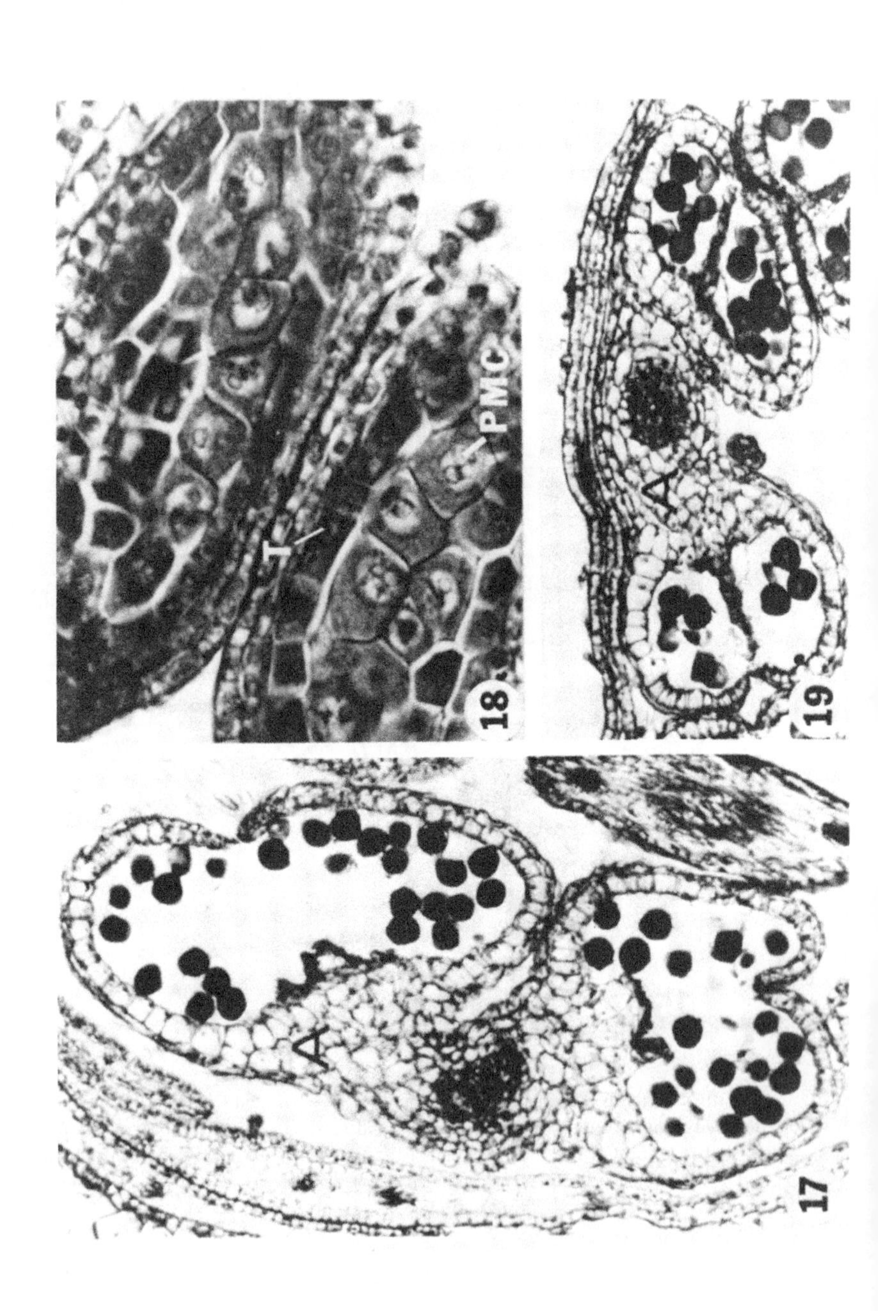
PMC
T
18
A
19
A
17

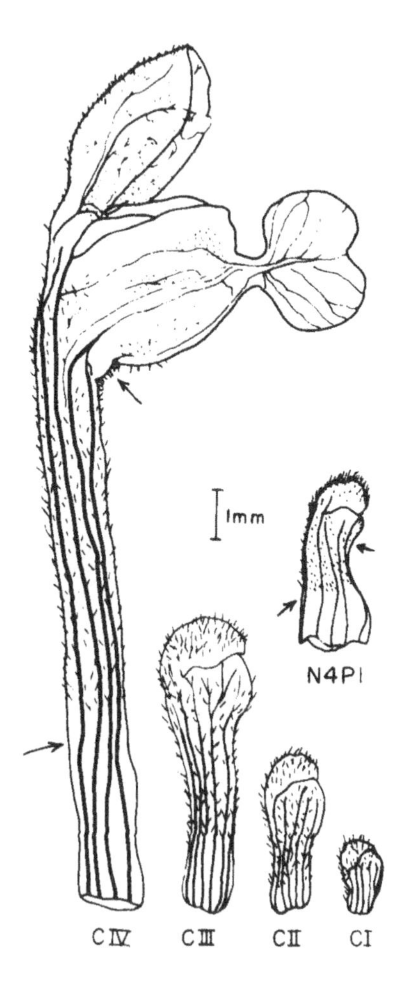

FIGURE 20. Mature CL corolla (N4P1) and stages (CI to CIII) in the development of the mature CH corolla (CIV) of *L. amplexicaule*. (From Lord, E. M., *Am. J. Bot.*, 67, 1430—1441, 1980. With permission.)

(Figure 26). Chlormequat (CCC), a GA inhibitor, can be shown to prevent CH flower production under inductive conditions (Figure 27). In addition, excision of flowers reduces internodal growth, and application of GA_3 reverses this tendancy (Figure 28). We postulated that the underlying developmental phenomenon of increasing anther size from lower- to upper-node flowers had local control of corolla expansion via GA produced in the anthers. Pollen is a known source of GA,[18] and since this hormone in anthers has been implicated in the expansion of floral parts in a number of species,[19-23] it is possible that the smaller anther size of the CL flowers results in GA levels too low to permit anthesis.

When morphologically and histologically undifferentiated primordia of the two flower types were placed on a defined growth medium, they developed into CL forms unless GA_3 was added (Figures 29 through 31 and Table 2).[24] With GA_3, CH flowers were induced which produced typical CL anthers (Figure 32) and normal ovules (Figure 33). The results suggest that GA is responsible for anthesis and organ elongation in the CH flowers; but GA_3 fails to induce a complete CH flower with respect to cell number in the corolla and anther size. Controls on early developmental processes responsible for divergence in form in the two flowers are more complex than the action of a single hormone.

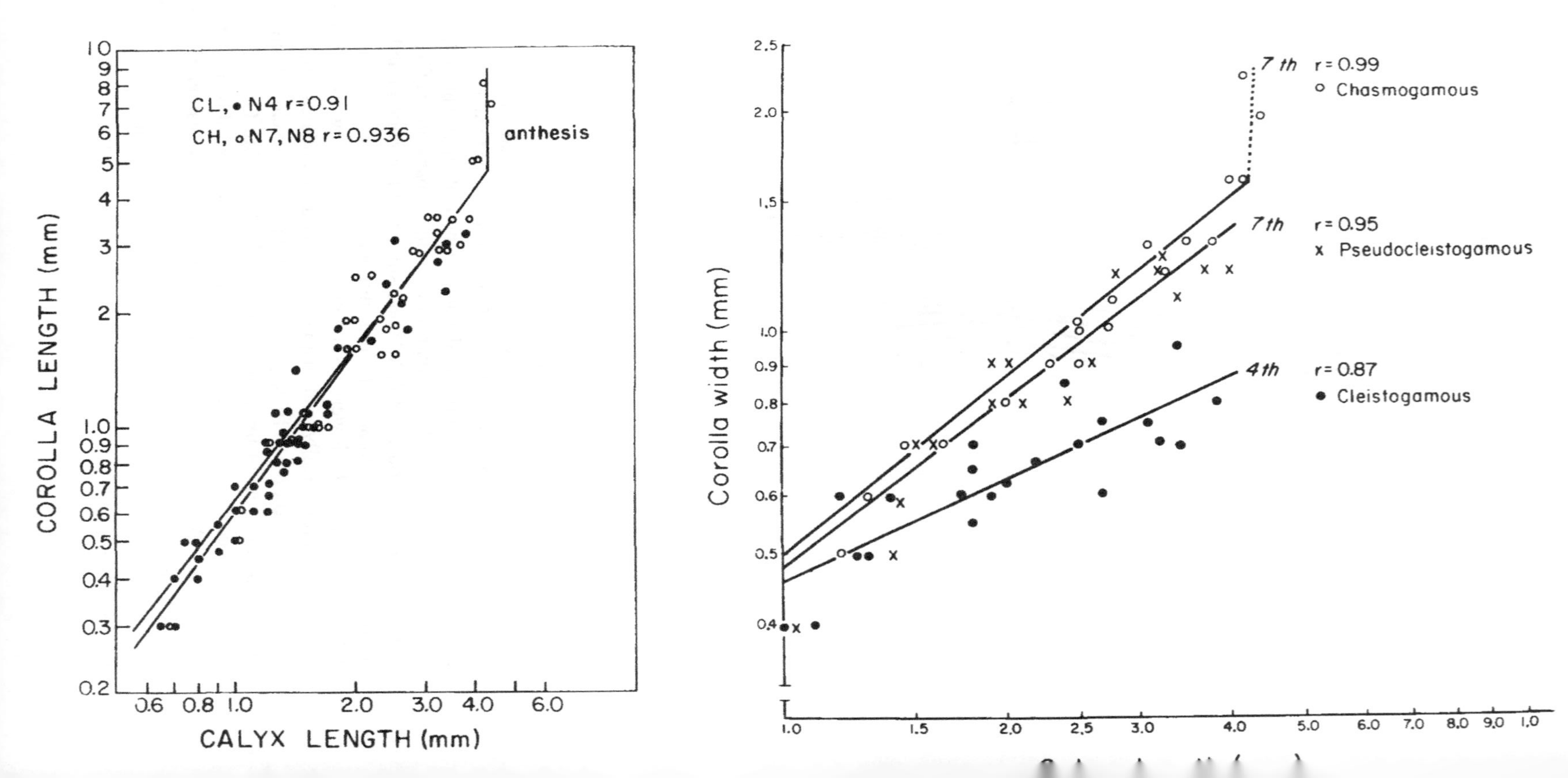
CL, • N4 r = 0.91
CH, o N7, N8 r = 0.936
anthesis
COROLLA LENGTH (mm)
CALYX LENGTH (mm)
7 th r = 0.99
o Chasmogamous
7th r = 0.95
x Pseudocleistogamous
4th r = 0.87
• Cleistogamous
Corolla width (mm)

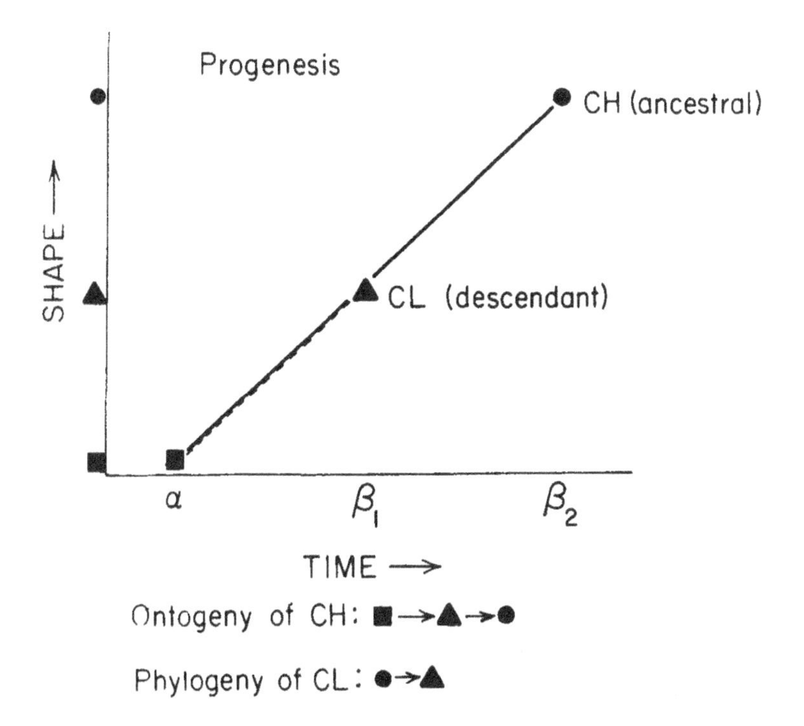

FIGURE 23. Model for the ontogeny and phylogeny of the CL floral form (descendent) from the CH floral form (ancestral). (After Alberch et al.[26]) Signs represent developmental stages; α = meiosis, β = pollination. (From Lord, E. M., *Bot. Gaz. (Chicago)*, 143, 63—72, 1982. With permission.)

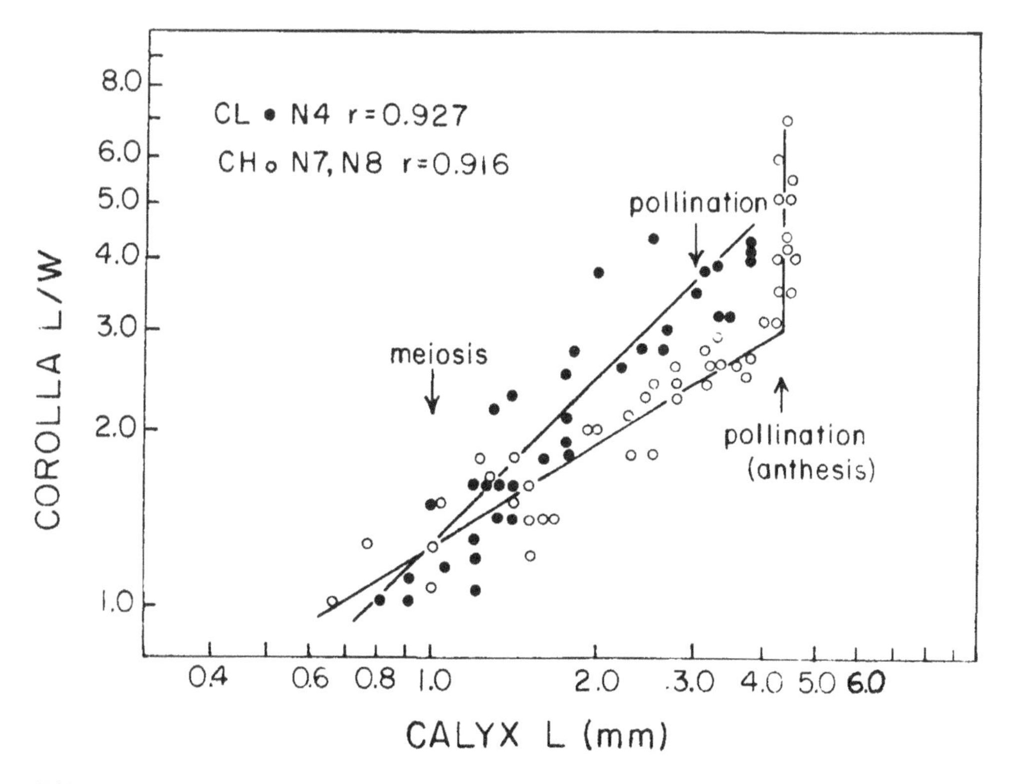

FIGURE 24. Comparative allometric growth plot of corolla length/width ratio vs. calyx length in CL and CH flowers of *L. amplexicaule*. (From Lord, E. M., *Bot. Gaz. (Chicago)*, 143, 63—72, 1982. With permission.)

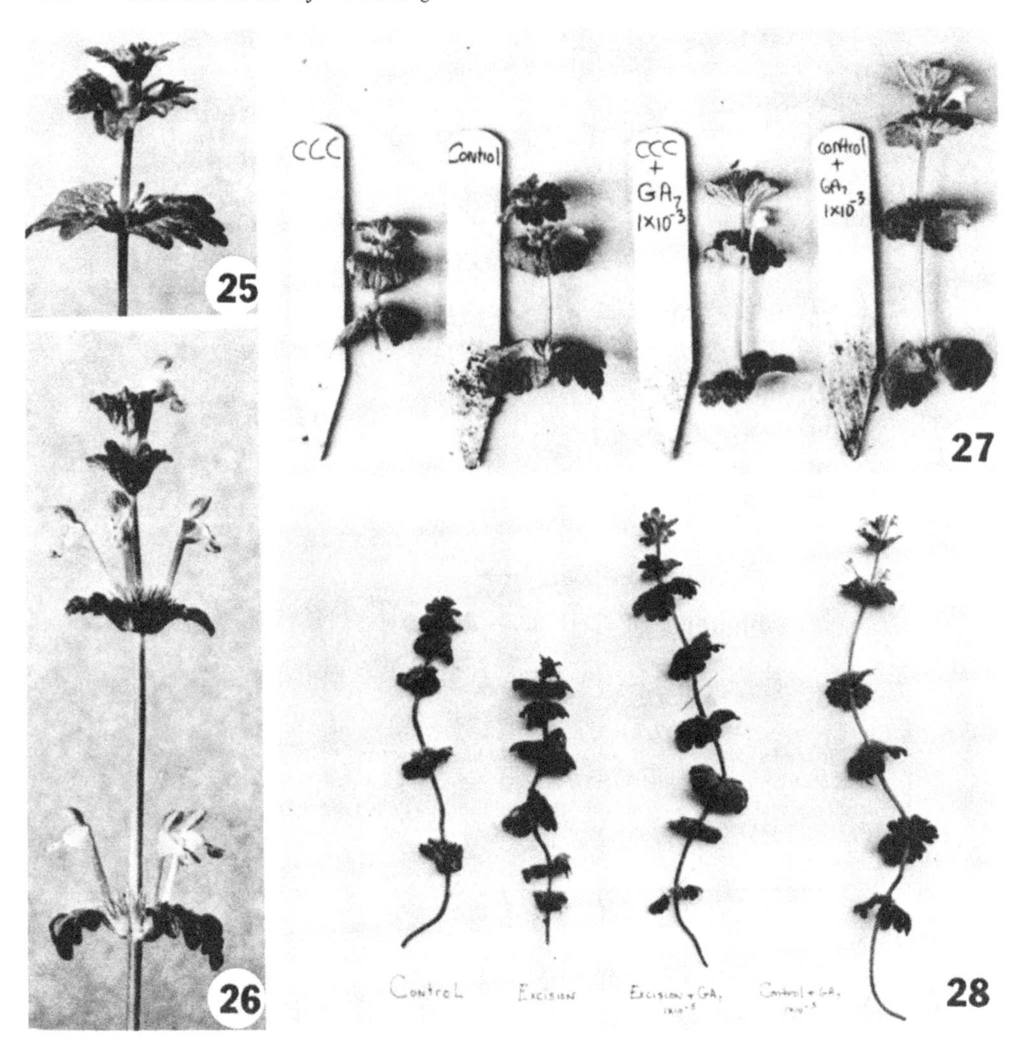

FIGURES 25 to 28. Figure 25. Control plant grown under SD bearing only closed flowers. (Magnification × 0.7.) Figure 26. GA-treated (100 μ*M* solution applied in 20-μℓ drops to shoot apex every 2 weeks) plant growing under SD conditions. Note all corollas have expanded. (Magnification × 0.7.) Figure 27. Results of a chlormequat (CCC) experiment with GA application (chlormequat, 0.06%, applied to soil: 4 days later GA, 100 μ*M* solution applied weekly in 20-μℓ drops to shoot apex). Plants grown under SD conditions. Note open flowers produced in GA-treated. (Magnification × 0.35.) Figure 28. Main shoots of a flower excision experiment. Note decrease in internode lengths in excision shoots (flowers removed). (Magnification × 0.3.) (From Lord, E. M., *Ann. Bot.*, 44, 757—766, 1980. With permission.)

To conclude, the phenomenon of cleistogamy provides a useful comparative tool for the study of floral morphogenesis, function, and evolution. In addition, there are agronomically important species that exhibit this type of breeding system, especially in the grass and legume families.[25] Since cleistogamy is an efficient means to ensure autogamous seed set, it represents a useful character for which to breed in crop plants.

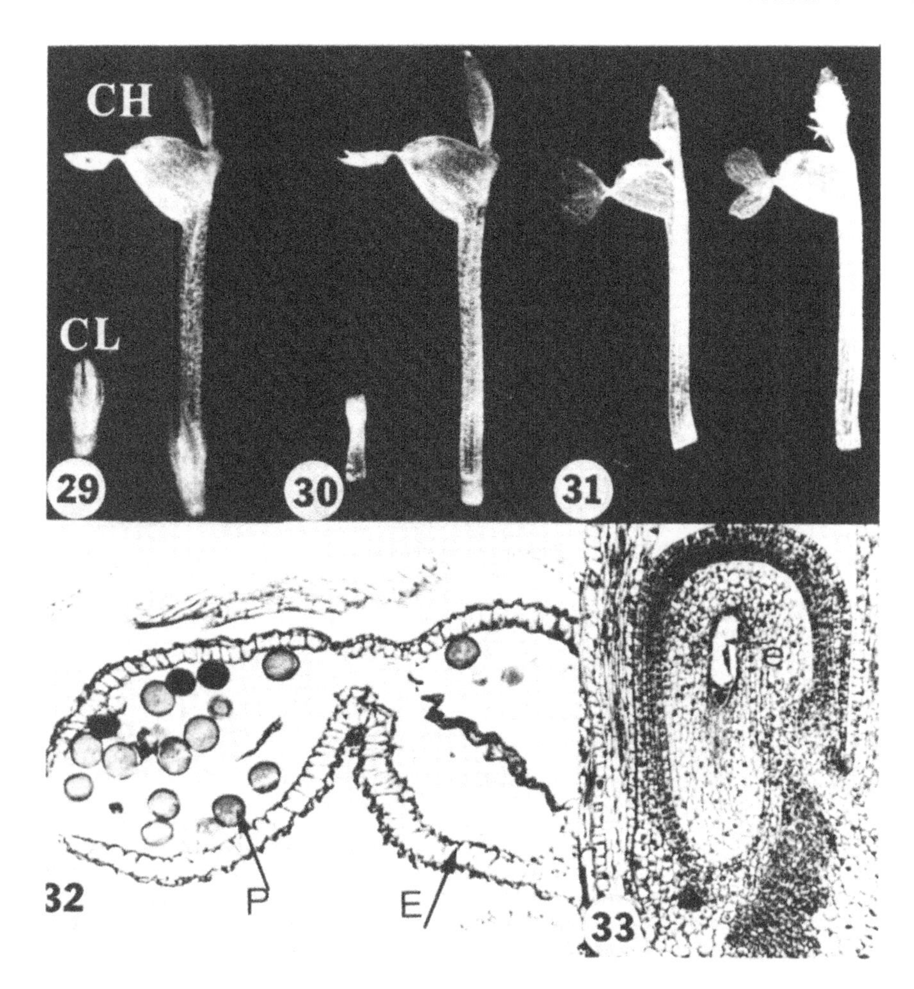

FIGURES 29 to 33. In vivo and in vitro experiments on flower opening in *L. amplexicaule*. Figure 29 shows mature CL and CH flowers. (Magnification × 3.6.) Figure 30. CL (left) corolla and GA_3-treated CL corolla (right). (Magnification × 3.6.) Figure 31. Corollas from CL (left) and CH (right) flowers grown in vitro on KGA_3 medium (K = kinetin). (Magnification × 3.6.) Figure 32. Section through an anther from CH flower cultured on KGA_3 medium. (Magnification × 180.) Figure 33. Section through an ovary from a CH flower cultured on KGA_3 medium. (Magnification × 126.) P = pollen, e = embryo, E = endothecium. (From Lord, E. M. and Mayers, A. M., *Ann. Bot.*, 50, 301—307, 1982. With permission.)

Table 2
ORGAN CULTURE OF NODE 7, 8, CH FLOWERS AND NODE 4, CL FLOWERS OF *LAMIUM*

Experiment no.	Node	Growth medium[a]	Buds developed/ buds cultured	Open corollas
1	7, 8	K	32/40	1
		KGA_3	31/38	31
	4	K	16/20	0
		KGA_3	17/25	17
2	7, 8	K	6/10	0
		KGA_3	6/10	6
	4	K	8/10	0
		KGA_3	10/10	10
3	7, 8	K	27/31	0
		KGA_3	18/29	16

[a] K = kinetin; KGA_3 = kinetin plus gibberellic acid.

From Lord, E. M. and Mayers, A. M., *Ann. Bot.*, 50, 301—307, 1982. With permission.

REFERENCES

1. **Gams, H.,** Labiatae, in *Illustrierte Flora von Mittel-Europa,* Vol. 5, Hegi, G., Ed., J. F. Lehmans, Munich, 1927, 2255—2548.
2. **Gulyas, S.,** Zusammenhang zwischen Struktur und Produktion in den Nektarien einiger *Lamium* Arten, *Acta. Biol., (Szeged),* 13, 3—10, 1967.
3. **Bernstrom, E.,** Cleisto- and chasmogamous seed setting in di- and tetrapolid *Lamium amplexicaule, Hereditas,* 36, 492—506, 1950.
4. **Bernstrom, E.,** Cytogenetic intraspecific studies in *Lamium.* I, *Hereditas,* 38, 163—220, 1952.
5. **Correns, C.,** Genetische Untersuchungen an *Lamium amplexicaule* L. IV, *Biol. Zentralbl.,* 50, 7—19, 1930.
6. **Darwin, C.,** *The Different Forms of Flowers on Plants of the Same Species,* Appleton, New York, 1877, 352.
7. **Lord, E. M.,** Cleistogamy: a tool for the study of floral morphogenesis, function and evolution, *Bot. Rev.,* 47, 421—449, 1981.
8. **Uphof, J. C. T.,** Cleistogamic flowers, *Bot. Rev.,* 4, 21—49, 1983.
9. **Lord, E. M.,** The development of cleistogamous and chasmogamous flowers in *Lamium amplexicaule* (Labiatae): an example of heteroblastic inflorescence development, *Bot. Gaz.* (Chicago), 140, 39—50, 1979.
10. **Allard, H. A.,** Flowering behavior of the hog peanut in response to length of day, *J. Agric. Res.,* 44, 127—137, 1932.
11. **Lord, E. M.,** Effect of daylength on open flower production in the cleistogamous species *Lamium amplexicaule* L., *Ann. Bot.,* 49, 261—263, 1982.
12. **Lord, E. M.,** Cleistogamy: a comparative study of intraspecific floral variation, in *Contemporary Problems in Plant Anatomy,* White, R. A. and Dickison, W. C., Eds., Academic Press, New York, 1984.
13. **Lord, E. M.,** The Development and Function of Chasmogamy in *Lamium amplexicaule* L., Ph.D. thesis, University of California, Berkeley, 1978.
14. **Lord, E. M.,** Floral morphogenesis in *Lamium amplexicaule* L. (Labiatae) with a model for the evolution of the cleistogamous flower, *Bot. Gaz. (Chicago),* 143, 63—72, 1982.
15. **Lord, E. M.,** An anatomical basis for the divergent floral forms in the cleistogamous species, *Lamium amplexicaule* L., (Labiatae), *Am. J. Bot.,* 67, 1430—1441, 1980.
16. **Goebel, K.,** Die kleistogamen Blüten und die Anpassungstheorien, *Biol. Zentralbl.,* 24, 637—697, 737—753, 769—787, 1904.

17. **Lord, E. M.,** Physiological controls on the production of cleistogamous and chasmogamous flowers in *Lamium amplexicaule* L., *Ann. Bot.*, 44, 757—766, 1980.
18. **Letham, D. S., Goodwin, P. B., and Higgins, T. J. V.,** *Phytohormones and Related Compounds: A Comprehensive Treatise,* Vol. II, Elsevier/North Holland Press, Oxford, 1978, 542.
19. **Greyson, R. I. and Tepfer, S. S.,** Emasculation effects on the filaments of *Nigella hispanica* and their partial reversal by gibberellic acid, *Am. J. Bot.*, 54, 971—976, 1967.
20. **Murakami, Y.,** The role of gibberellins in the growth of floral organs of *Mirabilis jalapa, Plant Cell Physiol.*, 14, 91—102, 1975.
22. **Plack, A.,** Effect of gibberellic acid on corolla size, *Nature (London),* 182, 610, 1958.
23. **Zieslin, N., Biran, I., and Halevy, A. H.,** The effect of growth regulators on the growth and pigmentation of 'Baccara' rose flowers, *Plant Cell Physiol.*, 15, 341—349, 1974.
24. **Lord, E. M. and Mayers, A. M.,** Effects of gibberellic acid on floral development *in vivo* and *in vitro* in the cleistogamous species, *Lamium amplexicaule* L., *Ann. Bot.*, 50, 301—307, 1982.
25. **Frankel, R. and Galun, E.,** *Pollination Mechanisms, Reproduction and Plant Breeding,* Springer-Verlag, Berlin, 1977, 281.
26. **Alberch, P., Gould, S. J., Oster, G. F., and Wake, D. B.,** Size and shape in ontogeny and phylogeny, *Paleobiology,* 5, 296—317, 1979.

LITCHI CHINENSIS

En. Litchi, lychee; Fr. Litchi; Ge. Litchi; Sp. Litchi

A. J. Joubert

TAXONOMY

The litchi (*Litchi chinensis* Sonn.) belongs to the family Sapindaceae and is indigenous to southern China. It was described as early as 1059 A.D. by Tsai Hsiang, and has been cultivated for its delicious fruit ever since. The first detailed description of the litchi, including the botanical name, was published by Pierre Sonnerat in 1782, in his book *Voyage aux Indes Orientales et a la Chine*. The genus name *Litchi* is a romanization of the Chinese name for the fruit, which is written with two Chinese characters pronounced as ''lay'' and ''chee'', meaning ''to separate the fruit from the branches with knives.''[1]

The tree is a medium to large, much-branched evergreen tree growing to a height of up to 12 m or more and having an equal spread when grown under favorable conditions.

The litchi is also being cultivated successfully in north-central India, South Africa, Hawaii, and Florida (U.S.), Queensland (Australia), Taiwan (R.O.C.), Mauritius, Israel, and to a lesser extent in other countries. Various cultivars, mostly of early Chinese origin, are grown commercially in these countries. The most popular cvs are 'San Yue Hong' (''third month red''), 'Nuo-mi-ci' or 'Lo-mai-tzi' (''glutinous rice''), 'Hak Ip' or 'Hei yeh' (''black leaf''), 'Kwai Mi' or 'Kuei wei' (''cinnamon flavor''), and 'Gui Wei', which are grown on the mainland of China and Taiwan. In Hawaii, the cvs 'Kwai Mi' and 'Hak Ip' are very popular while ''Brewster'' (''Chen family purple'') is grown in Florida. In South Africa, cv 'Mauritius' (probably similar to 'Kwai Mi') is the most popular and widely grown. The most well-known cultivars from India are 'Muzaffarpur' and 'Bengal'.

ENVIRONMENTAL LIMITATIONS OF CULTURE

The Canton delta to which the litchi is indigenous is crossed by the Tropic of Cancer and is a subtropical region with dry, cool but frost-free winters and hot, wet summers.

FLOWER MORPHOLOGY AND DEVELOPMENT

Bearing Habits

The inflorescence of the litchi tree is a well-branched terminal panicle formed on the new growth, arising from branches which did not set fruit during the previous fruiting season. Flower structure and development have been reviewed recently.[23]

The numerous individual flowers are small, apetalous and cream-colored. The flowers of a panicle do not open simultaneously, but many flowers of different ages are found on one panicle at a given time.

The panicles appear at the end of winter, and the first flowers to open during early spring are male. This cycle of male flowers is terminated by the appearance of imperfect hermaphrodite flowers functioning as female flowers. The cycle of functionally female flowers is followed by another cycle of imperfect hermaphrodite flowers, this time functioning as male flowers. These three basic types of flowers have been designated chronologically by Mustard[2] as Type I, Type II, and Type III (Figure 1). The same cyclic flowering pattern was reported from India, Israel, and South Africa.[3-5] There is an overlap between the flowering cycles of Type I and Type II as well as between Type II and Type III flowers on a single panicle and between different panicles on one tree, with the result that dichogamy does not occur.

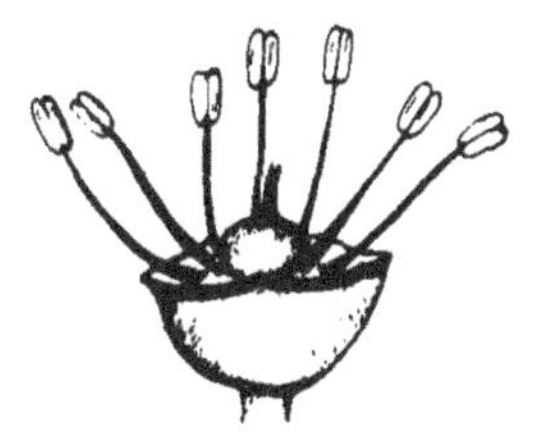

FIGURE 1. Litchi flowers, Type I (left), Type II (middle), and Type III (right).

It is generally accepted that insects, and in particular honeybees, serve as pollinators.[6] No studies have as yet been undertaken on the subject of self- or cross-incompatibility but it can be assumed that it does not occur, due to the fact that there are large orchards comprising only one cultivar in China and South Africa.[4] Campbell and Malo[7] stated that the litchi is self-fruitful and does not require plantings of mixed cvs for cross-pollination. Data from Israel indicate that certain cvs are, to some extent, incompatible.[8]

Onset of Flowering: Juvenile and Vegetative Periods

The litchi tree is slow to come into flowering, whether from seedlings, after completion of the juvenile period, or from grafted or air-layered trees after completion of the vegetative period.

Juvenile period — Litchi seedlings are not grown for commercial fruit production. The juvenile period of such seedlings is long, and it is known that they may take up to 8 years and longer to reach flowering stage.

Vegetative period — Litchi trees propagated by air layers or by grafting onto seedlings will flower about 4 years after planting. The trees are long-lived and a recent article reported that a 1200-year-old litchi tree in the eastern province of Fukien, China was still growing and producing well.[9]

Floral Differentiation

The development of the functionally female litchi flower has been studied by Banerji and Chaudhuri,[10] Mustard,[2] and Joubert.[11] The 2-lobed ovary contains two anatropous ovules. In the ovule, the megaspore mother cell can be identified 10 days before the embryo sac reaches maturity. Through meiotic divisions, the megaspore mother cell forms a linear tetrad of megaspores, of which the chalazal megaspore elongates into a one-nucleate embryo sac. After three mitotic divisions, the mature monosporic eight-nucleate embryo sac of the *Polygonum* type is formed. At maturity, the antipodals degenerate and cannot be observed after fusion of the two polar nuclei (Figure 2). During embryo sac development, the nucellus is absorbed except for the nucellar epidermis and nucellar cap at the micropylar end. Two integuments envelop the mature embryo sac. On the ventral side of the anatropous ovule the outer integument is poorly developed, and in cross section it appears only as a protuberance. An obturator is present, which, according to Joubert,[12] consists of three to four cell layers of enlarged stigmatoid tissue on the outer integument protuberance, forming a ring around the micropylar end of the mature ovule. The obturator in the litchi can be regarded as a pro-aril since it gives rise to the aril after fertilization.

The ovary wall consists of undifferentiated paranchyma, vascular tissue, and epidermal cell layers on the inner and outer surfaces. The distal surface is made up of closely associated protuberances consisting of 15 to 20 radial, and 12 to 15 tangential cell layers.

Only one ovule develops after fertilization to form the single-seeded fruit.[11]

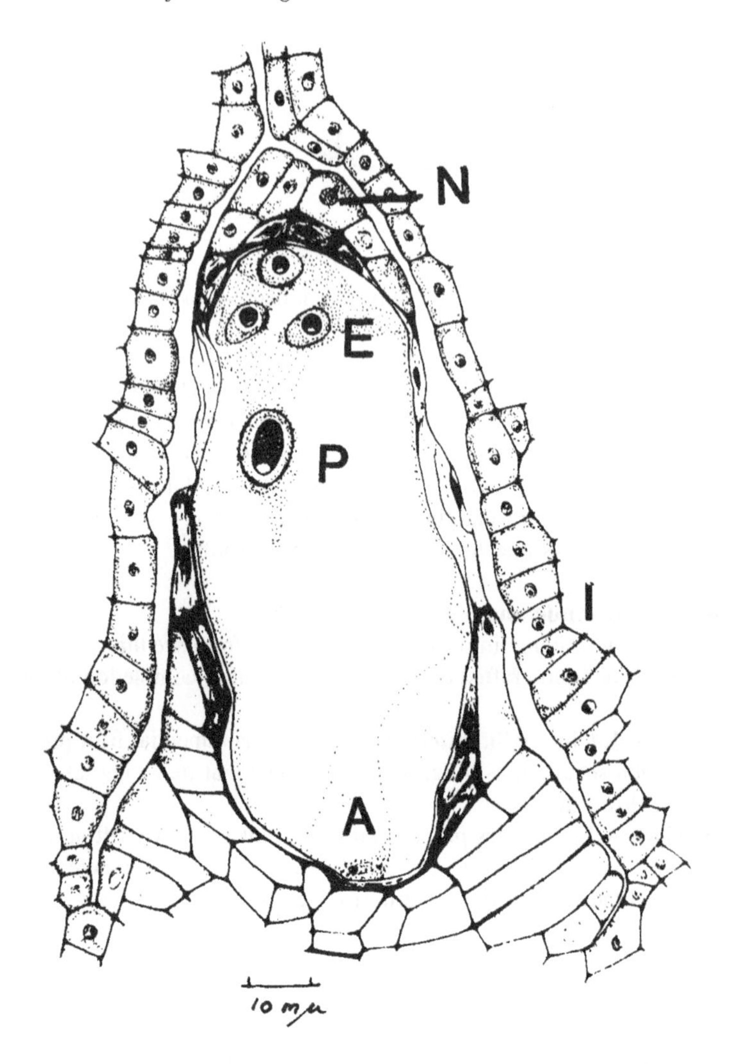

FIGURE 2. Mature embryo sac at full bloom. N. Nucellar cap; E. Egg apparatus; P. Primary endosperm nucleus; I. Integument; A. Degenerated antipodals.

FACTORS INFLUENCING FLORAL DIFFERENTIATION

Dormancy

According to the growth habit and flowering pattern of the tree, it can be accepted that the litchi requires a period of dormancy for floral initiation. Under field conditions, the periods of low winter temperatures and low soil moisture coincide and are the main factors inducing dormancy. Experiments were done to induce dormancy with growth regulators, cincturing, and root pruning, but with varying and limited success.[15] None of these methods are used commercially to induce flowering.

Temperature

It is generally accepted that the litchi tree requires a period of low temperature and/or low soil moisture during the pre-flowering period to provide the physiological changes required for flower initiation. Temperatures below 0°C, however, damage the flowers and leaves of the tree.

If the climatic data of the litchi-growing areas in Florida (U.S.), Canton (China), Mauritius, and South Africa are compared, it is evident that the average monthly minimum temperatures in winter drop below 14°C for a period of between 1 and 3 months. These cool winter months are generally rainless and such climatic conditions can be regarded as being favorable for successful flowering.[13]

A survey of world litchi-growing areas by Batten[14] revealed that the mean minimum temperature for the coldest months is approximately 10°C or lower, but the areas are frost-free and have a dry period coinciding with the low temperature period. These guidelines are the most reliable indicators of the climatic requirements for flowering of litchi trees.

Rainy weather during the flowering period in spring may restrict insect activity, resulting in reduced pollination and fruit set.

Moisture Stress and Nutrition

It is generally accepted that high soil moisture in autumn is inducive to vegetative growth and suppresses flowering, while low soil moisture restricts vegetative growth and promotes flowering.[15] This moisture stress period generally coincides with the winter period of low temperatures during the dormant season. The direct result of water stress on floral initiation has not yet been established.

Nitrogen is the most important mineral element in the nutrition of litchi trees.[16] In view of the importance of inducing dormancy during the winter, it is recommended that care should be taken to time nitrogen applications so that they do not encourage vegetative growth at the time of flower initiation. Split applications of fertilizer during spring and summer are recommended.

Photoperiod

According to Nakata and Watanabe[17] the litchi is a day-neutral plant with respect to floral initiation.

Cincturing and Root Pruning

The literature on the effect of cincturing and root pruning was reviewed by Menzel.[15] According to the literature, cincturing promoted flowering of 'Brewster' litchi trees in Hawaii, but in Florida the response to cincturing has been less spectacular. Experiments were conducted in Queensland, Australia to induce early cropping. The response to cincturing seems to depend on several factors, including plant vigor, pruning treatments, time of cincturing, genotype, temperature, and moisture conditions. Continuous cincturing can lead to retarded growth, alternate bearing, small fruit, leaf scorching and branch and tree die-back.[15]

Cincturing is not recommended as a general horticultural practice to induce flowering. Further, no experimental evidence on the effect of root pruning is available to draw any conclusions on its suitability as a method to induce floral initiation.

Growth Substances

It is reported from Florida and Hawaii that irregular bearing is a major problem in the commercial production of litchis, and attention has been given to the use of growth regulators to increase flowering and subsequent fruiting of litchi trees.[18,19] Ledin[20] investigated the use of sodium naphthalene acetate (SNA) at a concentration of 150 mg/ℓ to control dormancy and increase yield. He reported success in inducing vegetative dormancy, but found no subsequent increase in fruit yield. Investigations were made by Mustard et al.,[18] using SNA at a concentration of 100 mg/ℓ, NAA at a concentration of 30 mg/ℓ, and dinitro-*o*-cyclohexylphenol (DNO) at a concentration of 60 mg/ℓ during three consecutive seasons. Their results indicated that neither DNO, NAA, nor SNA at the concentrations used affected the vegetative condition of the trees. From Hawaii, Nakata[19] reported that SNA inhibited vegetative growth and promoted flowering only when there had been heavy rainfall during the pre-flowering period and adequate water available during the flowering period. Applications of SNA did not result in increased floral initiation and yield during a year with dry autumn months and on trees which yielded heavily during the previous year. Young[21] stated that sprays with SNA at 150 mg/ℓ and GA at 50, 250, 500, and 1000 mg/ℓ did not influence flowering or fruit set in 'Brewster' litchis at different localities in Florida. He also tested branch ringing and banding, i.e., removal of a band of bark or wrapping the branch with heavy butyl rubber bands during autumn. The ringed or banded branches flowered better and produced significantly more fruit than untreated trees for several consecutive seasons. However, he came to the conclusion that these methods did not offer a practical solution to the yield problem of litchis.

Taking all these results into account, it is obvious that climate is the determining factor in the checking of vegetative growth, promotion of dormancy, and the initation of flowers in litchi trees.

FLOWERING

The Litchi Flower

Female flower — The imperfect hermaphrodite flower functioning as a female flower has a 2-lobed, 2-locular superior ovary with a single style and bifurcated stigma. The 6 to 8 stamens are short and the anthers do not dehisce. The flowers are apetalous (Figure 3).

Male flowers — The functional male flowers of the first and second cycles differ very little morphologically. Both flower types are apetalous, and the 6 to 8 stamens are well developed, having long filaments. Both have nonfunctional ovaries. The rudimentary ovary of a first cycle male flower is smaller than that of a second cycle male flower (Figure 4).

Time and Duration of Anthesis

The litchi tree flowers in spring, i.e, March to April (northern hemisphere) and September to October (southern hemisphere). In a study with 14 different cvs during one season, it was found that the onset of flowering varied between March 1st to 26th, covering the range for early and late flowering cvs.[22] The flowering period of a single cv can be up to 6 weeks long. The first cycle of male flowers can last up to 3 weeks, overlapping with the female flowering period which is relatively short. The latter is reported for different cvs, to range from 4 to 6 days in Israel,[4] 5 to 10 days in Florida,[22] and 10 to 20 days in South Africa.[5] The last cycle of functionally male flowers also overlaps with the female flowering period and can also last for up to 3 weeks.

The dates of full female bloom were reported to vary between March 14th and April 24th (northern hemisphere) for 15 cvs during one season,[22] and between September 15th and October 4th (southern hemisphere) for one cv during 5 seasons.[5]

Blossom Longevity

No information is available concerning the length of the receptive period of the female flower. However, it is assumed that the shiny white stigmatic surface of a newly opened flower is an indication of the receptivity of the stigma. It was observed by the author that under cloudy, rainy conditions the stigma faded after 2 days. Under less favorable conditions of high temperatures and low humidity, the stigmatic surface remained receptive during 1 day only.

Abscission of Floral Buds and Defective Flowers

Shedding of litchi flowers occurs as a natural phenomenon after anthesis. All male flowers

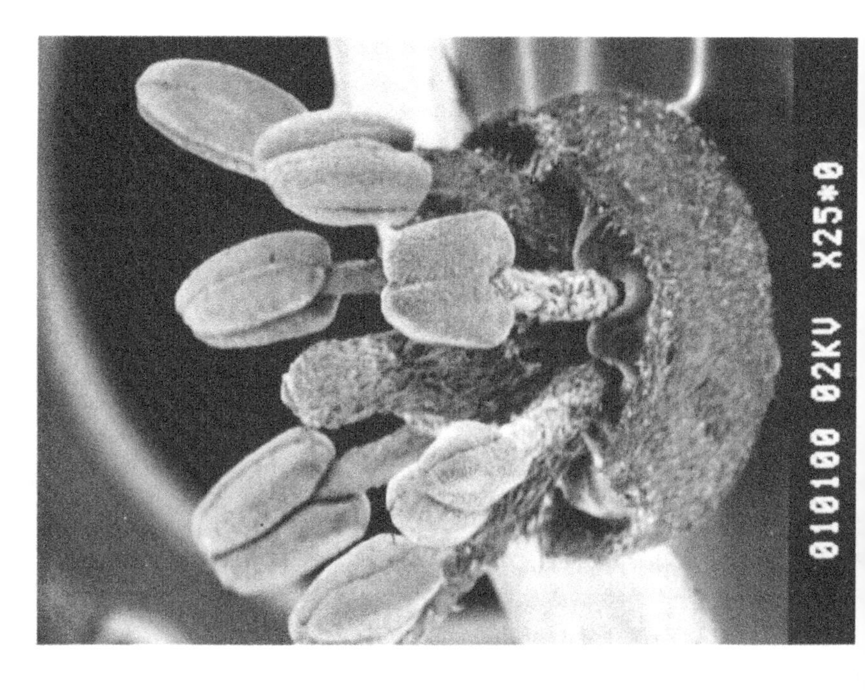
010100 02KV X25*0

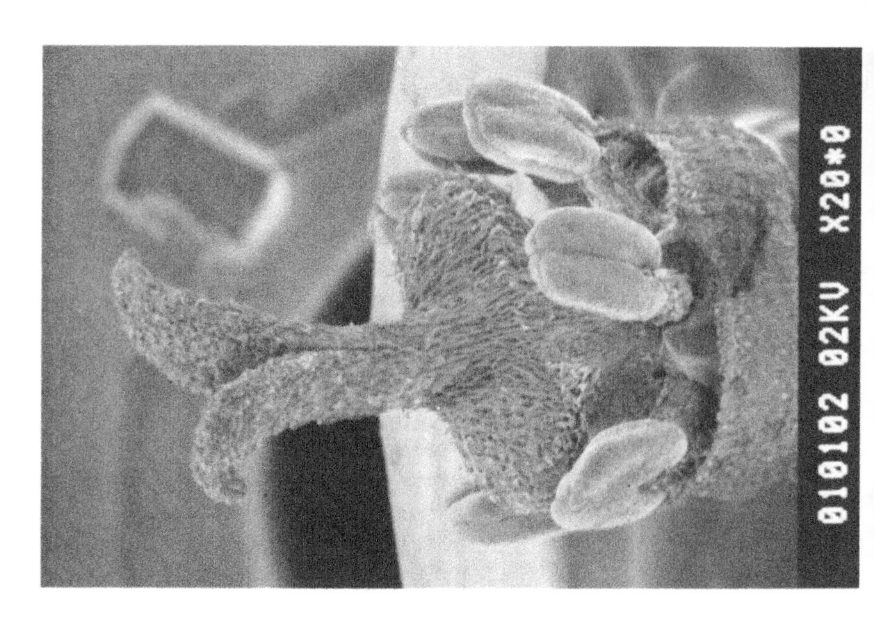
010102 02KV X20*0

are shed, as well as the unfertilized female flowers. Reference to the shedding of floral buds before anthesis or the presence of abnormal or malformed flowers could not be found in the literature, and it is assumed that abscission of floral buds is not a problem in litchi cultivation.

REFERENCES

1. **Groff, G. W.,** *The Lychee and Lungan,* Orange Judd Co., New York, 1921.
2. **Mustard, M. J.,** Megagametophytes of the lychee (*Litchi chinensis* Sonn.), *Proc. Amer. Soc. Hortic. Sci.,* 75, 292—304, 1960.
3. **Singh, L. B. and Singh, U. P.,** *The Litchi,* Superintendent Printing and Stationery, Lucknow, 1954.
4. **Pivovaro, S. Z.,** Studies of the Floral Biology and the Influence of Growth Regulators on Fruit Set and Drop of *Litchi chinensis* Sonn., M.Sc. (Agr.) thesis, Hebrew University, Rehovot, 1974.
5. **Joubert, A. J.,** *The Litchi,* Bull. 389, Gov. Printer, Pretoria, 1970.
6. **Butcher, F. G.,** Bees pollinate lychee blooms, in *Proc. Fla. Lychee Growers Assoc.,* 1956, 3, 59.
7. **Campbell, C. W. and Malo, S. E.,** The Lychee, Fruit Crops Fact Sheet 6, University of Florida, Gainesville, 1968.
8. **Monselise, S. P.,** personal communication, 1983.
9. **Anon.,** A 1200-year-old litchi tree flourishes in China. *Intern. Fruit World,* 37, 194—197, 1979.
10. **Banerji, I. and Chaudhuri, K. L.,** A contribution to the life-history of *Litchi chinensis* Sonn., in *Proc. Indian Acad. Sci.,* Sec. B., 19, 19—27, 1944.
11. **Joubert, A. J.,** Die bloei, Embriosak-, Embrio- en Vrug-ontwikkeling van *Litchi chinensis* Sonn. Cultivar HLH Mauritius, M.Sc. thesis, University of the Witwatersrand, Johannesburg, 1967.
12. **Joubert, A. J.,** 'n Studie van die arillus by die lietsjie (*Litchi chinensis* Sonn.), *Tydskrif Natuurwet.,* 9, 242—250, 1969.
13. **Joubert, A. J. and Bredell, G. S.,** Climatic Requirements for Tropical and Subtropical Crops Grown in South Africa, Report, Citrus and Subtropical Fruit Research Institute, Nelspruit, South Africa, 1982.
14. **Batten, D. J.,** Potential limitations to litchi (*Litchi chinensis* Sonn.) cultivation in Australia, in *Abstr. 20th Int. Hortic. Congr., Sydney, 1978,* [1591].
15. **Menzel, C. M.,** The control of floral initiation in lychee: a review, *Scientia Hortic.,* 21, 201—215, 1983.
16. **Koen, T. J., Langenegger, W., and Swart, G.,** Stikstofbemesting van lietsjiebome, *Subtropica,* 2(7), 9—11, 1981.
17. **Nakata, S. and Watanabe, Y.,** Effects of photoperiod and night temperature on the flowering of *Litchi chinensis, Bot. Gaz.,* 127, 146—152, 1966.
18. **Mustard, M. J., Nelson, R. O., and Goldweber, S.,** Exploration study dealing with the effect of growth regulators and other factors on the fruit production of the lychee, in *Proc. Fla. Lychee Growers Assoc.,* 3, 33—38, 1956.
19. **Nakata, S.,** Floral initiation and fruit-set in lychee, with special reference to the effect of sodium naphthaleneacetate, *Bot. Gaz.,* 117, 126—134, 1955.
20. **Ledin, R. B.,** Sub-tropical Crops of Minor Importance, Annual Report, Florida Agricultural Experiment Station, Gainesville, 1954.
21. **Young, T. W.,** An appraisal of the lychee in Florida, in *Proc. Fla. Lychee Growers Assoc.,* 6, 18—20, 1964.
22. **Mustard, M. J., Liu, S., and Nelson, R. O.,** Observations of floral biology and fruit setting in lychee varieties, in *Proc. Fla. State Hort. Soc.,* 1954, 66, 212—220.
23. **Menzel, C. M.,** The pattern and control of reproductive development in lychee: a review, *Scientia Hortic.,* 22, 333—345, 1984.

MANGIFERA INDICA

En. Mango; Fr. Laubi manguier; Ge. Mangobaum; It. Mangifera-commune

K. L. Chadha and R. N. Pal

INTRODUCTION

Even though the mango originated in India, it is now being grown in at least 87 countries, with India, Brazil, Pakistan, Mexico, the Philippines, and Bangladesh the chief producers. Other important regions growing mango are Indonesia, Thailand, Burma, Malaysia, Sri Lanka, Egypt, southeast Africa, South Africa, Israel, tropical Australia, U.S. (Hawaii, California, and Florida) and the West Indies; but nowhere is it so greatly valued as in India, where it occupies about 1 million ha or 70% of the area devoted to fruit crops. Mango occupies the same position in India as the apple in temperate climates and grapes in "Mediterranean" areas. India is the largest producer, far exceeding the other leading mango-growing countries, and contributes about 65% of the total world production.[5,10] India, however, exports only a very negligible proportion (0.02%) of its product, as fresh fruit mainly absorbed in the Gulf States. The U.S. imported 8000 tonnes in 1975, most of it from Mexico and Haiti. Of the total 15,000 t trade in world mango products, India accounts for a share of 11,000 t (1969 to 1970). At present, mango juice (nectar, etc.) constitutes the major item of export, followed by mango chutney and pickles and mango slices in brine. The Philippines, Thailand, Mexico, and India are the major exporters of fresh mangoes.

TAXONOMY

The mango belongs to the dicotyledonous family Anacardiaceae, which consists of 64 genera, mostly of trees and shrubs. All the cultivated Indian mangoes belong to the single species *Mangifera indica* L.; other well-known plants of the family are the cashew nut (*Anacardium occidentale* L.) and pistachio nut (*Pistacia vera*). A few other edible species of *Mangifera* are *M. odorata* Grif., *M. foetida* Lour, and *M. caesia* Jack, which are being cultivated in the Malaysian region. In India, only two species are found, *M. indica* L. and *M. sylvatica* Roxb. However, fruits of no other species are as good in quality as those of *M. indica*. Most of the members of the family Anacardiaceae are characterized by resin canals. The genus *Mangifera* consists of 41 valid species. However, most of them are wild and economically unimportant. Almost all grafted varieties of mango have been selected from naturally occurring seedlings, based upon their earliness or lateness and superior fruit quality. There are about 1000 varieties of mango, out of which only about 25 to 40 are of commercial importance. In India, almost all the commercial cultivars are monoembryonic, but important varieties of mango grown in other countries, e.g., Carabao, Pico, etc., are polyembryonic.

INFLORESCENCE

The mango inflorescence is primarily terminal, but axillary and multiple panicles may also arise from axillary buds quite frequently. The panicles consists of a main axis bearing many branched secondary axes (Figures 1 to 5). The secondary branches may bear a cyme of three flowers, or tertiary branches may again arise on them which bear a cyme of three flowers, each flower borne in bracteate pedicels. The flowers are closely clustered towards the apices of each branch or main axis and are either male or hermaphrodite. The total

FIGURE 1. Just-emerged panicle.

FIGURE 2. Half-developed panicle.

FIGURE 3. Panicle consists of a main axis bearing many branched secondary axes.

FIGURE 4. Terminal inflorescence.

FIGURE 5. Axillary and mixed panicles (inflorescence having flowers and leaves).

number of flowers in a panicle may vary from 300 to 3000, depending on the cv (Figure 6). The panicle of mango also varies in length from a few cm to 60 cm.[60,69] It is the hermaphrodite flowers which, after proper pollination and fertilization, set fruits. The percentage of hermaphrodite flowers in a panicle is subject to appreciable variation depending upon the early or late emergence of the panicles and the variety. The percentage of perfect flowers in the panicles of North Indian cvs, e.g., Dashehari and Langra, was found to be 30.6 and 69.8, respectively. In the South Indian cvs, it varies from 16.41 in 'Neelum' to 3.17 in 'Allampur Baneshan.' It was also found that the percentage of perfect flowers was less in early-emerged than in late-emerged panicles of 'Dashehari'.

TIME OF FLOWER BUD DIFFERENTIATION

Various studies have been done on the time of flower bud differentiation in mango. Mustard and Lynch[38] reported October to be the time of flower bud differentiation in mango under Florida conditions, but Reece et al.[50] reported that differentiation of the inflorescence begins within a very short period before the expansion of the terminal buds from December to February in the cv Haden, and the process continues during the period of bud expansion. In India, too, the time of flower bud differentiation has been reported by various workers to be from October to December. Masahib-ud-din[37] had reported August to be the flower bud differentiation time under Punjab conditions, which appears to be too early a period since there is no dormancy between fruit bud differentiation and inflorescence elongation. In the cv Baramasi, however, differentiation occurs sometimes twice a year and in certain years only once. The differentiation period in this cv is generally during May to June and

FIGURE 6. Profusely flowering 'Langra' mango tree.

September to October, which appears to be a genetic character. Fruit bud differentiation in 'Alphonso' mango, an irregular or erratic-bearing cv grown under the mild tropical climate of Dharwad (Karnataka), was initiated in early October and reached a peak by mid-November.[49] Singh[64] reported the last week of December to be the critical time for flower bud differentiation. This may vary depending on fluctuation in temperature and the previous load of fruit on the tree. Further, he did not find any period of dormancy between the time of flower bud differentiation and the time of inflorescence expansion.

TIME AND DURATION OF FLOWERING

The time of flowering in different regions is mainly governed by local weather conditions, and to some extent it may also vary in different cvs grown under the same climatic conditions.[60] Flowering may start as early as November or usually during December in Rayalseema (Andhra Pradesh) and South Konkan on the west coast of India. In northern India, mango flowers from February to March, and the period of full bloom may be sometime during the second fortnight of March. Thus, under the milder climatic conditions of southern and western India, mango may start flowering from December, whereas under the extreme climatic conditions of the north, the flowering time is comparatively more precise and late (February to March). The bloom period in eastern India is earlier as compared to the north.[69] The flowering time in the U.S. is from January to March, whereas in the Philippines the mango flowers during December to January. In Java, the local cvs flower from June to August and bear fruits to maturity during October and November. The flowering periods in Queensland and South Africa are during June to July and August to September, respectively. In Jamaica the local cvs flower from February to March, whereas in Egypt and Israel flowering occurs in February and March, respectively.

Some cvs develop all their flowers within 10 days after the first bud opens, whereas others may take several weeks or even months. There are certain cvs of mango in India such as Rumani, Bangalora, Neelum, and Alipasand, which put forth flower flushes two or three times a year, particularly when these are grown under Kanyakumari conditions. Baramasi is yet another erratic cv which may flower once, twice, or three times a year, even under north Indian conditions. The duration of flowering is for about 20 to 25 days in northern India.

FACTORS AFFECTING FLOWERING

There are several factors which affect flowering in mango.

Growth Pattern

New shoots arise mostly as laterals from axillary buds around the stump of the fruited twigs of the previous year. Such growth either remain unextended or make further extension growths in subsequent months, largely depending on the variety. Terminal growth is always in the form of an extension of shoots already produced. Growth in mango is produced in different flushes which vary from variety to variety and under different environmental conditions. Under north Indian conditions, March to April and May to June are the most important periods for the emergence of new shoots. However, stray shoots and sporadic extension growth may emerge at any time between July to October.[64] Under south Indian conditions, two active flushes occurring from February to June and October to November were reported. Three main growth flushes in February to March, March to April, and October to November were reported under western Indian conditions. Only one or rarely two periods of active growth in the dry zone of Ceylon were reported by Buell.[9] However, he reported two to six growth flushes in the wet zone, which was attributed to the irregular fruiting found in that zone. Nakasone et al.[39] reported that under Hawaiian conditions, the vegetative flushes in 'Pirie' mango are scattered throughout the year. An average of 18 months was considered necessary by them between a vegetative flush and subsequent flowering.

Most of the earlier workers strongly felt that early initiation and cessation of growth, followed by a definite dormant period, will help the shoots to attain proper physiological maturity which is essential for fruit bud initiation. However, now it is more-or-less established that growth of shoots in mango is a varietal characteristic and their fruit bud differentiation in regular bearing cvs is an annual feature. In biennial-bearing varieties, "on" and "off" year phases of the trees govern the flower bud differentiation rather than age and cessation of growth of shoots. The shoot, depending upon the cv, may stop putting forth extension growth after May or continue until September or later and the potential of these shoots to form flower buds will depend on the floriferous conditions of the tree, which in turn will be determined by the amount of fruit load carried by the tree in the previous year.[68]

At the Indian Institute of Horticultural Research, Bangalore, it was observed that when 'Langra' mango tree was sprayed with ethephon for flower induction, it resulted in shoot extension which after about 15 days produced a terminal inflorescence. This shows that maturity of wood is not essential for flowering.[14]

Nitrogen and Carbohydrate Reserves

Kraus and Kraybill[27] were the first to propose that the concentration of sugars should be greater than that of nitrogenous compounds for flowering to take place. Many workers have studied the seasonal changes in carbohydrate and nitrogen content of the shoots in relation to flowering.[54,56,60,66] Except for cv 'Baramasi,' it was found that higher starch reserve, total carbohydrate, and C/N ratio in the shoots favored flower initiation in mango. Flowering in mango in the "off" years and increased flowering in "on" years could be successfully induced by ringing of branches, a practice known to increase the C/N ratio of shoots.[30] However, it is now known that wounding stem tissues may also produce ethylene and other hormonal factors.

In Israel, Gazit[24] found that in "on" trees of the 'Haden' in December, polysaccharides were highest in the wood of shoots with enlarged apical buds and lowest in shoots of both "on" and "off" trees where the apical bud was absent. Polysaccharide levels were intermediate in shoots of the "on" trees where apical buds were dormant. However, in "off" trees where lateral flower buds were preparing to break, polysaccharide content was almost as high as in shoots of "on" trees with enlarged apical buds. Thus, shoots actually forming flowers have a high starch content; April samples (at the usual flowering time) gave similar results.

Recently many studies have been done on biochemical constituents in relation to flower bud differentiation in mango. High levels of RNA, phenols, glucose, fructose, nitrogen, phosphorus, iron, and zinc and lower levels of carbohydrate and amino acids were associated with flower bud differentiation in 'Langra', but leaf protein level appeared not to be involved with this process.[34,35] Rao et al.[45] reported that amino acids, particularly arginine and glutamic acid, increase both qualitatively and quantitatively during fruit bud differentiation in 'Alphonso' mango. Chacko and Ananthanarayanan[13] reported greater accumulation and metabolism of carbohydrates (particularly sucrose), protein, and amino acids in the bark tissue at the time of flower initiation and an almost 4-fold enhancement in the specific activities of amylase and protease in mature (10-year old) mango trees as compared with juvenile (3-year old) nonflowering trees. Suryanarayana and Rao[74] reported that glucose and fructose levels significantly decreased in 4 cvs during flower bud formation but then rose sharply, reaching a maximum during inflorescence development and panicle emergence. The sucrose level gradually increased to reach a maximum during flower bud formation and declined thereafter. A similar trend was also noted with regard to starch content. Veera and Rao[82] reported higher dry matter and total carbohydrate content at the time of flower bud differentiation, but no correlation was noted between C/N ratio and flowering in mango. Higher levels of reducing sugars and calcium and low levels of insoluble carbohydrates, nonreducing sugars, magnesium, and potassium at the time of flower bud differentiation in 'Alphonso' mango were also reported.[49]

The role of the C/N ratio, protein, and amino acids in induction of flowering was further confirmed by studies made in mango grown at Kanyakumari. Although a few cvs of mango are known to flower during the off season, production of fruits outside the normal season (April to June) is invariably exhibited by all cvs grown at Kanyakumari (8 N, 77.5 E) and the surrounding areas situated at the extreme southern tip of the Indian peninsula. This peculiar behavior is not seen anywhere else in India and the reason for this is not known. Suryanarayana[77] compared the levels of endogenous constituents of shoots in relation to off-season bearing at two places in southern latitudes at Kanyakumari (8 N, 77.5 E) as against single flowering at Coimbatore (11.2 N, 76.6 E), 500 km distant. The C/N ratio, protein, and amino acid values were generally higher at Kanyakumari. He concluded that high metabolic activity and continuous production of a number of vegetative flushes which mature at different times of the year thereby make off-season flowering possible.

In many instances, no correlation was found between the flower bud differentiation and nitrogen and carbohydrate composition of shoots; but, nitrogen and carbohydrate reserves play an important role in flower bud initiation, though they are not the primary factor. Accumulation of these compounds may create a favorable condition for the synthesis and action of substances actually responsible for flower induction in mango.

Effect of Phosphorus Content

The investigations carried out at the Indian Agricultural Research Institute (IARI) on cv Dashehari had indicated a high level of phosphorus in shoots to be favorable for flower bud initiation. During the month of August. there was not much difference in the level of different fractions of phosphorus in potential flowering and nonflowering shoots. However, as the time of flower bud initiation approached, i.e., November to December, the level of phosphorus fractions, except nucleic acids, became quite high in the flowering stems. The leaves from flowering shoots also registered a high level of phosphorus, but differences were not as marked as in stems.

It appears that flower bud formation in mango is not under the control of a single factor but is govered by proper adjustment of a number of factors including high level of phosphorus, especially of organic fractions.[67]

Crop Load as Related to Flower Formation

In mango, fruiting is an exhaustive process. The number of fruits retained until harvest is a varietal feature. However, the total number of fruits that are harvested is important because of its adverse effects on production of new shoots and their subsequent fruit bud differentiation. Thimmaraju[80] reported that when the number of fruits harvested was optimum, a lower number of shoots was produced which in the following year failed to differentiate fruit buds in 'Dashehari' mango. In the ''off'' year when there were no fruits, a large number of new shoots was produced which successfully differentiated fruit buds in the following year. Thus, the fruit load on the tree appears to be the main factor governing production of shoots, their fruit bud differentiation, and ultimately the biennial bearing.

In contrast, regular bearing cvs like Neelum, Bangalora, and Baneshan produce a large number of shoots during the time of fruiting and immediately after harvest which would in turn differentiate flower buds in the coming season. In such cvs even unextended shoots which carried fruits up to maturity during the previous season, differentiate fruit buds in the succeeding year.[65] Singh,[65] after a comprehensive study of various factors contributing to flower bud initiation in a number of regular and biennial cvs, concluded that when the necessary stimulus is present, fruit buds can differentiate and develop from any point of the tree irrespective of the size and nature of shoots or sometimes even without new leaves; otherwise, not.

Hormonal Control of Flower Formation

It has long been speculated that some flowering stimulus is synthesized in the leaves which is responsible for flowering in mango. Reece et al.[50] were the first to show that the flower-inducing stimulus in mango is synthesized in the leaves. Singh[59] demonstrated that the flower-inducing stimulus can be transmitted from a mature tree to juvenile mango seedlings through grafting, resulting in flowering of the young stalk. However, he observed that the donating scion shoots failed to induce flowering in the nondefoliated seedling stock. He proposed that some substances (growth promoting or inhibiting) produced in the leaves of the accepting seedling counteracted the action of the flowering hormone donated by the scion, resulting in lack of flowering. It has been further shown through grafting experiments that the response to flowering in the accepting seedling was the same, bearing no relation to the fruiting habit (regular or biennial) of the variety involved. Thus, the nature and action of the flower-inducing hormone is the same in both regular and biennial-bearing cvs. The newly emerged leaves in the shoots of regular bearing cvs such as Neelum were capable of synthesizing the flower-inducing hormone.

In alternate-bearing cvs such as Dashehari and Langra, the shoots in the "off" year arise during March to April and most of them flower in the following "on" year. However, in the "on" year the new vegetative shoots generally arise in May to June but do not flower in the following year. A few of the vegetative shoots which sprout even during March to April during the "on" year either completely fail to flower the following year or only a few of them may flower. Thus, it is the "on" and "off" year condition of the tree that governs whether a shoot arising early or late in the season will differentiate flowers or remain vegetative.[60] In regular-bearing cvs such "on" and "off" year conditions do not exist, possibly because of the production of the flower-inducing hormone even in young leaves.

Auxins

Lal and Ram[29] reported that shoot tips of cv Dashehari contain 2 acidic and 1 nonacidic auxins. In December, which corresponds with the period of flower bud differentiation, shoot tips of 'Dashehari' in an "on" year contained about 35 times more of auxin than those in "off" years. Chacko[12] has shown that the shoots from 'Dashehari' during the "on" year and 'Totapuri Red Small' trees contained higher levels of growth-promoting substances during the period of flower bud initiation than the shoots of 'Dashehari' "off" year trees, which remained vegetative. Thus, auxins appeared to play a role in the induction of flowering in mango.

Gibberellins

Pal and Ram[43] isolated 8 GA-like substances from mango shoot-tips, tentatively identified as A_1, A_7, A_9, A_4, A_3, A_5, A_6, and one unidentified GA X. The endogenous level of GAs was higher in "off" year shoot-tips than in "on" year shoot-tips. This suggests that the failure of flowering in the "off" year may be accompanied by higher levels of GAs in the shoot tips. The exogenous application of GA_3 (10^{-1}, $10^{-2}M$), applied on the buds of "on" year 'Dashehari' mango trees, just before flower bud differentiation, inhibited flowering by 95 and 75%, respectively.[26] Lower concentrations (10^{-3} and $10^{-4}M$) inhibited flowering to a much lesser extent, but delayed the emergence of panicles by nearly 2 weeks. Shawky et al.[57] and Bakr et al.[4] also reported delay and inhibition of flowering in mango by GA_3 applications.

Cytokinins

Studies were made of the endogenous cytokinins in the shoot-tips of mango cv Dashehari between September and February in "on" and "off" years.[1] Eleven cytokinin-like substances were isolated in the "on" year, including zeatin riboside and zeatin. Cytokinin levels at the time of flower bud differentiation (December to February) were higher in the "on" year than in the "off" year. These results suggested that flowering in mango shoot-tips is associated with high levels of endogenous cytokinins. The effect of endogenous growth substances on fruiting in mango was also reported from the Republic of China.[19]

Lipids

Eleven endogenous lipid-like factors were isolated from mango shoot-tips. Higher levels of lipids were observed in the "on" year in comparison to the "off" year,[28] suggesting their probable role in flowering of mango. A lipid fraction was present in traces only in fruiting shoots of 'Dashehari', as reported by Singh and Singh.[62]

Inhibitors

While studying inhibitory substances in the shoot, Chacko et al.[18] reported relatively higher levels of inhibitor (similar to abscisic acid) in "on" year shoots of 'Dashehari' and 'Totapuri Red Small' during fruit bud initiation than in "off" year shoots. They concluded that growth-inhibiting substances may be involved in the flowering of mango. Singh and Singh[62] found that there was a high level of growth-inhibitory substances in the shoots when the fruits were at the pea stage, and their levels then gradually decreased to a low level as the fruit developed. Chowdhury et al.[20,21] were of the opinion that in cv Langra, the level of inhibitor was low during the active stage of shoot growth (March) and increased sharply during cessation of growth from April to August. The level also remained high in the shoot till fruit ripened but became low at the time of flower bud differentiation.

Ethylene

Saidha et al.[52] studied the seasonal changes in the leaf ethylene levels as influenced by pruning and thinning treatments in 'Mulgoa', 'Neelum', and 'Bangalora' cvs in relation to flowering. There was a gradual increase in leaf ethylene levels in all cvs from June to November, with a sharp increase in December coinciding with inflorescence emergence. The ethylene levels in all cvs were higher during 1979 (normal year) than during 1978 (lean year). Flowering shoots showed 3 to 5 times higher ethylene levels compared to vegetative shoots.

Environmental Factors

Environmental factors are also associated with the biennial bearing of mango trees. Frost, high temperature accompanied by low humidity, and hail storms directly damage the fruit bud and developing fruits, respectively, and thereby reduce the crop considerably. Cloudy weather and rains during the flowering period reduce the crop indirectly by creating favorable conditions for the spread of diseases and pests. Besides, mango trees in extremely humid places and under milder climatic conditions may not flower at all, owing to their increased tendency towards vegetative growth. Even if they flower, owing to excessive humidity, pollen is never in a suitable condition to be transferred by insects for pollination purposes. When the flowers are damaged completely by environmental factors, an "on" year may be converted into an "off" year. The following year, which would have been normally an "off" year becomes an "on" year in which an excessive number of fruit buds are laid down, resulting in heavy fruiting. Withholding water for 2 to 3 months preceding flowering induces vigorous flowering. Excess soil moisture after October may even retard formation of blossoms.

MEASURES TO OVERCOME BIENNIAL FLOWERING

Deblossoming and Thinning of Fruits

Earlier work on control of biennial bearing in mango related mainly to deblossoming in the "on" year. It has been observed that partial or complete removal of flowers or young fruits in the "on" year increased flowering during the next year, though only to a slight extent.[53] Singh and Khan[58] suggested that deblossomed shoots in mango tend to bloom the following year and partial defloration is of importance in regulating the crop. Singh[60] reported that deblossoming response was precise in 'Dashehari' but not so in the case of 'Langra'. 'Dashehari' trees, deblossomed either half or branch-wise, yield a fair crop every year, but no such tendency is noticeable in 'Langra'. The deblossoming response appears to be less marked in overvigorous or vigorous trees. Singh[67] is of the view that deblossoming of a few individual shoots is of no consequence in mango. However, branch-wise regulation of crop in varieties like 'Dashehari' is quite possible by resorting to deblossoming only once in 10 years. Thus, instead of taking a full load of crop in one year, it can be distributed over 2 years for better profit. Some de-fruiting experiments were also conducted at IARI, New

FIGURE 7. Effect of ethephon on flowering in mango. (A) A juvenile plant which flowered after spraying with 1000 ppm ethephon; (B) a

FIGURE 8. Profuse flowering in mature mango tree after spraying with ethephon.

FIGURE 9. Flowering panicles emerging from dormant buds on main branches which have been sprayed with ethephon.

FIGURE 10. Cauliflory observed in mango after spraying with ethephon.

Delhi, and it was found that de-fruiting at the pea stage was relatively more effective than at other stages of fruit growth. Further, emergence of shoots early in the season does not ensure flowering in such shoots in the following year. However, owing to a huge amount of fruit drop, especially at the initial stages of fruit growth and the exhaustion caused by the fruitlets when these are still in the pea stage of growth, de-fruiting at a later stage may not be a feasible practice. The tree takes 1 year to recoup, thus resulting in biennial-bearing in mango.

Recently a number of abscission-promoting chemicals such as ethephon, cycloheximide, MB 25105, Pik-off, carbaryl, Gebutox, etc. were tried for chemical deblossoming in cvs Dashehari and Mallika.[11,41] Very good results were obtained with ethephon (500 ppm) and cycloheximide (250 and 500 ppm) treatments.

Pruning

Mango, being an evergreen tropical species, is seldom pruned. However, pruning in mango gave flowering and fruiting every year in old 'Mulgoa' trees.[44,45] The trees were pruned either in the second fortnight of August or the first fortnight of September under Tamilnadu conditions, i.e., 4 to $4^1/_2$ months before the expected time of flowering. First-year pruning in the case of large, densely growing trees consisted of removal of a few branches all around the tree and in the center in such a way as to open the tree to admit more light and air. The terminal clusters of shoots and young flushes of leaves which are normally 3 to 5 in number were thinned out to 1 or 2 healthy ones. This was considered to be more important than pruning of large branches. This was done all around the tree on the periphery and inside the canopy. Dried twigs and branches were completely removed.

Manures were applied 1 month after pruning. Pruning in subsequent years consisted of mere thinning of the terminal whorls of shoots, particularly young flushes. There was no need to prune large branches every year. It was observed that in addition to the crop on the terminal ends of the shoots, the fruits were produced directly from old branches in pruned trees.[46] It was postulated that removal of 15 to 20% of vegetative growth by pruning may increase the ratio of root to shoot which is expected to bring about an optimum cytokinin-gibberellin balance to promote flowering every year.

Pruning has also been tried in South Africa on cv Peach of mango.[2] Average yields were highest (304 kg per tree) with benefit pruning, compared with 99, 40, and 226 kg per tree with selective, severe, and no pruning, respectively.

Smudging

Smudging or smoking mango trees to induce out-of-season flowering and fruiting is a unique feature of mango growing in the Philippines.[25,81] A mango tree is ready for smudging if it has an appearance of suspended growth. At this stage, the leaves are dull green or greenish brown in color and are brittle when crushed in the hand. The terminal buds should be dormant but well formed. The smudging operation consists of building a smoky fire below the tree canopy and directing the smoke to pass through the foliage. The common practice is to make the first smudge big and hot. Later, green grass, rice hulls, etc. are placed on top of the smudge to produce smoke. Some smudgers keep the first smudge burning day and night for quick results. Others smudge only during the day. Smudging is often done in December and early January to advance flowering and fruiting by a few weeks. However, it is also done at any time of the year whenever a tree becomes ready for smudging. Through this practice, mango fruits are now available almost the whole year round and out-of-season fruits command a premium price. However, under Indian conditions, smudging was not found to be effective in inducing flowering in mango.

Effect of Potassium Nitrate

Recent studies in the Philippines have shown that polyembryonic cvs are dramatically influenced by KNO_3 sprays.[3,6-8] Potassium nitrate at 10 to 40 g/ℓ induced flowering of 4.5- to 8.5-month old 'Carabao' mango shoots. The oldest shoots (8.5 months) required only 10 g KNO_3 per liter to produce the best flowering response, whereas younger (4.5 to 7.5 month) shoots were most responsive to 20 g/ℓ. At all stages of maturity, higher concentration, i.e., 40 g KNO_3 per liter reduced the percentage of flowering, panicle length, number of flowers, primary panicle branches, and sex ratio. Studies were also conducted on other polyembryonic varieties, e.g., 'Pahutan' and 'Pico'. Thus, KNO_3 can modify the flowering behavior of mango and makes it possible to induce flowering and fruiting at any time of the year, breaking the biennial bearing habit of mango. However, under Indian conditions, KNO_3 did not induce flowering in mango. Pal et al.[42] studied the effect of potassium nitrate, calcium nitrate, and magnesium nitrate at 10 g/ℓ each on cv Dashehari continuously for 5 years from 1975 to 1979. Whole trees were sprayed once on January 1, i.e., 1 month before actual flowering. Potassium nitrate and other nitrates failed to induce significantly more flowering as compared to control trees when the data were pooled for 5 years. The differences may be due to the fact that growth under Philippine conditions is continuous, but under Indian conditions occurs in distinct flushes in mango. The commercial cvs of the Philippines are polyembryonic whereas Indian cvs are monoembryonic.

Use of Plant Growth Regulators

Various growth retardants have been tried for induction of flowering in mango. Maiti and Sen[33] and Maiti et al.[32] reported beneficial effects from both daminozide and chlormequat on flower induction in 'Langra'. Chlormequat at 2000 ppm was found to be better than other treatments. Chlormequat not only increased the percentage of flowering shoots but also induced the axillary buds on the treated shoots to flower along with the apical buds and thereby significantly increased the average number of panicles per shoot as compared to the control. This was considered to be due to partial loss in apical dominance brought about by the repeated sprays of chlormequat. Das and Panda[22] reported that treatment with daminozide increased the flowering to 66% in shoots of 'Banganpalli' mango. Mukhopadhyay[36] reported that chlormequat at 5000 ppm significantly increased the production of panicles and hermaphrodite flowers in 'Langra' and 'Baramasi' mango. Daminozide (5000 ppm) and L-methionine (300 ppm) did not exhibit any increasing effect on production of panicles per shoot but significantly increased the percentage of hermaphrodite flowers. Rath and Das[48] reported that in "off" year 'Langra' trees, chlormequat (300 ppm) on ringed shoots produced the greatest number of flower panicles (62.3%) as compared to control (8%). Ethephon at 400 ppm also enhanced the flowering time by 16 days.

Suryanarayana[71-73,75,76,78,79] studied the effect of foliar application of chlormequat and daminozide each at 5000 ppm at monthly intervals from April to December on flowering and on different biochemical parameters. Flowering was early and heavier in treated trees. The activity of proteolytic enzymes was lower in growth-retardant-treated trees as compared to controls, which may have been responsible for retardation of protein degradation. Both retardants also reduced the rate of respiration but increased the levels of RNA, proteins, sucrose, starch, chlorophyll, carotene, xanthophyll, and ascorbic acid which ultimately lead to enhanced flowering in mango. However, no quantitative relationship between amino acid content and flowering was established.

Chacko et al.[15-17] have shown the possibility of using ethephon — an ethylene-releasing compound — for flower induction in 'Langra' trees during the "off" years. Heavy uniform flowering and high yields were obtained in the "off" year by 5 applications of 200 ppm of ethephon at fortnightly intervals starting from the middle of September. Panicles in the treated trees emerged directly from the axillary buds of the previous year's fruited stump. Also, a large number of panicles were also reported to emerge even from dormant buds located in the woody branches of the treated trees, indicating the extreme floriferous conditions induced by chemical treatment.

Mango, like most trees, exhibits a distinct juvenile phase. Normally, it takes about 6 years or more for mango seedlings to start flowering. Five spray applications (1000 ppm) of ethephon at weekly intervals during November and early December, induced heavy flowering in ringed and nonringed juvenile mango seedlings.[14] The treated plants produced flower buds which opened by the end of December, while the control and ringed seedlings produced only a new flush of vegetative growth. This may be used for early evaluation of hybrid seedlings in mango.

Dutcher[23] reported that weekly or twice-monthly spray applications of 125 to 200 ppm ethephon resulted in flowering in 'Carabao' mango in the Philippines within 6 weeks from initial treatment. Histological studies of buds showed that floral differentiation of treated trees was induced at least 4 weeks earlier than the control. Pandey et al.[44] also reported the usefulness of ethrel in inducing flowering in the "off" year in 'Dashehari' mango.

Spraying 1000 ppm ethrel three times at 2-week intervals starting 1 month before the initiation of flowering, and once at 500 ppm 1 month before the normal date of flowering, increased the number of panicles by 55 to 40%, respectively, in 10-year-old 'Haden' mango trees.[39] Ethephon treatment at 200 ppm, applied in October, combined with girdling in September, increased flowering in "on" year trees of the two mango cvs Langra and Ewaise; whereas TIBA at 50 ppm in October stimulated flowering in "off" year trees.[70]

Sen et al.[55] reported that ethephon at 250 and 500 ppm inhibited the vegetative growth and promoted flowering in "on" year in Langra and Bombai cvs but failed to induce flowering in any of the concentrations tested (100, 200, 400, and 800 ppm) during the "off" year. They concluded that efficacy of ethephon as a flower-inducing chemical in the "off" year is doubtful. The chemical may promote flowering and fruiting only when conditions for bearing are not otherwise unfavorable.

Pal et al.[42] studied the effects of ethephon (100 and 200 ppm) and chlormequat and daminozide at 500 ppm each on induction of flowering in 'Dashehari' mango continuously on the same trees for 5 years (1975 to 1979). Whole trees were sprayed with different plant growth regulators five times at monthly intervals beginning from the middle of September. Considering the 5 years' data together, none of the chemicals were found to be effective in inducing flowering as well as yield in mango.

It must be emphasized that most of the work on the effect of ethephon and other plant growth regulators on induction of flowering in mango has been carried out for 1 or 2 seasons only. Further, usually the experiments have been done on a shoot basis, which is not very desirable in perennial fruit trees like mango. The experiment should be done on a whole tree basis, so that useful results may be recommended to the growers to overcome the problem of biennial bearing. Ethephon may increase the flowering in "on" years or may promote flowering initially for 1 or 2 years, but experiments should be continued for at least 4 years (i.e., 2 "on" and 2 "off" years) and only then can some meaningful conclusions be drawn on the effect of ethephon and other plant growth regulators in inducing flowering in mango.

Possibilities of Growing Regular Bearing Cultivars

Most of the commercial cvs of north India, such as Dashehari, Langra, Chausa, and Bombay Green are biennial in bearing habit. Some of the south Indian cvs such as Neelum, Bangalora, and Totapuri Red Small show a pronounced tendency towards regular bearing. Neelum and Bangalora bear regularly even under north Indian conditions. However, trees of Neelum, instead of being vigorous or semi-vigorous, become dwarfed in north India. In particular, regular bearing is responsible for this dwarfing effect. Although inferior in fruit quality, Neelum and Bangalora can be recommended for obtaining regular fruiting and for making fruits available late in the season. Some cvs of West Bengal such as Himsagar, Anupam, and Safdar Pasand which show comparatively less bienniality, need popularization in areas where these can suitably be grown. Further, a thorough search of the existing germplasm of mango should be made to locate more regular-bearing cvs having good fruit quality.

According to Singh[69] any hybrid which shows shoot growth on the fruited twig in the same season of fruiting, and such growths are able to differentiate flower buds in the following season, can be termed a potentially regular-bearing hybrid. Inheritance of this essentially regular-bearing characteristic has considerably shortened the time involved in assessment of the hybrids in regard to their bearing tendency. Thus, by raising a large number of hybrids of desirable combinations, it would be possible to evolve a regular-bearing hybrid. Recently 2 regular-bearing and dwarf mango cvs Mallika and Amrapalli were evolved at IARI, New Delhi. These are crosses between the regular bearing Neelum and biennial-bearing Dashehari cvs.

Cultural Practices

A number of cultural practices have been tried in mango to reduce the intensity of biennial flowering and fruiting. Sen[53] observed that regularity of bearing in mango is a cultural problem which can be corrected by influencing nutritive conditions of the tree, but other workers reported the opposite view. Singh[61] has shown that liberal manuring, irrigation, and ploughing given to 20-year-old biennial-bearing trees failed to induce regular cropping in them. Thus, it appears that while nutritional requirements must be met for optimum performance, biennial bearing cannot be overcome by addition of nutrients.

Roy[51] recommended 700 g of nitrogen for a 12-year tree. It should be doubled during the "on" year to avoid the exhaustion of the tree. The recommended mineral requirement of mango was NPK 1.1:0.27:1.0.

CONCLUSIONS

Flowering and fruiting in mango is complex, and it is not possible to pinpoint a single factor for biennial bearing. The floriferous condition of the tree is probably obtained by combination of factors.[68] However, high levels of endogenous auxins, cytokinins, ethylene, and inhibitor (similar to ABA) and quite low levels of GAs are vital for a floriferous shoot in mango. The fact that some cvs of mango do flower and fruit every year under environmental conditions similar to conditions under which others fail to do so indicates the genetical nature of the biennial-bearing problem in mango. The work on mango hybridization has shown that the regular-bearing habit can be transmitted to the F_1 hybrids. There are, thus, more chances of tackling the problem of "biennial" bearing through hybridization rather than any other means. More systematic work needs also to be done on endogenous plant growth substances and their role in flowering of mango.

REFERENCES

1. **Agarwal, A., Ram, S., and Garg, G. K.,** Endogenous cytokinins of mango shoot tips and their significance in flowering, *Indian J. Exp. Biol*, 18(5), 504 — 509, 1980.
2. **Anon.,** Mango pruning methods evaluated, Information Bull. Citrus and Sub-Tropical Fruit Research Institute, *South Africa Bull. No. 80*, pp. 8—9, 1979.
3. **Astudillo, E. O. and Bondad, N. D.,** Potassium nitrate induced flowering of 'Carabao' mango shoots at different stages of maturity, *Phillipp. J. Crop Sci.*, 3(3), 147—152, 1978.
4. **Bakr, E. I., Abdalla, K. M., Meligi, M. A., and Ismail, I. A.,** Floral differentiation in mango as affected by growth regulators, ringing and defoliation, *Egypt J. Hortic.*, 8(2), 161—166, 1981.
5. **Bondad, N. D.,** World mango production and trade, *World Crops*, 32(6), 160—168, 1980.
6. **Bondad, N. D. and Apostol, C. J.,** Induction of flowering and fruiting in immature mango shoots with potassium nitrate, *Curr. Sci.*, 48(13), 591—593, 1979.
7. **Bondad, N. D. and Linsangan, E.,** Flowering in mango induced with potassium nitrate, *HortScience*, 14(4), 527—528, 1979.
8. **Bondad, N. D., Mercado, E. L., Apostol, C. J., and Astudillo, E. O.,** Smudging and potassium nitrate spray of mango—comparative effects and costs, *Indian J. Hortic.*, 36(4), 369—375, 1979.
9. **Buell, E. P.,** Flowering and fruiting habits of the mango in the wet zone, *Trop. Agric.*, 110, 280—284, 1954.
10. **Chadha, K. L.,** Mango industry in India. Ad-hoc government consultation on the improvement and development of the mango industry in Asia and the Far-East, Bankok, October 13 to 15, 1975, pp. 1—58.
11. **Chadha, K. L., Pal, R. N., and Sahay, R. K.,** Effect of abscission promoting chemicals on de-blossoming of mango, *Indian J. Hortic.*, 36, 238—243, 1979.
12. **Chacko, E. K.,** Studies on the Physiology of Flowering and Fruit Growth in Mango, Ph.D. thesis, Indian Agricultural Research Institute, Delhi, 1968.
13. **Chacko, E. K. and Ananthanarayanan, T. V.,** Accumulation of reserve substances in *Mangifera indica* L. during flower initiation, *Z. Pflanzenphysiol.*, 106(3), 281—285, 1982.
14. **Chacko, E. K., Kohli, R. R., Swamy, R. D., and Randhawa, G. S.,** Effect of 2-chloroethyl phosphonic acid on flower induction in juvenile mango *(Mangifera indica)* seedlings, *Physiol. Plant.*, 32, 188—190, 1974.

15. **Chacko, E. K., Kohli, R. R., and Randhawa, G. S.,** Studies on the effect of 2-chloroethyl phosphonic acid (Ethrel) on mango, I. Flower induction in 'off' year in Langra trees, *Indian J. Hortic.*, 29(1), 1—4, 1972.
16. **Chacko, E. K., Kohli, R. R., and Randhawa, G. S.,** Investigations on the use of 2-chloroethylphosphonic acid (Ethephon, CEPA) for the control of biennial bearing in mango, *Scientia Hortic.*, 2, 389—398, 1974.
17. **Chacko, E. K. and Randhawa, G. S.,** Towards an understanding of the factors affecting flowering in mango, *Andhra Agric. J.*, 18(6), 226—236, 1971.
18. **Chacko, E. K., Singh, R. N., and Kachru, R. B.,** Studies on physiology of flowering and fruit growth in mango. VII. Naturally occurring auxins and inhibitors in the shoots of flowering (on) and vegetative (off) mango trees, *Indian J. Hortic.*, 29(2), 115—125, 1972.
19. **Chen, W. W.,** Physiological studies of fruiting in mango trees. II. Effect of endogenous growth substances on fruiting, *Proc. Natl. Sci. Counc. Repub. China*, 5, 49—55, 1981.
20. **Choudhuri, J. M., Basu, R.N., and Sen. P. K.,** Changes in endogenous growth regulating substances in the *Mangifera indica* L. shoot in relation to growth and development, *Acta Hortic.*, 24, 196—205, 1972.
21. **Choudhuri, J. M. and Rudra, P.,** Physiological studies on chemical control of growth and flowering in mango, *Indian Agric.*, 15, 127—135, 1971.
22. **Das, G. C. and Panda, J.,** Study on the effect of B-nine and maleic hydrazide on vegetative shoots of late occurrence in mango, *Orissa J. Hortic.*, 4(1 and 2), 33—37, 1976.
23. **Dutcher, R. D.,** Induction of early flowering in 'Carabao' mango in the Philippines by smudging and ethephon application, *HortScience*, 7, 343, 1972.
24. **Gazit, S.,** Initiation and Development of Flower Bud in Various Mango Varieties (in Hebrew), Ph.D. Dissertation, Hebrew University of Jerusalem, Israel, 1960.
25. **Gonzales, L. G.,** The smudging of mango trees and its effects, *Philipp. Agric.*, 12, 15—27, 1923.
26. **Kachru, R. B., Singh, R. N., and Chacko, E. K.,** Inhibition of flowering in mango *(Mangifera indica* L.) by gibberellic acid, *HortScience*, 6(2), 140—141, 1971.
27. **Kraus, E. J. and Kraybill, H. R.,** Vegetation and reproduction with reference to the tomato, *Bull. Oreg. Agric. Exp. Stn.*, 149, 1918.
28. **Kumar, A. and Ram, S.,** Endogenous lipids of mango shoot and their significance in flowering, personal communication, 1982.
29. **Lal, K. and Ram, S.,** Auxins of mango shoot tip and their significance in flowering, *Pantnagar J. Res.*, 2(1), 31—35, 1977.
30. **Mallik, P. C.,** Inducing flowering in mango by ringing the bark, *Indian J. Hortic.*, 8(1), 1—10, 1951.
31. **Maiti, S. C., Basu, R. N., and Sen, P. K.,** Chemical control of growth and flowering in *Mangifera indica* L., *Acta Hortic.*, 24, 192—195, 1972.
32. **Maiti, S. C., Mukhopadhyay, A. K., and Sen, P. K.,** Effect of growth regulants on flowering and apical dominance of mango, *Curr. Sci.*, 40, 388, 1971.
33. **Maiti, S. C. and Sen, P. K.,** Effect of growth retardants on flowering and fruiting of Langra mango, *Curr. Sci.*, 37(19), 566—567, 1968.
34. **Misra, K. A. and Dhillon, B. S.,** Carbohydrates and mineral composition of leaves in relation to fruit bud differentiation in Langra mango, *Indian J. Agric. Sci.*, 48(1), 46—50, 1978.
35. **Misra, K. A. and Dhillon, B. S.,** Ribonucleic acid, proteins, phenols and amino acids in the leaves in relation to fruit bud differentiation in Langra mango, *Indian J. Agric. Sci.*, 51(6), 447—449, 1981.
36. **Mukhopadhyay, A. K.,** A note on the effect of growth retardants and L-methionine on flowering of mango, *Haryana J. Hortic. Sci.*, 5(3 and 4), 169—171, 1976.
37. **Musahib-ud-din,** A note on fruit bud differentiation in mangoes in the Punjab, *Punjab Fr. J.*, 10, 30—31, 1946.
38. **Mustard, M. J. and Lynch, S. J.,** Flower-bud formation and development in *Mangifera indica* L., *Bot. Gaz.*, 108, 136—140, 1946.
39. **Nakasone, H. T., Bowers, F. A. I., and Beaumont, J. H.,** Terminal growth and flowering behaviour in the 'Pirie' mango in Hawaii, *Proc. Am. Soc. Hortic. Sci.*, 66, 183—191, 1955.
40. **Nunez Elisea, R., Becerriland, A. E., and Martinez Garza, A.,** The effect of ethrel on the flowering of mango cv. Haden, *Chapingo*, 23/24, 43—49, 1980.
41. **Pal, R. N. and Chadha, K. L.,** Deblossoming mangoes with cycloheximide, *J. Hortic. Sci.*, 57(3), 331—332, 1982.
42. **Pal, R. N., Chadha, K. L., and Rao, M. R. K.,** Effect of different plant growth regulators and other chemicals on flowering behaviour of mango, Paper presented at Mango Workers Meeting held at Panaji, Goa, May, 2nd to 5th, 1979, pp. 331—336.
43. **Pal, S. and Ram, S.,** Endogenous gibberellins of mango shoot-tips and their significance in flowering, *Scientia Hortic.*, 9, 369—379, 1978.
44. **Pandey, R. M., Singh, R. N., and Sinha, G. C.,** Usefulness of ethrel in regulating flower bearing in mango, *Sci. Cult.*, 39(3), 148—150, 1973.

45. **Rao, M. M., Ravishanker, H., and Bojappa, K. M.,** Amino acid composition of shoots of Alphonso mango at pre-, during, and post-fruit bud differentiation stages, *South Indian Hortic.*, 30(1), 1—3, 1982.
46. **Rao, V. N. M.,** A note on pruning as a remedy for irregular bearing in mango, *Andhra Agric. J.*, 18(6), 242—245, 1971.
47. **Rao, V. N. M. and Abdul Khader, J. B. M. M.,** More about mango pruning, *Indian Hortic.*, 23(4), 2—5, 1979.
48. **Rath, S. and Das, G. C.,** Effect of ringing and growth retardants on growth and flowering in mango, *Scientia Hortic.*, 10(1), 101—103, 1979.
49. **Ravishanker, H., Rao, M. M., and Bojappa, K. M.,** Fruit bud differentiation in 'Alphonso' and 'Totapuri' mango under mild tropical rainy conditions, *Scientia Hortic.*, 10(1), 95—99, 1979.
50. **Reece, P. C., Furr, J. R., and Cooper, W. C.,** Further studies of floral induction in the 'Haden' mango, *Am. J. Bot.*, 36, 734—740, 1949.
51. **Roy, R. S.,** Study of Irregular bearing of mango, *Indian J. Hortic.*, 10(4), 157—160, 1953.
52. **Saidha, T., Rao, V. N. M., and Santhanakrishnan, P.,** Internal leaf ethylene levels in relation to flowering in mango, *Indian J. Hortic.*, 40(3 and 4), 139—145, 1983.
53. **Sen, P. K.,** The bearing problem of mango and how to control it, *Indian J. Hortic.*, 1, 48—71, 1943.
54. **Sen, P. K.,** You can get a full crop of mango every year, *Punjab Fr. J.*, 10, 31—34, 1946.
55. **Sen, P. K., Bandopadhyay, M., Roy, S. S., and Basu, R. N.,** Use of ethrel in controlling non-uniform bearing of mango, *Indian Agric.*, 17(3), 285—288., 1973.
56. **Sen, P. K., Sen, S. K., and Guha, D.,** Carbohydrate and nitrogen contents of mango shoots in relation to fruit bud differentiation in them, *Indian Agric.*, 7, 133—138, 1963.
57. **Shawky, I., Zidan, Z., El-Tomi, A., and Dahshan, D. I.,** Effect of GA sprays on time of blooming and flowering malformations in Taimour mango, *Egypt. J. Hortic.*, 5(2), 123—132, 1978.
58. **Singh, L. and Khan, A. A.,** Forcing mango trees to bear regularly, *Indian Farming*, 1, 380—383, 1940.
59. **Singh, L. B.,** Movement of flowering substances in mango leaves, *Hortic. Adv.*, 3, 20—27, 1959.
60. **Singh, L. B.,** *The Mango. Botany, Cultivation and Utilization*, Leonard Hill (Books) Ltd., London, 1960.
61. **Singh, L. B.,** Biennial bearing in mango as affected by cultural operations, weather conditions and tree vigour, *Hortic. Adv.*, 5, 17—24, 1961.
62. **Singh, Ranvir and Singh, R. N.,** Lateral bud growth in *Mangifera indica* L. in relation to auxin and inhibitor content of shoots and fruits, *Acta Hortic.*, 24, 175—184, 1972.
63. **Singh, Ranvir and Singh, R. N.,** Changes in the organic phosphorus content of deblossomed and fruiting mango shoots during fruit growth and flower bud formation, *Indian J. Hortic.*, 30(1 and 2), 357—363, 1973.
64. **Singh, R. N.,** Studies in the differentiation and development of fruit buds in mango. II. Morphological and histological changes, *Hortic. Adv.*, 2, 37, 1958.
65. **Singh, R. N.,** Studies in the differentiation and development of fruit buds in mango *(Mangifera indica* L.) varieties. III. Mango shoots and fruit bud differentiation, *Hortic. Adv.*, 3, 28—40, 1959.
66. **Singh, R. N.,** Studies in the differentiation and development of fruit-buds in mango *(Mangifera indica* L.). IV. Periodical changes in the chemical composition of shoots and their relation with fruit bud differentiation, *Hortic. Adv.*, 4, 48—59, 1960.
67. **Singh, R. N.,** Studies in the differentiation and development of fruit buds in mango. V. Effect of defoliation, decapitation and de-blossoming on fruit bud differentiation, *Indian J. Hortic.*, 18, 1—11, 1961.
68. **Singh, R. N.,** Biennial bearing in fruit trees—accent on mango and apple, *Indian Tech. Bull. (Agric)*, ICAR No. 30, pp. 1—47, 1971.
69. **Singh, R. N.,** Mango, Indian Council of Agricultural Research, New Delhi, India, 1978, 1—99.
70. **Stino, G. R., Fayek, M. A., Khattab, M. M., and Bastawrous, M. B.,** Physiology of biennial bearing of mango trees, *Ann. Agric. Sci. Moshtohor*, 16, 193—207, 1981.
71. **Suryanarayana, V.,** Proteolytic enzyme changes in mango shoots as affected by growth retardants in relation to flowering, *Curr. Sci.*, 46(4), 127—128, 1977.
72. **Suryanarayana, V.,** Seasonal changes in ribonucleic acid and protein contents in mango shoots in relation to flowering, *Plant Biochem. J.*, 5(1), 9—13, 1978.
73. **Suryanarayana, V.,** Amino acid changes in mango shoots in relation to flowering, *Plant Biochem. J.*, 5(1), 50—57, 1978.
74. **Suryanarayana, V. and Rao, V. N. M.,** Studies on certain endogenous constituents of shoots in relation to flowering in mango. I. Changes in sugars and starch, *Orissa J. Hortic.*, 4(1 and 2), 1—12, 1976.
75. **Suryanarayana, V. and Madhava Rao, V. N.,** Ascorbic acid changes in shoots of mango cv Mulgoa as affected by growth retardants in relation to flowering, *Indian J. Plant Physiol.*, 20(1), 88—90, 1977.
76. **Suryanarayana, V. and Madhava Rao, V. N.,** Effect of growth retardants on certain bio-chemical changes in relation to flowering, *Indian J. Plant Physiol.*, 21(1), 1—6, 1978.
77. **Suryanarayana, V.,** A comparative study of some endogenous constituents in mango shoots in relation to 'off' season flowering in the southern latitudes, *Plant Biochem. J.*, 7(1), 72—77, 1980.

78. **Suryanarayana, V.,** Amino acid changes in mango shoots as affected by growth retardants in relation to flowering, *Plant Biochem. J.*, 7(1), 78—82, 1980.
79. **Suryanarayana, V.,** Influence of growth retardants on respiration, chlorophyll and carotenoid pigments in mango leaves in relation to flowering, *Indian J. Hortic.*, 38(1 and 2), 29—34, 1981.
80. **Thimmaraju, K. R.,** Studies on the Biennial Bearing of Mango, Ph.D. thesis, Indian Agricultural Research Institute, New Delhi, 1966.
81. **Valmayor, R. V.,** The Philippine mango industry — its problems and progress, *Acta Hortic.*, 24, 19—23, 1972.
82. **Veera, S. and Rao, V. N. M.,** Studies on certain endogenous constituents of shoots in relation to flowering in mango. II. Changes in dry matter, total carbohydrates, total nitrogen and C:N ratio, *Orissa J. Hortic.*, 5(1 and 2), 24—34, 1977.

OENOTHERA

En. Evening primrose, Sundrops; Fr. Oenothère; Ge. Nachtkerze

Atsushi Takimoto

INTRODUCTION

Oenothera belongs to the family Onagraceae (Oenotheraceae), and consists of 15 subgenera originally distributed widely throughout North and South America.[7] Although native only to the New World, it was carried to Europe in the 17th century, and has now spread throughout Europe and other areas, including Japan. The subgenus *Oenothera (Onagra* or *Euoenothera)* includes *O. lamarckiana* with which de Vries worked extensively and developed his famous mutation theory.[18] This plant has played a distinctive role in the field of cytogenetics, cytotaxonomy, and evolution.

Since most of the work on flowering in *Oenothera* has been done with the subgenus *Oenothera,* particularly with *O. biennis* and *O. lamarckiana,* this section will deal mainly with this subgenus which originated in North America.

Because the chromosomes of this genus regularly form one or more closed circles in diakinesis, the chromosome configuration of a variety of plants of this subgenus has been examined, and has been found to consist of many strains with different segmental arrangements; indeed, each species consists of a variety of strains. *O. biennis,* often used for the study of flowering, is the most widespread species of *Oenothera* in Europe consisting of various strains; and *O. lamarckiana* is now believed to have been synthesized in Europe as the result of crosses between *O. biennis* and *O. hookeri.*

The seed oil of *Oenothera* is now being used as a dietary supplement of essential fatty acids, and has been used for a very wide range of ailments.[5] The unique quality of this oil is that it contains gamma linolenic acid which is rarely found in other plants.[12]

MORPHOLOGY OF FLOWER

Oenothera has yellow flowers that open in the evening and wither the following morning. It is usually a biennial plant growing in a rosette form during the first year. Its stem elongates (bolting) in spring reaching 50 to 150 cm in height, the size varying with the species. Lanceolate-oblong leaves are arranged in an alternate phylotaxy, and an inflorescence develops at the top of the shoot. Each flower has a bract at the base without a peduncle. The 4 sepals are connate until flower opening, when they separate, often splitting into 2 parts each with 2 connate sepals. Each flower has 2 petals, 4 stigmatic lobes, 8 stamens, and an inferior ovary covered with fine hairs. The capsule splits into 4 parts, each holding many small seeds.

VERNALIZATION AND PHOTOPERIODISM

The subgenus *Oenothera* consists of many species with various chilling and photoperiodic requirements for flowering.[3] Many of the species are biennial, requiring both vernalization and subsequent LD, but some are annual and may or may not require LD. Table 1 summarizes the requirement for vernalization and photoperiod of various species reported by Chouard.[3]

Oenothera biennis has an obligate requirement for chilling followed by LD if it is to flower.[3,10] Seed vernalization is not effective, and only those plants (rosette plants) with 10 or more leaves when grown under LD conditions, and with 20 or more leaves when grown

Table 1
CHILLING AND PHOTOPERIODIC REQUIREMENT FOR FLOWERING OF *OENOTHERA* SPECIES[3]

Species	Requirement for vernalization	Requirement for photoperiod
O. biennis	Obligate	Obligate
O. lamarckiana	Obligate	Obligate
O. parviflora	Obligate	Obligate
O. suaveolens	Facultative	Obligate
O. longiflora	Facultative	Obligate
O. rosea	None	Facultative
Some others	None	None

under SD conditions, respond to chilling. The sensitivity to chilling treatment increases with advance of plant age, and plants having 60 to 70 leaves show maximum sensitivity. The minimum length of the chilling period necessary for floral induction in plants at the most sensitive stage was 10 to 11 weeks for plants grown under LD conditions, and 13 to 14 weeks for those grown under SD.[10]

Muller-Stoll and Hartmann[6] reported that in *O. biennis* (wild species in Berlin area), sensitivity to vernalization appeared at 30 days of age and showed two maxima, one at 56 days and the other at 80 days, although such two maxima were not observed in Picard's experiment.[10] Alternation of chilling and moderate low temperature, such as 3 and 11° C is more effective than chilling at a constant temperature.[9,10] Photoperiod during the chilling period has no effect at all.[10]

When vernalized plants of *O. biennis* are kept under SD (18 to 22° C) for more than 15 days, they are almost completely devernalized, losing their ability to respond to LD. Plants exposed to SD for a long period after vernalization can not be vernalized again.[3]

O. lamarckiana and *O. parviflora* also require both chilling and subsequent LD qualitatively, but the requirements are satisfied much more easily. Exposure to 3 to 4° C for 1 to 2 months is generally sufficient to induce flowering, and alternate exposure to 3° C and 10 to 12° C is more effective. Plants exposed to SD after the end of vernalization are devernalized, but they are revernalized when exposed to chilling again.[3]

In *O. suaveolens, O. longiflora,* and *O. stricta,* the chilling requirement is not obligate, although flowering is promoted by chilling treatment. These plants flower under LD even in the absence of previous chilling. *O. rosea* is a facultative LDP without any vernalization requirement.[3]

Many grafting experiments have been done with plants with an obligate vernalization requirement *(O. biennis),* but the effect of vernalization could not be transmitted to the partner.[10]

EFFECT OF GIBBERELLIN AND GROWTH RETARDANTS ON FLOWERING

GA_3 applied to nonvernalized rosette plants of *O. biennis* under LD conditions causes some shoot elongation, but the elongation stops and raised rosettes are formed without producing flowers.[2,8,10] GA_3 applied to partly vernalized rosette plants *(O. biennis),* however, caused bolting and flowering.[8,10] For instance, GA_3 applied to plants chilled for 7 weeks can induce flowering, whereas plants require more than 10 weeks of chilling for flowering without GA_3 application.

2-(Chloroethyl) trimethylammonium chloride (chlormequat) given after chilling treatment has no effect on flowering and significantly promotes stem elongation when a small amount

(50 to 250 μM) is applied to the apical bud in *O. biennis*. By contrast, *N*-dimethylamino-succinamic acid (daminozide) applied after chilling, particularly when applied after suboptimum chilling treatment, retards both stem elongation and flowering, and this effect is reversed by simultaneous application of GA_3. If the retardation of flowering by daminozide treatment exceeds a certain limit, the plant is devernalized.[11]

In *O. lamarckiana* and *O. parviflora*, repeated application of GA_3 to nonvernalized plants induces shooting and finally flowering under LD conditions, and sometimes even under SD though slowly.[3]

FLOWER OPENING

Oenothera is peculiar in its behavior in flower opening, i.e., its flower buds open at a definite time in the evening. However, only a few experiments have been made on the factors determining the time of flower opening of this plant.

In the 1930s, Sigmond[15,16] reported some experimental data on the effect of the light-dark cycle on flower opening of *Oenothera*, but at that time he considered that high humidity in the evening rather than the light-dark cycle is the main factor in determining the time of flower opening. In 1959, Arnold,[1] working with *O. berteriana* and *O. campylocalyx*, criticized Sigmond's conclusion, and emphasized the importance of the light-dark cycle in the determination of flower-opening. He found that these plants flowered in the evening, even when kept under continuous light, suggesting that the time of flower opening is determined by an endogenous rhythm. When exposed to 12-hr light (6:00 to 18:00) and 12-hr dark (18:00 to 6:00) cycles, plants flowered toward the end of the light period, i.e., about 12 hr after the beginning of the 12-hr photoperiod. If the phase of the 12L:12D cycle was reversed, i.e., if light was given from 18:00 to 6:00 instead of from 6:00 to 18:00, the flower buds still opened in the evening (around 18:00) during the first 2 days, but from the 3rd day on, they opened at the end of the day in the new L:D cycles. With the extension of the photoperiod per day beyond 12 hr, flower buds opened slightly earlier than the time of light-off, and with the shortening of the photoperiod they opened slightly later. The bud itself was found to be the site of photoperception in the control of flower-opening.

Later, the effect of light on flower opening of *O. lamarckiana* was extensively investigated by the late Dr. M. Saito (who met with an untimely death before publishing her complete data), and partly by Yamaki et al.[17] The following information was obtained from Saito's doctoral thesis[13] and partly from personal communication, unless otherwise mentioned.

The buds of *Oenothera lamarckiana* under natural conditions on a fine day in Tokyo (35° 35′ N) open 20 to 30 min after sunset in June to August, and somewhat earlier in October, November, and May. The time of flower opening is affected by the weather of the day the bud opens and also, to a lesser extent, by that of the previous day.

When the plant with buds expected to open at around 19:00 is transferred to darkness at 12:00, flower opening occurs at around 17:00. By repeating this procedure for 3 consecutive days, buds can be made to open 20 to 50 min after light-off. The plant may also be made to flower later in a similar manner by delaying the time of light-off.

The first splitting of the green connate sepals occurs on one of the four connate lines 2 to 3 hr before flower opening. The reflection of the connate sepals occurs immediately after the secondary split has proceeded at the lower part of the remaining three lines, resulting in the release of petals, i.e., flower opening (Figure 1). Light given only to the lower one third of the sepals inhibits this secondary splitting and this part is considered to be a site of photoperception. The stronger the intensity of light, the greater the delay of flower opening. Light stronger than 15 klx given from 30 min before the expected time of flower opening completely suppressed flower opening, but the light of 5 klx delayed it by only about 30 min.

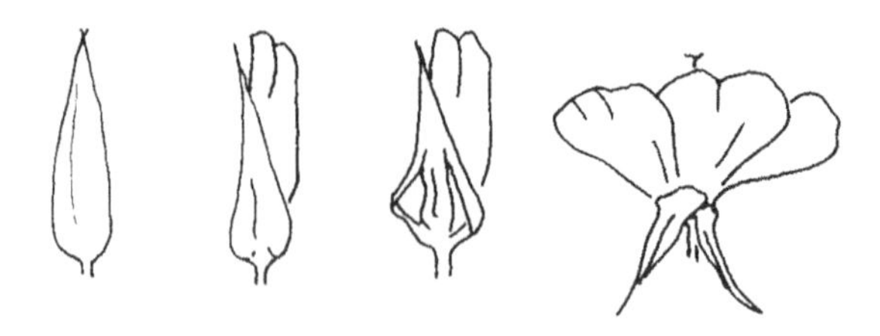

FIGURE 1. Flower opening of *Oenothera*.

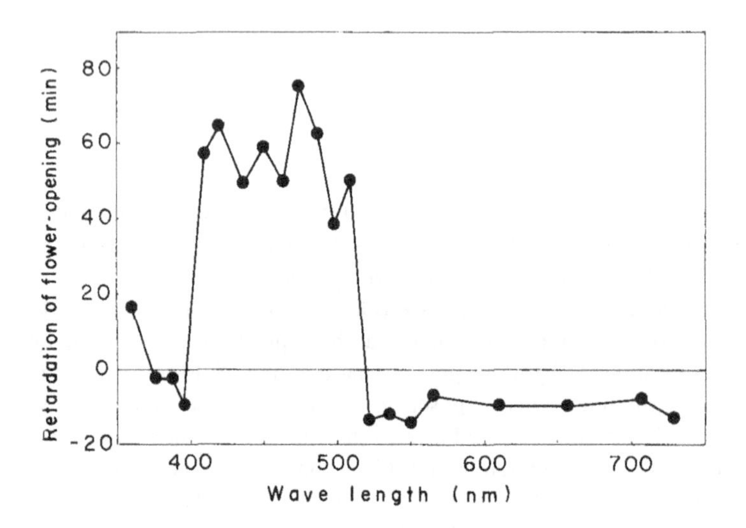

FIGURE 2. Action spectrum for suppression of flower opening of *O. lamarckiana*. (Reprinted by permission from Saito, M. and Yamaki, T., *Nature*, 214, 1027, 1967. Copyright ©1967, Macmillan Journals Limited.)

The action spectrum for the suppression of flower opening of *O. lamarckinana* showed that only the blue and green region of the spectrum is active.[14,17] Interestingly, the action spectrum had a peak at 510 nm which is usually inactive for photomorphogenesis in plants (Figure 2).

In Saito's experiments,[13] *O. lamarckiana* exposed to continuous light blossomed at various times of the day, but only between 15:00 and 3:00, whereas in Arnold's experiments, *O. berteriana* and *O. campylocallyx* blossomed at a definite time of the day even under continuous light. This might be caused by the difference in the species used, but more probably by the difference in the light intensity they used: 10 klx in Saito's experiment and 5 klx in Arnold's experiment. Under continuous darkness, even *O. lamarckiana* blossomed at definite time every day, strongly suggesting the participation of an endogenous rhythm in flower-opening.

Saito exposed the plants of *O. lamarckiana* to various light-dark cycles (12- to 48-hr cycles), each with various light and dark periods. The flower-opening rhythm synchronized well with the light-dark cycles of 14 to 34 hr, blossoming soon or several hours after the end of each light period. In 12-hr cycles (6L:6D), however, they blossomed every other cycle, and in a 48-hr cycle, twice in each cycle.

Six hours of darkness or 4 hr of light included in each cycle was enough to synchronize the flower-opening rhythm. However, the longer the light period in each cycle, the shorter the time between the light-off and flower opening, results similar to the tendency seen in *O. berteriana* by Arnold.

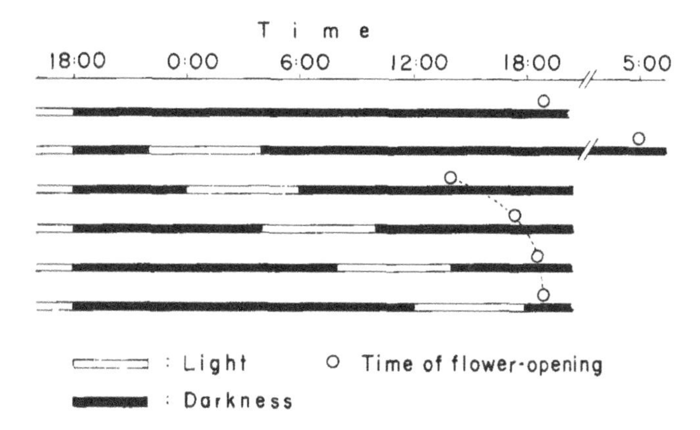

FIGURE 3. Time of flower opening of *O. lamarckiana* in continuous darkness during which a 6-hr light period was inserted at various times.

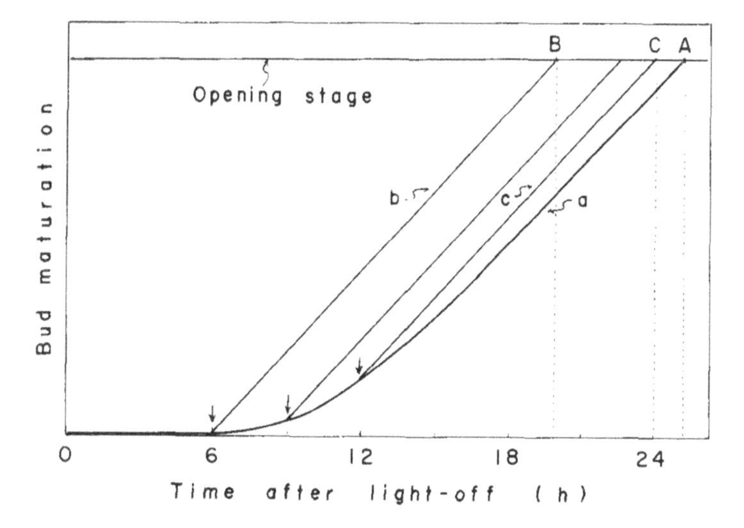

FIGURE 4. Maturation curve of flower buds of *O. lamarckiana* (after Saito[13]). a, The maturation curve in darkness (standard curve). b, c, The maturation curves when exposed to light after 6- and 12-hour dark period, respectively. A, B, C, Time when the maturation curves a, b, and c attain the opening stage. Arrows = onset of light period.

Saito's most interesting discovery is that whenever the plants were transferred into continuous darkness, the buds expected to open the next day opened about 25 hr after the light-off. She considered that some bud maturation process starts at the light-off, and that it takes about 25 hours in darkness to reach the opening stage. If the buds placed in darkness are exposed to a light period of 4 hr or more, provided it is not earlier than the 6th hour of the dark period, this process is accelerated and the buds reach the opening stage earlier than in continuous darkness. When the light period is terminated before reaching the opening stage, the buds open as soon as the opening stage is attained. On the contrary, when the buds were still kept under light when they reached the flower-opening stage, they could not open until darkness had commenced.

Figure 3 summarizes the time of flower opening of the plants placed under continuous darkness, a 6-hr light period being inserted at various times. From this and many other experiments in which various lengths of light periods were inserted in continuous darkness at various times, Saito proposed the model shown in Figure 4.

Curve *a* in Figure 4 is the maturation curve which proceeds after a 6-hr lag period in darkness. This is the basic curve (standard curve of flower-opening in Saito's paper), and the buds open when the curve attains the opening stage (time A) in darkness. When plants are exposed to light (for longer than 4 hr), starting at the 6th hour of the dark period, bud maturation proceeds along the straight line *b* in Figure 4, reaching the opening stage at time B. If plants are exposed to light starting at the 12th hour, the maturation curve proceeds along the straight line *c* reaching the opening stage at time C. This model fits the experimental results very well.

The relationship between this bud maturation process and the endogenous rhythm mentioned above remains to be investigated in detail, but it is clear that light controls the time of flower-opening at least at three different steps.

1. Light-off of the day before opening gives a signal to start the bud maturation process.
2. Light given 6 or more hours after the light-off accelerates the maturation process.
3. Light given at the time when the maturation process is at the opening stage suppresses the opening; relatively high-intensity light is required for this suppression.

Under natural conditions, usually, the plants receive a dark period of longer than 6 hr, and the maturation process reaches the opening stage a few or several hours before sunset. That is, the buds under natural conditions are ready to open before sunset, but their opening is suppressed by high-intensity light until dusk. This may be why the time of flower-opening is always related to dusk under natural conditions.

REFERENCES

1. **Arnold, C. G.,** Blütenoffnung bei *Oenothera* in Abhängigkeit vom Licht-Dunkelrhythmus, *Planta,* 53, 198—211, 1959.
2. **Chouard, P.,** Presentation de quelques plantes en cours d'experimentation sur les facteurs de la floraison, *Bull. Soc. Bot. Fr.,* 105, 135—136, 1958.
3. **Chouard, P.,** Vernalization and its relations to dormancy, *Annu. Rev. Plant Physiol.,* 11, 191—238, 1960.
4. **Cleland, R. E.,** *Oenothera: Cytogenetics and Evolution,* Academic Press, London, 1972.
5. **Graham, J.,** *Evening Primrose Oil,* Thorsons Publishing Ltd., Wellingborough, England, 1984.
6. **Muller-Stoll, W. R. and Hartmann, W.,** Vernalisation-phases und Stoffwechselerscheinungen bei zweijährigen Pflanzen, *Proc. 9th Int. Bot. Congr. Montreal,* Abst. Vol. I, 275—276, 1959.
7. **Munz, P. A.,** Onagraceae, in *North American Flora,* Ser. II, Part 5, N. Y. Botanical Garden, 1965, 1—231.
8. **Picard, C.,** Remarques sur l'action de l'acide gibberellique sur *Oenothera biennis* L., *Compt. Rend. Physiol. Vegetale,* 247, 2184—2187, 1958.
9. **Picard, C.,** Remarques sur l'exigence de températures froides alternées pour la vernalisation d' *Oenothera biennis* L., *Compt. Rend. Physiol. Vegetale,* 250, 573—575, 1960.
10. **Picard, C.,** Contribution à la connaissance de la vernalisation, de ses particularités et de sa signification chez *Oenothera biennis* L. var. *subfurea* de Vries, *Ann. Sci. Nat. Bot.,* 12th Sér., 6, 197—314, 1965.
11. **Picard, C.,** Action du CCC et du B995 sur la mise à fleur d'une plante bisannuelle, l' *Oenothera biennis, Planta,* 74, 302—312, 1967.
12. **Riley, J. P.,** The seed fat of *Oenothera biennis* L., *Jour. Chem. Soc.,* 2728—2731, 1949.
13. **Saito, M.,** Effect of Light on Flower-opening of *Oenothera lamarckiana,* Doctoral thesis, Tokyo University, Japan, 1976.
14. **Saito, M. and Yamaki, T.,** Retardation of flower-opening in *Oenothera lamarckiana* caused by blue and green light, *Nature,* 214, 1027, 1967.
15. **Sigmond, H.,** Die Entfaltung der Blütenknospen zweier Oenothera-Arten. II, *Beih. Bot. Zbl.,* 46, 476—488, 1930.
16. **Sigmond, H.,** Die Entfaltung der Blütenknospen zweier Oenothera-Arten. II, *Beih. Bot. Zbl.,* 47, 69—138, 1931.
17. **Yamaki, T., Gordon, S. A., and Chorney, W.,** On the inhibition by green light of flower bud opening in *Oenothera,* Annu. Rep. Argonne Natl. Lab., Biol. Med. Div., ANL-7278, 1966, 289—290.
18. **de Vries, H.,** *Die Mutationsteorie,* Veit, Leipzig, 1901—1903.

PANICUM MILIACEUM

En. Proso, proso millet, common millet, hog millet, broom-corn millet, Hershey millet, Cheena millet, Cheno, Vari, Gajro, Sava Chinee, Varagu, Chirwa, Bansi, Phikar, Rali; Fr. Millet commun; Ge. Echte Hirse, Rispenhirse; Sp. Mijo; It. Miglio comune

Lenis A. Nelson and Jerry D. Eastin

INTRODUCTION

Proso (*Panicum miliaceum* L., Gramineae) is a short-season species which has more tolerance to warm weather than most cereal grains. It is a grass plant with a C-4 photosynthesis cycle. It produces a seed that is slightly oblong and about 4 to 5 mm long. The seed can be used for human food but has a rather strong flavor. It is used for hog, poultry, cattle, and bird feed. There is some discrepancy as to the place of origin of proso. Matz[10] indicated that proso is probably native to Egypt and Arabia, while Lysov and Gudimovich[8] suggested that China was the place of origin. Proso is grown in most parts of the world as a minor crop. It is best adapted to places of low rainfall and short growing season.

MORPHOLOGY

Proso has a panicle type of inflorescence. There are three main types of panicles: (1) compactum, which is compressed; (2) contractum, which droops to one side; and (3) effusum, which is loosely spreading. Both primary and subsequent orders are arranged on the axis of the inflorescence in a spiral. The branches are arranged along the axis separately and can develop up to the fifth order. The spikelets are borne on a pedicel and are usually awnless and biflorate.[7]

The spikelets are oval and acuminate and 4 to 5 mm long. The lower glume is broad, clasping, acuminate, 5- to 7-nerved, and from half to two thirds the length of the spikelet. The upper glume is as long as the spikelet, oblong, subacuminate, concave at the base, and up to 13-nerved. There are 2 flowers, the lower sterile, the upper perfect. The sterile flower is represented by its lemma and palea, the former resembling the second glume, but 7- to 13-nerved, the latter much reduced. The fertile upper flower has a broad obtuse lemma, 7-nerved, becoming coriaceous, and smooth and shining. The palea is broad, 3-nerved, becoming coriaceous, smooth and shining, and light brown in color. There are two lodicules which are short, broad, and truncate. The stamens are three in number and the ovate ovary has two styles ending in pluminose stigma.[5] At times the sterile lower floret will also be fertilized and set seed. This fertilization may be in response to favorable environmental conditions but has also been reported to be under genetic control as a recessive character. When all spikelets have contained two fertilized florets, the yield was reduced from the normal condition.[9]

The mature grain of proso consists of a lemma and palea that are tightly attached to the seed. The lemma and palea determine the color of the seed, which can range from nearly white to yellow, red, gray, and dark brown. The seed is yellow once the lemma and palea are removed.

FLOWER INDUCTION AND DEVELOPMENT

Proso is generally considered a SDP and is also a short-season summer annual plant. As planting is delayed, the time from emergence to flowering is generally hastened due to decreasing daylengths. Although there are varietal differences, usually an 8- to 10-hr day gives the most rapid development of flowering and a photoperiod as low as 4 hours promotes more rapid flowering than a long photoperiod. One week of SD (10 hr) exposure during the 1st or 5th week after emergence had the greatest effect on hastening flowering. Plants which were hastened to maturity by SD had the lowest dry weight.[2]

Anthesis within a panicle occurs over 12 to 25 days depending on the source (Russia, China, etc.).[1] The first florets to open are at the top and progress basipetally. The intensity of anthesis is regulated by heat, humidity, and mechanical bumping once the floret is physiologically ready. At the time of anthesis, the lemma and palea are pushed open by the lodicules exposing the anthers and the stigma. During the period that the floret is open (2 to 24 min but usually less than 10) the anthers dry and dehisce, spreading pollen into the air.[1] The stigma are receptive to pollen at the same time; thus there is a reasonable chance of cross pollination during the time of floret opening.

Quality and intensity of light also influence the time of flowering. In general, light of predominantly shorter wave length (B) hastens flowering while light rich in longer wave lengths (R) delays flowering.[11] Proso grown at light intensities of 70 klx flowered sooner than proso grown at 6 klx.[4]

Some male sterility has been observed in proso, especially when grown in the greenhouse in the winter months. Generally, this sterility does not occur in the field. Genetic male sterility useful for hybridization has not been reported in proso at this time.

GROWTH REGULATORS

Phenolic compounds and GA have also been demonstrated to affect the flowering period in proso. GA_3, salicylic acid, β-naphthol, catechol, resorcinol, chlorogenic acid, and tannic acid hasten the emergence of panicles, increase the number of branches and panicles, and increase the number of seeds, seed weight, and 1000-seed weight of proso. GA_3 can also cause flower formation in proso under continuous light when it would normally not initiate flowers.[3,6]

REFERENCES

1. **Agafonov, N. P. and Podvez'ko, YE. S.,** The biology of flowering in proso millets of different ecologo-geographical groups under conditions of Poltava province *Byull. Vses. Inst. Rastenievod. N. I. Vavilova*, 51, 43—49, 1975.
2. **Chailakhyan, M. Kh.,** Internal factors of plant flowering, *Ann. Rev. Plant Phys.*, 19, 1—36, 1968.
3. **Datta, K. S., Kumar, S., and Nanda, K. K.,** Effects of some phenolic compounds and gibberellic acid on flowering and yield characters of cheena millet (*Panicum miliaceum* L.), *J. Agric Sci.*, 91, 731—735, 1978.
4. **Fedorov, A. K.,** Effect of various durations of day on development of proso, *Tr. Akad. Nauk. SSSR Inst. Genet.*, 30, 129—135, 1963.
5. **Hector, J. M.,** *Introduction to the Botany of Field Crops*, Vol. 1, *Cereals*, Central News Agency, Johannesburg, 1936, 478 pp.
6. **Kumar, S., Datta, K. S., and Nanda, K. K.,** Gibberellic acid causes flowering in short day plants *Panicum miliaceum* L., *P. miliare* Lamk., and *Setaria italica* (L.) P. Beauv., *Planta*, 134, 95—96, 1977.
7. **Kuperman, F. M. and Rostovtseva, Z. P.,** On panicle formation in millet, *Dokl. Vses. Akad. Skh. Nauk.*, 16(3), 16—18, 1951.
8. **Lysov, V. N. and Gudimovich, M. A.,** Proso, *Leningrad. Vses. Inst. Rastenievod.*, B. 11, 60—65, 1962.
9. **Maksimcuk, I. H.,** Development of new forms and characters in millet on hybridization, in *Plant Breeding Abstr.*, 27, 2716, 1957.
10. **Matz, S. A.,** *Chemistry and Technology of Cereals*, AVI Publishing, Westport, Conn., 1959.
11. **Pavlov, P.,** The effect of light on the growth, development and yield of millet, *Akad. Nauk. Inst. Rastenievod. Izv.*, 12, 209—238, 1961.

PERILLA

J. A. D. Zeevaart

INTRODUCTION

Perilla is a strongly aromatic annual plant which belongs to the Labiatae and has a square stem with decussate leaves. It is native to Asia. As a garden escapee, it has become widely distributed in some areas of the U.S.[31] Two types of *Perilla* can be distinguished: the red-leaved varieties grown as ornamentals for their colorful foliage, and the green-leaved ones. The latter produce valuable seed oil which is used as drying oil for paints and lacquers. *Perilla* plants are poisonous due to the presence of perilla ketone and other 3-substituted furans which cause acute emphysema in laboratory animals.[31] It has therefore been suggested[31] that the use of *Perilla* in oriental foods and medicines poses a potential health hazard to humans.

Perilla is a SDP which has been studied extensively since the early 1930s. The history of the physiology of flowering in *Perilla* has been reviewed before.[33] Although Garner and Allard,[13] working in Washington, D.C., discovered quite early that *Perilla* is a typical SDP, relatively little work has been done with this plant in the U.S.[15-17,34,36] In the Soviet Union, Chailakhyan[7] and his associates have worked extensively on flowering in *Perilla* for almost half a century. Lona, working in Italy, contributed for some time actively to our knowledge on flowering in *Perilla*. A summary of his work has appeared in English.[20] At Wageningen, the Netherlands, Wellensiek[29,30] and students[11,28,32] have worked with red *Perilla*. More recently, Allot-Deronne and Blondon[1-3,10] in the Phytotron at Gif-sur-Yvette, France, have done extensive work on the mechanism of photoperiodic induction in green and red *Perilla*.

This chapter is not meant to be a comprehensive review on flowering in *Perilla*. For the older literature the reader is referred to a previous review by the author.[33] Since *Perilla* has been used most extensively for studying transmission of the floral stimulus, this aspect has been emphasized in the present chapter, particularly since some interesting new results in comparative studies with green and red *Perilla* have appeared recently.[1]

The nomenclature of *Perilla* is confusing and has been discussed before.[33] Wherever possible, the material used in a particular study will be referred to in this chapter as green or red *Perilla*, depending on the color of the leaves. The green *Perilla* has been variously described as *P. ocymoides* L., *P. frutescens* (L.) Britt., or *P. crispa* (Thunb.) Tanaka var. *ocymoides* L., whereas the red-leaved variety has been called *P. nankinensis* (Lour.) Decne., *P. ocymoides* (L.) var. *nankinensis* (Lour.) Voss, *P. frutescens* Britt. var. *crispa*, or *P. crispa* (Thunb.) Tanaka.[33] Some of the varieties of red *Perilla*, such as *laciniata*,[1] have wrinkled leaves that are irregularly lobed.

GROWTH HABIT AND FLOWERING RESPONSE CRITERIA

Uniform batches of plants are most readily grown from seed, although propagation via cuttings is also possible. The plants can be kept vegetative in the greenhouse by supplementing daylight with light from incandescent bulbs or from fluorescent tubes to give a photoperiod of 20 hr. Once the fourth leaf pair is fully expanded, maximal photoperiodic sensitivity has been reached, the flower buds will appear 18 to 20 days after transfer to an 8-hr photoperiod and a night temperature around 20°C. If the plants are kept under continuous SD, they will flower, produce seeds, senesce, and then die. However, if after 10 to 20 SD the plants are returned to LD, the terminal inflorescence will produce some flowers and fruits, and then revert to vegetative growth (Figure 1). The first indication of reversion is

FIGURE 1. Red *Perilla* exposed to 20 SD and subsequently grown in LD for 2 months. After fruits had formed on the main stem, complete reversion to vegetative growth took place. Upper axillary branches also exhibit reversion (From Zeevaart, J. A. D., in *The Induction of Flowering. Some Case Histories,* Evans, L. T., Ed., Macmillan of Australia, Melbourne, 1969, 116—155. With permission of Macmillan of Australia, Melbourne).

enlargement of the bracts. A gradual transition to leaves takes place until a completely vegetative shoot growing out of an inflorescence is ultimately formed.[1,33] Reversion to vegetative growth in *Perilla* in suboptimally induced plants is due to the fact that the terminal shoot apex splits off axillary flower buds, but is never transformed to a terminal flower bud itself.[33]

Detached leaves of red *Perilla* can be easily grown in vials with nutrient solution.[32] Roots are formed readily at the base of the petiole, but shoots are never regenerated.

Perilla is extremely suitable for grafting and for demonstrating the flower-inducing stimulus. Various grafting techniques have been used.[1,32,33]

The inflorescence of *Perilla* is a verticillaster with inconspicuous flowers; the color is pinkish in the red strains, and white in the green ones.

The following criteria have been used to measure the flowering response in *Perilla:*

1. The percentage of plants with flower buds.
2. The numbers of days from the start of a treatment to appearance of flower buds or until anthesis. A response in only 50% of the plants has also been used.[1,10]
3. A scale from 0 to 3 has been used to score excised apices cultured on agar, 0 being vegetative, and stage 3 shoots having open flowers.[24]

JUVENILE PHASE

Young seedlings of red *Perilla* are much less sensitive to SD treatment than plants at a later stage of development. For example, it took 51 SD to produce flower buds in 12-day-old plants, but only 24 SD in 75-day or older plants.[32] Likewise, it was observed in green *Perilla* that the older the plants when transferred to SD, the fewer the number of SD before inflorescences appeared.[15] When green *Perilla* was given SD for about 3 weeks from germination until the second leaf pair was mature, no flowering took place in subsequent LD,[1] indicating that the first two leaf pairs are quite insensitive to SD induction.

The leaf pairs of red *Perilla* show distinct morphological differences. When numbered in order of appearance, the first two pairs are small and oval-shaped, with little anthocyanin. The fourth pair and higher ones are ovate, much larger in area than the lower ones, and have a dark purple color due to the presence of anthocyanin.[32]

As established in grafting experiments with red *Perilla*, leaves on the second node required 46 SD before they could function as donors, whereas leaves on the fourth node required only 26 SD.[32] This suggests that the difference in photoperiodic sensitivity between the different leaf pairs is due to their position on the stem, and not to their physiological age. This was further demonstrated as follows. In two groups of plants, sown at different times, the fifth and second leaf pairs had just fully expanded. The blades of these two leaf pairs were reduced to 25 cm^2 and grafted onto LD stocks above the fourth node. Once the graft unions had been established, the grafted leaves were induced by daily darkening with light-tight bags. Leaves from the fifth node needed 14 SD to cause flowering, whereas those from the second node required 28 SD.[32] Since the leaf area and physiological age of the donor leaves were kept the same, it can be concluded that the difference in photoperiodic sensitivity was due to their original position on the plant. These results explain why young seedlings need more SD for flowering than older plants. The first leaves are quite insensitive to induction, but as new leaves appear, sensitivity increases until it levels off in the fourth or fifth leaf pair. Thus, the physiological basis for the juvenile phase in *Perilla* is the inability to produce the floral stimulus at optimal rates in the first two or three leaf pairs. Related to these results may be the finding that when excised leaf discs were cultured on agar, flower bud formation occurred only in explants from the higher leaf positions on the mother plants.[27] Thus, a gradient in the capacity to form flower buds exists along the stem.

NUMBER OF INDUCTIVE CYCLES

In red *Perilla*, 9 SD were needed for a 100% flowering response,[34] but in a different strain of the red variety only 6 SD were required.[1] A requirement for at least 9 SD has also been reported for green *Perilla*,[16,25,36] but other workers[1,10] found that flower formation took place after exposure to only 6 SD (Table 1). The differences found by various workers are probably due to the use of different strains and differences in environmental conditions,

Table 1
COMPARISON OF THE FLOWERING RESPONSES IN THE SDP GREEN AND RED *PERILLA*[a]

Response	*P. ocymoides* (green)	*P. nankinensis* (red)
Photoperiodic type	Quantitative SDP	Obligate SDP
Critical daylength	17 hr	14 hr
Minimal number of SD for flowering	6 9[15,25,36]	6 9[33,34]
Minimal duration of SD (22° C) induction for transmission of floral stimulus (days)	8	12
Minimal number of days at 5° C, CL for flowering	45	50
Minimal duration of induction at 5° C, CL for transmission of floral stimulus (days)	45	120
Interspecific transmission of floral stimulus	Green→red >>>	Red→green
Indirect induction	Yes	No
Excised apices cultured in vitro flower in LD	Yes	Yes

[a] Data from Allot-Deronne.[1]

such as temperature during the inductive dark period, and light intensity. Although few inductive cycles are necessary to cause the formation of flower buds in *Perilla,* it must be emphasized that 3 to 4 weeks of SD are required for abundant flowering and seed production.

CRITICAL DAYLENGTH

Allot-Deronne,[1] working in a controlled environment with a photoperiod of 9 hr and a night temperature of 22° C, found that the critical photoperiod was 14 and 17 hr for red and green *Perilla,* respectively (Table 1). Other workers[5,15,22,26] also reported that 14 hr of light was the longest daylength under which red *Perilla* flowered. Different numbers have been reported for green *Perilla,*[1,12,15] but as a whole it is clear that the critical daylength of this species is at least 2 hr longer than that of the red one. Jacobs[15] found that under his experimental conditions red *Perilla* behaved as an obligate SDP, whereas green *Perilla* responded as a quantitative SDP which eventually flowered in continuous light 77 days after sowing.

LIGHT INTENSITY

Takimoto and Ikeda[26] exposed red *Perilla* to 12 hr light/12 hr dark cycles. Weak supplementary light from fluorescent lamps was given during the first or last 3 hr of the dark period to determine the light intensity necessary to suppress flowering. While an intensity of 200 lux did not completely inhibit flowering when given during the first part of the dark period, no flowering took place with 10 lux given at the end of the night. This indicates that *Perilla* is more sensitive to light at the end than at the beginning of the dark period.

Red *Perilla* is able to produce flower buds under LD provided the light intensity is low.[11] However, under these conditions the flowering response is very slow. For example, under continuous fluorescent light of low intensity ($3\ W\ m^{-2}$) flower buds appeared after 70 days[11] compared to 20 days under SD conditions. Flowering in other *Perilla* strains in continuous light of low intensity has also been observed by others.[1,12]

EFFECTS OF TEMPERATURE

The optimal temperature for the inductive dark period (15 hr) was found to be between 22 and 27°C in both green and red *Perilla*. At lower as well as at higher night temperatures the flowering response was much delayed.[1,10]

Wellensiek[29] exposed red *Perilla* to different temperatures during the 8-hr light and 16-hr dark periods. Darkness given at 5°C completely prevented flower formation, whereas low temperature given during the light delayed flowering only slightly. This suggests that the processes of photoperiodic induction in *Perilla* take place in the dark and are of a biochemical nature. Wellensiek[30] also gave differential temperature treatments during light and darkness to red *Perilla* plants in LD. Plants exposed to 5°C during the photoperiod (16 hr) and 20°C during the dark period (8 hr), produced flower buds after 8 weeks. Thus, light was less effective in preventing flowering when the long photoperiods were given at low temperature. This observation was further explored by Deronne and Blondon.[1,10] These workers found that flowering in both types of *Perilla* will occur in continuous light (CL) at 5°C. However, this flowering response was relatively weak. For example, the minimal duration for SD induction in green *Perilla* was 6 days at 22°C (Table 1), but 45 days under CL at 5°C.[10] Successive treatments with the two inducing factors, each at a subthreshold level, also resulted in flowering. For example, 4 SD + 15 CL at 5°C, or 20 CL at 5°C + 4 SD resulted in 100% flowering. In other words, the effects of the two subthreshold treatments were additive which suggests that the two factors acted through a common mechanism. Both the inductive effects of SD and of low temperature could be transmitted via leaf grafts to LD receptor stocks.[1,10] However, flower buds appeared more rapidly on the receptors after grafting with SD-induced leaves (after 20 days) than with leaves induced in CL at 5°C (after 50 days).[1,10] See below.

THE FLORAL STIMULUS

General Characteristics

Perilla was one of the first plants in which evidence for a transmissible stimulus was obtained: by giving SD to the leaf blades only, flower formation occurred in the axillary shoots, thus implying the movement of a flower-inducing stimulus from the leaves to the shoot tips.[21] This type of experiment is generally taken as evidence that the leaf is the site of photoperiodic perception. However, it should be added that mature plants of red *Perilla* without any leaves or buds, except the terminal shoot apex, respond to SD with flower formation just as rapidly as intact plants.[33] Thus, in contrast to many other photoperiodic plants, *Perilla* stems can perceive the relative length of day and night and produce the floral stimulus.

Whereas in experiments on localized induction (see above) the inductive treatment and its expression are separated in space, SD induction and subsequent flowering can be separated in both space and time by use of the grafting technique. Results of such grafting experiments with red *Perilla* were first reported in 1937.[9] In this instance, leaves on donor stocks could transmit the floral stimulus to receptor shoots, even though the SD induction was completed before the grafting was performed, and the entire graft combination was kept under a photoperiod unfavorable for flowering. Since these original experiments, grafting has been used extensively for studying the nature of photoperiodic induction in *Perilla*. For single leaves, both cleft-[32] and approach-grafting[1] have been used. Transmission of the floral stimulus has been demonstrated in various types of graft combinations: a shoot, a leaf, or part of a blade could function as donors. In double-grafts, it was possible to graft one scion on two different stocks, or two different scions on one stock.[32] By the use of approach-grafting up to 8 leaves could be grafted per internode.[1]

Table 2
RESULTS OF RECIPROCAL GRAFTING EXPERIMENTS BETWEEN GREEN AND RED *PERILLA* INDUCED BY EITHER 5°C IN CL OR SD[a]

	Donor			
	Green		Red	
Receptor	5° C, CL	SD	5° C, CL	SD
Green	48 (>80)	20 (>80)	—[b]	32 (>80)
Red	45 (56)	12 (20)	47 (>80)	22 (40)

Note: Numbers indicate days until appearance of flower primordia in 50% of the receptors. In brackets: days until anthesis in 50% of the receptors.

[a] Data from Allot-Deronne,[1,2]
[b] Grafting not performed.

Transmission of the floral stimulus from donor to receptor can be demonstrated most readily when axillary shoots and buds are removed from the donor and the receptors are defoliated. However, there are several examples in which noninduced leaves were present on the receptor, and flowering did occur, albeit somewhat delayed.[1,17,32]

Grafts between Green and Red *Perilla*

Green and red *Perilla* each need a minimum of 6 SD for flower formation (Table 1), but exhibit more abundant flowering after exposure up to 25 SD. However, when reciprocal graft combinations between green and red *Perilla* were established, the green species was a much more effective donor than the red one (Table 2). For example, a SD-induced red donor caused appearance of flower buds after 22 days in red receptors, whereas flower buds were visible after 12 days when grafted with green donor leaves. A similar trend was discernable when the donors were induced by low temperature under continuous light, although in that case the flowering response was generally weaker than after induction of the donors by SD (Table 2).

Leaf grafts in all four possible combinations between green and red *Perilla* have also been performed by Zeevaart.[36] With green donor leaves, flower buds on both green and red receptor shoots were visible after 12 to 13 days, and the first flowers opened 22 to 24 days after grafting (Figure 2). This is a more rapid flowering response than when *Perilla* plants are normally induced by SD. With red leaves as donors, flower buds appeared on red and green receptors after 16 and 24 days, respectively, while flowering started after approximately 40 days.[36] It is clear from these results that induced leaves of green *Perilla* are much more effective donors than those from red plants. This suggests that green *Perilla* produces more floral stimulus per unit leaf area than the red one. In addition, the red receptor shoots appear to be more sensitive to the floral stimulus than green receptors (Table 2).

Duration of Induction

For red *Perilla*, it was found that 9 SD were necessary for 100% flowering, whereas at least 12 SD were required before transmission of the floral stimulus could be demonstrated with leaf grafts.[34] In green *Perilla*, 8 SD or 45 days at 5°C in CL were necessary to obtain

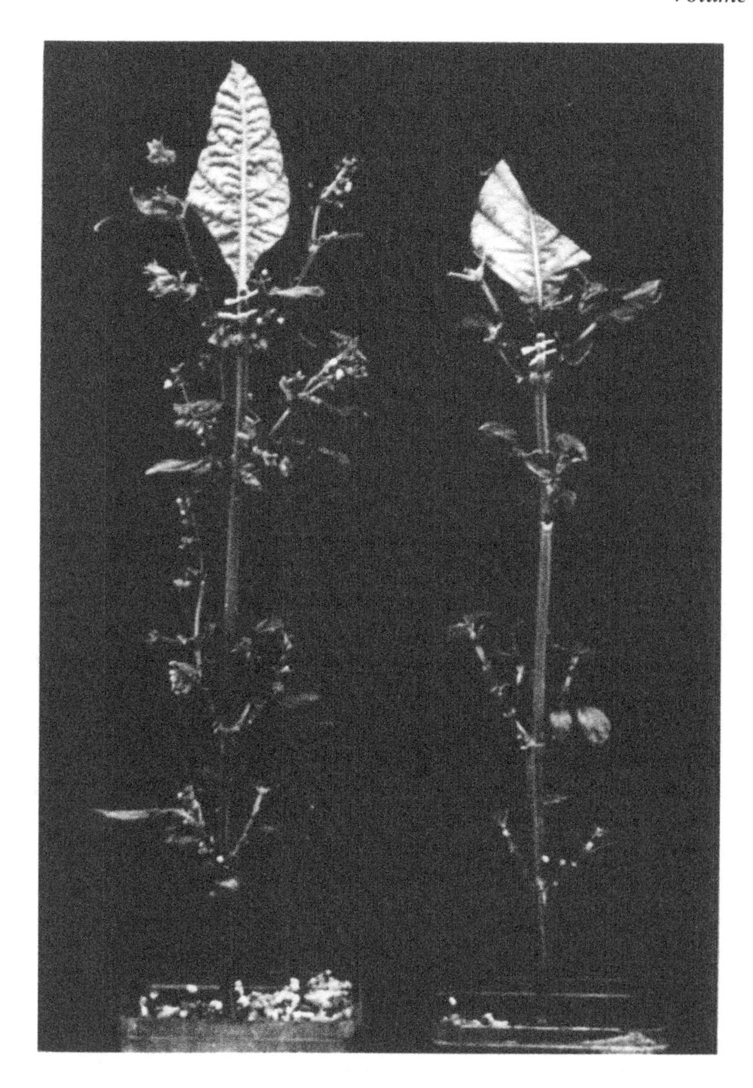

FIGURE 2. Donor leaves (30 cm^2) of green (left) and red (right) *Perilla* grafted on red receptor stocks. Receptor shoots on the left show flowering and fruiting; in plant on the right only flower buds are visible. Photograph 36 days after grafting.[38]

100% flowering in the receptors;[1] for red *Perilla*, these numbers were 12 and 120 days, respectively (Table 1). In general, the response with cold-induced donor leaves was much weaker than with SD-induced ones. This is also reflected in the minimal area of donor leaves necessary for transmission of the floral stimulus: 4 and 35 cm^2 for SD- and cold-induced leaves, respectively.[1] However, when four cold-induced leaves were grafted on one receptor stock, flower buds appeared at the same time as after grafting with a single SD-induced donor leaf.[1]

As shown in Table 1, the minimal number of SD for flowering in both green and red *Perilla* is less than the minimal duration of SD induction necessary for transmission of the floral stimulus. A pertinent question in this context is, at what time during induction does production of the floral stimulus start? It can be expected that a certain threshold level of floral stimulus must accumulate before flower formation can take place, but does production of the stimulus already begin during the first inductive cycle? This question was investigated[1] by taking advantage of the finding that red receptors are more responsive to the stimulus than green ones (see above). Green *Perilla* plants were exposed to 3, 4, and 5 SD, or to 15, 20, and 30 days at 5°C under CL. None of these treatments resulted in flowering, but donor leaves from these plants induced flower formation in red *Perilla* receptors. In all treatments, flower buds appeared on the receptors after approximately 80 days. It was concluded, therefore, that a certain amount of floral stimulus was produced under conditions of subthreshold induction which was inadequate for flowering in the green *Perilla* donor plants, but which was sufficient to cause flowering in red *Perilla* receptors.[1,2] However, green *Perilla* will ultimately produce flower buds in LD.[15] It is possible, therefore, that the subthreshold induction in combination with the gradual acquisition of the induced state in the donor leaves under LD resulted in production of the floral stimulus and ultimate flowering of the receptor shoots (see also below).

Duration of Graft Contact

Under favorable conditions, a contact period of 6 days between red donor leaves and receptor buds was sufficient for 100% flowering,[33] whereas in green *Perilla* a contact period of 5 days already resulted in a 100% flowering response.[1] However, the longer the contact period between donor and receptor, the more flower buds were produced.[32] Transport studies with radioactive sucrose from donor leaves to receptor shoots also indicated that 6 to 7 days of contact were necessary before label could move across the graft union. It was concluded, therefore, that a functional phloem connection between donor and receptor is required for transfer of the stimulus.[32]

Movement of the Floral Stimulus

The floral stimulus in *Perilla* can move in the stem both in the upward and downward direction, depending on the position of the induced leaves and the receptor buds.[33] For transmission of the floral stimulus to occur, it is essential that translocation of assimilates can take place from the donor to the receptor (see below). This, combined with the requirement for a functional phloem connection in grafts (see above), indicates that phloem is the only pathway for long-distance movement of the floral stimulus.

Inhibitory Effect of Noninduced Leaves

Perilla leaves in noninductive conditions are most inhibitory to flowering when positioned between donor leaves and receptor shoots. This inhibitory effect of noninduced leaves could be due either to a specific inhibitor of flowering produced under LD, or to interference with movement of the floral stimulus. With respect to the former possibility, it should be pointed out that flowering can occur with LD leaves present on the receptors. For example, a grafted donor leaf of 30 cm^2 caused flowering in axillary shoots with noninduced subtending leaves with an area of 100 cm^2 or more present.[17] Thus, per unit leaf area the flower-promoting effect coming from induced leaves is much stronger than the flower-inhibiting effect from leaves in LD. Support for the second possibility has come mainly from studies in which distribution of ^{14}C-assimilates was correlated with movement of the floral stimulus as evident from the flowering response. In a qualitative study[8] with red *Perilla,* it was shown that noninduced leaves in close proximity to the receptors prevent the products of SD leaves from reaching the shoots in adequate amounts. In a more quantitative approach,[17] export and distribution of ^{14}C-labeled assimilates from donor or LD leaves were determined in grafted plants of red *Perilla* (Figure 3). When the donor leaf was labeled with $^{14}CO_2$, radioactive assimilates were exported to receptor shoots on nodes 1 and 3 which flowered, but not to the shoots on nodes 2 and 4 which remained vegetative (Figure 3).[17] In *Perilla,* with opposite leaves, both the floral stimulus and assimilates are translocated in specific orthostichies of the stem. Since the floral stimulus moves in parallel with the assimilates in the phloem, the pattern of assimilate distribution from different leaves also determines the

FIGURE 3. Induced leaf (30 cm^2) grafted on LD receptor stock of red *Perilla*. LD leaf pair present at node 2 (nodes numbered from the graft union in basipetal direction). Axillary shoots flowering and fruiting at nodes 1 and 3, vegetative at nodes 2 and 4. Photograph 60 days after grafting. (From King, R. W. and Zeevaart, J. A. D., *Plant Physiol.*, 51, 727—738, 1973. With permission of the American Society of Plant Physiologists.)

movement of the floral stimulus. Inhibition of flowering by *Perilla* leaves in LD is, therefore, the result of competition between assimilate streams originating in the induced and noninduced leaves. Thus, there is no evidence that a specific transmissible inhibitor is involved in the control of flowering in *Perilla*.[17] Bhargava[4] showed that a flower inhibitor could be produced in red *Perilla* leaves, but only after the plants had received a subthreshold induction, followed by LD. This is clearly different from the LD inhibition discussed above where leaves were always kept in LD.

The Induced State and the Floral Stimulus

When the duration of induction is suboptimal, *Perilla* plants will revert to vegetative growth (see above). There are two possible explanations for this phenomenon: (1) the floral stimulus accumulates during SD treatment and is used up in subsequent LD, so that vegetative growth resumes, or (2) induced leaves have acquired an irreversible state and continue to produce floral stimulus in LD, but leaves developing in LD do not, and ultimately cause reversion to vegetative growth. This question of reversible vs. permanent induction was investigated in red *Perilla* by the use of leaf grafts.[32] Induced leaves were grafted on receptor stocks and at regular intervals regrafted on new groups of receptor plants. In the most extensive experiment,[32] leaves were grafted successively on seven different groups of stocks. Flowering was obtained in all receptors, whereas control receptor plants grafted similarly with noninduced leaves remained invariably vegetative. In the final grafting, performed 97 days after the first one, the flowering response was the same as in the beginning. It was concluded, therefore, that red *Perilla* leaves, once induced, continue to produce the floral stimulus throughout the rest of their life span, regardless of the photoperiodic conditions. On the other hand, flowering receptors, or leaves from such shoots, failed to cause flowering in receptor shoots.[32] Thus, the phenomenon of indirect induction of flowering does not take place in red *Perilla*. Photoperiodic induction in red *Perilla* involves therefore the following steps:[32]

1. During SD induction the induced state gradually builds up. The induced state is irreversible, and stictly localized in the leaves that have been exposed to a minimal number of inductive cycles.
2. Leaves that have acquired the induced state produce the floral stimulus which is mobile and transmitted to receptor buds. The stimulus is apparently metabolized, since it does not accumulate in receptor shoots as is evident from the negative results with grafted leaves.

The situation with respect to these phenomena is somewhat different in green *Perilla*. While induced leaves can be regrafted and will continue to supply the floral stimulus under LD, they senesce much sooner than those of red *Perilla*.[1,38] However, the most striking difference between green and red *Perilla* is that indirect induction can be obtained in the green species, but not in the red one.[1,3,10] SD-induced donor leaves of green *Perilla* caused appearance of flower buds on receptor shoots after 21 days. When leaves from these shoots were grafted on vegetative stocks, it took 43 days until flower buds were visible on the receptors. In a similar experiment with red *Perilla*, no indirect induction was observed.[1,10] Also, when induced leaves of green *Perilla* were grafted on red stocks, the leaves of the red receptors did not function as donors.[1,3] These observations have been confirmed:[36] leaves taken from a flowering green receptor shoot induced flower formation in both green and red receptors, regardless of whether the original donor leaf was green or red. However, LD control grafts (which were not included by Allot-Deronne[1,3,10]), also caused flowering, albeit later than with leaves from flowering receptor shoots. For example, leaves from flowering and vegetative receptors caused formation of flower buds on receptor shoots after 50 and 80 days, respectively.[38] This suggests that the induced state gradually builds up in leaves of green *Perilla* even when grown under continuous LD. This process is clearly accelerated when the shoots develop under the influence of a SD-induced donor leaf. Thus, the phenomenon of indirect induction in green *Perilla* is only a quantitative one, and is not as clear cut as in other plants, such as *Bryophyllum, Silene armeria,* and *Xanthium* (see respective chapters in Volumes 2 and 4[21a,24a,30a,39]).

Detached *Perilla* leaves without roots or meristems can be induced, since they cause flowering after grafting on receptor stocks.[1,32,33] Thus, acquisition of the induced state in red *Perilla* under SD is a noncorrelative phenomenon. No treatments or conditions have been found so far that will destroy the induced state and nothing is known about the underlying biochemistry.

Nature of the Floral Stimulus

Although the presence of the floral stimulus can be easily demonstrated in physiological experiments, its chemical nature remains unknown. The various possibilities with respect to the nature of the floral stimulus have been discussed.[37] In *Perilla*, long-distance transport and the continued production of stimulus in induced leaves under various conditions make unlikely the idea that the floral stimulus is a balance of several substances, or represents the sequential arrival of several compounds at the apex. The possibility that the floral stimulus is a single flower-inducing substance remains therefore a viable hypothesis.[37]

Since the floral stimulus moves in the phloem, several workers have attempted to collect phloem exudate for bioassay by treatment of the petioles with chelating agents.[18] Purse[24] incorporated the entire exudate in the medium on which excised shoot tips of red *Perilla* were cultured. In some experiments, the exudate obtained from induced leaves caused flowering, but it was not possible to obtain these results consistently. In this author's own experiments,[38] phloem exudate or fractions thereof were applied to vegetative *Perilla* plants in the transpiration stream via a stem flap, or directly to the shoot tips. Again, no consistently positive results were obtained. In further attempts to characterize the floral stimulus, phloem exudates from vegetative and induced leaves were analyzed and compared for possible differences in the neutral, basic, and acidic fractions. No significant differences were detected when exudates from SD-induced leaves and from leaves which has received SD with an interrupted night were compared.[6] Thus, the floral stimulus remains a physiological concept, and has not (yet) become a chemical reality.

Tissue Contact with Other Species

Several attempts have been made to use *Perilla* as a donor for other species, or to induce flower formation in *Perilla* in intergeneric grafts. All these experiments have yielded negative results and have been reviewed before.[33] Since then, transfer of the floral stimulus has been claimed from red *Perilla* to the LDP *Silene armeria*.[28] Since no functional union was established in these grafts, the receptor scions were regrafted after 5 weeks on vegetative *Silene* stocks to observe the flowering response. Most of the receptors ultimately produced flower buds, a result which was taken as evidence that cell-to-cell movement of the floral stimulus from *Perilla* to *Silene* had taken place.[28] However, it is also possible that flower formation was the consequence of the removal of the roots from the receptors, since in *Silene* this causes flowering under SD conditions.[28] Thus, there is at present no conclusive evidence for exchange of the floral stimulus between *Perilla* and the species of other genera or families.

EVENTS IN THE SHOOT APEX

Photoperiodic induction is perceived in the leaves, but the ultimate expression, differentiation of flower primordia, occurs in the apical meristems. The changes that occur in the apex of red *Perilla* following induction have been reviewed before.[33] Arrival of the stimulus in the apex affects specifically the central zone: the cells in this zone show an increase in mitotic activity, RNA content, and number of ribosomes. It is of further interest that application of GA which does not cause flowering in *Perilla* under LD,[14,33] has no effect in the central zone, but stimulates mitotic activity in the peripheral zone (more rapid leaf formation) and in the subapical pith region (causing extra stem growth).

Although histochemical studies of the shoot apex indicate that arrival of the stimulus causes an increase in RNA synthesis in the central zone, the question is whether more of the same RNA is made, or if new "reproductive" m-RNA is synthesized. This latter possibility implies that qualitatively different proteins would appear during the transition of the meristems from the vegetative to the reproductive state. Chailakhyan's group[19,23] has investigated this problem in red *Perilla* by the use of immunochemical methods. A specific reproductive protein was detected in floral primorida and bracts which was absent in the leaves of both vegetative and induced plants. This protein could be detected in the shoot apex after 6 SD and increased with longer SD induction. It was suggested[19] that the appearance of this new protein was associated with the action of the floral stimulus in the apex.

CONCLUDING REMARKS

All *Perilla* species are SDP which require several SD before flower buds will be produced. The green-leaved *Perilla* is less strict in its photoperiodic requirement than red *Perilla*.[1,15] This, at least in part, appears to be the reason that the phenomenon of indirect induction occurs only in green *Perilla*.[1,36]

During SD treatment of *Perilla* the induced state gradually builds up in the leaves and is of a permanent nature. The evidence obtained in interspecific grafts indicates that the floral stimulus is already being produced before the threshold of induction for flowering in green *Perilla* has been reached.[1,38]

The presence of a floral stimulus in *Perilla* is obvious from physiological experiments, but attempts to identify this material have failed so far. Possible reasons for this failure, such as low levels and/or lability of the stimulus outside the plant, as well as lack of a suitable assay, have been discussed before.[33,35] Clearly, the chemical characterization of the floral stimulus remains a pivotal problem in physiology of flowering.

The floral stimulus acts specifically in the cells of the central zone of *Perilla*, and indications are that new genes are activated, resulting in the appearance of a "reproductive" protein. The role of this protein in the transition from the vegetative to the reproductive state remains to be determined.[19]

ACKNOWLEDGMENT

My research since 1965, which is discussed in this review, was supported by AEC/ERDA/DOE. The preparation of this chapter was supported by Contract DE-AC02-76ERO-1338.

REFERENCES

1. **Allot-Deronne, M.,** Feuille et induction florale. Analyse sur une plante-modèle: le *Perilla*, Thèse de Doctorat d'État, University of Paris VI, Paris, 1983, 1—170.
2. **Allot-Deronne, M. and Blondon, F.,** Mise en évidence, par greffes interspécifiques, de l'édification progressive de l'état induit dans la feuille du *Perilla ocymoides* L., *C. R. Acad. Sci. Ser. D*, 286, 41—44, 1978.
3. **Allot-Deronne, M. and Blondon, F.,** Acquisition d'une induction florale indirecte dans la feuille du *Perilla:* comportement opposé du *Perilla ocymoides* L. et du *Perilla nankinensis laciniata* Decne, *C. R. Acad. Sci. Ser. D*, 291, 785—788, 1980.
4. **Bhargava, S. C.,** A transmissible flower bud inhibitor in *Perilla crispa*, *Proc. K. Ned. Akad. Wet. Ser. C*, 68, 63—68, 1965.
5. **Bouillenne, R.,** Recherche de la photopériode critique chez diverses espèces de jours longs et de jours courts cultivée en milieu conditionné, *Bull. Acad. Roy. Belg., Classe Sci.*, 49, 337—345, 1963.

6. **Boyer, G. L. and Zeevaart, J. A. D.,** Analysis of EDTA-enhanced phloem exudate obtained from flowering and vegetative *Perilla,* in *Plant Research '81,* Ann, Rep. MSU-DOE Plant Research Laboratory, Michigan State University, East Lansing, 1982, 125—126.
7. **Chailakhyan, M. Kh.,** Forty years of research on the hormonal basis of plant development — some personal reflections, *Bot. Rev.,* 41, 1—29, 1975.
8. **Chailakhyan, M. Kh. and Butenko, R. G.,** Translocation of assimilates from leaves to shoots during different photoperiodic regimes of plants, *Sov. Plant Physiol.,* 4, 450—462, 1957.
9. **Chailakhyan, M. Kh. and Yarkovaya, L. M.,** New facts in support of the hormonal theory of plant development, I, *C. R. (Dokl.) Acad. Sci. URSS,* 15, 215—217, 1937.
10. **Deronne, M. and Blondon, F.,** Mise en évidence chez le *Perilla ocymoides* L., plante de jours courts typique, d'un autre facteur de l'induction florale: les températures basses. Études de l'état induit acquis par la feuille en jours courts ou au froid, *Physiol. Veg.,* 15, 219—237, 1977.
11. **DeZeeuw, D.,** Flower initiation and light intensity in *Perilla, Proc. K. Ned. Akad. Wet. Ser. C,* 56, 418—422, 1953.
12. **Gaillochet, J., Mathon, C. C., and Stroun, M.,** Nouveau type de réaction et changement du type de réaction au photopériodisme chez le *Perilla ocymoides* L., *C. R. Acad. Sci.,* 255, 2501—2503, 1962.
13. **Garner, W. W. and Allard, H. A.,** Effect of abnormally long and short alterations of light and darkness on growth and development of plants, *J. Agric. Res.,* 42, 629—651, 1931.
14. **Gukasyan, I. A., Chailakhyan, M. Kh., and Milyaeva, E. L.,** Effect of daylength and gibberellin on the growth, flowering and apex differentiation in red *Perilla, Sov. Plant Physiol.,* 17, 63—70, 1970.
15. **Jacobs, W. P.,** Comparison of photoperiodic sensitivity of green-leafed and red-leafed *Perilla, Plant Physiol.,* 70, 303—306, 1982.
16. **Jacobs, W. P. and Raghavan, V.,** Studies on the floral histogenesis and physiology of *Perilla.* I. Quantitative analysis of flowering in *P. frutescens* (L.) Britt., *Phytomorphology,* 12, 144—167, 1962.
17. **King, R. W. and Zeevaart, J. A. D.,** Floral stimulus movement in *Perilla* and flower inhibition caused by non-induced leaves, *Plant Physiol.,* 51, 727—738, 1973.
18. **King, R. W. and Zeevaart, J. A. D.,** Enhancement of phloem exudation from cut petioles by chelating agents, *Plant Physiol.,* 53, 96—103, 1974.
19. **Kovaleva, L. V., Milyaeva, E. L., and Chailakhyan, M. Kh.,** Clarification of a specific protein in stem apices of the short-day plant red *Perilla* with transition from the vegetative to the reproductive state, *Dokl. Akad. Nauk SSSR,* 260, 1513—1516, 1981.
20. **Lona, F.,** Results of twelve years of work on the photoperiodic responses of *Perilla ocymoides, Proc. K. Ned. Akad. Wet. Ser. C,* 62, 204—210, 1959.
21. **Lubimenko, V. N. and Bouslova, E. D.,** Contribution à la théorie du photopériodisme. II, *C. R. (Dokl.) Acad. Sci. URSS,* 14, 149—152, 1937.
22. **Mathon, C. C. and Stroun, M.,** Mise à fleur de *Perilla nankinensis* Voss., plante de jours court typique, en jour continue, *C. R. Soc. Biol.,* 155, 1387—1388, 1961.
23. **Milyaeva, E. L., Kovaleva, L. V., and Chailakhyan, M. Kh.,** Formation of specific proteins in stem apices of plants in transition from vegetative growth to flowering, *Sov. Plant Physiol.,* 29, 253—260, 1982.
24. **Purse, J. G.,** Phloem exudate of *Perilla crispa* and its effects on flowering of *P. crispa* shoot explants, *J. Exp. Bot.,* 35, 227—238, 1984.

24a. **Salisbury, F. B.,** *Xanthium strumarium,* in *CRC Handbook of Flowering,* Vol. 4, Halevy, A. H., Ed., CRC Press, Boca Raton, Fla., 1985, 473—524.

25. **Schwabe, W. W.,** Studies on long-day inhibition in short-day plants, *J. Exp. Bot.,* 10, 317—329, 1959.
26. **Takimoto, A. and Ikeda, K.,** Effect of twilight on photoperiodic induction in some short-day plants, *Plant Cell Physiol.,* 2, 213—229, 1961.
27. **Tanimoto, S. and Harada, H.,** Hormonal control of morphogenesis in leaf explants of *Perilla frutescens* Britton var. *crispa* Decaisne f. *viridi-crispa* Makino, *Ann. Bot.,* 45, 321—327, 1980.
28. **Van de Pol, P. A.,** Floral induction, floral hormones and flowering, *Meded. Landbouwhogesch. Wageningen,* 72(9), 1—89, 1972.
29. **Wellensiek, S. J.,** Photoperiodism and temperature in *Perilla, Proc. K. Ned. Akad. Wet. Ser. C,* 55, 701—708, 1952.
30. **Wellensiek, S. J.,** The inhibitory action of light on the floral induction of *Perilla crispa, Proc. K. Ned. Akad. Wet. Ser. C,* 62, 195—203, 1959.

30a. **Wellensiek, S. J.,** *Silene armeria,* in *CRC Handbook of Flowering,* Vol. 4, Halevy, A. H., Ed., CRC Press, Boca Raton, Fla., 1985, 320—330.

31. **Wilson, B. J., Garst, J. E., Linnabary, R. D., and Channell, R. B.,** Perilla ketone: a potent lung toxin from the mint plant, *Perilla frutescens* Britton, *Science,* 197, 573—574, 1977.
32. **Zeevaart, J. A. D.,** Flower formation as studied by grafting, *Meded. Landbouwhogesch. Wageningen,* 58(3), 1—88, 1958.
33. **Zeevaart, J. A. D.,** *Perilla,* in *The Induction of Flowering. Some Case Histories,* Evans, L. T., Ed., Macmillan of Australia, Melbourne, 1969, 116—155.

34. **Zeevaart, J. A. D.,** Lack of evidence for distinguishing florigen and flower hormone in *Perilla, Planta,* 98, 190—194, 1971.
35. **Zeevaart, J. A. D.,** Physiology of flower formation, *Annu. Rev. Plant Physiol.,* 27, 321—348, 1976.
36. **Zeevaart, J. A. D.,** The floral stimulus in red and green *Perilla,* in *Plant Research '78,* Annual Report MSU-DOE Plant Research Laboratory Michigan State University, East Lansing, 1979, 32—33.
37. **Zeevaart, J. A. D.,** Perception, nature and complexity of transmitted signals, in *La Physiologie de la Floraison,* Champagnat, P. and Jacques, R., Eds., Coll. C. N. R. S. No. 285, Paris, 1979, 59—90.
38. **Zeevaart, J. A. D.,** unpublished data, 1980.
39. **Zeevaart, J. A. D.,** *Bryophyllum,* in *CRC Handbook of Flowering,* Vol. 2, Halevy, A. H., Ed., CRC Press, Boca Raton, Fla., 1985, 89—100.

PERSEA AMERICANA

En. Avocado; Fr. Avocatier; Ge. Avocatobirne; Sp. Abogado, Aguacate, Palta

B. O. Bergh

INTRODUCTION

The avocado, *Persea americana* (Mill.), apparently originated in the region from Mexico through Central America[31]. Kopp[19] reduced the probable area of origin to a Chiapas-Guatemala-Honduras center. The Spanish conquest found the fruit being widely consumed from Mexico south to Peru. The Aztec name *ahuacatl* became the Spanish *aguacate* — the general appellation in post-conquistador Mexico and Central America. This, in turn, was modified to the English *avocado* and to similar words in other European languages, such as *avocat* in French and *abogado* in European Spanish. Other avocado-growing regions developed their own local names, notably *palta* along the west coast of South America, and *custard apple* in West Africa. English-speaking consumers in different parts of the world often referred to them as "pears", from the similarity of their usual shape and color to the deciduous fruit; this was sometimes given the distinguishing modifier "alligator" pear. Indeed, for a time, there was a debate concerning the preferred English designation, among avocado, alligator pear, and aguacate, with the first-listed triumphing and now accepted nearly universally.

The pre-Columbian Indian tribes did not practice asexual propagation, but, evidently by seed selection over many generations, they developed types of high horticultural quality. Smith[47] found evidence of avocado consumption in Puebla state, Mexico, 9000 years ago. His evidence for selection, beginning about 6000 years ago, is not convincing, but prehistoric horticultural improvement over the small-fruited, large-seeded wild forms[5] was so great that major cvs like 'Fuerte' and 'Nabal' were taken directly from Mexican-Central American seedlings to found important commercial enterprises in other countries and on other continents. In 28 years of "scientific" avocado breeding, this author has made apparently significant cultivar improvements, but no dramatic advances.

The avocado has been important in the native diets of Mexico, Central America, and South America for at least hundreds of years. It has become important as a world crop only the last few decades. Production in different countries has increased rapidly in the past 15 years, and further considerable expansion is possible.[18] The most recent available estimates[1,37] showed Mexico to be the top producer with about 300,000 MT. Brazil and California came next with less than half that amount (California is now producing over 200,000 MT, but Mexican production is increasing even much more rapidly). For most Caribbean and Central and South American countries, production figures are uncertain and contradictory, but there are estimates in very roughly the 50,000 MT range for the Dominican Republic, Haiti, Cuba, Guatemala, El Salvador, Costa Rica, Colombia, Venezuela, Peru, and Chile, with considerable production also in other countries in the whole region. More accurate figures[1] placed Florida production at about 27,000 MT, Israel at 20,000 MT, and South Africa at 15,000; all three of these have also been expanding rapidly. Other countries with sizeable industries include Zaire, Morocco, the Philippines, and Australia. Only Israel and South Africa have major export industries, both to Europe, chiefly France. California grower returns increased rapidly to $112 million 5 years ago, but have stabilized at about that figure as ever-increasing production has caused sharply falling prices; retail value is about four times greater.

The avocado fruit ripens (softens to edibility) several days after it is picked. Unlike most fruits, its flesh is neither acid nor sweet, but with a unique rather bland flavor. It is never cooked, being eaten with salt or lemon or lime juice, mashed with other ingredients, or as a sweetened dessert. Most adults only gradually acquire a liking for it, but then it often becomes a prized delicacy. For most people in the developed world, its high calorie content is considered a drawback; however, its nutritional density is such that the calorie recommended daily allowance (RDA) percent is only about half the RDA percent of vitamins A, C, E, and several Bs, and iron and magnesium.[1] Also, its protein content is unusually high for a fruit, its high fat content is largely unsaturated, and it has digestive tract benefits. For regions where the problem is too few rather than too many calories, the high avocado calorie content is an advantage. Hume[17] believed that "perhaps the most important contribution of the New World to the human diet has been the avocado" — certainly a questionable assumption considering maize and beans, but doubtless true of tree fruits. According to Ruehle,[36] "The avocado possesses nutritive values far exceeding those of other fresh fruits;" and Purseglove[32] rated it "the most nutritious of all fruits."

BOTANY

The avocado is in the largely tropical family Lauraceae, with the classic laurel *(Laurus)* of Greek antiquity honor and the commercial camphor and cinnamon trees *(Cinnamomum* spp.). The only other member of the family significant for edible fresh fruit is *Persea schiedeana,* rather like the avocado on the outside, but with flesh that is quite different and inferior, to most people's taste; it is sometimes cultivated on a small scale, such as near Orizaba, Mexico.

For the avocado, *P. americana,* the genus name, *Persea,* is of unknown source — the avocado has no known connection with Persia. The species name, *americana,* reflects its New World origin, and was found to have precedence over the *P. gratissima* Gaertn. designation commonly used some decades ago.

Three horticultural races of *P. americana* have been recognized for a hundred years or more. In order of decreasing tropical adaptation, they are designated: West Indian (or "Antillean"), Guatemalan, and Mexican. Samson[37] therefore described them, respectively, as tropical, semitropical, and subtropical, although the latter two terms are usually used interchangeably. The three races differ in various tree and fruit horticultural and botanical traits.[4,31] The Guatemalan fruit matures considerably later than that of the other two races and is superior in terms of smaller and tight seed. For "Mediterranean" climates like Israel, South Africa, west Australia, California, and the highlands of Mexico and of Central and South American, cultivars of Guatemalan origin or with varying admixtures of Mexican germplasm are successful. For more tropical regions, West Indian cultivars and their hybrids with Guatemalan are adapted. The three races have sometimes been separated into different species, with the Mexican race given individual species status earlier, and the Guatemalan race placed in a separate species more recently. But hybridization results and morphological, geographical, and biochemical analyses indicate that the three are best regarded as botanical varieties of a single species.[8] Recent isozyme studies have strongly supported that conclusion (Bergh, B. O. et al.[8a]). So the West Indian, Guatemalan, and Mexican races become, respectively, *P. americana* var. *americana,* var. *guatemalensis,* and var. *drymifolia,*

The avocado fruit is a berry. Fruit color varies from yellow green through deep green or reddish to purple or black. Fruit size varies from under 50 g to almost 2 kg. Preferred commercial size is about 250 to 300 g, with some successful tropical cultivars considerably larger. The mesocarp becomes the edible portion, consisting of an oily pulp. Oil content varies from under 5% in many West Indians to 30% or more in many Mexican lines; it is intermediate in Guatemalans, but closer to Mexicans. Oil content becomes higher in a cooler climate.

FIGURE 1. Avocado inflorescence, showing a terminal panicle, with a small axillary panicle developing below it, and a larger axillary panicle grown out from farther back on the stem. (From Bergh, B. O., in *Outlines of Perennial Crop Breeding in the Tropics*, Ferwerda, F. P. and Wit, F., Eds., Miscellaneous Papers No. 4. Landbouwhogeschool [Agricultural University], The Netherlands, 1969, 23—51. With permission.)

Seedling trees may reach a height of 20 m. Grafted tree height varies with cultivar and degree of crowding, from about 4 m up. Commercial trees over about 10 m are often cut back to facilitate harvesting, a procedure that reduces flowering and fruit set. Compared with citrus under the same conditions, the avocado[12] tends to grow from apical buds at the end of branches rather than from upright "watersprouts" in older wood; to have thicker, but spongier and weaker wood; to grow less continually and with more distinct flushes — commonly two each summer for young trees, three or more for older trees; and to accumulate more starch between flushes — therefore, again somewhat resembling deciduous trees rather than typical tropical evergreens, at least as grown in California.

The flowers are usually borne on the current season's growth, typically on the first early spring growth flush. Thus, they are around the periphery of the tree.

FLOWER FORM AND FUNCTION

The Inflorescence

The typical avocado flower grouping (Figure 1) is a panicle of cymes.[34] These panicles may number in the hundreds, with up to thousands of flowers each, so that the total flower number per tree may literally be in the "millions;"[44] perhaps a million flowers would be a more average number for a mature tree. Some cvs, especially Guatemalans, tend to defoliate when they bloom heavily (Figure 2). Only one or two fruits are likely to mature from each panicle. A good average fruit crop is about 200 fruit per tree, so perhaps 0.02% of the flowers would be enough for heavy bearing. "Nevertheless, bloom that is less than normal for that cultivar indicates a physiological condition not conducive to good set, while genetically sparse bloom may indicate unusually good fruit-setting ability, perhaps because

FIGURE 2. Avocado tree in heavy bloom, largely defoliated. (Courtesy of C. A. Schroeder.)

superabundance of flowers is physiologically exhausting.''[4] Unlike most fruit trees, avocados can seriously deplete photosynthate reserves by their heavy flower production.[12]

Most panicles are indeterminate, ending in a vegetative bud.[34] The secondary lateral inflorescence axes arise alternately, and bear individual flowers on tertiary or higher-order axes. The flowers are commonly in groups of three, with the terminal (central) bud maturing first. Inflorescences arise on either terminal or lateral shoots (Figure 1). About 5 to 20% of all inflorescences are determinate,[40] the proportion increasing with heavier blooming. Exceptional cvs, such as 'Topa Topa', may have 90% determinate inflorescences.

Reece[34] made bud sections of the Guatemalan × West Indian hybrid 'Lula' and found that the earliest flower primordia could be identified in January. These developed rather slowly until the warmer weather of March induced rapid anthesis. Similar studies by Schroeder[41] in California, on three cvs varying from pure Mexican to pure Guatemalan, gave essentially similar results: floral primordia were recognizable no earlier than 2 months before anthesis and usually only 4 or 5 weeks. Gazit[14a] found flower primordia possibly identifiable considerably longer before anthesis in Israel. That at least some avocados can differentiate new floral primordia and then bloom within a matter of weeks is shown by lines with Mexican-race germplasm, which, when the original inflorescences failed to set fruit, produced secondary inflorescences where none had previously been identifiable.[4,41]

Occasional seedlings have had adventitious inflorescences arising from large branches or even the trunk;[7a] the effect was reminiscent of the jaboticaba *(Eugenia cauliflora)*. These inflorescences were small and were never observed to set fruit.

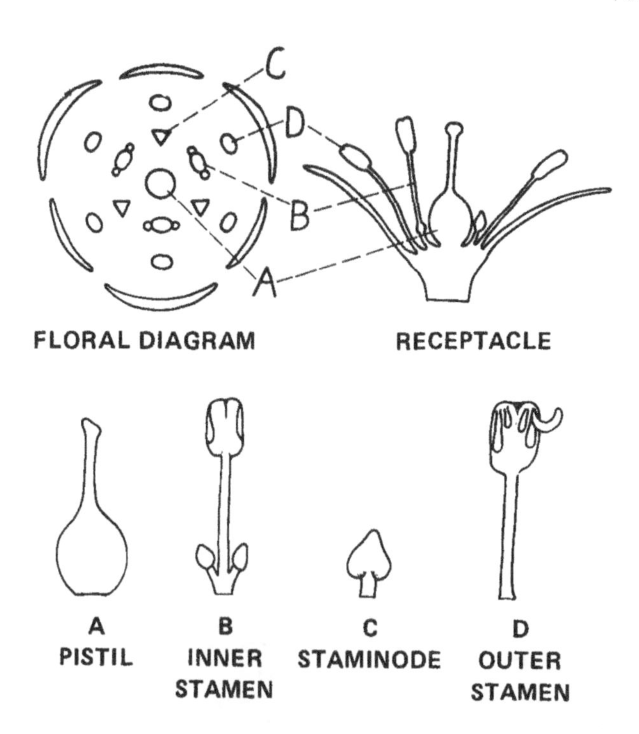

FIGURE 3. Avocado floral diagram, vertical section, and sexual organs. (From Bergh, B. O., in *Advances in Fruit Breeding*, Janick, J. and Moore, J. N., Eds., Purdue University Press, West Layfayette, Ind., 1975, 541—567. With permission.)

The Flower

The avocado flower is pubescent, regular, complete, and trimerous (Figures 3 and 5). The 6 greenish-yellow, partly united perianth segments are half sepals and half petals as shown by their traces arising at two levels.[33] The 3 sepals average about 5 mm in length — slightly shorter than the petals — and reflex slightly more on the average. Inside the 2 perianth whorls are 3 whorls of stamens and then 1 of staminodes, all 6 whorls having 3 members each. All stamens are quadrilocular. The 2 outer stamen whorls have introrse pollen sac valves, while the inner whorl stamens are extrorse or lateral (Figure 4C). Each of these inner 3 stamens has a pair of nectaries attached near the base of its filament. The short staminodes secrete nectar at the first or female flower opening; the nectaries secrete at the second opening (Figures 4D and 5). The simple pistil has a superior ovary with 1 ovule. The ovule is anatropous according to Schroeder,[42] who made a detailed study of sporogenesis and related developments. The style is rather slender, the stigma papillate (Figure 4B).

The staminodes, the more inward of the two outer stamen whorls, and the petals, are opposite each other. So are the inner stamens, the outermost stamen whorl, and the sepals (Figure 3).

Floral abnormalities are quite common in the avocado. The three most frequent types reported[39] were abnormal organs, extra organs, and conversion of one organ into another. Respective examples of these were (1) up to 8% of naked ovules in different cultivars; (2) a flower with 36 perianth parts, 62 stamens, and 13 pistils (instead of 6, 9, and 1); and (3) extra stamens, evidently transformed from perianth lobes or staminodes which were deficient in number. Frequent partial conversions were also noted. The same three types of abnormalities were later observed in the 'Bacon' cv.[7a] Here, the supernumerary organs were sometimes consistently so, so that the normal trimerous condition became quadripartite (6

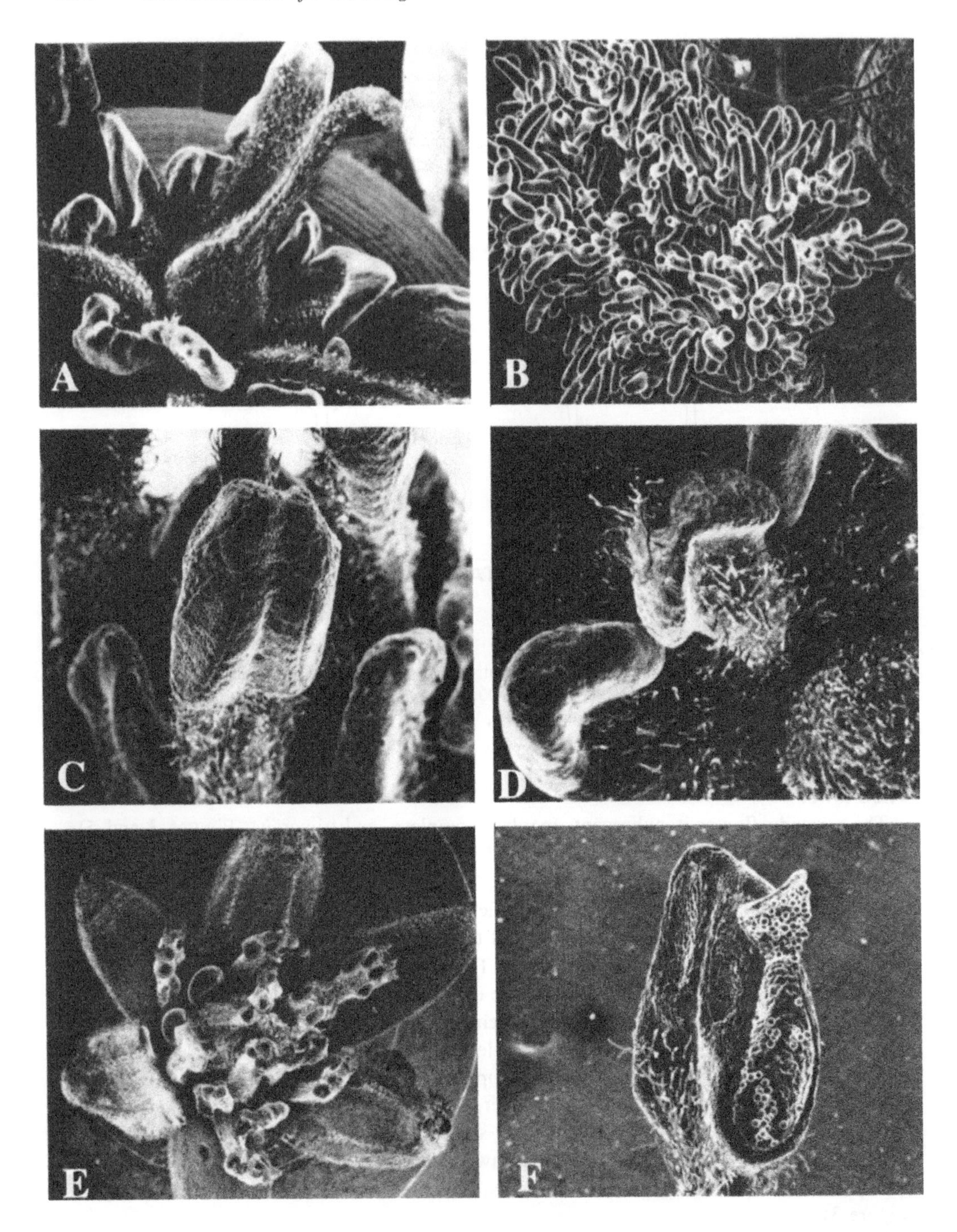

FIGURE 4. Scanning electron microscope views of avocado flower parts. A through D, female-function stage: (A) pistil, (B) papillae of stigma surface, (C) stamen, and (D) two nectaries flanking a staminode with much nectar. E through H, male-functioning stage: (E) dehisced anthers, (F) open pollen chamber, (G) anther valve with clumped pollen grains, and (H) enlarged pollen grain. (From Scholefield, P. G., *Scientia Hortic.*, 16, 263—272, 1982. With permission.) Courtesy of CSIRO Division of Horticultural Research.

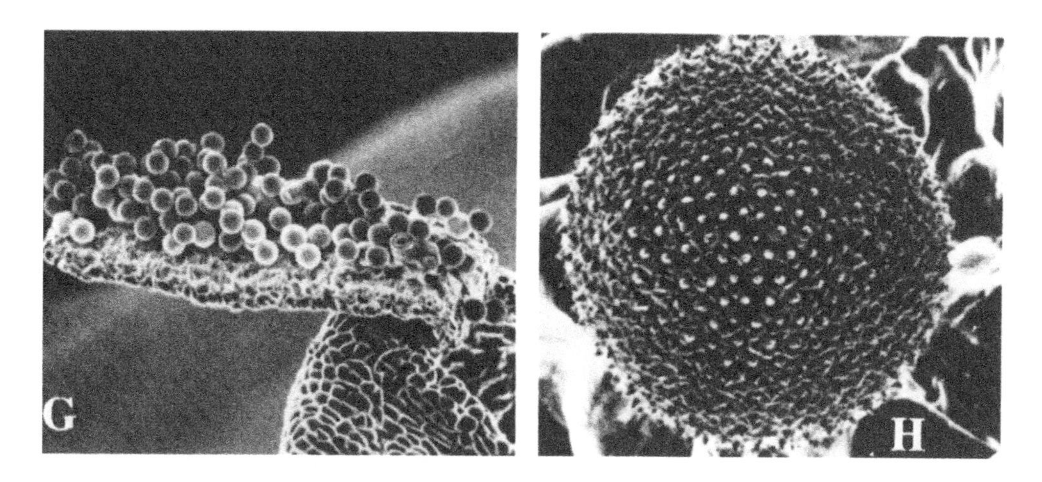

FIGURE 4 (continued)

whorls of 4 members each). When only one organ was increased in number, in 'Bacon' this was usually the nectaries; sometimes outer stamens or even perianth parts had vestigial or functional nectaries along their margins, or there was a varying number of "free" nectaries. The most common conversion here was a reversion of staminodes into apparently normal stamens.

Dichogamy

The unique avocado behavior can be described as "protogynous dichogamy with synchronous daily complementarity."[4]. It was discovered by Nirody[29] and by Stout;[48,49] it has been confirmed and illustrated by later researchers.[2-4,6,30,35,38,52] Each flower opens twice, the first time functioning as female (pistil-receptive; Figures 4A, 5A, and B), the second time functioning as male (pollen-shedding; Figures 4E-H, 5C, and D). The two open periods are on consecutive days. A flower of type "A" cultivars or seedlings is female in the morning and male the following afternoon. Type "B" flowers are just the reverse. Thus, each type complements the other in terms of female and male performance. Nature thereby encourages cross-pollination, bringing about the genetic variability on which natural selection acts to enhance adaptation. The avocado flower is structurally bisexual, but functionally unisexual. Regardless of rootstock, many hectares of a given cultivar will have all flowers on all trees typically entering and leaving both female and male stages more or less synchronously.

The above-described behavior, with strict synchronous complementarity, is commonly observed under the warm weather conditions prevalent in the avocado's Central American regions of origin. A large body of publications from different countries around the world has confirmed the basic situation and noted significant exceptions to it.[2] The most common cause of dichogamy modification, with some male-female overlap permitting self-pollination, is cooler weather or rapid weather changes. Thus, large plantings of a single cultivar in California may set quite well — although, even here, provision for cross-pollination has consistently increased fruit set. In quite cool weather, the sexual timing of both "A" and "B" types may be reversed from that outlined above. Still cooler weather may extend the 2-day double opening to 5 days[11] or even longer,[26] or result in female stages opening at night for "A" cultivars and not at all for "B" cultivars examined.[44]. Even without cool weather, there may, at times, be a little overlap of the two sexual stages within one cultivar, especially of "A" type, and also puzzling night openings.[30] Sedgley[54] has recently found that, except for either continuous dark or continuous light, normal floral cycling occurred

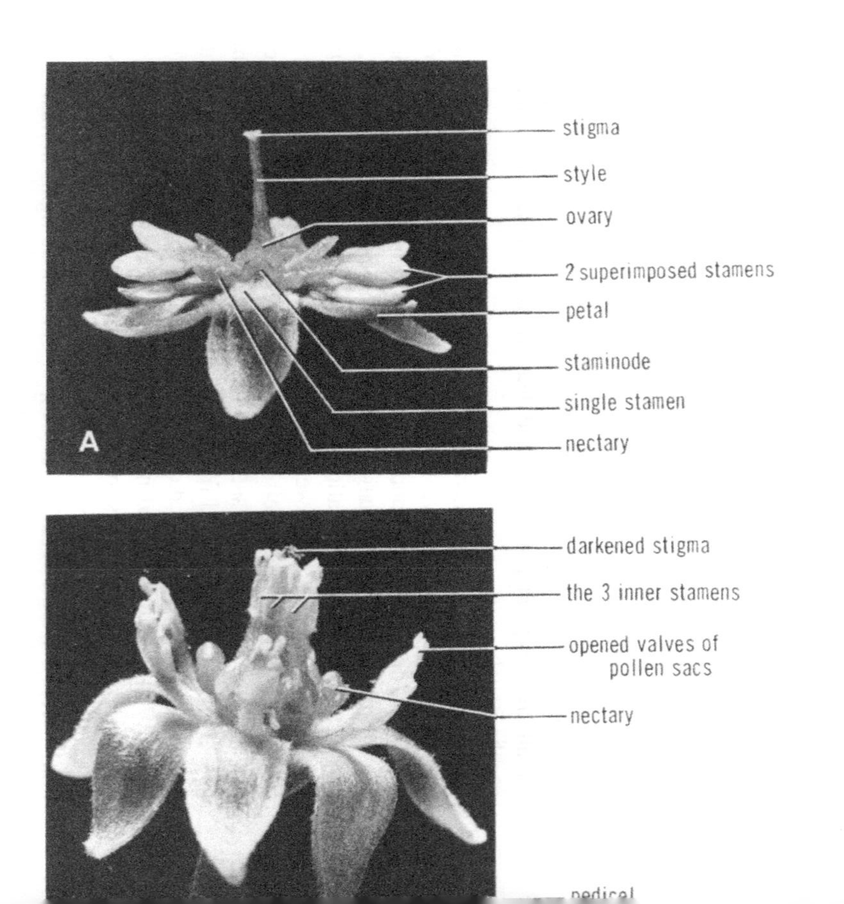
stigma
style
ovary
2 superimposed stamens
petal
staminode
single stamen
nectary
A
darkened stigma
the 3 inner stamens
opened valves of pollen sacs
nectary

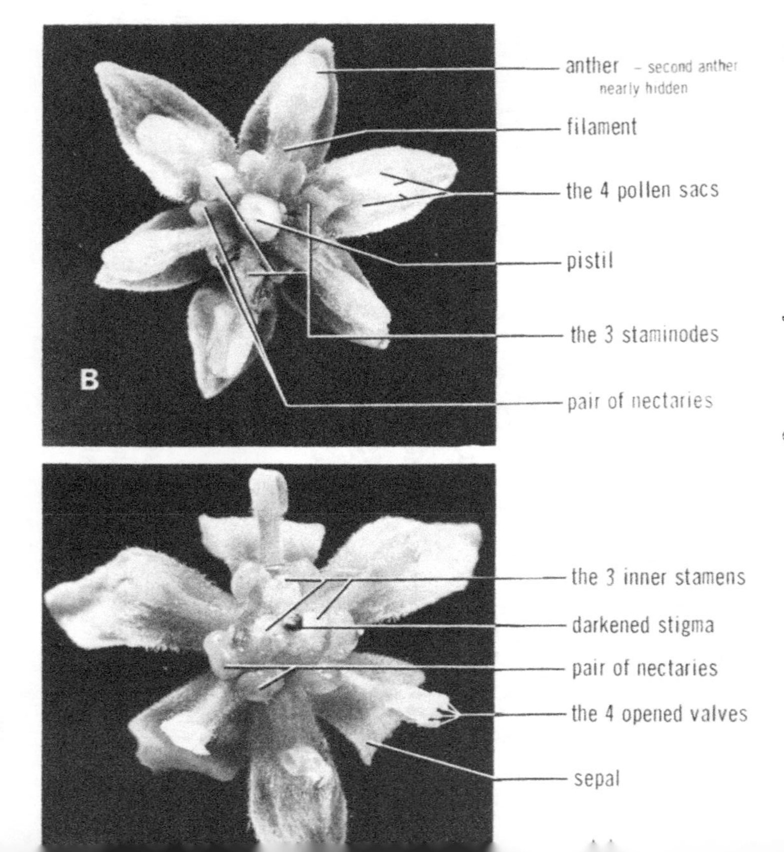
anther – second anther nearly hidden
filament
the 4 pollen sacs
pistil
the 3 staminodes
pair of nectaries
B
the 3 inner stamens
darkened stigma
pair of nectaries
the 4 opened valves
sepal

at various controlled daylengths. Emasculation at the first flower opening prevented the second (male) opening in a "B" cultivar but not in an "A" type.

FACTORS AFFECTING FLOWER DEVELOPMENT

Varietal Effects

Climatic Inhibition

The different races (botanical varieties) respond differently to different climates. The West Indian race is much the most tender to cold, and so its cultivars are severely injured by the occasional California freezes. But after several frost-free years here, West Indian cultivars that have reached 10 m or more in height, that appear perfectly healthy, and that are far past the stage of expected fruiting, still have not bloomed. Indeed, this author has never found a single flower on any pure West Indian cultivar regardless of age at either of the University of California's two avocado stations: inland, where it is colder in winter and hotter and drier in summer (Riverside); or nearer the ocean, where the temperatures are more moderate (there has not been a real freeze in over 30 years) and the humidity is higher (South Coast Field Station), where one might expect West Indian types to be better adpated, and where all the available West Indians have been tried. West Indian hybrids with either of the other two races have bloomed and set well at both locations.

The reason for the bloom failure of pure West Indians is not known. An "antiflorigen" could be involved. One might suspect a photoperiodic limitation since the West Indian race is from the tropics and the author grows it here at about 34° latitude north. But West Indians bloom and bear well in Israel at nearly the same latitude. The climates of the two countries are considered quite similar, with hot, dry summers and cool, rainy winters. However, the Israeli climate is more favorable to tropical plants, as is shown by its ability to produce bananas commercially — possible nowhere in California. Is this simply a reflection of warmer average winter temperatures? Or are there more subtle, significant differences? Precise climatic comparisons between the two countries are unknown to this author.

Moreover, Serpa[46] has reported the converse situation in tropical parts of Venezuela. There, Mexican race cultivars — and by implication apparently even some Guatemalans — "would not even produce flowers;" more detailed observations, preferably in conjunction with experiments in environmentally controlled growth chambers, should throw light on interesting differences in race physiology. There may well turn out to be complex interactions involving both photoperiod and climate. Certainly, all three races can bloom (and bear) well at the same location, as shown not only in Israel, but also in Morocco,[13] in Florida (where Mexican types flower and produce fruit in even the more tropical regions, but are inferior because of anthracnose fungus susceptibility, uneven fruit ripening, and limitations of small fruit size and thin skin), and probably elsewhere.

Seasonal Effects

There are differences between cultivars of a given race, but the three races consistently differ in average time of both beginning and ending anthesis. Observations in Florida[36] are typical: each cultivar bloomed about 2 months; a Mexican and a Mexican-Guatemalan hybrid were earliest, beginning in early to mid-January; pure West Indians came next, beginning mid-January to mid-February; finally, pure Guatemalans started blooming from mid-Feburary to mid-March. West Indian × Guatemalan hybrids varied widely. Flowering periods of the very earliest and the very latest were discrete, but most cultivars had much bloom overlap, centering in March. Similar observations have been made elsewhere, for example in Peru by Morin.[28] He, however, found that the respective dates for beginning bloom were spread over a considerably longer period, whereas length of bloom of any cultivar was not extended; hence, there was much less bloom overlap.

California differs from the standard of Florida and other more tropical countries in two respects. First, West Indian lines ordinarily do not flower. Second, the difference in time of beginning anthesis between Mexicans and Guatemalans is increased. This latter comes about by a somewhat puzzling climatic effect. Cool weather delays the onset of flowering, so that bloom tends to begin later with an increase in either elevation or latitude. But, California's climate causes a reversal of this effect: Mexican lines (and some of its hybrids with Guatemalans such as cvs 'Fuerte' or 'Pinkerton') may actually bloom considerably earlier here, beginning about November. Similar behavior has been found in similar climates elsewhere.[11] Again, the reason is not known, but photoperiod or the sharper seasonal difference in climate (or a combination of the two) are the likely suspects. Evidence against the photoperiod hypothesis may be the fact that in Corsica, at the considerably greater latitude of 42°, this early-blooming effect was not found;[26] Mexicans again started blooming earlier than Guatemalans, and all cvs bloomed for 2 or 3 months. In California, while lines with Mexican germplasm finished flowering at about the same time as in Corsica (roughly the end of April, depending on cv and weather), their total bloom period had been as long as 6 or 7 months. Rarely, such largely Guatemalan cvs as 'Hass' will also have some autumn bloom; this is more common in the cooler climate of New Zealand (personal observation of the author). I have never seen such on late-blooming pure Guatemalans like 'Nabal'.

Autumn bloom will occasionally set a small crop of fruits. This "offbloom" fruit has sometimes had a marketing advantage; but, on the whole, the advanced anthesis of Mexican-type avocados, in climates like California's, represents a considerable wastage of photosynthate due to bloom at seasons too cold for fruit set and maintenance, and must be regarded as a maladaptive physiological consequence of growing the tree outside of its region of evolutionary development.

Unlike Mexicans, Guatemalan cvs in California bloom a little later than Guatemalan types in Florida, beginning about March 15; in Corsica, they are a little later yet, starting about early April. These trends are the expected reflection of cooler climates. In both of the less tropical regions, the Guatemalan period extends a similar 2 months or so (sometimes longer in California), and usually overlaps considerably with the bloom of Mexican types and hybrids.

Other Climatic Effects

Bloom Period

In subtropical regions, most avocados bloom in the midwinter through spring season. This may reflect floral initiation (see "The Inflorescence" under "Flower Form and Function" above) resulting from the impeding of vegetative growth and perhaps cytokinin accumulation by the lower temperatures and perhaps the shorter days of autumn. Alternatively, floral initiation may be abetted directly by cold, or photoperiod change, or both. Growth chamber studies of cv 'Fuerte'[10] grafts found that, at a daylength of 15 hr, daytime temperatures of 25°C or higher inhibited bloom; bloom was abundant after 25 weeks at 20°C. At 9 hr daylength and 20°C, bloom was 8 weeks earlier, though less.

In tropical areas, the avocado does not have this spring bloom periodicity — a single tree can have flowers as well as fruits of all sizes.[37] In Kenya, at a nontropical high altitude but very near the equator, 'Fuerte' trees had various stages from flowers to mature fruit at one time (personal observation of the author). Trinidad, at about 10° latitutde, has an intermediate location, and intermediate flower behavior was reported:[30] different trees had bloom periods ranging from 2 to 8 months, from one bloom period to repeated flushes. Presumably, the reduced distinct periodicity of bloom in more tropical-equatorial regions reflects the reduced periodicity of one or more of: tree growth, temperature, and photoperiod.

Drought Dormancy

Especially in near-tropical regions without irrigation, drought may occur regularly in winter and thereby delay bloom for up to several months. This is true even in part of Trinidad,[47a] where West Indian types may not bloom until May or June after the rains resume; the fruit then develops during the hottest part of the year, requiring as little as 3 months from set to maturity. This compares with about 6 months for the typcial West Indian maturation period, with Mexicans about the same, Guatemalans up to 12 months (to 18 months in a climate like that of California). Storey[47a] noted that a few flowers might open and set just before the drought began, thus maturing fruits as the main blooming period was coming on.

Illumination

Lahav[20] studied flowering (and fruiting) in cv 'Ettinger' on 150 branches in two orchards over 3 years, comparing branches differing as to maturity, angle of growth, wind exposure, and illumination. His major finding was: "Better illumination produced earlier [more precocious] flowering." Hence, flowering on the top of the tree was earlier than on its lower branches and on the northern side. And so, because productivity was his ultimate concern, he concluded that "the lower part of the tree gives little fruit because of lack of illumination, and the same is true of the interior part of the tree." This would have practical applications in such practices as choosing a south slope where feasible (in the northern hemisphere), avoiding shading from windbreaks, and avoiding tree crowding.

Temperature

In a comparison of three different temperature levels, Sedgley[43] found that 'Fuerte' "reproductive growth was inhibited in favor of vegetative growth at 33/28 [°C day and night, respectively, the highest of her three levels], as exhibited by smaller floral parts and abscission . . . ." A later study of 'Hass' at the same three temperatures[45] found the latter cultivar to be much more tolerant of the high temperature; still, at 33/28°C, less than half as many flowers opened as at the two lower levels. The length of flowering period was about six times as long at the lowest temperature level (17/12°C) as at the highest — 85.3 days vs. 15. Lahav and Trochoulias[24] suggested from their own data involving different temperature regimes that the decline in reproductive development at high temperatures is due to increased competition from enhanced vegetative growth.

Promoting Bloom

Optimum Care

This author's repeated experiences with cooperators growing breeding seedlings and, also, with growers of commercial grafted trees, have demonstrated that adequate tree care is the major factor in bringing about the earliest possible flowering and fruiting. It is true that tree stress can sometimes induce shortening of the juvenile period and thereby earlier flowering. My experience with the avocado, however, has been that this is much more likely to have just the opposite effect and to delay flowering. Perhaps a certain level or type of stress combined with otherwise good care would be still better; if so, we do not know what that level or type is — the best possible care has given earliest average flowering. This involves such obvious measures as protection from wind and pathogens to maximize photosynthesizing surface.

Optimum irrigation and fertilizers require expert knowledge. Excess water quickly injures the avocado, which is exceptionally susceptible to harm from "wet feet". Conversely, more frequent irrigation, which presumably lowers water stress, has been associated with more precocious flowering.[23] Optimum fertilizer level varies with cv, but the most attractive lush, deep green foliage reflects a level of nitrogen too high for maximum fruiting.[14] These researchers did not attempt to determine whether the excess nitrogen reduced fruit set by reducing flowering, or by reducing the proportion of flowers maturing fruit; but Embleton[13a] believed from his field observations that it was the former — a reduction in flower formation.

Grafting

Apparently, no good study has been made to determine if grafting avocado seedlings causes them to bloom when younger. An approximate rule of thumb in California[7b] is that selections when grafted will bloom in about half the number of years required by the original own-root seedling, but this could be entirely a result of the grafting wood now being taken from a nonjuvenile, fruiting tree. Results from grafting small seedlings into cut-back mature trees have not been encouraging; here one might expect that the maturity of the stock would add to any effect of the grafting itself in inducing greater precocity, but several thousand seedlings grafted this way showed no obvious overall gain in precocity and the procedure has been abandoned.[7b] Earlier blooming in the avocado from grafting is, at best, undemonstrated.

Girdling

The removal of a narrow strip of bark around a limb or limbs has sometimes strikingly increased avocado fruit set in California[24a] and in Israel.[21,50] The usual result is for the girdled limb(s) to bear much more than nongirdled limbs on that tree, but for such nongirdled limbs to bear significantly less than limbs on nongirdled control trees.[7]

"Detailed work needs to be carried out initially to establish whether girdling enhances bud differentiation and/or fruit set."[53] Ticho[50] had stated that the girdling is done "in order to increase fruit bud differentiation. . .," but he gives no supporting evidence. Tomer's[51] detailed analysis found that under his conditions "girdling did not affect the number of flowers on the girdled part of the tree, but decreased the number of flowers on the ungirdled part of girdled trees as compared with control trees." It appears that there may be different effects on bloom under different circumstances and, unfortunately, flowering observations have been neglected in favor of the commercially important criterion of ultimate fruit production. Thus, Bergh[7] found that young 'Fuerte' trees set about three and one half times as much fruit when about one third of each tree was girdled, but he made no bloom comparisons. A number of observations are anecdotally interesting, but statistically meaningless. For example, Ticho[50] reported a 12-year-old avocado seedling that "started flowering and fruiting only after scaffold branch was ringed.... The remainder of the tree continued with its juvenile habit." And an introduction from Guatemala, G755C, is of much interest because of resistance to *Phytophthora cinnamomi,* but it has been of no value to the breeding program here because not one flower has been observed on trees in California; last fall we girdled the trunk of a G755C tree — this spring it is blooming well.

Girdling may be useful for inducing precocity in seedlings generally, and also in commercial trees, if tests of physiological readiness for anthesis can be developed. Girdling of about 2000 seedlings of various parentages of an age when some of them would be expected to start blooming the following spring resulted in a higher proportion blooming that spring than of the check seedlings,[7c] but the larger proportion suckered heavily from below the girdle and, in the judgement of this author, many that would have bloomed on their own a year later had flower initiation actually delayed for at least one additional year by the girdling. I could not differentiate these negative-responders on the basis of pregirdling morphology or other appearances. Perhaps anatomical or biochemical studies could identify young avocado trees that are physiologically ready to bloom more precociously when girdled. Desirable material for such an investigation would be genetically identical twin seedlings obtained by splitting the embryos into two (or fairly easily into more) viable sections, or cvs grafted onto identical clonal rootstocks.

Bending Down of Branches

This has induced more precocious blooming in some plants, but this author's avocado experience in California and in Israel has not been encouraging. Whether side branches or main stems were bent over, the result was to encourage juvenile vertical shoots to break from buds along the more or less horizontal stem. My judgement was that blooming was, on the whole, delayed rather than advanced.

Chemicals

Maleic hydrazide[8b] produced significantly more precocious flowering of seedling avocados in the greenhouse. There was some plant injury at the concentrations used, which, with reports of chemical carcinogenicity, discouraged further trials. Paclobutrazol [(2RS, 3RS)-1-(4-chlorophenyl)-4,4-dimethyl-2-1,2,4-triazol-1-1-yl-1 pentan-3-ol] seems quite promising, and trials are beginning.

Boswell et al.[9] found that both NIA-10637 (ethyl hydrogen 1-propyl-phosphonate) and NIA-10656 (1-propylphosphonic acid) at 1250 and 2500 ppm caused a stunting of new growth to one fourth or less (at 1% level of significance) as compared with the check 'Bacon'-grafted trees, but no observations were made as to subsequent flowering differences. Similarly, McCarty et al.[27] found that TIBA at preferably 250 or 500 ppm broke apical dominance and caused wide-angled side-branch development on 'Bacon' seedlings at an age where the check trees had no axillary shoot development; but again there was no follow-up observation on flowering.

In Spring 1983, this author placed several dozen small grafted trees of various cvs, on various rootstocks, from both commercial and University nurseries, in an outdoor canyard, in two groups separated by a large block of seedlings. All trees were for planting out in the field in the spring of 1984 and, in the meantime, they were left in the routine care of others. Examination early in the spring revealed an unexpected situation: the seedlings had grown normally, but all grafted trees, regardless of location, scion, stock, or source, were severely dwarfed and covered with a frequently astonishing mass of bloom — most would not have been expected to produce any flowers until they were a year or, more likely, 2 years older. The most reasonable explanation seems to be drift from some chemical spray, which has not yet been identified. An added puzzle is the apparent immunity of own-rooted young seedlings. At least this incident seems to indicate that avocado flowering is susceptible to manipulation (chemical?) to a degree that could be highly important for commercial production.

Genetic Selection

Many avocado seedlings, even in its native Central American habitat, will apparently never fruit. Those that do bloom and set fruit have required about 6 years to begin doing so in tropical or near-tropical regions,[15,36,37] and somewhat longer on the average in less tropical climates like California (Bergh, unpublished). Over the past 25 years, the University of California avocado breeding program has sharply reduced this average time to flowering. Formerly, seedling plantings were left for 10 years, with the reasoning that trees that had not yet fruited, but would do so subsequently, were probably of inferior productivity; now (Bergh, unpublished), seedling plantings are left only 5 years, with the same reasoning. One seedling group has actually had over half the trees flower (and set fruit that matured) just 2 years from planting — demonstrating precocity superior to that of most grafted commercial cvs. Very occasional seedlings are now blooming a year from planting, although none of these has as yet matured fruit from that early bloom.

This increased precocity has resulted from the combination of two related procedures: self-fertilization that has reduced tree vigor and so tipped the physiological balance from the vegative toward the reproductive;[4] and concomitant selection for more precocious bloom (and heavier fruit set). Unlike the experience with many other plants, the avocado has had no commercially harmful side effects from this selfing-selection program.

Modifying Time of Bloom

Girdling

Although it is not yet clear if girdling is likely to increase the amount of bloom, when carried out at the usually most effective time — in autumn before the spring chief blooming period — girdling apparently always advances the time of flowering. For 'Fuerte', the advance is 60 to 85 days.[22] As girdling is practiced closer to spring, bloom advance becomes less and less, as usually does also fruit set increase.

Previous Crop

A heavy fruit set carried to maturity both delays bloom and reduces bloom. 'Fuerte' set that reached the equivalent of 30,000 kg/ha resulted in very late and very little bloom on the heavy bearing trees the following spring.[7d] Thus, avocado trees tend to be alternate bearers. Conversely, unfruitful trees, whether seedlings or grafts, are usually outstanding for the earliness and heaviness of their flowering. Early fruit harvesting may increase both the earliness and amount of the subsequent bloom.[16]

Weather

Earlier, we noted the interaction between race and beginning of bloom time in terms of climate, and also the delayed bloom from regular winter drought in some locations. For any given location, cooler weather in the months preceding normal bloom time can be expected to delay the onset of anthesis. The degree of delay varies with degree of coolness; 1 month is about the maximum that I have observed. The chief factor appears to be daytime highs. Hence, maximum bloom retardation has been observed from stretches of cloudy, cool days, even though nighttime lows are higher under cloudy conditions. In a cool winter climate like that of California, greenhouse avocados usually bloom considerably earlier than those out of doors. Another way to manipulate blooming time with temperature is to store budwood that has differentiated floral buds at about 4°C for up to 4 months and then graft it.[44]

Chemicals

In less tropical regions like California, Israel, South Africa, Chile, New Zealand, southern Australia, and much of Mexico, much of the avocado bloom ordinarily occurs when the weather is still too cool for best fruit set.[4] Hence, anything that would delay bloom period could be beneficial. The weather (see preceding subsection) is helpful only sometimes. Rootstocks have not been found to make an appreciable difference. In Israel, Levin et al.[25] tested "numerous" substances for bloom retardation. Some had no effect and others had destructive side effects. However, sodium dikegulac delayed bloom (and vegetative growth) for a month or longer, with apparently acceptable injury levels. They recommended deferring treatment until flower buds can be identified. No suggestions were published as to concentration.

REFERENCES

1. **Ahmed, E. M. and Barmore, C. R.,** Avocado, in *Tropical and Subtropical Fruits: Composition, Properties, and Uses*, Nagy, S. and Shaw, P. E., Eds., AVI, Westport, Conn., 1980, chap. 2.
2. **Bergh, B. O.,** Avocado *(Persea americana* Miller), in *Outlines of Perennial Crop Breeding in the Tropics*, Ferwerda, F. P. and Wit, F., Eds., Misc. Pap. No. 4, Landbouwhogeschool (Agricultural University), Wageningen, The Netherlands, 1969, 23—51.
3. **Bergh, B. O.,** The remarkable avocado flower, *Calif. Avocado Soc. Yearb.*, 57, 40—41, 1974.
4. **Bergh, B. O.,** Avocados, in *Advances in Fruit Breeding*, Janick, J. and Moore, J. N., Eds., Purdue University Press, West Lafayette, Ind., 1975, 541—567.
5. **Bergh, B. O.,** Avocado: *Persea americana* (Lauraceae), in *Evolution of Crop Plants*, Simmonds, N. W., Eds., Longman, London, 1976, 148—151.

6. **Bergh, B. O.,** Avocado breeding and selection, in *Proc. 1st Int. Tropical Fruits Short Course: The Avocado,* Sauls, J. W., Phillips, R. L., and Jackson, L. K., Eds., University of Florida, Gainesville, 1977, 24—33.
7. **Bergh, B. O.,** Factors affecting avocado fruitfulness, in *Proc. 1st Int. Tropical Fruits Short Course: The Avocado,* Sauls, J. W., Philipps, R. L., and Jackson, L. K., Eds., University of Florida, Gainesville, 1977, 83—88.

7a. **Bergh, B. O.,** unpublished, 1977.

7b. **Bergh, B. O.,** observations over several years.

7c. **Bergh, B. O.,** unpublished observations, 1980.

7d. **Bergh, B. O.,** unpublished observations, 1970.

8. **Bergh, B. O., Scora, R. W., and Storey, W. B.,** A comparison of leaf terpenes in *Persea* subgenus *Persea, Bot. Gaz.* (Chicago), 134, 130—134, 1973.

8a. **Bergh, B. O., Torres, A. M., Zentmyer, G. A., and Ellstrand, N. C.,** unpublished, 1985.

8b. **Bergh, B. O. and Hield, H. Z.,** unpublished, 1973.

9. **Boswell, S. B., Burns, R. M., and McCarty, C. C.,** Chemical inhibition of avocado top regrowth, *Calif. Avocado Soc. Yearb.,* 55, 113—116, 1972.
10. **Buttrose, M. S. and Alexander, D. McE.,** Promotion of floral initiation in 'Fuerte' avocado by low temperature and short daylength, *Sci. Hortic.,* 8, 213—217, 1978.
11. **Calvino, E. M.,** Biologia fiorale della *Persea drymifolia* (aguacate) in Sanremo, *Stazione Sperimentale di Floricoltura 'O. Raimondo'* (San Remo, Italy) N. 31, (Transl.), *Calif. Avocado Soc. Yearbk.,* 1939, 79—86.
12. **Chandler, W. H.,** The avocado, in *Evergreen Orchards,* 2nd ed., Lea & Febiger, Philadelphia, 1958, chap. 7.
13. **Chavanier, G.,** New observations on avocado growing in Morocco, *Calif. Avocado Soc. Yearbk.,* 51, 111—113, 1967.

13a. **Embleton, T. W.,** private communication, 1984.

14. **Embleton, T. W. and Jones, W. W.,** Development of nitrogen fertilizer programs for California avocados, *Calif. Avocado Soc. Yearb.,* 55, 152—155, 1972.

14a. **Gazit, S.,** private communication, 1984.

15. **Higgins, J. E., Hunn, C. J., and Holt, V. S.,** The avocado in Hawaii, *Hawaii Agric. Exp. Stn. Bull.* No. 25, 1911.
16. **Hodgson, R. W.,** The avocado — a gift from the middle Americas, *Econ. Bot.,* 4, 253—293, 1950.
17. **Hume, E. P.,** Growing avocados in Puerto Rico, *U.S. Dep. Agric. F. Exp. Stn. Puerto Rico Circ.* No. 33, 1951.
18. **Knight, R., Jr.,** Origin and world importance of tropical and subtropical fruit crops, in *Tropical and Subtropical Fruits: Composition, Properties and Uses,* Nagy, S. and Shaw, P. E., Eds., AVI, Westport, Conn., 1980, chap. 1.
19. **Kopp, L. E.,** A taxonomic revision of the genus *Persea* in the Western Hemisphere (*Persea*-Lauraceae), *Mem. N.Y. Bot. Gard.,* 14(1), 1966.
20. **Lahav, E.,** Localization of fruit on the tree, branch girdling, and fruit thinning, in *Div. Subtrop. Hortic. Volcani Inst. Agric. Res. 1960—1969,* 1970, 60—61.
21. **Lahav, E., Gefen, B., and Zamet, D.,** The effect of girdling on the productivity of the avocado, *J. Am. Soc. Hortic. Sci.,* 96, 396—398, 1971.
22. **Lahav, E., Gefen, B., and Zamet, D.,** The effect of girdling on fruit quality, phenology and mineral analysis of the avocado tree, *Calif. Avocado Soc. Yearbk.,* 55, 162—168, 1972.
23. **Lahav, E. and Kalmar, D.,** Water requirements, in *Div. Subtrop. Hortic. Volcani Inst. Agric. Res.,* 1960—1969, 1970, 69—72.
24. **Lahav, E. and Trochoulias, T.,** The effect of temperature on growth and dry matter production of avocado plants, *Aust. J. Agric. Res.,* 33, 549—558, 1982.

24a. **Lee, B. W.,** unpublished, 1978.

25. **Levin, A., Adato, I., and Gazit, S.,** Effect of growth substances on avocado flowering and vegetative growth, in *Volcani Cent. Isr. Inst. Hortic. Dept. Subtrop. Hortic. Spec. Publ.* No. 222, 86—87, 1983.
26. **Lichou, J. and Vogel, R.,** Biologie florale de l'avocatier en Corse, *Fruits,* 27, 705—717, 1972.
27. **McCarty, C. D., Boswell, S. B., and Burns, R. M.,** Chemically induced sprouting of axillary buds in avocados, *Calif. Avocado Soc. Yearbk.,* 55, 117—119, 1972.
28. **Morin, D.,** El palta, in *Cultivo de Frutales Tropicales y Menores,* Editorial Juridica, Lima, Peru, 1965, chap. 2.
29. **Nirody, B. S.,** Investigations in avocado breeding, *Calif. Avocado Assoc. Annu. Rep. 1921—1922,* 65—78, 1922.
30. **Papademetriou, M. K.,** Some aspects of the flower behavior, pollination, and fruit set of avocado (*Persea americana* Mill.) in Trinidad, *Calif. Avocado Soc. Yearb.,* 60, 106—152, 1976.
31. **Popenoe, W.,** Central American fruit culture, *Ceiba,* 1, 304—310, 1952.

32. **Purseglove, J. W.,** Avocado, in *Tropical Crops: Dicotyledons,* Vol. 1, Longman, London, 1972, 192—198.
33. **Reece, P. C.,** The floral anatomy of the avocado, *Am. J. Bot.*, 26, 429—433, 1939.
34. **Reece, P. C.,** Differentiation of avocado blossom buds in Florida, *Bot. Gaz. (Chicago),* 104, 323—328, 1942.
35. **Robinson, T. R. and Savage, E. B.,** Pollination of the avocado, *U.S. Dep. Agric. Circ.*, No. 397, 1926.
36. **Ruehle, G. D.,** The Florida avocado industry, *Univ. Fla. Agric. Exp. Sta. Bull.*, No. 602, Gainesville, 1963.
37. **Samson, J. A.,** Avocado, in *Tropical Fruits,* Longman, London, 1980, 196—204.
38. **Scholefield, P. G.,** A scanning electron microscope study of flowers of avocado, litchi, macadamia and mango, *Sci. Hortic.*, 16, 263—272, 1982.
39. **Schroeder, C. A.,** Floral abnormality in the avocado, *Calif. Avocado Soc. Yearb.*, 1940, 36—39.
40. **Schroeder, C. A.,** The avocado inflorescence, *Calif. Avocado Soc. Yearb.*, 1944, 39—40.
41. **Schroeder, C. A.,** Flower bud development in the avocado, *Calif. Avocado Soc. Yearb.*, 36, 159—163, 1951.
42. **Schroeder, C. A.,** Floral development, sporogenesis, and embryology in the avocado, *Persea americana, Bot. Gaz. (Chicago),* 113, 270—278, 1952.
43. **Sedgley, M.,** The effect of temperature on floral behavior, pollen tube growth, and fruit set in the avocado, *J. Hortic. Sci.*, 52, 135—141, 1977.
44. **Sedgley, M. and Alexander, D. M.,** Avocado breeding research in Australia, *Calif. Avocado Soc. Yearb.*, 67, 129—135, 1983.
45. **Sedgley, M. and Annells, C. M.,** Flowering and fruit-set response to temperature in avocado cultivar 'Hass', *Sci. Hortic.*, 14, 27—33, 1981.
46. **Serpa, D.,** Avocado culture in Venezuela, *Calif. Avocado Soc. Yearb.*, 52, 153—168, 1968.
47. **Smith, C. E., Jr.,** Archaeological evidence for selection in avocado, *Econ. Bot.*, 20, 169—175, 1966.
47a. **Storey, W. B.,** private communication, 1980.
48. **Stout, A. B.,** A study in cross-pollination of avocados in Southern California, *Calif. Avocado Assoc. Annu. Rep. 1922—1923,* 29—45, 1923.
49. **Stout, A. B.,** The flowering behavior of avocados, *Mem. N.Y. Bot. Gard.*, 7, 145—203, 1927.
50. **Ticho, R. J.,** Girdling, a means to increase avocado fruit production, *Calif. Avocado Soc. Yearb.*, 54, 90—94, 1971.
51. **Tomer, E.,** The Effect of Girdling on Flowering, Fruit Setting and Abscission in Avocado Trees, Ph.D. thesis, Hebrew University, Rehovot, 1977 (in Hebrew, English summary).
52. **Torres, J. P.,** Some notes on avocado flower, *Philipp. J. Agric.*, 7, 207—227, 1936.
53. **Trochoulias, T.,** Girdling, in Proc. Aust. Avocado Res. Workshop, Binna Burra Lodge, Lamington National Park, South Queensland, October 10 to 14, 1977, 66—70.
54. **Sedgley, M.,** Some effects of daylength and flower manipulation on the floral cycle of two cultivars of avocado, *(Persea americana* Mill., Lauraceae), a species showing protogynous dichogamy, *J. Exp. Bot.*, 36(166), 823—832, 1985.

PINACEAE*

Hormonal Promotion of Flowering

Richard P. Pharis and Stephen D. Ross

INTRODUCTION

Prior to 1975 the promotion of flowering for commercially important conifers within the Pinaceae family was limited to traditional cultural practices, of which water stress, root-pruning, N fertilizing, girdling, and high temperature were most commonly employed.[28,29a,46,64a] However, these techniques were usually only effective on older trees that had already begun to flower naturally; even then successful increases in seed and pollen cones were sporadic and subject to the vagaries of nature. GA_3, which had been found to promote profuse flowering even in very young seedlings of conifers within the Cupressaceae and Taxodiaceae,[28,43,52] did not seem to work on conifers of the Pinaceae family.[43,52] Why this should be so was a perplexing question, for members of all three families were known to respond to similar cultural treatments.

It is now of course apparent that the GA of choice, GA_3, was the wrong choice for use on Pinaceae conifers. Successful promotion of flowering by a mixture of GA_4 and GA_7 was first reported in 1974 for *Pinus contorta*[58] and in 1973 through 1976 for *Pseudotsuga menziesii*.[44—46,52,56,66,76a,77] These initial successes have been expanded to 17 species in 5 genera of the Pinaceae (Table 1). It is apparent from the 60 plus research papers listed in Table 1 that hormonal promotion of flowering in Pinaceae family conifers by $GA_{4/7}$ is not an isolated event. It seems likely that most conifer species will be amenable to manipulation of flowering through the judicious timing of $GA_{4/7}$[13,23,37,40a,72,80,89] applied at a proper concentration. This response to $GA_{4/7}$ can be enhanced, often synergistically, in the presence of adjunct cultural treatments such as girdling, nitrate-N fertilization, water stress, root-pruning and high temperatures (Figure 1 and see References 2,4,12a,17,19,21,22,24,25,27,29,35, 38,39,44—46,52—55,58—60,62—64,67,68,73,74,77,79,80,82—85,87—89,92, and 93).

Because most of the reports dealing with hormonally induced flowering in Pinaceae species have been largely empirical in nature, some workers[19,41,84,94] have taken a conservative view in discussing the relevance of $GA_{4/7}$-promoted flowering to flowering in nature. Dunberg and Oden[19] in their comprehensive review of the subject, cite several lines of evidence which led them to conclude that, "The GA effect on flowering in the Pinaceae is of undetermined physiological significance." Although uncontrovertible proof is lacking for the hypothesis[47,79a,82] that endogenous GAs of a less-polar nature (e.g., mono-hydroxylated GAs such as GA_4 and GA_7) play an important regulatory role in the flowering of Pinaceae family conifers, the questions posed by others equally, in our opinion, do not negate this hypothesis. We will attempt to examine the physiological relevance of $GA_{4/7}$-promoted flowering in the Pinaceae and go on to discuss in a somewhat speculative way how exogenous $GA_{4/7}$ (and by implication endogenous GAs) might be promoting flowering in Pinaceae species.

THE METHODOLOGY OF GROWTH REGULATOR APPLICATION

The early successes of GA_3 (and other GAs) in promoting flowering in Cupressaceae and Taxodiaceae conifers (see Reference 52 and references cited therein) generally utilized aqueous sprays to drip-off, with varying amounts of a surfactant. This approach had been

* Supplement to the chapter "Temperate Conifer Forest Trees" published in Volume I.

Table 1
CHECKLIST FOR SPECIES OF PINACEAE FAMILY CONIFERS THAT WILL FLOWER IN RESPONSE TO APPLICATION OF A MIXTURE OF THE PLANT HORMONES GIBBERELLIN $A_{4/7}$

Species	Ref.
Larix leptolepis (Japanese larch)	3
L. decidua (European larch)	3
Picea abies (Norway spruce)	11,12a,14,17,35,38
P. engelmannii (Engelmann spruce)	73
P. glauca (white spruce)	9,40,40a,52,57a
P. mariana	24a
P. sitchensis (Sitka spruce)	59—61,85—90
Pinus banksiana (jack pine)	6—8
P. caribaea	77a
P. contorta (lodgepole pine)	52,56,58,93
P. densiflora (Japanese red pine)	28,31—34,52
P. elliottii (slash pine)	25,26,52
P. palustris (longleaf pine)	26,27
P. radiata (Monterey pine)	52,75,84
P.sylvestris (Scots pine)	10,12,12a,13,36—39
P. taeda (loblolly pine)	21—26,33,34,76
P. thunbergii (Japanese black pine)	31,32,52
Pseudotsuga menziesii (Douglas fir)	2,29,41,42b,44—48,53,54a,55,56,57,65,66, 67—72,76a—78,81—83,92
Tsuga heterophylla (Western hemlock)	4,53,62—64,64b,80,82

tested on Pinaceae family conifers with GA_3, and on occasion with $GA_{4/7}$ mixtures, but with generally limited success (see References 28,52,79 and 82 and references cited therein).

The reasons for the lack of success with early treatments of plant-growth regulators on Pinaceae family conifers appear to include:

1. Use of the relatively inactive (in terms of flowering) GA_3.
2. Poor timing of application. Unlike most Cupressaceae and Taxodiaceae family conifers, which initiate conebuds on previously vegetative meristems over a long period of the growing season, Pinaeceae family conifers generally differentiate conebuds from previously initiated (but undifferentiated) lateral primordia, the differentiation period for each of male and female conebuds taking place over a discrete (and sometimes different) period of varying length and time (depending on species and location).
3. The use of only one or several applications of growth regulator over a relatively short period of time, often outside of, or barely bracketing the period of sexual differentiation.
4. A lack of use of adjunct cultural treatments which were known to promote flowering (e.g., girdling, water stress, etc.) together with the $GA_{4/7}$ treatment — these have now been shown to be highly synergistic when used with $GA_{4/7}$.
5. Use of surfactants or application methods which did not give sufficient uptake of hormone. Successful methods of applying $GA_{4/7}$ now include topical application in ethanolic solution, with or without a surfactant, an especially useful technique for initial probes with new species or very young individuals;[3,3a,45,46,55,77] stem injection in aqueous solution;[41,53,80,82,83] sprays of aqueous or ethanolic solutions with the surfactant AROMOX C-12W;[3,3a,70,74,79,82] and an especially promising approach using ultra-low-volume (ULV) sprays in an appropriate oil carrier.[3a,73,74,79] In each case success has been enhanced by timing the application to coincide with that period when sexual differentiation of male and/or female conebuds is expected to occur. Finally, for best results — especially with very young seedlings and recently grafted/rooted

FIGURE 1. An example of abundant flowering which can be induced in a conifer of the Pinaceae family by $GA_{4/7}$ applied in conjunction with appropriate cultural practices. This potted 6-year-old *Picea engelmannii* grafted propagule (sexually mature scion on seedling rootstock) was sprayed weekly to wetness with 400 mg/ℓ $GA_{4/7}$ (in aqueous solution containing 0.05% AROMOX C-12/W surfactant) and subjected to high temperatures within a 30°c day/20°C night polyethylene house for 3 weeks the previous year during that period (toward the end of shoot elongation) when conebud differentiation normally occurs. Although only seed cones are shown in this photograph, taken in August, abundant male conebuds were also produced and provided more than adequate pollen for supplemental pollinations. Treated trees averaged 45 seed cones, compared to only 3 cones each for noninduced trees outdoors; and each cone contained an average of 100 seeds of which 50% or more were filled.

propagules, or on sites unfavorable for flowering (i.e., moist, cool, with relatively low solar insolation during the differentiation period) — the $GA_{4,7}$ should be combined with other growth regulators and with cultural treatments that have a proven history of at least occasionally promoting flowering (e.g., water stress, girdling, high temperature, rootpruning, etc.).

THE RELEVANCE OF $GA_{4/7}$-INDUCED FLOWERING IN CONIFERS TO FLOWERING IN NATURE

Why Does the Flowering Response in Pinaceae Species Require Such a High Dosage of $GA_{4/7}$, Relative to GA_3 in the Cupressaceae?

Dunberg and Oden[19] have commented that "The amount of GA that must be applied to a single shoot to have any effect on flower bud initiation is...many thousand times higher than the quantities required to produce a comparable response in the Cupressaceae." The idea that all conifers of the Cupressaceae and Taxodiaceae are highly "GA efficient", whereas all those of the Pinaceae family are highly "GA efficient", is, we believe, a misconception. Thus, *Tsuga heterophylla* is a highly "GA inefficient" Pinaceae family conifer that, by all accounts (see Table 1 for references), is more efficient than either *Chamaecyparis nootkatensis*[72a] or *Sequoia sempervirens*,[52c] although not as efficient as, for example, *Cupressus arizonica*[50] or *Thuja plicata*.[52b,72a] In one study,[74] *T. heterophylla* rooted propagules grown in containers (1.5 to 2.5 m tall) within a greenhouse produced an average of 1348 conebuds per plant (71% male, 29% female) in response to six spray applications (weekly intervals) of 200 mg/ℓ $GA_{4/7}$, together with the adjunct cultural treatments of $Ca(NO_3)_2$ fertilization and modest water stress. Although containerization and these adjunct cultural treatments are required to maximize the response to $GA_{4/7}$ in *T. heterophylla*,[4,29,62,63,80,83a] they are generally ineffective by themselves in promoting flowering[4,62,63,80,82] in this species. Such adjunct treatments are unnecessary, although usually beneficial, for Cupresseaceae species.[28,52]

The ability of exogenous GA_3 to induce flowering at an extremely early age in Cupressaceae species is well known, the earliest being 88 days from germination for *C. arizonica* (see Reference 52). There are no reports as yet of such an early induction of flowering in the Pinaceae. Even so, early flowering by normal standards has been achieved for several Pinaceae conifers in response to $GA_{4/7}$ plus adjunct cultural treatments. In *Tsuga heterophylla*[4] and *Pinus banksiana*[6] flowering was promoted in the second year from seed (application of $GA_{4/7}$ at an age of about 6 to 10 months for *P. banksiana)*, and for *Pseudotsuga menziesii*,[73] in the third year from seed. It seems likely that these ages will be reduced even more as the $GA_{4/7}$ treatment is optimized with adjunct cultural regimes.

Although the literature on flowering in Cupressaceae and Taxodiaceae family species is replete with examples of early and profuse flowering by application of GA_3 (usually applied by spray to drip-off, in solutions ranging from 50 to 500 mg/ℓ), there are few examples where direct calculations of flowering response to varying amounts of injected GA_3 can be made. However, for *Cupressus arizonica* the single injection of 50 ng to 2.5 μg of GA_3 into the upper branches of approximately 1-year-old seedlings yielded a significant flowering response.[50] By calculation, ratios of 2 to 22 ng of GA_3 were injected per male conebud produced at the lowest and highest concentrations of GA_3, respectively. This estimate, however, does not take into account the fact that the GA_3 treatment must be continued if significant conebud abortion is to be prevented, and if normal growth and development to pollen shedding the next spring is to occur (see references cited in References 46 and 52).

McMullan[41] injected 250 μg of $GA_{4/7}$ over a 10-week period to each shoot for Douglas fir clones that were known to be "poor producers", yielding an average of 41 conebuds (male and female) per shoot. The injection was just behind the shoot; hence the amount available to any one differentiating lateral primordium was probably far in excess of amounts in other studies (see below), which may explain the relatively favorable "applied $GA_{4/7}$ dosage: conebuds produced" ratio of 6 μg to one female conebud. McMullan noted subsequent twig toxicity on one family, and thus considered the $Ga_{4/7}$-induced flowering to be a "stress crop" (e.g., not necessarily related to a normal physiological role played by the hormone). However, the apparent absence of toxicity on the other five families, and the continued success (Table 1) of $GA_{4/7}$ to promote flowering in Pinaceae family conifers without appreciable toxicity (in most studies) at effective dosages leads us to conclude that $GA_{4/7}$ in her study[41] was exerting a morphoghenic effect on the sexual differentiation of lateral primordia over and above any "stress effect".

McMullan's[41] bioassay estimates of endogenous GAs yielded, at the most, only 10 ng per shoot. The number of shoots making up her samples (which were as high as 20 g d.w. at later stages of elongation) are not given, but if one assumes that a single shoot weighs 1 to 2 g d.w. (at least in the middle to late phases of elongation), her estimates are 10- to 20-fold less than the 115 ng of GAs per g d.w. obtained from other rapidly growing, 30-year-old Vancouver Island Douglas fir trees.[1,16] Thus, her 5,000:1 (applied $GA_{4/7}$ to endogenous GAs) ratio may be too conservative, and values of 250:1 or 500:1 may be more likely (see below for further discussion of the relevance and calculation of such a ratio).

In another study with Douglas fir[83] 72 mg of the $GA_{4/7}$ mixture were injected over a 7-week period into the main stem of 2- to 4-m tall saplings (9 to 10 years old), resulting in 700 female conebuds (103 μg $GA_{4/7}$ applied per female conebud produced) about 10 months following injection. Another study with western hemlock, where dosage of hormone was varied over a wide concentration, gave a flowering promotion roughly proportional to the log of hormone concentration.[80] A similar proportionality to the log of hormone concentration exists for $GA_{4/7}$-promoted flowering in loblolly pine *(Pinus taeda)*[23] and Douglas fir,[72] and also for the GA_3-induced flowering in Cupressaceae species[50] (see also the literature cited in Reference 52).

Recently Ross[74] used "containerized" western hemlock rooted propagules (1.5 to 2.5 m tall) in a greenhouse, spraying them to "wetness" weekly for 6 weeks with $GA^{4/7}$ at 200 mg/ℓ in aqueous solution, with AROMOX C-12W surfactant. Adjunct cultural treatments were also used: modest water stress and $Ca(NO_3)_2$ fertilization. By themselves these treatments generally have no or a nominal effect on flowering in western hemlock,[4,80,83a] although they effectively synergize the $GA_{4/7}$ treatment.[4,80] Ross's sprays to "wetness" for hemlock[74] are analogous to many of the studies in the early literature of GA_3-induced flowering in Cupressaceae species[28,52] in terms of concentration, frequency, and mode of application.

Across seven clones, Ross's western hemlock plants, when surveyed 10 months later, averaged 1348 conebuds per plant (71% males, 29% female) over and above controls.[74] Some individual clones produced as many as 3735 conebuds per plant in response to the hormone treatment. Thus, although the amount of $GA_{4/7}$ actually applied to the hemlock propagules could not be calculated, the "spray to wetness" mode of application and concentration was comparable to earlier methods used on Cupressaceae species. And, most important, response in terms of conebud number was in the thousands per plant for trees of both families, a roughly comparable response.

Although sprays of $GA_{4/7}$ to hemlock seedlings yielded flowering responses per plant roughly comparable to GA_3 sprays to Cupressaceae species, the Pinaceae species required "containerization", water stress, and $Ca(NO_3)_2$ adjunct treatments to synergize the $GA_{4/7}$ effect. Such adjunct treatments are unnecessary (although usually beneficial) for Cupressaceae species.

Injections of $GA_{4/7}$ into the main stem of Douglas fir have thus yielded ratios of about 6 μg to 200 μg of $GA_{4/7}$ applied, to bring conebuds through a 10-month period to pollen shedding/anthesis-like development. Injections of GA_3 into Cupressaceae species yielded ratios as low as 2 ng injected to bring 1 male conebud to about day 35 development. Unfortunately, the amount of GA_3 required to bring a *Cupressus* male conebud to the stage of pollen shedding is not calculable from published work, since appropriate studies have used "sprays to drip-off".

We are thus left with a number of differences between the highly "GA efficient" Cupressaceae species, and the apparently much less "GA efficient" Pinaceae species. Namely:

1. Cuppressaceae calculations are made using the slowly metabolized GA_3,[49,49a] whereas Pinaceae calculations are made using the less polar, mono-hydroxylated $GA_{4/7}$ mixture, of which GA_4, at least, is very rapidly metabolized in conifers[49,57,91] as well as other plants (the rate of metabolism of GA_7 has not been examined in higher plants). Additionally, high dosages of GA_4 may disproportionately increase conjugation, relative to low dosages.[49]
2. Cupressaceae calculations are made for male conebuds, which are smaller than female conebuds, especially following differentiation, and may thus require less GA for maintaining differentiation and continued development.
3. Cupressaceae calculations are made only for the presence of conebuds shortly after injection of GA_3, and do not take into account the continued requirement for GA_3 to bring the conebud through development to pollen shedding the next spring (see literature cited in References 46 and 52).
4. Cupressaceae calculations do not take into account the very much shorter period of initiation/differentiation required to evoke a permanent change in the vegetative apex of *Cupressus* (about 22 days), relative to the much longer period (6+ weeks) to yield a similar change in the differentiating lateral primordia of Douglas fir.
5. Cupressaceae calculations do not take into account the fact that the distribution of the injected GA_3 occurred over a very few small branchlets, weighing perhaps 100 to 200 mg d.w., relative to very much larger tissue amounts in Pinaceae experiments.
6. Cupressaceae calculations do not take into account the fact that species in this family may have 10 to 1000 times the number of potentially reproductive apices/lateral primordia, relative to Pinaceae family species.

While the first four items in this list are imponderables, they should cause one to pause before totally discarding the possibility that applied $GA_{4/7}$ in Pinaceae species is acting to supplement endogenous levels of less-polar GAs in a manner similar to that postulated for Cupressaceae species[43,46,52] (see also discussion below), where exogenous GA_3 appears to be supplementing endogenous (and suboptimal) levels of an endogenous GA-like substance.

Why Are Levels of $GA_{4/7}$ Applied So Much Higher Than Endogenous Levels of GAs in Pinaceae Family Conifers?

It may be worthwhile to digress briefly and consider just how much of the 72 mg of $Ga_{4/7}$ injected between the fourth and fifth whorls into the Douglas fir samplings[83] (that produced in excess of 700 female conebuds on the entire sapling the following spring) might have found its way to the differentiating lateral primordia that will become next year's "flower". On the 7 whorls of branches (4 above and 3 below the injection point) should be in excess of 250 elongating new shoots of sufficient vigor to bear seed cones. Each of those shoots may have 20 undifferentiated lateral primordia or approximately 5,000 primordia per sapling. If all of the 72 mg of $GA_{4/7}$ moved in the xylem and/or phloem only to those primordia, then each primordium would receive 16 μg. However, such a calculation is patently unreasonable. The elongating new shoots should attract only a small proportion of the water moving through the xylem, since they do not constitute a major component of the transpiring surface of the tree. Similarly, although the elongating shoot is a strong sink for photosynthate, and would thus mobilize that $GA_{4/7}$ moving in the phloem, there are other demands (woody tissue and roots). An unknown proportion of the injected $GA_{4/7}$ would be therefore expected to remain in woody tissue in the main stem, in older branches, and in older needles. Metabolism of the mono-hydroxylated GA_4 is known to be rapid,[57,91] and over any given period of time, as the $GA_{4/7}$ moves through living xylem/phloem/shoot tissue, only a proportion (unknown, but probably small) of the precursor $GA_{4/7}$ will remain, and many of the metabolites can be expected to be inactive (e.g., C-2 hydroxylated in the case of GA_4 metabolites, and also GA glucosyl conjugate-like substances[91]). Finally, the approximately 20 differentiating primordia along each elongating shoot make up only a very small part of the dry matter in that shoot, and unless they are disproportionately adept at secondering $GA_{4/7}$ and/or its "active" metabolites, we would expect the hormone to be relatively uniformly distributed on the basis of dry matter and/or transpiring surface.

In a (sister) study[57] $[^3H]GA_4$ (1.0 Ci/mmol) was injected in early June into the woody tissue subtending the elongating shoot in a manner similar to that accomplished by McMullan.[41] Although approximately 2.6% of the 6.3 μCi (approximately 2.0 μg GA_4) of injected radioactivity moved into the elongating terminal shoot over 192 hr, only 0.05% of this 2.6% (e.g., 0.0013% of the injected radioactivity) was associated with any one primordium and its immediately adjacent shoot tissue (which could not be readily separated from the primordium). At a lower specific activity (e.g., 5 mCi/mmol) of $[^3H]GA_4$ (which would promote flowering) total uptake would probably be reduced by perhaps 2/3 due to increased applied GA_4 amounts, yielding a theoretical uptake into the elongating shoot of about 3.5 μg GA_4.

The same Douglas fir shoot tissue analyzed above[57] with regard to uptake of $[^3H]GA_4$ at 1.0 Ci/mmol was also examir :he Tan-ginbozu microdrop bioassay (in serial dilution) after purification on reverse phase C_{18} HPLC,[57] and yielded 57 ng of less-polar GA-like substances (retention time similar to GA_4, GA_7, GA_9, and GA_{36}). These shoots were comparable to those on "sister" trees in the Ross et al.[83] study where feeding 72 mg of $GA_{4/7}$ between the 4th and 5th whorl had yielded around 700 female conebuds per sapling (e.g., 103 μg of $GA_{4/7}$ gave one female conebud when applied in this manner, compared to 6 μg of $GA_{4/7}$ per female conebud in McMullan's study,[41] where the hormone was applied immediately subtending the elongating shoot).

A comparison of $GA_{4/7}$ applied in the Ross et al.[83] study to endogenous GA levels per shoot[57] can be obtained by dividing the total amount of hormone applied (72 mg) by the approximate number of potentially conebud-bearing shoots (approximately 250 per sapling). This yields a very approximate ratio of 5000:1 ($GA_{4/7}$ applied per shoot to endogenous less-polar GA-like substances present in that shoot). From the McMullan work[41] a similar ratio can be obtained, although her endogenous GA levels were very low (10 ng per shoot for total GA-like substances), and other studies[1,16,57] show endogenous total GA levels of about 100+ ng per shoot for Douglas fir (see discussion in the preceding subsection).

But are these ratios of "applied to endogenous" GAs relevant? Actual calculations where 2.0 μg of GA_4 (e.g., $[^3H]GA_4$ at 1.0 Ci/mmol) was applied to Douglas fir shoots in the woody tissue subtending an elongating shoot showed[57] that the ratio of extracted GA_4/GA_4 metabolites to endogenous less-polar GA-like substances was approximately 1:1, even 192 hr after the application was made. If the injected dosage of GA_4 were increased 200$\times$ to 402 μg (e.g., 5 mCi/mmol $[^3H]GA_4$) per shoot, estimates of the ratio of extracted GA_4/GA_4 metabolites to endogenous less-polar GAs found in the shoots might be around 60:1 (e.g., movement of GA_4 may be reduced to approximately one third to increased carrier GA_4). But, metabolism of precursor $[^3H]GA_4$ will also be slowed, and instead of the 10% of precursor $[^3H]GA_4$ remaining after 192 hr (1.0 Ci/mmol, 2 μg GA_4 dosage), we might expect 30% to remain unmetabolized at the higher (402 μg) dosage. If GA_4 (and/or GA_7) per se is the "effector" exogenous GA with regard to promotion of flowering in Douglas fir, then the amount of exogenously applied GA_4 present in the shoot after 192 hr at the 402 μg GA_4 dosage would be approximately 1 μg, thus yielding a ratio of 18:1 for GA_4 to endogenous GAs.

Which then is the most relevant calculation? Gross amounts applied relative to endogenous levels, or amounts of exogenously applied GA found in the shoot during the period of lateral primordia differentiation, relative to endogenous levels?

One can also examine, for comparison, a Cupressaceae species. *Cupressus arizonica* contains 100 to 300 ng of endogenous GA_3-like substance per g d.w. of vegetative tissue under optimal (e.g., LD, high light intensity) conditions.[1,43] The injection of 50 ng of GA_3 into tissue with a dry weight of near 100 to 200 mg yielded 25 male conebuds.[50] By rough calculation the injection (application) amount was thus equal to, or several times in excess of the expected endogenous level. But, in the same study[50] an injection of 2500 ng of GA_3 into tissue with a similar d.w. yielded 114 male strobili. By rough calculation the injection amount was 42 to 125 times the expected endogenous level. One can see that if the tissue is somewhat low in endogenous GAs, (e.g., 100 ng per g d.w.), and the dosage level is at the high end of the "log response curve" (see above), then one can have application amounts that vary from one to over a hundred times the expected endogenous levels.

Movement of [^{14}C]GA_3 was not charted in those studies on *C. arizonica* where flowering efficacy was followed, but no other work indicates that movement from the point of injection into tissue just above and near the injection point might be as high as 20 to 60%,[52a] and metabolism of GA_3 is slow[49,49a] relative to that of GA_4.[49,91]

Thus, in examining the relevance of exogenous application of hormone to endogenous events, we think that one should look at the probable fate of a finite number of GA molecules that are sprayed onto, or injected into trees weighing, in some cases, up to 10 kg or more d.w. Then, contrast the number of surviving effector GA molecules that might find their way to a differentiating primordium with the expected endogenous GA levels. When approached in this manner, the proportion of exogenous GA that finds its way into tissue of relevance, relative to endogenous GA amounts, no longer appears to be so excessive.

Are the Effects of $GA_{4/7}$ in Promoting Flowering in the Pinaceae Unique?

According to Dunberg and Oden, "...the effect of applied GAs is not unique. Other chemicals and many cultural treatments produce comparable effects. Kubitz (1967) has reported increases in female flowering after application of KCN, IAA and guanine to *Pinus contorta.*"[19]

There are several points that should be made in reference to this statement. First, no repeat of the Kubitz results (cited in Reference 19) has found its way into the literature. In fact, many chemicals have been tested, but only the less-polar GAs have consistently promoted flowering in Pinaceae species. It is, however, worthwhile examining examples where chemicals other than GAs have been applied to conifers in order to promote flowering. There are several reports that the synthetic auxin NAA can enhance both male and female flowering in response to $GA_{4/7}$ in *Pseudotsuga menziesii*,[54a,55,66] *Pinus contorta*,[93] and male flowering in *Picea sitchensis*,[88] although positive results were not obtained with NAA in other studies on *Tsuga heterophylla*,[80] nor were the results consistent on *Pinus banksiana*.[6] Only in *Larix leptolepis* has NAA been reported to be effective in the absence of applied $GA_{4/7}$ (see literature cited in References 28). Although results of Ross and Pharis[77] indicated that the cytokinin, ^{6}*N*-benzyladenine slightly enhanced the response to $GA_{4/7}$ in *P. menziesii*, this result is not readily repeatable for other conifer species.[84] Similarly, whereas certain growth retardants will promote flowering in some woody angiosperms, these chemicals have not proved successful for conifers.[29a,51,82,95] Where successes have occurred — for example, with CCC in *Picea abies*[11,1b] and ABA and chlorflurenol in *Pinus radiata*[84a] — it has been where the growth retardant or growth regulator was applied together with one or more GAs. The positive results with CCC + GA_3 on *P. abies*[11,1b] are not always repeatable,[11] and the work with *P. radiata* using ABA and chlorfurenol + $GA_{4/7}$ has not been repeated.

Several studies have related increased flowering to elevated levels of arginine,[1a,19b,84b] thus leading some workers to speculate that this amino acid plays a specific role in conebud differentiation.[19a,19b,83b] However, attempts to promote flowering in *Pinus radiata*,[84b] *P. elliottii*, and *P. taeda*,[25] and *P. menziesii*[41] by applications of arginine all have been unsuccessful, in contrast to the stimulatory effect of $GA_{4/7}$ in these same species (Table 1). Recently, Hare[26] reported increases in flowering from soil applications of nitrapyrin in *P. taeda*, but this compound is probably acting to increase available soil nitrogen, and fertilization with NH_4NO_3 is known to promote flowering (albeit variably) in sexually mature southern pines.[25,64a]

The second point to be emphasized is that while there are a number of cultural treatments which will promote flowering in conifers, they generally are not as consistently effective as $GA_{4/7}$.[79,82] Root pruning in *P. menziesii* may be an exception (see Reference 83 and the literature cited therein), as may high temperature in *Picea*,[14,61,73,88] although high temperature alone does not consistently promote flowering in *Tsuga heterophylla*.[62,73] Furthermore, there is evidence (see later) that these and certain other cultural practices, such as nitrate-N fertilization, water stress, and girdling, may be promoting flowering in part through effects on GA metabolism. However, the fact that these cultural practices frequently interact synergistically with $GA_{4/7}$ application suggests that they may influence conebud differentiation by other mechanisms as well[79a], mechanisms that may involve reduced shoot growth and/or increased availability of assimilate nutrients,[75] reduced root activity and export therefrom of possible flowering inhibitors,[2,16a,60] and/or altered pattern of N metabolism.[19a,83b]

The fact, however, that the less polar GAs are not unique in their ability to promote flowering in Pinaceae species should not be regarded as evidence against the hypothesis that endogenous less-polar GAs play an important regulatory role in the process. Ross and Pharis[79a] pointed out that conebud initiation/differentiation in conifers involves a continuum of interacting processes that occur over a relatively long (6 to 12+ weeks) period of time. A variety of actors — not only the concentrations of appropriate GAs, but also of other growth regulators, possibly inhibitors, carbohydrates, and nitrogenous substances — undoubtedly must remain highly favorable during the entire conebud initiation and differentiation phases.

Does the Effect of Applied GAs in Promoting Flowering Imply a Similar Role by Endogenous GAs?

To quote Dunberg and Oden, "...the effects of applied GAs must not necessarily be interpreted as evidence for a flowering-stimulatory function of endogenous GAs. It appears equally possible that applied GAs block the synthesis or action of endogenous GAs that are, as in woody angiosperms, inhibitory to flowering."[19]

Certainly, one should not assume that, merely because a substance causes flowering, it mimics the action of endogenous substance(s). However, we know of no evidence in the literature that applied GAs block the synthesis or action of endogenous GAs in conifers. Nor is there any evidence that endogenous GAs in conifers are "inhibitory to flowering."

The postulate that endogenous GAs inhibit flowering in woody angiosperms has been critically examined primarily in biennial bearing cultivars of apple, pear, *Citrus*, *Salix*, and black currant (see literature cited in References 42, 51, and 95 and chapters "Rosaceae — Deciduous Fruit Trees" in Volume 1, "*Citrus*" in Volume 2, and "*Ribes*" in Volume 4). Even in ivy, where GA_3 will cause reversion of flowering plants to the juvenile state, the evidence that the juvenile state (and absence of flowering) is maintained by endogenous GAs is not unequivocal (see literature cited in Reference 95 and chapter "*Hedera*" in Volume 3). The inhibitory effect of exogenous GAs on flowering in woody angiosperms is not straightforward (see References 42, 51, and 95 and literature cited therein). For apple, the inhibitory effect is obtained for flowering of spurs, but not on flowering in new wood.[42] And, some fruit must remain on the tree for GA_3 to be an effective inhibitor of next year's flowers in one cultivar of apple; flowering in completely deblossomed trees is not inhibited by GA_3.[20,42] Additionally, extractable GAs can remain high in tissues of trees which will flower well the next year, relative to trees on which flowering will be inhibited by the presence of fruit (see literature cited in References 42, 51, and 95).

We would answer the question posed in the above subtitle with a somewhat *equivocal* "yes" since only a few studies have examined endogenous levels of GAs, or GA metabolism, under circumstances where valid comparisons can be made between flowering and non-flowering trees. However, we do think it is unlikely that the effects of exogenous GAs in promoting flowering in the many species of conifers thoughout the Cupressaceae, Taxodiaceae, and Pinaceae families will prove to be only of a "pharmacological nature". For the Cupressaceae there is evidence that the exogenous GA is supplementing endogenous levels (see discussion in References 43,45,46,52, and 82), and that at least one cultural treatment (low N nutrition) changes the endogenous GA status (less-polar GAs are increased in trees that flower, more polar GAs reduced; see data presented in Reference 30). Similar evidence exists for Pinaceae family conifers (see References 15,18,19,47,57, and discussion below).

What Is the Evidence that Endogenous GAs Are Involved or Regulate the Flowering in Pinaceae Conifers?

A recent review[94] questions a conclusion[56] put forward in 1976 that "...the flowering state was associated with an increased level of GAs less polar than GA_3, and with a decrease in polar GAs," noting that the work by McMullan[41] found "...endogenous GAs as measured by the dwarf rice bioassay changed greatly with stage of development, but were similar in good and poor cone-producing clones of *Pseudotsuga.*" Data were provided in 1977[47] supporting the conclusion[56] using two cultural treatments (drought and NO_3 nitrogen fertilization). Additional confirmatory work on *Picea abies,* using the cultural treatment of high temperature, was published in 1982[15] and 1983;[18] this work is discussed in Ross et al.[82] and below.

McMullan[41] had compared the endogenous GAs of tissue taken from abundantly flowering vegetatively propagated trees of a single cone (parent tree #16), with endogenous GAs of tissue taken from poor flowering propagules of two parent trees (#24 and 25, tissue bulked). Although her bioassay-detectable GAs were low (relative to other studies with Vancouver Island Douglas fir shoots[1,16]), suggesting that inhibitors may have been a problem, the two clones compared did exhibit similar developmental trends during shoot elongation. However, it may be expecting too much to assume that the single genotype that was chosen to represent "good flowering" would have significantly increased endogenous GAs, relative to bulked tissue from two "poor flowering" genotypes. Lack of sufficient amounts of endogenous GAs of a "promotory nature" may not be the only factor limiting flowering in "normal" years for "poor flowering" genotypes. And, in such comparisons, it would be useful to include sufficient genotypes of each "flowering type", so as to ascertain the variability within each population. Or, if the logistics of such an approach are too formidable, then tissue from a number of genotypes of each "flowering type" should be bulked in a statistically sound way to reduce the number of replicates required — this would tend to average out individual genotype variability.

A similar comparison by Sweet[84] did show differences in endogenous GAs from bulked tissue samples of trees that had a high probability of being "good flowerer", "nondeveloper", and "mainly branch buds". The only trend apparent, however, was that trees which would be expected to produce mainly "branch buds" had slightly more of a GA_3-like substance, and appreciably more of a nonpolar GA-like substance that eluted from SiO_2

partition columns between GA_9 and $GA_{4/7}$ (C_{20} GAs without a hydroxyl might be expected to elute here). Thus neither Sweet's results nor those of McMullan confirm the trends noted above (and discussed below), where cultural treatments were used to promote flowering. However, a reexamination of endogenous GAs of "good" vs. "poor" flowering clones using improved chromatographic methods (reverse phase C_{18}HPLC), and with better identification (GC-MS/GC-SIM) is warranted, and should now be feasible.

The review by Ross et al.[82] discusses investigations on endogenous GAs where cultural treatments of low N nutrition *(Cupressus arizonica)*,[30] water stress,[47] NO_3-N fertilization,[47] and high temperature[15,18] had been used as promotory treatments. By necessity, the harvest of tissue is a destructive event, and for the Pinaceae species only in the case of the NO_3-N fertilization treatment on large trees of Douglas fir[47] could the changes in GAs be compared directly with the "flowering promotive ability" of the treatment. However, the other cultural treatments are known to promote flowering (albeit sporadically), and to act in a highly synergistic manner with applied $GA_{4/7}$. The conclusion from all of these studies was that cultural treatments known to promote flowering yielded increased levels of less-polar GAs (e.g., chromatographing with, or near GA_9 and $GA_{4/7}$), and reduced levels of more-polar GAs. One study[18] examined the metabolism of high-specific-activity [^{3}H]GA_4 (a component of the active $GA_{4/7}$ mixture) in *Picea abies,* and noted that the cultural treatment of high temperature reduced the rate of GA_4 metabolism, relative to controls, thereby implying that this cultural treatment may promote the retention of endogenous GAs of a less-polar nature by inhibiting their oxidative metabolism.

Since the Ross et al. review,[82] there has been additional work on Douglas-fir using the cultural treatment of rootpruning.[83,92] Rootpruning was more effective than $GA_{4/7}$ in promoting flowering in this study. When $GA_{4/7}$ was applied to rootpruned trees, the combination was highly synergistic (discussed above in the initial subsection). Within the trees there could be distinguished two groups. One group of trees had a "history of no or poor flowering", and another group had "a history of average or good flowering". Control (e.g., untreated) trees with a history of "no or poor flowering" tended to have less vegetative shoot growth than did control trees from families with a "good flowering history". Trees from families with a history of no or poor flowering did not respond well to $GA_{4/7}$ alone (no rootpruning) in terms of flowering, but they did respond significantly in terms of vegetative growth. Conversely, trees from families with a "good flowering history" flowered much better in response to $GA_{4/7}$ (no rootpruning), but they did not respond significantly to $GA_{4/7}$ in terms of vegetative growth. These results are consistent with a hypothesis that exogenous *and* endogenous GAs are used preferentially for vegetative growth processes, with increased flowering occurring only after a threshold concentration of the effector GA is reached.

In simplistic terms, trees with a "history of no or poor flowering" had diminished vegetative shoot growth, and appeared to be "GA deficient". These trees appeared to use exogenous $GA_{4/7}$ first for vegetative growth, then for flowering. Conversely, trees with a "history of flowering" appeared to have sufficient levels of endogenous GAs for vegetative growth. Thus, they did not respond to exogenous application of $GA_{4/7}$ by growing more vegetatively. Instead they utilized the exogenous $GA_{4/7}$ for flowering. (Indeed, $GA_{4/7}$ alone was a much more effective flowering stimulator for trees from families with a "good flowering history" than for trees from families with a "poor flowering history".)

In a sister study[57] to that described above, [^{3}H] GA_4 of high specific activity (e.g., low amounts of GA_4) was administered to control trees (e.g., no treatment), and to trees given the cultural treatment of rootpruning. Shoot tissue harvested during the period of lateral primordia sexual differentiation from trees which flowered in response to rootpruning was analyzed for trends in [^{3}H] GA_4 metabolism, and for levels of endogenous GA-like substances, relative to tissue from control trees which did not flower. Preliminary results[57] on a limited number of comparisons indicate that rootpruning (which caused flowering, and which will significantly synergize $GA_{4/7}$-promotion of *flowering*[82,83]) caused the following:

1. A reduction in the rate of metabolism of the precursor [^3H] GA_4
2. A decrease in the metabolism of [^3H] GA_4 into a [^3H] GA_{34}-like compound (GA_{34} is a C-2 hydroxylated GA of low biological activity)
3. A tendency for decreased metabolism of [^3H] GA_4 into GA glucosyl conjugate-like metabolites (GA glucosyl conjugates are also presumed to be biologically inactive unless the GA moiety is freed by hydrolysis)
4. An increase in amounts of bioassay active endogenous GA-like substances in shoot (but not needle) tissue, especially those GAs with a retention time similar to $GA_{4/7}$, GA_{36}, and GA_9

HOW MIGHT EXOGENOUS $GA_{4/7}$ BE PROMOTING FLOWERING?

Understanding of the mechanism of GA action in the extensively studied barley aleurone and lettuce hypocotyl systems still remains at a level that can best be termed, "unresolved". It is thus presumptuous to believe that research of only a few years on the complex system of sexual differentiation in Pinaceae family conifers can provide more than a glimpse into the "role", or the possible "mechanism" by which the GA accomplished sexual differentiation.

The example[50] alluded above in the initial subsection of "The Relevance of $GA_{4/7}$-Induced Flowering in Conifers to Flowering in Nature", where GA_3 injections of just 50 ng yielded male conebud initiation in *C. arizonica* (e.g., 2 ng injected per conebud produced), indicates that the promotion of flowering in this conifer, at least, can require very nominal amounts of hormone. This would seem to rule out nonspecific pharmacological effects or toxic effects that "inhibit" vegetative growth, thereby "causing" flowering. One concept for any flowering mechanism is that the hormone promotes the initiation/differentiation/growth of the flower by diverting additional nutrients (photosynthate and other hormones, perhaps even essential mineral elements) to the apex, or to the appropriate lateral primordium, thereby creating a sink, and causing, in a yet unexplained way, flower/conebud initiation/differentiation (see literature cited in References 22, 75, 84, and 89).

An attempt to examine the concept of nutrient diversion as a mechanism in the $GA_{4/7}$-induced flowering of *Pinus radiata* has been made by Ross et al.[75] Carefully matched "paired" shoots were given carrier solvent only or $GA_{4/7}$ at a level known to promote flowering. One week later, needles subtending the shoot were fed $^{14}CO_2$, and analyses of ^{14}C assimilate and dry matter distribution made 24 to 48 hr later (e.g., days 8 to 9 after $GA_{4/7}$ administration). The $GA_{4/7}$ had, within that period, significantly increased movement of both ^{14}C assimilate and dry matter into lateral long-shoot buds (which are potential seed cone or lateral branch buds). This was accomplished at the expense of the structural tissue and apical dome. In a second experiment, $GA_{4/7}$ was compared with GA_3, again relative to a 0-GA control, using dry-matter distribution throughout the sexual differentiation period as an index of "nutrient diversion". Both GA_3 and $GA_{4/7}$ promoted significant diversion of dry matter to lateral long-shoot buds, but only $GA_{4/7}$ caused significant flowering. Ross et al.[75] thus concluded the $GA_{4/7}$ was acting in a morphogenic manner, in addition to any role that it might play as a "diverter of nutrients".

The mechanism by which the morphogenic effect of the hormone is brought about, however, still remains a matter of speculation, and a subject of future research.

CONCLUSIONS

The results[57] noted earlier in the first four subsections of "The Relevance of $GA_{4/7}$-Induced

Flowering in Conifers to Flowering in Nature'' on the effects of rootpruning on GA metabolism and endogenous GAs are thus similar to results in the literature[15,18,30,47] for the other cultural treatments known to promote flowering. The responses of trees with "good" and "poor" histories of past flowering to exogenous $GA_{4/7}$ are also indicative of a role for endogenous GAs in both vegetative growth[92] and flowering,[83] and thus are confirmatory of similar responses in Cupressaceae conifers (see References 43, 52, and 82, and the literature cited therein).

It is certainly wise to be conservative when speculating about mechanisms by which exogenous substances cause flowering, and by which flowering in nature is brought about. This is especially true when the past work is primarily empirical (e.g., utilizes exogenous application of GAs under a variety of conditions), and when the methods of endogenous hormone extraction, purification, chromatography, and assay are relatively crude, and are subject to the vagaries of inhibitor to promoter ratio, limited replication, and the absence of appropriate internal standards.

However, even though such problems have bedeviled all work of this nature, they did not prevent the noting of the trends of "increased less-polar GAs" being associated with the flowering state in conifers (see References 30, 47, and 76a). Indeed, these early trends inspired an applied series of tests[44,46,52,55,56,58,66,77,78,93] using applications of less-polar GAs on Pinaceae family conifers. These initial tests led to successful procedures by which flowering in 17 species of Pinaceae family conifers can now be promoted (see references cited in Table 1).

The scientific method should eventually tell us whether the discovery of $GA_{4/7}$-induced flowering in Pinaceae conifers was serendipitous, and whether endogenous GAs do indeed play a regulatory role in normal, precocious, or enhanced flowering in conifers. The answers to date, however, are not unequivocal.

ACKNOWLEDGMENTS

The authors gratefully acknowledges assistance from C.S.I.R.O. Division of Plant Industry, Canberra, and the Research School of Chemistry, Australian National University, for technical assistance in the preparation of this manuscript. An International Collaborative Research Grant to R.P.P. from the Natural Sciences and Engineering Research Council of Canada is also gratefully acknowledged.

REFERENCES

1. **Aoki, H., Crozier, A., and Pharis, R. P.,** Endogenous growth regulators in forest trees, *Zesz. Nauk. Uniw. Mikolaja Kopernika Toruniu Nauki Mat. Przyr. Biologia*, 13, 114—121, 1970.

1a. **Barnes, R. L. and Bengston, G. W.,** Effects of fertilization, irrigation, and cover cropping on flowering and on nitrogen and soluble sugar composition of slash pine, *For. Sci.*, 14, 172—180, 1968.

1b. **Bleymüller, H.,** Investigations on the dependence of flowering in spruce *(Picea abies* (L). Karst.) upon age and hormone treatment, *Silvae Genet.*, 25, 83—85, 1976.

2. **Bonnet-Masimbert, M.,** Influence de l'etat l'activate des racines sur la floriason induit par des gibberellines 4 et 7 chez *Pseudotsuga menziesii* (Mirb.) Franco, *Silvae Genet.*, 31, 178—182, 1982.

3. **Bonnet-Masimbert, M.,** Effect of growth regulators, girdling and mulch on flowering of young European and Japanese larches under field conditions, *Can. J. For. Res.*, 12, 270—279, 1982.

3a. **Bower, R. C. and Ross, S. D.,** Evaluation of ULV sprays for seed orchard application of $GA_{4/7}$, in *Proc. 20th Meet. Can. Tree Improvement Assoc.*, Laval Univ., Quebec, 1985, in press.

4. **Brix, H. and Portlock, F. T.,** Flowering response of western hemlock seedlings to gibberellin and water-stress treatments, *Can. J. For. Res.*, 12, 76—82, 1982.

5. **Buban, T. and Faust, M.,** Flower bud induction in apple trees: internal control and differentiation, *Hortic. Rev.*, 4, 174—203, 1982.

6. **Cecich, R. A.**, Applied gibberellin $A_{4/7}$ increases ovulate strobili production in accelerated growth Jack pine seedlings, *Can. J. For. Res.*, 11, 580—585, 1981.
7. **Cecich, R. A.**, Flowering of jack pine seedlings two years after application of $GA_{4/7}$ in, Proc. 2nd N. Cent. Tree Improvement Conf. 1981, Rocky Mountain Forest and Range Experimental Station, U.S. Forest Service, 1982, 176—182.
8. **Cecich, R. A.**, Flowering in a jackpine seedling seed orchard increased by spraying with gibberellin $A_{4,7}$, *Can. J. For. Res.*, 13, 1056—1062, 1983.
9. **Cecich, R. A.**, White spruce *(Picea glauca)* flowering in response to spray application of gibberellin $A_{4/7}$, *Can. J. For. Res.*, 17, 170—174, 185.
10. **Chalupka, W.**, Effect of growth regulators on the flowering of Scots pine *(Pinus sylvestris* L.) grafts, *Silvae Genet.*, 27, 62—65, 1978.
11. **Chalupka, W.**, Effect of growth regulators on flowering of Norway spruce [*Picea abies* (L.) Karst.] grafts, *Silvae Genet.*, 28, 125—127, 1979.
12. **Chalupka, W.**, Regulation of flowering in Scots pine *(Pinus sylvestris)* grafts by gibberellins, *Silvae Genet.*, 29, 118—121, 1980.
12a. **Chalupka, W.**, Influence of growth regulators and polythene covers on flowering Scots pine and Norway spruce grafts, *Silvae Genet.*, 30, 142—146, 1981.
13. **Chalupka, W.**, Time of $GA_{4/7}$ application may affect the sex of Scots pine flowers initiated, *Silvae Genet.*, 33, 173—174, 1984.
14. **Chalupka, W., and Giertych, M.**, The effect of polyethylene covers on the flowering of Norway spruce [*Picea abies* (L.) Karst.] grafts, *Arbor. Kornickie*, 22, 185—192, 1977.
15. **Chalupka, W., Giertych, M., and Kopecewicz, J.**, Effects of polythene covers, a flower inducing treatment on the content of endogenous gibberllin-like substances in grafts of Norway spruce, *Physiol. Plant.*, 54, 79—82, 1982.
16. **Crozier, A., Aoki, H., Pharis, R. P., and Durley, R. C.**, Endogenous gibberellins of Douglas fir, *Phytochemistry*, 9, 2454—2459, 1970.
16a. **Dunberg, A.**, Flower induction in Norway spruce, in *Proc. IUFRO Working Parties in Norway Spruce Provenances and Norway Spruce Breeding*, International Union of Forest Research Organizations, Escherode, F. R. G., 1979, 139—157.
17. **Dunberg, A.**, Stimulation of flowering in *Picea abies* by gibberellins, *Silvae Genet.*, 29, 51—53, 1980.
18. **Dunberg, A., Malmerg, G., Sassa, T., and Pharis, R. P.**, Metabolism of tritiated gibberellin A_4 and A_9 in Norway spruce, *Picea abies* (L.) Karst. Effects of a cultural treatment known to enhance flowering, *Plant Physiol.*, 71, 257—262, 1983a.
19. **Dunberg, A. and Oden, P.**, Gibberellins and conifers, in *The Biochemistry and Physiology of Gibberellins*, Vol. 2, Crozier, A., Ed., Praeger, New York, 1983b, 221—296,
19b. **Ebell, L. F.**, Cone production and stem-growth response of Douglas fir to rate and frequency of nitrogen fertilization, *Can. J. For. Res.*, 2, 327—338, 1972.
19b. **Ebell, L. F. and McMullan, E. E.**, Nitrogenous substances associated with differential cone production responses of Douglas fir to ammonium and nitrate fertilization, *Can. J. Bot.*, 48, 2169—2177, 1970.
20. **Fulford, R. M.**, Flower initiation, effect of gibberellin sprays, *East Malling Res. Stn. Report for 1972*, 1973, 93—94.
21. **Greenwood, M. S.**, Flower stimulation techniques for lobolly pine *(Pinus teada* L.), in *Proc. 3rd World Consult. Forest Tree Breeding, March, 1977, Vol. 2*, CSIRO, Canberra, Australia, 1978, 1031—1042
22. **Greenwood, M. S.**, Reproductive development loblolly pine, II. The effect of age, gibberellin plus water stress and out-of-phase dormancy on long shoot growth behavior, *Am. J. Bot.*, 68, 1184—1190, 1981.
23. **Greenwood, M. S.**, Rate, timing and mode of gibberellin application for female strobilus production by grafted loblolly pine, *Can. J. For. Res.*, 12, 998—1102, 1982.
24. **Greenwood, M. S., O'Gwynn, C. H., and Wallace, P. G.**, Management of an indoor, potted loblolly pine breeding orchard, in Flowering and Seed Development in Trees: A Symposium, May 1978, U.S. Forest Service, New Orleans, 1979, 94—98.
24a. **Hall, P.**, Genetic improvement of black spruce in Newfoundland, in *Proc. Canadian Tree Improvement Assoc.*, Part 1, Aug. 1983, Canadian Forestry Service, Ottawa, 1984, 38—41.
25. **Hare, R. C.**, Promoting flowering in loblolly and slash pine with branch, bud and fertilizer treatments, in Flowering and Seed Development in Trees: A Symposium, May 1978, U.S. Forest Service, New Orleans, 1979, 112—121.
26. **Hare, R. C.**, Application method and timing of gibberellin $A_{4/7}$ treatments for increasing pollen conebud production in southern pines, *Can. J. For. Res.*, 14, 128—131, 1984.
27. **Hare, R. C., Snyder, E. B., and Schmidtling, R. C.**, Longleaf pine flowering in response to nitrogen fertilization, branch girdling, growth substances and cultivation, in Proc. 13th Lake States Forest Tree Improvement Conf., August, 1977, Gen. Tech. Rep. NC-50, U.S. Forest Service, St. Paul, Minn., 1979, 11—16.
27a. **Harrison, D.**, unpublished research results, 1985.

28. **Hashizume, H.,** Studies on flower bud formation, flower sex differentiation and their control in conifers, *Bull. Tottori Univ. For. Tottori Daigaku Nogakubu Enshurin Hokoku*, 7, 1—139, 1973.
29. **Heussler, C. and Ross, S. D.,** Container and seed orchard research — 1980/81, in *Proc. 18th Meet. Can. Tree Improvement Assoc., Part 1*, Canadian Forest Service, Chalk River, Ontario, 1981, 29—33.
29a. **Jackson, D. I. and Sweet, G. B.,** Flower initiation in temperate woody plants, *Hortic. Abstr.*, 42, 9—24, 1972.
30. **Kamienska, A., Pharis, R. P., Wample, R. L., Kuo, C. C., and Durley, R. C.,** Gibberellin in conifers, in *Plant Growth Substances 1973*, Tamura, S., Ed., Hirokawa, Tokyo, 1974, 305—313.
31. **Kanekawa, T. and Katsuta, M.,** Promotion of strobilus production in *Pinus thunbergii* Parl. and *P. densifolora* Sieb. et Zucc. by gibberellins, *J. Jpn. For Soc.*, 64, 101—106, 1982.
32. **Katsuta, M.,** Recent advances in the promotion of strobilus production by gibberellins in Japanese conifers, in *Proc. Symp. Flowering Physiology, 17th IUFRO World Congr. Kyoto, Japan*, Krugman, S. and Katsuta, M., Eds., Japan Forest Tree Breeding Assoc., Tokyo, 1981, 68—74.
33. **Kawamura, K., Uetsuki, K., and Ida, S.,** Promotion of flowering in pines by gibberellins, *Annu. Rep. Kansai For. Tree Breed. Inst.*, 14, 39—42, 1978 (in Japanese); as cited by **Kanekawa, I. and Katsuta, M.,** Promotion of Strobilus production in *Pinus thunbergii* Part. and *P. densiflora* Sieb. et Zucc. by gibberellins, *J. Jpn. For. Soc.*, 64, 101—106, 1982.
34. **Kawemura, K., Uetsuki, and Ida, S.,** Promotion of flowering in pines by gibberellins, *Annu. Rep. Kansai For. Tree Breed. Inst.*, 13, 83—84, 1982 (in Japanese); as cited by **Kanekawa, T. and Katsuta, M.,** Promotion of strobilus production in *Pinus thunbergii* and *P. densiflora* Sieb. et Zucc. by gibberellins, *J. Jpn. For. Soc.*, 64, 101—106, 1982.
34a. **Kubitz, H.,** Zur Wirkung von Wirkung von Wachstums-und Stoffwechselregulatoren auf die Blühinduktion von *Pinus contorta* Dougl. *Wiss. Z. Univ. Rostock. Math. Naturwiss. Reihe*, 16, 539—540, 1967; as cited by **Dunberg, A. and Oden, P.,** Gibberellins and conifers, in *The Biochemistry and Physiology of Gibberellins*, Vol. 2, Crozier, A., Ed., Praeger, New York, 1983b, 221—296.
35. **Luukkanen, O.,** Hormonal treatment increases flowering in Norway spruce grafts grown in plastic greenhouse, *Finn. Fdn. For. Tree Breed. Annu. Rep.*, Helsinki, 1979, 20—26.
36. **Luukkanen, O.,** Promotion of flowering in Scots pine grafts by means of gibberellins, in Proc. IUFRO Working Party S2.03.05 Symp.'Scots Pine Forestry of the Future', Kornik, Poland, 1980.
37. **Luukkanen, O.,** Effect of gibberellins A_4 and A_7 on flowering in Scots pine grafts, *Silva Fenn.*, 15, 359—365, 1981.
38. **Luukkanen, O. and Johansson, S.,** Flower induction by exogenous plant hormones in conifers, *Silva Fenn.*, 14, 95—105, 1978.
39. **Luukkanen, O. and Johansson, S.,** Effect of exogenous gibberellins on flowering in *Pinus sylvestris* L. grafts, *Physiol. Plant.*, 50, 365—370, 1980.
40. **Marquand, R. D. and Hanover, J. W.,** Sexual zonation in the crown of *Picea glauca* and flowering response to exogenous $GA_{4/7}$, *Can. J. For. Res.*, 14, 27—30, 1984.
40a. **Marquard, R. D. and Hanover, J. W.,** Relationship between GA_{4+7} concentration time of treatment and crown position on strobilus production of *Picea glauca, Can. J. For. Res.*, 14, 547—553, 1984.
41. **McMullan, E. E.,** Effect of applied growth regulators on cone production in Douglas-fir, and relation of endogenous growth regulators in cone production capacity, *Can. J. For. Res.*, 10, 405—422, 1980.
42. **Monselise, S. P. and Goldschmidt, E. E.,** Alternate bearing in fruit trees, *Hortic. Rev.*, 4, 129—173, 1982.
42b. **Owens, J. N., Webber, J. E., Ross, S. D., and Pharis, R. P.,** Interaction between gibberellin $A_{4/7}$ and rootpruning on the reproductive and vegetative processes in Douglas-fir. III. Effects on anatomy of shoot elongation and terminal bud development, *Can. J. For. Res.*, 15, 354—364, 1985.
43. **Pharis, R. P.,** Precocious flowering in conifers: the role of plant hormones, in *Toward the Future Forest: Applying Physiology and Genetics to the Domestication of Trees*, Ledig, F. T., Ed., *Yale University School of Forestry Bulletin* No. 85, 1974, 51—80.
44. **Pharis, R. P.,** Method and Composition for Treating Trees, New Zealand Patent 180378, 1976; Australian patent, 514194, 1976; U.S. Patent 4110102, 1978; U.K. Patent 1547571, 1979; Canadian Patent 1078203, 1980.
45. **Pharis, R. P.,** Promotion of flowering in conifers by gibberellins, *For. Chron.*, 51, 244—248, 1975.
46. **Pharis, R. P.,** Manipulation of flowering in conifers through the use of plant hormones, in *Modern Methods in Forest Genetics*, Miksche, J. P., Ed., Springer-Verlag, Berlin, 1976, 265—282.
47. **Pharis, R. P.,** Interaction of native or exogenous plant hormones in the flowering of woody plants, in *Regulation of Developmental Processes in Plants*, Schutte, H. R., and Gross, D., Eds., Academy of Science, Halle, G.D.R., 1978, 343—360.
48. **Pharis, R. P.,** Promotion of flowering in the Pinaceae by hormones — a reality, *U.S.D.A. For. Serv. Gen Tech. Rep. NC-50 (Proc. 13th States Tree Improv. Conv. 1977)*, 1979, 1—10.
49. **Pharis, R. P., Koshioka, M., and Takeno, K.,** unpublished research results, 1984.
49a. **Pharis, R. P. and Bottini, G.,** unpublished research, 1984.

50. **Pharis, R. P., Crozier, A., Reid, D. M., and Kuo, C. C.,** Activity of gibberellin in A_{24} in 10 plant bioassays, *Can. J. Bot.*, 47, 815—820, 1969.
51. **Pharis, R. P. and King, R. W.,** Gibberellins and reproductive development in higher plants, *Annu. Rev. Plant Physiol.*, 36, 517—568, 1985.
52. **Pharis, R. P. and Kuo, C. G.,** Physiology of gibberellins in conifers, *Can. J. For. Res.*, 7, 299—325, 1977.

52a. **Pharis, R. P., Kuo, C. C., and Glenn, J. L.,** Gibberellin, a primary determinant in the expression of apical dominance, apical control and geotropic movement of conifer shoots, in *Plant Growth Substances, 1970*, Carr, D. J., Ed., Springer-Verlag, Basel, 1972, 441—448.

52b. **Pharis, R. P. and Morf, W.,** Experiments on the precocious flowering of western red cedar and four species of *Cupressus* with gibberellin A_3 and $A_{4/7}$ mixture, *Can. J. Bot.*, 65, 1519—1524, 1967.

52c. **Pharis, R. P. and Morf, W.,** Precocious flowering of coastal and giant redwood with gibberellins A_3, $A_{4/7}$ and A_{13}, *BioScience*, 19, 719—720, 1969.

53. **Pharis, R. P. and Ross, S. D.,** Gibberellins: their potential uses in forestry, *Outlook Agric.*, 9, 82—87, 1976.
54. **Pharis, R. P. and Ross, S. D.,** Progress in hormonal cone induction in Pinaceae conifers, in *Proc. Canadian Tree Improvement Assoc., Aug., 1983*, Canadian Forest Service, Ottawa, in press, 1984.

54a. **Pharis, R. P. and Ross, S. D.,** Flowering of Pinaceae family conifers with gibberlin $A_{4/7}$ mixture: how to accomplish it, mechanisms, and integration with early progeny testing, in Conifer Tree Seed in the Inland Mountain West, August 5 to 7, 1985, U.S. Forest Service, Missoula, Mont., in press, 1985.

55. **Pharis, R. P., Ross, S. D., and McMullan, E. E.,** Promotion of flowering in the Pinaceae by gibberellins. III. Seedlings of Douglas-fir, *Physiol. Plant.*, 50, 119—126, 1980.
56. **Pharis, R. P., Ross, S. D., Wample, R. L., and Owens, J. N.,** Promotion of flowering in conifers of the Pinaceae by certain of the gibberellins, *Acta Hortic.*, 56, 155—162, 1976.
57. **Pharis, R. P., Ross, S. D., Webber, J., and Owens, J. N.,** unpublished research results, 1984.

57a. **Pharis, R. P., Tomchuk, D., Beall, F . D., Rauter, R. M., and Kiss, G.,** Promotion of flowering in white spruce (*Picea glauca*) by gibberelin $A_{4/7}$ alone, and with auxin (naphthaleneacetic acetic acid), girdling and $Ca(NO_3)_2$ as adjunct cultural treatments, *Can. J. For. Res.*, in press, 1985.

58. **Pharis, R. P., Wample, R. L., and Kamienska, A.,** Growth, development and sexual differentiation in *Pinus* with emphasis on the role of the plant hormone, gibberellin, in *Management of Lodgepole Pine Ecosystems*, Baumgartner, D. M., Ed., Pullman, Wash., 1974, 106—134.
59. **Philipson, J. J.,** Flowering of Sitka spruce, in *Report on Forest Research, 1980*, Her Majesty's Stationery Office, London, 1981, 30—31.
60. **Philipson, J. J.,** The role of gibberellin $A_{4/7}$, heat and drought in the induction of flowering in Sitka spruce, *J. Exp. Bot.*, 34, 291—302, 1983.
61. **Philipson, J. J.,** Flowering stimulated in Sitka-FC research, *For. Br. Timber*, Feb., 1983, 21.

61a. **Phillipson, J. J.,** The promotion of flowering in large field grown Sitka Spruce by girdling and stem injections of gibberellin $A_{4/7}$, *Can. J. For. Res.*, 15, 166—170, 1985.

62. **Pollard, D. F. W. and Portlock, F. T.,** Effects of temperature on strobilus production in gibberellin-treated seedlings of western hemlock, *Can. For. Serv. Res. Notes*, 1, 21—22, 1981.
63. **Pollard, D. F. M. and Portlock, F. T.,** Response of strobilus production to gibberellin and fertilizer treatment in a young western hemlock clone bank on western Vancouver Island, *Can. For. Serv. Res. Notes*, 1, 27—28, 1981.
64. **Pollard, D. F. W. and Portlock, F. T.,** Timing and duration effects of gibberellin and fertilizer treatment on strobilus production in young western hemlock, *Can. For. Serv. Res. Notes*, 3, 3—5, 1983.

64a. **Puritch, G. A.,** Cone Production in conifers, *Can. For. Serv. Pac. For. Res. Cent. Inf. Rep.*, BC-X-65, 1972, 94 pp.

64b. **Pollard, D. F. W. and Portlock, F. T.,** The effects of photoperiod and temperature on gibberellin $A_{4/7}$-induced strobilus production of western hemlock, *Can. J. For. Res.*, 14, 291—294, 1984.

65. **Puritch, G. A., McMullan, E. E., Meagher, M. D., and Simmons, C. S.,** Hormonal enhancement of cone production in Douglas-fir grafts and seedlings, *Can. J. For. Res.*, 9, 193—200, 1979.
66. **Ross, S. D.,** Differential flowering responses by young Douglas-fir grafts and equisized seedlings to gibberellins and auxin, *Acta Hortic.*, 56, 163—167, 1976.
67. **Ross, S. D.,** Influences of Gibberellins and Cultural Practices on early Flowering of Douglas-fir Seedlings and Grafts, *Weyerhaeuser Forestry Research Technical Report 042/3001/77/7*, Centralia, Wash., 1977, 1—9.
68. **Ross, S. D.,** Influences of gibberellins and cultural practices on early flowering of Douglas-fir seedlings and grafts, in *Proc. 3rd World Consult. Forest Tree Breeding, March, 1977*, Vol. 2, CSIRO, Canberra, Australia, 1978, 997—1107.
69. **Ross, S. D.,** Efficacies of Gibberellins Alone and with Auxins NAA and 2,4,5-TP for Promotion of Male and Female Flowering on Douglas-fir Cuttings, Weyerhaeuser Forestry Research Technical Report 042-3201/79/29, Centralia, Wash.,1979, 1—7.

70. **Ross, S. D.,** Evaluation of Foliar Spray Formulations for Seed Orchard Application of Gibberellins, Weyerhaeuser Forestry Research Tech. Rep. 042-3001/79/10, Centralia, Wash., 1979, 1—12.
71. **Ross, S. D.,** Early Flower Induction in Douglas-fir: 1976 Trials at Sequim, Washington. I. Hormone Treatments and Application Regimens, Weyerhaeuser Forestry Research Technical Report 042-3201/79/45, 1979, 1—13.
72. **Ross, S. D.,** Enhancement of shoot elongation in Douglas fir by gibberellin $A_{4/7}$ and its relation to the hormonal promotion of flowering, *Can. J. For. Res.*, 13, 986—994, 1983.
72a. **Ross, S. D.,** Optimization of Gibberellin A_3 Treatments for Enhancing Yields in Seed Orchards of Western Red Cedar and Yellow Cedar, *B.C. Min. of Forests Progress Rept.* E. P. 935, Victoria, 1983, 26 pp.
73. **Ross, S. D.,** Promotion of flowering in potted *Picea engelmannii* (Perry) grafts: effects of heat, drought, gibberellin $A_{4/7}$, and their timing, *Can. J. For. Res.*, 15, 618, 624, 1985.
74. **Ross, S. D.,** Container seed orchard and crown management research, in Proc. Can. Tree Improvement Assoc., Part 1, Aug. 1983, Canadian Forest Service, Ottawa, 1984, 192—195.
75. **Ross, S. D., Bollman, M. P., Pharis, R. P., and Sweet, G. B.,** Gibberellin $A_{4/7}$ and the promotion of flowering of *Pinus radiata:* effects of partitioning of photoassimilate within the bud during primordia differentiation, *Plant Physiol.*, 76, 326—330, 1984.
76. **Ross, S. D. and Greenwood, M. S.,** Promotion flowering in the Pinaceae by gibberellins. II. Grafts of sexually mature and immature *Pinus taeda* L., *Physiol. Plant*, 45, 207—210, 1979.
76a. **Ross, S. D. and Pharis, R. P.,** Gibberellin-induced flowering of Douglas-fir grafts, *Plant Physiol.*, 51, 36, 1973.
77. **Ross, S. D. and Pharis, R. P.,** Promotion of flowering in the Pinaceae by gibberellins. I. Sexually mature non-flowering grafts of Douglas-fir, *Physiol. Plant*, 36, 182—186, 1976.
78. **Ross, S. D. and Pharis, R. P.,** Progress in the promotion of early flowering in Douglas-fir by gibberellins, *Weyerhaeuser Forestry Research Technical Report* 50/3001/76/32, Centralia, Wash., October, 1976, 1—8.
79. **Ross, S. D. and Pharis, R. P.,** Recent developments in enhancement of seed production in conifers, in *Proc. 18th Meet. Can. Tree Improvement Assoc., 1981, Part 2,* Pollard, D. F. W., Edwards, D. G. W., and Yeatman, C. W., Eds., Canadian Forest Service, Ottawa, 1982, 26—38.
79a. **Ross, S. D. and Pharis, R. P.,** Promotion of flowering in crop trees: different mechanisms and techniques with special reference to conifers, in *Attributes of Trees as Crop Plants,* Cannell, M. G. R., Jackson, J. E., and Gordon, J. C., Eds., Institute of Terrestrial Ecology, Monks Wood Experiment Station Abbots Ripton, Huntingdon, U.K., 1985, 387—401.
80. **Ross, S. D., Piesch, R. F., and Portlock, F. I.,** Promotion of cone and seed production in rooted ramets and seedlings of western hemlock by gibberellins and adjunct cultural treatments, *Can. J. For. Res.*, 11, 90—98, 1980.
81. **Ross, S. D., Pharis, R. P., and Heaman, J. C.,** Promotion of cone and seed production in grafted and seedling Douglas-fir seed orchards by application of gibberellin $A_{4/7}$ mixture, *Can. J. For. Res.*, 10, 464—499, 1980.
82. **Ross, S. D., Pharis, R. P., and Binder, W. D.,** Growth regulators and conifers: their physiology and potential uses in forestry, in *Plant Growth Regulating Chemicals,* Nickell, L. G., Ed., CRC Press, Boca Raton, Fla., 1983, 35—78.
83. **Ross, S. D., Webber, J. E., Pharis, R. P., and Owens, J. N.,** Interaction between gibberellin $A_{4/7}$ and rootpruning on the reproductive and vegetative process in Douglas-fir. I. Effects on flowering, *Can. J. For. Res.*, 15, 341—347, 1985.
83a. **Ross, S. D., Eastham, A. M., and Bower, R. C.,** Potential for container seed orchards, in *Proc. Symp. Conifer Tree Seed Inland Mountain West,* University of Montana, 1985, in press.
83b. **Schmidtling, R. C.,** Fruitfulness in conifers: nitrogen, carbohydrate, and genetic control, in *Proc. 3rd N. Am. Forestry Biol. Workshop,* Reid, C. P. P. and Fechner, G. H., Eds., Colorado State University, Fort Collins, 1974, 148—164.
84. **Sweet, G. B.,** A physiological study of seed cone production in *Pinus radiata, N.Z. J. For. Sci.*, 9, 20—33, 1979.
84a. **Sweet, G. B.,** personal communication, 1983.
84b. **Sweet, G. B. and Hong, S. O.,** The role of nitrogen in relation to cone production in *Pinus radiata, N.Z. J. For. Sci.*, 8, 225—238, 1978.
85. **Tompsett, P. B.,** Studies of growth and flowering in *Picea sitchensis* (Bong.) Carr. I. Effect of growth regulator applications to mature scions on seedling rootstocks, *Ann. Bot.*, 41, 1171—1178, 1977.
86. **Tompsett, P. B.,** Studies of growth and flowering in *Picea sitchensis* (Bong.) Carr. II. Initiation and development of male, female and vegetative buds, *Ann. Bot.*, 42, 889—900, 1978.
87. **Tompsett, P. B.,** Effects of environmental and growth regulator treatments on the flowering of mature Sitka spruce, in *Monogr. Br. Crop Prot. Council Symp. 'Opportunities for Chemical Plant Growth Regulation'*, British Crop Protection Council, Croyden, Surrey, England, 21, 75—81, 1978.

88. **Tompsett, P. B. and Fletcher, A. M.,** Increased flowering of Sitka spruce [*Picea sitchensis* (Bong.) Carr.] in a polythene house, *Silvae Genet.*, 26, 84—86, 1977.
89. **Tompsett, P. B. and Fletcher, A. M.,** Promotion of flowering on mature *Picea sitchensis* by gibberellin and environmental treatments: the influence of timing and hormonal concentration, *Physiol. Plant*, 45, 112—116, 1979.
90. **Tompsett, P. B., Fletcher, A. M., and Arnold, G. W.,** Promotion of cone and seed production on Sitka spruce by gibberellin application, *Ann. Appl. Biol.*, 94, 421—429, 1980.
91. **Wample, R. L., Durley, R. C., and Pharis, R. P.,** Metabolism of gibberellin A_4 by vegetative shoots of Douglas-fir at three stages of ontogeny, *Physiol. Plant.*, 35, 273—278, 1975.
92. **Webber, J. E., Ross, S. D., Pharis, R. P., and Owens, J. N.,** Interaction between gibberellin $A_{4/7}$ and rootpruning on the reproductive and vegetative process in Douglas-fir. II. Effects on shoot growth, *Can. J. For. Res.*, 15, 348—353, 1985.
93. **Wheeler, N. C., Wample, R. L., and Pharis, R. P.,** Promotion of flowering in the Pinaceae by gibberellins. IV. Seedlings and sexually mature grafts of lodgepole pine, *Physiol. Plant*, 50, 340—346, 1980.
94. **Zeevaart, J. A. D.,** Gibberellins and flowering, in *The Biochemistry and Physiology of Gibberellins*, Vol. 2, Crozier, A., Ed., Praeger, New York, 1983, 333—374.
95. **Zimmerman, R. H., Hackett, W. P., and Pharis, R. P.,** Hormonal aspects of phase change and precocious flowering, in *Encyclopedia of Plant Physiology*, (n.s.), Vol. II, Pharis, R. P. and Reid, D. M., Eds., Springer-Verlag, New York, 1985, 79—115.

PSIDIUM GUAJAVA

En. Guava Hindi (Amrud); Fr. Goyave; Sp. Guayaba; Port. Goiaba; Brazil. Araca

K. L. Chadha and R. M. Pandey

INTRODUCTION

The common guava (*Psidium guajava* L., Myrtaceae) originated in tropical America. It is now grown throughout the tropics and subtropics and is of commercial importance in India, Florida and Hawaii (U.S.), Egypt, South Africa, Brazil, Colombia, and the West Indies.[12,13,17] In India, the guava plant was introduced and distributed by the early 17th century.[10] At present, it is the fourth most important fruit crop of India and is being grown on an area of 58,230 ha, producing 200,000 M tons of fruit annually.[26] Much of the interest in guava cultivation has been due to its extremely high ascorbic acid content (up to 979 mg/100 g),[26] high nutritive value, ease of its culture, and the popularity of processed guava products, especially jellies and jams. However, the average ascorbic acid content of guava fruits varies from variety to variety and under different climatic conditions. The average content is approximately about 300 mg/100 g pulp.

Guava is a very hardy tree and can tolerate a wide range of climates and soils of tropical and subtropical regions, up to 1500 m altitude, if winter is frost-free. However, the optimum temperature for successful crop production lies between 23 to 28°C, and the optimum rainfall is from 100 cm to 200 cm. Guava is one of the few tropical and subtropical fruit crops which have tolerance to salinity and can be grown on marginal land with less care. Soil pH for guava cultivation may range from 4.5 to 8.2.[5] It may survive to a great extent in adverse climatic and soil conditions. Guava wilt is one of the serious problems, which renders its cultivation less remunerative. In order to put the flourishing guava industry on a sound footing, a comprehensive breeding program is needed to select a wilt-resistant guava.

TAXONOMY

Guava is one of the most important fruits of the Myrtaceae family, to which also belongs the Jamun (*Syzygium cuminii* Skeels), a fruit indigenous to India. It is classified under the genus *Psidium*, which contains about 150 species, but only *Psidium guajava* L. has been exploited commercially. The haploid chromosome number of guava is 11.

The common guava is an arborescent shrub or small tree that attains a height of up to 10 m. The trunk is slender with greenish brown scaly bark. The young pubescent branchlets are quadrangular. The opposite leaves are simple, entire, oblong-elliptic to oval, 7 to 15 cm long, acute to rounded at the apex, finely pubescent on the lower surface, with pigmentation throughout in young leaves, and have short petioles. The prinicpal veins are prominent, exstipulate, and pinnately veined. Flowers are borne in the axils of branchlets of current growth, solitary or in cymes upon a slender peduncle.

The fruit of the common guava is round and ovoid or pyriform in shape having rough yellow skin. As these two types of fruits have many variations, Linnaeus classified them into different species, *Psidium pyriferum* for the ovoid or pyriform-shaped fruits and *Psidium pamiferum* for the round ones. At present these two "species" are considered to be the pear-shaped and round-shaped varieties of *Psidium guajava*.[17] The pear-shaped forms are often called "pear guava" and the round ones "apple guava". Similarly, a large white-fleshed guava previously kept under the species, *Psidium guinaense* in Florida and California and a round, red-fleshed variety introduced into California under the species *Psidium aro-*

FIGURE 1. A twig showing six floral buds at one node, occasionally found in guava.

FIGURE 2. Flowering behavior in guava. (Modified from Teaotia, S. S., et al., *Progressive Hortic.*, 2, 101—112, 1970.)

maticum are also horticultural forms of *Psidium guajava.*[17] Other guavas, though somewhat less valuable than the common guava for commercial production, are the strawberry guava (*Psidium cattleianum* Sabine), Costa Rican guava (*P. friedrichsthalianum* Ndz.), and Guisaro guava (*P. molle* Bertol.). The fruits of these guavas, with high ascorbic acid content, are used principally for jelly making, but sometimes they are also used as fresh fruits.

The strawberry guava is ordinarily a bushy shrub, but sometimes grows into a small tree upto 8 m tall. The bark is smooth, grey-brown in color. The leaves are elliptic to obovate in shape, somewhat glossy and deep green in color. The flowers, which are produced singly in the leaf axils, are white. The calyx is obscurely lobed. The corolla is composed of 4 orbicular petals. The fruit is obovate to round in shape with sweet and aromatic flavor similar to that of strawberry. The strawberry guava is also called by other names such as ''Cattley guava'', and ''Chinese guava''. In the United States, different horticultural forms of the *Psidium cattleianum* have been listed under several botanical names, including *Psidium lucidum, P. chinense, P. sinense, P. araca* and *P. acre*. More recently, however, these species have been included under *Psidium cattleianum.*[17]

The Costa Rican guava (*Psidium friedrichsthalianum* Ndz.) is an erect tree about 8 m tall, with slender trunk and branches. The branchlets are wiry, quadrangular, and reddish in color. The leaves are elliptic, oblong-elliptic, or oval in shape, acuminate at the apex, with pubescence on the lower surface. White, fragrant, and about 2.5-cm broad flowers are singly borne on slender peduncles. The fruit is round or oval in shape, sulfur yellow, with few seeds and soft white flesh of acid flavor.

The Guisaro guava (*Psidium molle* Bertol.) is a bushy plant, which rarely grows more than 3 m tall. The young branchlets, peduncles, and lower surfaces of the leaves are reddish-velvety. The leaves are oblong-oval, obtuse at the apex, and rather stiff. The flowers, in cymes of 3 flowers, are borne in the leaf axil like common guava. The fruit is round, yellowish green to pale yellow, with whitish flesh containing numerous hard seeds.

FLORAL MORPHOLOGY AND BIOLOGY

Guava produces flowers on the current season's growth in the axils of leaves. The shoots bearing the flowers are terminal as well as lateral.[1,4,22] However, it has also been reported that the fruit bud in guava is mixed and is terminal only.[8] The flowers are solitary or in cymes of 2 to 4 (Figures 1 and 2). The receptacle is hollow and united to the ovary. The

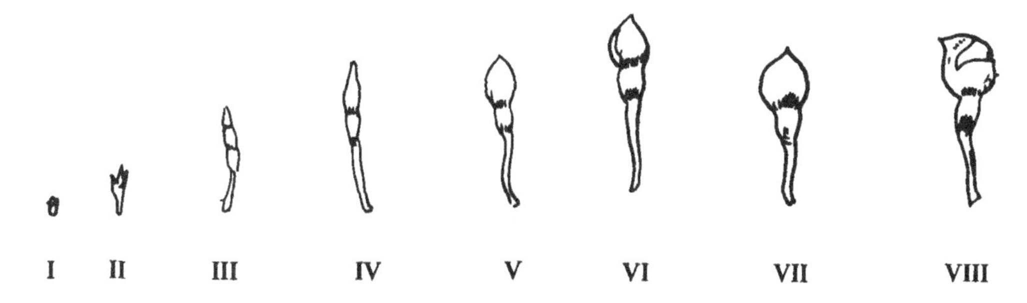

FIGURE 3. Stages in the development of the guava flower bud.

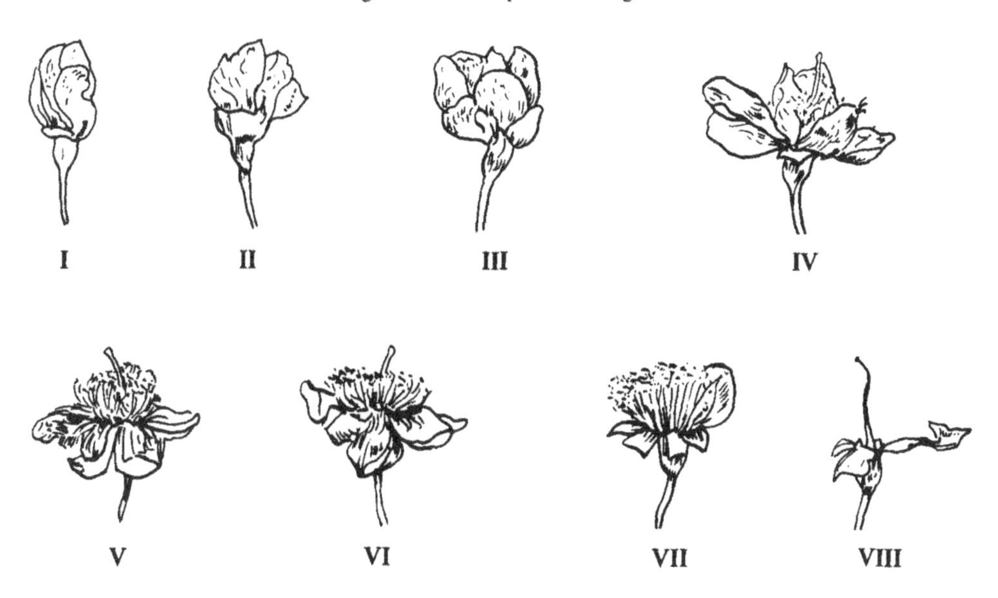

FIGURE 4. Different stages of flower development. I to V, stages of opening of the flower; VI, petals deflexed; VII, petals shed; VIII, filaments dropped, calyx cup intact.

flower is epigynous, cyclic, heterochlamydeous, actinomorphic, and hermaphrodite. The first pair of perianth leaves encloses the flower bud and is generally seen to crack slightly a day in advance of opening of flowers to expose the petals slightly (Figure 3). The calyx is obovate and is not separable into any definite number of sepals. On full emergence of the flower, the calyx is thrown off bodily as a cap or torn into 2 to 5 lobes which may subsequently drop away (Figure 4). The corolla consists of white, circular or oval, unequal-sized petals. The petals vary in number, from 5 to 12 to 1 or 2 whorls. The androecium consists of several thin filaments carrying bilobed anthers at the summit. In some cases, the filaments bear miniatrue petal-like structures, instead of anthers. The stamens are superior, of different sizes, and vary in number from 160 to 400. They are arranged in several rows in the receptacle over the ovary. The gynoecium consists of an inferior ovary, syncarpous, with axile placentation. The style is subulate, smooth, and bearded at the top. It is longer than the filaments and is bent over the stamens in the bud stage.

Floral buds emerge in the axils of different leaves after some period of shoot growth. The shoot growth period for emerging floral buds may vary with the variety and season of flowering.[1,24] Generally, the shoots can be 1 to 2 months old before they bear flower buds. It is not necessary that all the current seasons' shoots produce flower buds. Similarly, not all the buds in the axils of leaves on a shoot are floral. However, it has been confirmed by several scientists that the basal and terminal buds on the same shoot always remain vegetative.

FIGURE 5. Position of guava fruit in the shoot (the dark-shaded twig represents the previous season's growth). (Adapted from Sehgal, O. P. and Singh, R., *Indian J. Hortic.*, 24, 118—126, 1967.)

It is also a general consensus that the 2nd and 3rd basal buds in the axils of leaves on a floral shoot are flower buds. In guava, differentiation of a floral bud from a vegetative bud is not possible till two thin green bracteoles are developed from the flower buds. The fully developed floral buds are ovoid to oblong-oval in the adnate part, and the free part is also ovoid but more or less pointed. The terminal portion of the fully developed bud usually becomes yellowish. Often 2 to 4 minute dark ridges diverge from the apical point, marking the lines along which the calyx may split later.

Generally, guava flower buds take 30 to 45 days, depending upon the varieties, from visible differentiation of the flower bud to opening of the flower. Flowers usually open at from 5—5.30 a.m. to 7.30—8.00 a.m. the dehiscence of anthers starts normally 15 to 30 min prior to anthesis of the flowers and continues for about 2 hr. The stigma also becomes receptive within 2 to 3 hr after flower opening and remains receptive up to 48 hr.

The bearing twigs cease their growth as soon as fruit development starts (Figure 5). After fruit maturity, these twigs generally dry up from the terminal bud backwards. Pollen grains in guava morphologically are syncolporate, with 3, sometimes 4 to 5, colpi provided with a circular central opening (ora or endocolpium united at the poles). The exine is about 1 μ thick, and its outer surface is faintly granulated or almost smooth (psilate). The various layers of exine are hardly discernible. The pollen grain size and its frequency varies among the varieties. Normally the pollen grain size varies from 17.5 μ to 23 μ. There is an apparent correlation between the morphology and germination behavior of these grains. The big grains normally do not germinate except in a few varieties, e.g., Lucknow-49.[15] However, the pollen germination of the big grains is important for the production of better fruits, as these grains are possibly the result of polyploidy. Apart from the big grains, the 4 colpate and 5 colpate are also of possible significance in the improvement of varieties by means of selective fertilization using improved pollen grains.

CROPPING PATTERN

Guava bears more than one crop in a year. In tropical regions, if irrigation is given regularly, it may bear throughout the year though not always in large quantities. However, 3 distinct flowering seasons, viz., spring, rainy, and autumn have been well defined.[6,14,20,22,24] Flowering season varies with the locality and variety. In Florida and the West Indies the season of ripening guava fruit is late summer and autumn, but in India, guava is harvested in two or three seasons depending on the locality.

Flowering Season:

There are contradictory reports with regard to flowering season in northern India. According to Rathore and Singh,[20] the guava plant flowers 3 times in a year, viz., in summer, the rainy season, and in autumn, but only two seasons, summer and rainy, were important from the point of view of flowering and yield. Sehgal and Singh[22] and Smith[23] observed only two flowering seasons, viz., summer and rainy season. In south India also there are only two flowering seasons, spring and rainy.

It is also important to mention that the period of flowering and the duration of flowering in different seasons vary with the locality. In south India, flowering in the spring season commences in January and continues up to March and in the rainy season extends from September to October.[1,2] In Uttar Pradesh the two peaks of flowering are in February and June for the spring and rainy season crops, respectively.[6,23] In northern India, flowering takes place in April to May for the summer season and July to August for rainy season crop.[20,22] In terms of days, the duration of the flowering period under north Indian conditions for different varieties varies from 27 to 39 days in the summer season and from 28 to 46 days in the rainy season, whereas the corresponding figures for south Indian conditions are 49 to 63 days and 29 to 37 days for spring and rainy season flowering, respectively. Autumn season flowering in north India persisted for a longer period, i.e., October to December, without showing any peak. Commencement of early flowering in the spring season under south Indian conditions is due to the tropical climate.[10]

As far as the flower production in different flowering seasons is concerned, the highest flower production has been observed in summer season flowering. The production of flowers in greater or less quantity is mainly a physiological phenomenon. Since flowering and vegetative growth during winter are almost negligible, the plant accumulates sufficient food reserves, which are obviously responsible for the initiation of maximum new vegetative growth in spring and subsequent maximum flowering during summer season.

Though maximum flower production occurs during summer season flowering, fruit set and retention have been found to be highest in autumn flowering followed by rainy season flowering. Fruit set and retention, apart from physiological factors, are governed by various climatic factors, especially temperature and humidity. The hot summer season in north India, characterized by high temperature, low humidity, and dessicating winds, is responsible for poor set and retention of fruits of guava from summer season flowering. Autumn is the most favorable season for the set and retention of fruit.

It is a general view that the rainy season crop, developed from the summer season flowering, besides being higher in fruit yield, is attacked by many insect pests and diseases, and the fruit is rough, insipid, poor in quality, and less nutritive.[9,10,19,21,25] On the other hand, the winter season crop, developed from rainy season flowering, is superior in fruit quality, has more nutritive value, and is free from diseases and insect pests. As regards the net return in terms of cash, the winter season crop is more profitable than the rainy season crop due to high selling rates and less damage to the fruits by diseases and insect pests. Besides, winter season fruits have better storage life and can be transported over a long distance. Rathore and Singh[20] and Gupta and Nijjar[9] studied the various cropping patterns, and recommended a single winter crop in a year in order to harvest a highly economical crop of best quality fruits.

Regulation of Flowering

Flowering in guava may be regulated either by suspending the vegetative growth flush through cultural treatments (e.g., root exposure and pruning), withholding irrigation, fertilization, or by deblossoming by hand or by chemicals. The main objective in suspending the vegetative growth is to provide rest to the plant, which results in accumulation of food reserves in large quantity for enhancing flowering in the next season.

Root exposure is an old method practiced by guava growers in the west zone of India. It should be done carefully, so as not to cut or bruise the massive, thick roots from the major root system, while removing the upper 8 cm of soil from an area of 50 cm radius around the trunk. After these roots are exposed, the clusters of minute fiber-like roots borne on them are pruned by pruning shears. By root exposure, the water supply from soil to plant is checked, resulting in leaf fall, and ultimately plant rest. When leaves have fallen the exposed roots are again covered with the original soil. The root exposure practice is not suitable where plants are grown on sandy soils.

The same principle applies to the induction of rest during flowering season for accumulating food reserves in the tree, by withholding irrigating water to the plant. The fruit sets in the spring season and most of the leaves fall off; subsequently the plant enters a resting stage when irrigation of the trees is stopped from February to May. On resuming irrigation, from the end of May onward at intervals of 10 to 15 days until the monsoon sets in, the trees blaze into rainy season flowering. The excess of water in the soil during the flowering season has also been found to be helpful for normal flowering and setting of fruits.[7]

Besides cultural treatments to regulate the flowering season, deblossoming of the guava tree is another method. Guava plants can be deblossomed either by hand or by applying chemicals. Manual deblossoming is successful, but economically it is not acceptable and hence has not been popularized in practice.

Efforts have been made to find out some suitable chemicals, which may efficiently cause deblossoming.[3,11,16,18,25] However, a great variation exists in the response to chemicals tried for this purpose. This variation may be due to various factors such as cultivars, tree condition, soil type, and environment.

Some chemicals such as NAA, 2,4-D, NAD, ethephon and MH have been tried to achieve this objective. NAA has been found to be the most effective chemical for removal of the flower buds and subsequently in increasing the winter season crop.[3,16,18,25] However, there are varietal differences in response to this chemical. This is evident from the fact that 600 ppm NAA is the optimum concentration for the variety Lucknow-49 under Pantnagar (in north Uttar Pradesh) conditions;[16] Rathore[18] recommended 100 ppm for cv Safeda grown under Delhi conditions; and in Bihar, 50 ppm of NAD was recommended by Kumar and Hoda[11] for variety Allahabadi Safeda. Hence, still more comprehensive studies on deblossoming guava trees of various cvs under different local conditions, through chemical manipulation, are needed.

REFERENCES

1. **Balasubrahmanyam, V. R.,** Studies on blossom biology of guava (*Psidium guajava* L.), *Indian J. Hortic.*, 16, 69—75, 1959.
3. **Chundawat, B. S., Gupta, O. P., and Godra, N. R.,** Crop regulation of Banarasi Surkha, a guava (*Psidium guajava* L.) cultivar, *Haryana J. Hortic. Sci.*, *4*, 23—25, 1975.
3. **Chundawat, B. S., Gupta, O. P., and Godra, N. R.,** Crop regulation of Banarasi Surkha, a guava (*Psidium guajava* L.) cultivar, *Haryana J. Hortic. Sci.*, *4*, 23—25, 1975.
4. **Dasarathy, T. B.,** The guava, *Madras Agric. J.*, 38, 521—527, 1951.
5. **El Baradi, T. A., Guava,** *Abstr. Trop. Agric.*, 1, 9—16, 1975.

6. **Gandhi, S. R.,** The guava, in *Fruit Culture in India*, Indian Council for Agricultural Research, New Delhi, 1967, 143—152.
7. **Gandhi, S. R.,** The guava in India, *Farm Bull. 26*, Indian Council for Agricultural Research, New Delhi, 1959.
8. **Gardner, V. R., Bradford, P. C., and Hooker, H. D.,** *Fundamentals of Fruit Production*, McGraw-Hill, New York, 1952.
9. **Gupta, M. R. and Nijjar, G. S.,** Crop regulation in guava (*Psidium guajava* L.), *Indian J. Hortic.*, 35, 23—27, 1978.
10. **Hayes, W. B.,** *Fruit Growing in India*, Kitabistan, Allahabad, 1957, 286—303.
11. **Kumar, R. and Hoda, M. N.,** Crop regulation studies in Allahabad Safeda guava, *Indian J. Hortic.*, 34, 13—14, 1977.
12. **Luh, B. S.,** Tropical fruit beverages, in *Fruit and Vegetable Juice Processing Technology*, AVI Publishing, Westport, Conn., 1971.
13. **Malo, S. E. and Campbell, C. W.,** The guava, Fruit Crops Fact Sheet No. 4, Florida Agricultural Extension Service, Gainesville, 1968.
14. **Naik, K. C.,** *South Indian Fruits and Their Culture*, P. Vardachary and Co., Madras, 1949, 446—451.
15. **Nair, P. K. K., Balasubramanyam, V. R., and Khan, H. A.,** Palynological investigations of some guava varieties, *Indian J. Hortic.*, 21, 79—84, 1964.
16. **Pandey, R. M., Lal, S., and Kaul, G. L.,** Effect of chemicals and flower thinning on regulation of crop in guava, *Indian J. Hortic.*, 37, 234—39, 1980.
17. **Popenoe, W.,** *Manual of Tropical and Sub-tropical Fruits*, 3rd ed., Macmillan, New York, 1974, 272—311.
18. **Rathore, D. S.,** Deblossoming of rainy season crop of guava by NAA, *Prog. Hortic.*, 7, 63—65, 1975.
19. **Rathore, D. S.,** Effect of season on the growth and chemical composition of guava (*Psidium guajava* L.) fruits, *J. Hortic. Sci.*, 51, 41—47, 1976.
20. **Rathore, D. S. and Singh, R. N.,** Flowering and fruiting in the three cropping patterns of guava, *Indian J. Hortic.*, 31, 331—336, 1974.
21. **Sachan, B. P., Pandey, P. D., and Shanker, G.,** Influence of weather on chemical composition of guava fruits (*Psidium guajava* L.) var. Allahabadi Safeda, *Punjab Hortic. J.*, 9, 119—123, 1969.
22. **Sehgal, O. P. and Singh, R.,** Studies on the blossom biology of guava (*Psidium guajava*). I. Flowering season, flowering habit, floral bud development, anthers and dehiscence, *Indian J. Hortic.*, 24, 118—126, 1967.
23. **Smith, W. S.,** The guava, Bull. No. 8, Utar Pradesh Department of Agriculture, 1934.
24. **Syamal, M. M., Singh, R. K., and Chhonkar, V. S.,** Studies on growth and flowering in guava (*Psidium guajava* L.) *Indian J. Hortic.*, 27, 243—249, 1980.
25. **Teaotia, S. S. and Pandey, I. C.,** Crop regulation studies in guava (*Psidium guajava* L.) *Prog. Hortic.*, 1, 25—28, 1970.
26. **Vandendriessche, H.,** Case studies of the industries in developing countries, Tropical Fruit Processing Industry, Dev. Chtr. Organ. Econ. Co-op. and Dev., Paris, 1976.

[illegible]

17. Pounder, W. [illegible] and Kohl, C. E., [illegible], *[illegible]*, 39, [illegible], 1960.
18. Pucerne, W., *Manual of [illegible]*, Macmillan, New York, 19[illegible].
19. Rathore, D. S., [illegible], *[illegible]*, 6[illegible], 1[illegible].
20. Rathore, D. S., Effect of [illegible] on the growth and chemical composition of [illegible], *[illegible]*, [illegible], 1[illegible].
21. Rathore, D. S. and [illegible], [illegible] in the [illegible], *[illegible]*, [illegible], 19[illegible].
22. [illegible], R. C., and Sharma, [illegible], The influence of [illegible], *[illegible]*, [illegible].
23. [illegible]

[illegible]

*RUDBECKIA**

En. Coneflower; Fr. Rudbeckie; Ge. Sonnenhut

Vyacheslav G. Kochankov and Mikhail Kh. Chailakhyan

TAXONOMY, ECOLOGICAL, AND BOTANICAL CHARACTERISTICS

History of Use

The herbaceous *Rudbeckia* (coneflower) plants were named by Linnaeus in honor of the Swedish anatomist and botanist Olaus Rudbeck (1630 to 1702). *Rudbeckia* plants are native to North America[65] and represent one of the most widely distributed weeds throughout the U.S.[105-110]

Cultivated *Rudbeckia* species are used as ornamental plants. Especially widespread are *R. laciniata* L. (cv Golden Glow) and *R. hirta* L. (black-eyed or brown-eyed Susan). The latter, spreading by seeds, is becoming and in some places has already become a weed, for instance, in the European part of the U.S.S.R. and in western and eastern Europe.[72] Recently, the polyploid *R. hybrida* Hort. (gloriosa daisy) has gained wide use in ornamental horticulture. It consists of plants with strong branching shoots bearing large (15 to 19 cm in diameter) flower heads differing in color and the degree of doubleness.[87]

Other species of *Rudbeckia* are being introduced in botannical gardens of the Soviet Union. Apart from the above-mentioned species, *R. triloba L.*, *R. bicolor* Nutt., *R. speciosa* Wend., *R. grandiflora* Gmel., *R. deamii* Blake, and *R. sullivantii* Boynton et Beadle have been introduced and recommended for practical use in the Botanical Garden of the Academy of Sciences of the Armenian S.S.R. (Yerevan) and the Botanical Garden of the Academy of Sciences of the Uzbek S.S.R. (Tashkent).[7,72] The distinguishing feature in the introduction of *Rudbeckia* plants to gardening is that they are introduced as species, most likely due to the fact that almost all wild species of the genus appear to be beautifully blooming plants.[72] Variety Golden Glow, which belongs to the species *R. laciniata*, but is apparently its sterile mutant, has been an exception.

Rudbeckia plants can be used other than for ornamental purposes. *R.laciniata*, for instance, can be used as a forage directly or for silage, since some of its varieties can produce a vast silage mass with much higher yield than that of sunflowers harvested from the same area. The long flowering period of *Rudbeckia* plants (2 to 3 months) makes it possible to use them as nectariferous plants for several months because of the high amount and quality of their nectar.[72]

As an object of physiological investigations, *Rudbeckia* found use as a typical LDP as far back as 1924 *(R. bicolor)* in the studies of Allard and Garner, and in their later studies on (*R. bicolor* var. *superba* Nutt., *R. nitida* Nutt., *R. hirta* L., *R. laciniata*, and *R. newmanii* Loud).[4,5,6] In the period of extensive studies on photoperiodism of plants in the 1930s and early 1940s, the physiology of *Rudbeckia* flowering was primarily investigated in America (Murneek,[96,97] Austin,[8,9] Roberts and Struckmeyer,[111-113]) and to some extent in Germany (Knodel).[74] These studies dealt as a rule with the main photoperiodic characteristics of *Rudbeckia* (critical length of day, temperature effect of photoperiodic response, investigations into reduction-oxidation processes at various daylengths, etc.).

After World War II, *Rudbeckia* became the object of studies in a number of research centers in Europe. In the U.S.S.R., beginning in 1947 and up to the present time, *R. bicolor* has been one of the main objects of studies at the laboratory of M.Kh. Chailakhyan in the Timiryazev Institute of Plant Physiology of the U.S.S.R. Academy of Sciences in Mos-

* Supplement to *Rudbeckia* in Volume 4 (pages 239 to 242) of this *Handbook*.

FIGURE 1. Comparative effect of LD induction and treatment with GA_3 under SD on growth and flowering of *R. bicolor* plants. Plants treated with: (1) short days; (2) 24 μg GA by daily application of 1 drop of solution (50 mg/ℓ) to the apex under SD conditions for 12 days; (3) induction by 12 long photoperiods. Plants are shown 57 days after the beginning of induction. (From Kochankov, V. G., Dissertation for a Master of Biology, K.A. Timiryazev Institute of Plant Physiology, U.S.S.R. Academy of Sciences, Moscow, 1972, 145a. With permission.)

cow.[60,61,117-120] In western Europe, *Rudbeckia* has been studied at the Institute of Plant Physiology in Göttingen, F.R.G. (Bünsow, Harder),[16] the Phytotron Laboratory in Gif-sur-Ivette in France (Harada, Nitsch),[66,67] and at Liège University, Belgium (Bouilenne, Bernier, Jacqmard, and Pont Lezica et al.).[12-14,69-71,100-104]

The investigations carried out in different countries in the postwar years, particularly in the 1950s to 1970s, used the particular properties of *Rudbeckia* plants — their ability to form flower stems and to bloom only under conditions of long photoperiodic cycles, while under conditions of a short day they form only a rosette (Figure 1). With the discovery of the physiological effects of gibberellins, growth inhibitors, and retardants, various species of *Rudbeckia* plants were employed as models for studying the differentiation of the apical meristem, hormonal status, and the regulation of stem growth and flowering, *R. bicolor, R. hirta, and R. speciosa* being the most popular species with physiologists.

Taxonomy

The genus *Rudbeckia* L. belongs to the Compositae family, tribe Heliantheae. According to American floras and other literature this genus is native to North America[65] and comprises 25 to 30 species, variable in habit and pubescence.[65,72] Their leaves vary in shape, and the species may be grouped according to the degree of leaf division and lobing.[108]

Table 1
BIOLOGICAL CHARACTERISTICS OF *RUDBECKIA* SPECIES

Species	Life span (years)	Flowering in: First year	Flowering in: Subsequent years	Method of reproduction	Seed dormancy	Period of seed viability (years)
R. triloba	2—3,[72] P (short-living[71]	+[72]		Seeds[72]	2 months[72]	<3 years[72]
R. bicolor	⊙[7,72]	+[72]		Seeds[72]	—[72]	8 years[72]
R. hirta	⊙[72] P (short-living)[65]	+[65,72]		Seeds[72]	—[72]	8 years[72]
R. speciosa	P[72]		2[72]	Seeds, vegetative[72]	6 months[72]	<3 years[72]
R. grandiflora	P[65,72]		2[72]	Seeds, vegetative[72]	6 months[72]	<3 years[72]
R. maxima	P[65,72]			Vegetative[72]		
R. occidentalis	P[72]			Vegetative[72]		
R. laciniata	P[65,72]			Vegetative[72]		

Note: ⊙, monocarpic species; P, polycarpic perennial species; references superscripted.

In the present article we use the names of species described according to the *Illustrated Flora* of Gleason[65] and *Wild Flowers of the United States* by Rickett,[105-110] with a few exceptions.

First, the English name "coneflower" refers not only to the genus *Rudbeckia* L. and the genus *Echinacea* Raf. separated therefrom, but also to the genus *Dracopis* Cass. and the genus *Ratibida* Raf.[65,105-110] In this chapter, we confined ourselves only to *Rudbeckia* L. plants proper.

Second, *Rudbeckia* plants, as representatives of aggressive weeds spreading to new territories, are apparently very variable and some species differ only insignificantly.[108] Therefore, even the special botanical literature indicates that *R. hirta* is often confused with *R. serotina,* and *R. bicolor* is represented as the subspecies *R. serotina.*[108] *R. hirta* and *R. bicolor* are completely identical in the seedling phase.[73]

Third, Allard and Garner mention the species *R. nitida* (var. Autumn Sun)[4,6] and *R. newmanii* Loud.[6] which we have not found in American floras. It is likely that *R. newmanii* in the work of Allard and Garner[6] and *R. speciosa* in the work of Murneek[97] represent one and the same species. Therefore, citing these works, we give the Latin names in the same way as the authors apply them.

Ecological and Botanical Characteristics

Under natural conditions in North America, *Rudbeckia* species occur in all the geographical zones — from moderate to subtropical climates, and in the regions from continental to Atlantic-continental and Atlantic-monsoon, while their habitat appears to include prairies, forest-steppes, and mixed and hardwood forests of the subtropical and moderate zones.[72] They are, however, most widely distributed in the East and the central mountain regions, as well as in prairies. They bloom as a rule for several months.

There are annual and perennial plants among *Rudbeckia.* The perennials begin to flower, as a rule, in the second year (Table 1). The data presented in Table 1 on the duration and time of the first flowering relate to plants grown in soil under conditions of natural daylength. Some authors who planted, for instance, perennial *R. speciosa* in a greenhouse at first under SD conditions and transferred them to LD conditions were able to obtain flowering of the plants after several weeks.[66] *Rudbeckia* plants reproduce as a rule by seeds, but the offspring of some of them are reproduced only vegetatively. The period of seed viability is 3 to 8 years, and seed dormancy is very short, or not present at all.

Growing Techniques.

Rudbeckia plants may be grown both from seeds and rhizomes (Table 1). Since we experimented with *R. bicolor* reproduced by seeds, we will consider only reproduction by seeds.

Seeds of *R. bicolor* are small, 3 to 4 mm in length and 1 mm in width. They are sown in a greenhouse with a temperature of 20 to 26°C in February-March, and transplantation to clay flower pots and soil is carried out in May, when the external day temperature reaches 15 to 20°C and there is no risk of frost. From sowing up to the beginning of the experiments the plants are exposed to short 7 to 9 hr days (SD) so that they remain in the form of a rosette, since, for instance, *R. bicolor* apparently becomes photoperiodically sensitive with the emergence of the first true leaves.

The only detailed study on the morphology of seedlings of several *Rudbeckia* species was made by Khudzhaniyazova.[73] At the period of seed germination, a rupture appears on its micropylar part, through which a tiny radicle emerges, then forming a large number of root hairs. As soon as the seedling forms a root system, intensive growth of the hypocotyl begins. During the first 3 to 5 days, the roots of rudbeckias grow very rapidly.[73] By the time the cotyledons are released from the testa, the roots have penetrated 3 to 7 cm into the soil. In 8 to 10 days after the beginning of growth, the rapidly elongating main root produces lateral roots. With 5 to 9 days after the onset of growth of the main root, there begins unfolding and growth of the first true leaf of the seedling that is the first leaf of the rosette.

Shoots usually emerge 7 days after planting. The plants are transplanted at the age of 2 to 2.5 months after growing under SD when the plants have 13 to 16 true leaves.

Plants intended for obtaining seeds are exposed to a natural 16 to 17.5-hr day after transplantation into the soil. Following transplantation, the plants begin to flower after 1 to 1.5 months; they bloom for 1.5 to 2 months and form seeds after 3 months. The greatest number of viable seeds are formed in the apical heads on the main stems (that is, on the first-order axes). The quality of seeds also depends upon their location within the receptacle (torus) of the head. Seeds from the lower and middle parts of the heads have higher weight and better ripeness and germinability.[72]

Plants intended for investigations of photoperiodism and physiology of flowering usually mature at the age of 3 to 5 months, with a well-developed rosette. They require less induction by LD for stem growth and transition to flowering.[117]

In summer, the plants are given a long 17- to 18-hr day provided by natural light, and when the length of day decreases a similar LD is provided by additional illumination with reflected incandescent lamps or mercury-fluorescent lamps "DRL" (500 W).

The only pest for *R. bicolor* plants (especially old plants) appears to be a spider mite which is particularly dangerous in the autumn/winter period, when the plants are grown in a heated greenhouse and with additional artificial illumination.

Shoot Apex and Inflorescence Structure; Criteria of Flowering Response

Common features of species belonging to the genus *Rudbeckia* appear to be the presence of large heads on long flower stalks. *Rudbeckia* plants have heads with yellow, orange, or rarely red-brown peripheral, infertile ligulate (ray) flowers and greenish, yellow or brown, small, tubular, fertile flowers (disk flowers).[78,108] The English name "coneflower" comes from the shape of its inflorescences, composed of tubular flowers forming a spiral around a conic, hemispherical, or sometimes columnar receptacle.[65,72,108]

The transition of *R. bicolor* plants to flowering can be subdivided into several developmental phases depending on the method of observation, that is, macroscopic or microscopic. While the first method enables us to follow all stages of the transition from vegetative growth to flowering, the latter is concentrated on the very first stages of this transition.

The stages of transition to flowering of *R. bicolor* in visual observations are the following: (1) a flat rosette of leaves (in plants growing under short day conditions); (2) raising of the horizontal rosette leaves to a more vertical position; (3) beginning of stem elongation (bolting); (4) flower bud formation; and (5) flowering.

Under induction of stem growth and flowering the first visible response appears to be the raising of the rosette leaves in the daytime— first of the young, upper leaves after 1 to 2 days, and then of the middle ones. However, initially, this response is reversible; after transfer to SD conditions the leaves again become horizontal after 2 to 3 photoperiods, without visible morphological alterations. The same reversible raising of the leaves is also caused by shading of the plants or by exposure to continuous darkness, by transplantation of young plants, or by low temperatures (5 to 10°C). Even in plants growing under SD conditions the young leaves of the rosette ascend a little after the end of the dark periods, but when exposed to light they become horizontal again after 1 to 2 hr.

Stem growth becomes apparent as a clearance between the bases of adjacent leaves which can be seen when the young petioles are bent sideways; the beginning of flower bud formation can be seen in the form of flower primordia which are apparent when the young leaves are pulled apart. Both the transition to flowering of *Rudbeckia* and the flowering period itself are quite extended in time. In our experiments with *R. bicolor,* we regard the appearance of yellow ligulate flowers (rays) as the beginning of flowering.

Under continuous LD exposure, the onset of leaf erection in *R. bicolor* is observed after 1 to 3 days, stem elongation after 5 to 15 days, and flower bud formation 16 to 25 days following the beginning of LD exposure, depending on the age of the plant.

On the other hand, in order to characterize the first stages of transition from vegetative growth to reproductive development, cytological criteria, namely, changes in the structure of the apical meristem, are of interest. However, numerous data, reported by Samygin[114] for many plants and by Pont Lezica[102] for *R. bicolor* var. Etoile de Kelvedon, indicate that the first stages of development — changes in the structure of the shoot apices and formation of flower primordia — do not necessarily lead to flowering and fruit formation. Therefore the histological evaluation of flowering is a great deal less reliable as compared to a visual assessment of flowering at the end of an experiment.

Studies carried out by Bernier et al.[12,13,68] and Chailakhyan et al.[36,92] showed that the apices of *R. bicolor* plants consist of three main zones: (1) central, (2) lateral (peripheral), and (3) medullary (pith-rib, subapical), giving rise to flowers, leaves, and stem pith, respectively.

In vegetative *R. bicolor* plants, growing under SD, the apex is flat (Figure 2) and small with, depending on age, a diameter of several tenths of a millimeter. The central apical zone is characterized by larger cells and nuclei with small nucleoli, well-developed plastids and vacuoles, low RNA and protein content, and occasional cell divisions. The lateral zone surrounding the central one consists of smaller cells with large nucleoli and small vacuoles, and a high content of RNA and protein. Below the central zone is the medullary zone, characterzied by regular layers of flattened cells resembling the cambium, with large vacuoles and a low RNA content.[12,68,92,102]

In the transition to flowering, the shoot apex of *R. bicolor* passes through several developmental phases. Milyaeva et al.[94] distinguish three phases: (1) vegetative, (2) evocative, and (3) reproductive. The period of apical development from the moment the flowering hormones enter the apex to the beginning of flower primordia formation (after exposure to 2 to 4 long photoperiods) represents the phase of flowering evocation. At this period the mitotic activity in the medullary zone (Table 2), which is responsible for stem elongation, increases. The mitotic activity in the central zone, responsible for flower primordia formation, also rises.[92] However, such a slight induction is insufficient for inducing flowering in 2- to 3-month-old plants.[94,95] Only the beginning of flower primordia formation marks the tran-

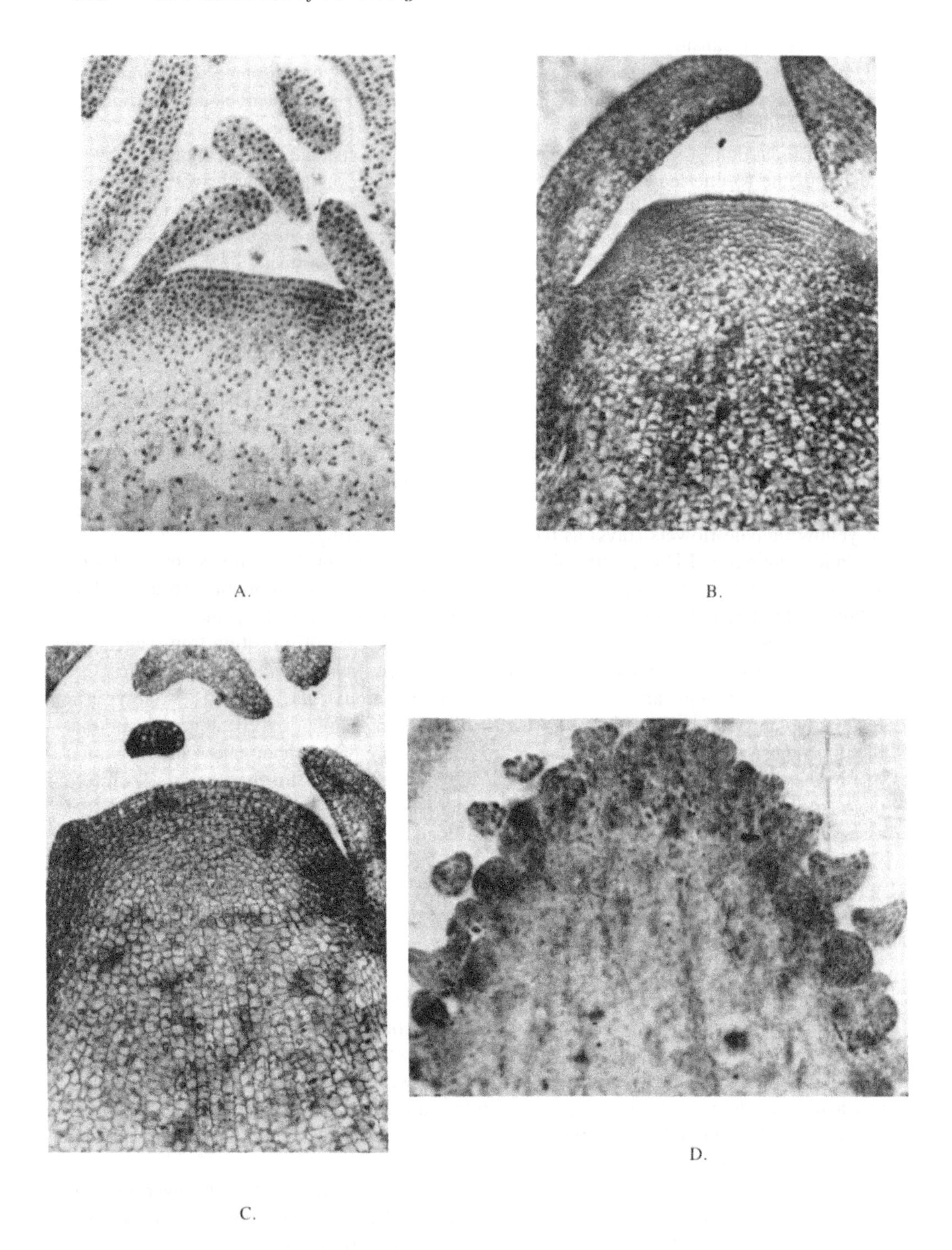

FIGURE 2. Longitudinal sections of shoot apices of *R. bicolor* plants exposed to SD (A) and induced by 4(B), 8(C), and 16(D) LD. (From Milyaeva, E. L., Kovalyova, L. V., and Chailakhyan, M. Kh., *Fiziol. Rast.*, 29(2), 253—260, 1982. With permission.)

Table 2
MITOTIC INDICES OF VARIOUS EPICAL ZONES IN *R. BICOLOR* PLANTS INDUCED BY LD OR TREATED WITH GA_3 UNDER SD CONDITIONS[92]

Number of long days	Mitotic indices		Number of days of treatment with GA_3	Mitotic indices	
	Central zone	Medullary zone		Central zone	Medullary zone
0 (control plants)	0.4 ± 0.15	0.2 ± 0.19	0 (control plants)	0.4 ± 0.15	0.2 ± 0.19
2	3.9 ± 0.39	4.0 ± 0.34	2	0.2 ± 0.19	3.5 ± 0.36
4	5.2 ± 0.32	4.3 ± 0.31	4	0.6 ± 0.42	5.0 ± 0.28
6	6.2 ± 0.29	3.3 ± 0.18	6	1.0 ± 0.31	6.2 ± 0.28
8	10.4 ± 0.26	1.0 ± 0.33	8	1.5 ± 0.16	8.0 ± 0.24

sition to the reproductive phase. At this period (after 6 to 8 long photoperiods) the mitotic activity in the medullary zone falls, whereas in the central zone it keeps increasing.[92]

Earlier, Bernier et al.[12,13] Jackmard,[68] and Pont Lezica[102] had distinguished three phases in the development of the apices, the vegetative, prefloral, and floral phases. In the prefloral phase the mitotic activity and RNA content in the central zone increase and leaf production stops. No difference is noted between the central and the lateral zone which together constitute a uniform outer tunica. The number of cells in the central zone rapidly increases, modifying the size and shape of the apex. The medullary zone shows stimulation of mitoses and this is linked with stem elongation. The apical meristem is gradually transformed into an inflorescence apex. In the floral phase, bracts are produced and an active development of floral organs takes place. the most important histochemical event in this phase is the decrease of DNA synthesis.

Studies on the effect of GA_3 on growth, flowering, and mitotic activity in the various zones of the shoot apex in plants, first carried out by Bernier et al.[12,13] and Jacqmard,[68,69] and then by Chailakhyan et al.,[36,92] showed an increase of the mitotic activity in the medullary zone (Table 2) resulting in an intensive elongation of the stem. GA_3 under SD caused a rise of the mitotic index in the central zone only 8 days after treatment, that is, much later than in flower induction by LD.[92] This shows that a LD activates the central zone to a greater extent as compared to the medullary one, while GA_3 activates the medullary zone more intensively as compared to the central one.

The diurnal pattern of mitotic activity in the central and medullary zones of the shoot apex during the first 4 days of photoperiodic induction (the phase of flowering evocation)[93] was of a periodic character both in induced and control plants. The maximal peaks of mitotic activity fall within the second half of the dark periods, while the minimal ones fall within the light periods. In the medullary zone, the increase of mitotic activity occurs earlier than in the central one (after 12 and 14 to 16 hr, respectively). Earlier, Jacqmard[70,71] had also established the periodic character in the mitotic activity in various zones of the apex under the action of GA_3 in plants of *R. bicolor* cv Etoile de Kelvedon.

The criterion of the flowering response is as a rule the number of days from the beginning of a treatment to the flowering in 100% of the plants.[6,63,117] Some authors indicate the percentage of flowering plants at the end of the experiment.[103,104] The number of days to flower bud formation in 100%[6,13,52,119] or 50% of the plants,[14,120] the number of days before and beginning of stem elongation in rosette plants,[61,119] and flower primordia formation or a change in the structure of the apical meristem[36,58,69,70,92] have also been regarded as the main or supplementary criteria.

Table 3
PHOTOPERIODIC RESPONSE OF *R. BICOLOR* PLANTS IN RELATION TO THEIR AGE[a,117]

Plant age at beginning of experiment (days)	Number of long photoperiods		
	7	16	Continuous LD
30	No response	No response	66
60	No response	No response	66
90	No response	64	51
120	52	54	45

[a] Number of days from beginning of experiment to flowering.

EFFECTS OF PLANT AGE AND JUVENILITY

The duration of the juvenile period in *Rudbeckia* plants has not been determined precisely. The influence of age on the transition to flowering under favorable long photoperiods was studied by Zhdanova,[117] but the minimum age of the *R. bicolor* plants used in this work was 30 days (Table 3).

The fact that flowering began earlier in more adult plants has been confirmed in other reports[36,82,103] from which it is clear that with the increase in age the plants require fewer inductive photoperiodic cycles for the inducing of flowering if they have been under SD conditions before.

One of the reasons for the increase in sensitivity to photoperiodic induction may be a higher GA activity with age even under unfavorable SD conditions,[82,101] though it does not reach the same level as under LD. The activity of ABA in this case keeps dropping.[83] Apparently, this can explain an age-related stimulation of the "ripeness to flower" state in plants when induction by as little as 3 LDs is capable of inducing flowering in 7-month-old plants.[82]

PHOTOPERIODIC RESPONSE

Obligatory Photoperiodic Response — Critical Daylength

Rudbeckia plants, at least those species studied by physiologists, are obligatory (qualitative) long-day plants (Table 4). At normal temperature (20 to 22°C) they do not start flowering (strictly speaking, this relates to annual species) under conditions of 8 to 9 hr SD for a period lasting 7 to 8 months.[14,103]

Under conditions of high (32°C) temperature (and after reaching a certain age)[97] of anaerobiosis (Figure 3)[49,50,55,56,85] or continuous darkness followed by SD,[47,48,61,82] *R. bicolor* may start flowering without having received a single inductive long photoperiod.

Table 4, based on data from Samygin's review,[114] with the addition of later results, illustrates the minimal critical length of day for *Rudbeckia* plants, ranging from 10 to 14.5 hr for various species. When the length of day is shortened below the critical limit, *Rudbeckia* plants do not flower, remaining in a vegetative state and forming a rosette.

The quantitative and qualitative influence of photoperiods of various duration on growth and flowering of a cultivated *R. bicolor* variety was investigated by Austin.[9] Immediately after germination, the plants were grown under 8-, 12-, 16-, or 20-hr photoperiods (Table 5). Under the 8-hr photoperiod they remained as rosettes up to the end of the experiment, 91 days after emergence of seedlings. Under the 12-hr photoperiod, after 85 days half the plants were developing stems, but by the end of the experiment some plants remained as rosettes.

Table 4
CRITICAL DAYLENGTH FOR SOME *RUDBECKIA* SPECIES

Rudbeckia species	Photoperiodic group	Flowering under SD	Critical daylength (hr)	Ref.
R. bicolor	LD	—[a]	—	8,62,64
	LD	—	—	16,63
	LD[b]	—	12—13	115
R. bicolor var. *superba*	LD	—	12	4
	LD[b]	—	12	97
R. bicolor var. Etoile de Kalvedon	LD	—	12	9
	LD	—	12	14
R. hirta	LD	—	14—14.5	5
	LD	—	10—12	6
	LD	—	—	118
	LD		—	16
R. laciniata	LD	—	14	5
	LD	—	—	74
	LD[b]	—	—	112,113
	LD	—	14—14.5	6
R. speciosa	LD	—	—	66,97
R. newmani	LD	—	12	6
R. nitida var. Autumn Sun	LD	—	12	4
R. nitida	LD	—	14—14.5	6

[a] Dash means no flowering under SD conditions; absence of dash indicates no data available.
[b] Photoperiodic response depends upon temperature.

FIGURE 3. The effect of anaerobiosis during darkness under SD on flowering of *R. bicolor* plants. (1) The control plant, under natural atmosphere during darkness (2) plant subjected to anaerobiosis during first 5 hr of dark period; (3) plant subjected to anaerobiosis during last 5 hr of dark period. (From Chailakhyan, M. Kh. and Konstantinova, T. N., *Dokl. Acad. Nauk S.S.S.R.*, 135(6), 1539—1542, 1960. With permission.)

The qualitative effect on growth was manifested by the fact that under exposure to relatively short photoperiods (8 hr) the plants remained in the form of a rosette, whereas relatively longer photoperiods (12 hr and more) gave rise to the elongation of the main stem. The quantitative effect on the growth of plants was shown by the fact that stem elongation started earlier under exposure to a longer photoperiod. It should be noted that the number of nodes

Table 5
THE EFFECT OF PHOTOPERIOD DURATION ON GROWTH AND FLOWERING OF *R. BICOLOR SUPERBA*[9]

Indices of development and growth	Duration of photoperiod (hr)			
	8	12	16	20
No. days from germination to:				
Initiation of stem elongation	Rosette	85(50%)	22.6±0.5	17.0±0.4
Flower bud formation	Rosette	—	40.2±0.3	35.6±0.3
Flowering	Rosette	—	66.9±0.3	58.7±0.3
Stem height (cm) after 91 days from germination	—	—	61.5±1.0	51.1±0.8

Table 6
THE EFFECT OF INDUCTION PERIOD DURATION ON THE DEVELOPMENT OF *R. BICOLOR* PLANTS[115]

Number of inductive cycles (days)	Number of days from beginning of inductive cycles to: Initiation of stem elongation (50%)	Flower bud formation (50%)
0	Rosette	Rosette
4	Rosette	Rosette
6	12	57
8	8	45
12	8	29
24	8	21
Continuous LD	8	21

Note: Plant age at beginning of experiment 130 days; LD, natural 16- to 18-hr day; period of observation, 57 days.

on the main stem was the same under different photoperiods and that photoperiod affected only the elongation of internodes.

The qualitative influence of daylength on flower bud formation was shown by the fact that no flower buds were observed during 91 days under relatively short photoperiods (8 and 12 hr). The quantitative effect was expressed in the fact that under exposure to photoperiods longer than the critical one, flower buds started to form earlier under longer photoperiods. Similar quantitative dependence of earlier flower bud formation under longer photoperiods was shown by Bouillenne[14] using *R. bicolor* plants treated with long photoperiods from the beginning of germination. Under exposure to continuous light, flower buds started to form after 50 days, as compared to 141 days under 13-hr photoperiods. A longer experiment (200 days) showed that *R. bicolor* plants could form flower buds under a 12-hr photoperiod (after 183 days).

The minimum photoperiod under which the plants were able to survive was 4 hours; under 2-hr photoperiods they died.[14]

Number of Inductive Cycles.

The number of inductive cycles needed for flowering depends on a number of factors, particularly the length of photoperiod (some researchers grew plants under conditions of a long day of 14 to 18 hr or under naturally changing long days), plant age, the criterion for the transition to flowering used (emergence of visible flower buds or microscopic investigation of flower primordia), and the duration of observations (Tables 3,6,7).

Table 7
THE EFFECT OF VARIOUS LD PERIODS ON THE GROWTH AND DEVELOPMENT OF *R. BICOLOR* PLANTS OF DIFFERENT AGE[82]

	Plant age 5 months			Plant age 7 months		
		No. days from beginning of exposure to			No. days from beginning of exposure to	
Period of exposure to LD (days)	Plant height on 47th day after exposure (cm)	Initiation of stem elongation	Flowering	Plant height on 35th day after exposure (cm)	Initiation of stem elongation	Flowering
0(SD)	0	Rosette	Rosette	0	Rosette	Rosette
3	—	—	—	9	6	45
4	6	7	54	—	—	—
5	—	—	—	28	6	42
6	11	7	53	—	—	—
7	—	—	—	44	7	39
8	20	8	53	—	—	—
9	—	—	—	70	6	31
10	30	8	56	—	—	—
11	—	—	—	81	6	36
12	37	10	49	—	—	—
13	—	—	—	105	8	26
14	70	7	49	—	—	—
15	—	—	—	95	8	28
16	60	9	51	—	—	—
17	—	—	—	101	8	29
18	74	10	48	—	—	—
20	82	9	45	—	—	—
Continuous	125	9	35	109	8	28

For *R. bicolor* plants, 1 to 3 LD are sufficient for the raising of the young leaves (a reversible response), 3 to 6 and more LD are enough for stem elongation, flower bud formation, and flowering, but flower bud formation and flowering in this event occur much later than in plants exposed to continuous LD. *R. bicolor* plants 5 to 7 months old require 13 or 14 inductive cycles to reach flower bud formation simultaneously with plants which have been under *LD all* the time.

Interruption of Light or Darkness — Alternation of Short Light and Dark Periods. Interruption of a light period in *R. hirta* and *R. laciniata* plants results in a more rapid transition to flowering.[6] If the plants are exposed to 10 hr of continuous light daily, they remain as rosettes and no transition to flowering is observed. When they are exposed to 5 hr of light, followed by 4 hr of darkness and again 5 hr of light, they have the same stem length and start to flower as they do under long (14 to 14.5 hr) photoperiods.

In *R. bicolor* plants, growing under 6-hr days (after 35 days), interruption of a 16-hr dark period in its middle by 2 hr of light from electric lamps led to the formation of a long stem[57,60] and flower buds.[57] The content of gibberellins and auxins in the leaves increased as compared to plants grown under SD conditions without interruption of the dark periods.[57]

Alternation of 6 hr light and dark periods so that the total duration of light was 12 hr within each 24 hr cycle and was equal to the critical length of day accelerated flower emergence in *R. bicolor* plants by 5 days. Alternation of light and dark periods (in the ratio of 1 to 1) from 5 sec to 1 hr, the total duration of light and darkness per day again constituting 12 hours, accelerated flowering to the same extent as continuous light.[62,63]

Table 8
THE EFFECT OF CONTINUOUS DARKNESS FOLLOWED BY SD ON THE GROWTH AND DEVELOPMENT OF *R. BICOLOR* PLANTS OF DIFFERENT AGE[82]

	Plant age 5 months			Plant age 7 months		
		No. days from beginning of exposure to			No. days from beginning of exposure to	
Period of continuous darkness (days)	Plant height on 58th day after exposure (cm)	Initiation stem elongation	Flowering	Plant height on 56th day after exposure (cm)	Initation of stem elongation	Flowering
0(SD)	0	Rosette	No	0	Rosette	No
2	0	Rosette	No	1	5	No
4	0	Rosette	No	2	5	No
6	3	9	No	3	7	No
8	3	8	No	14	8	53
10	3	10	No	15	7	50
12	2	10	No	20	7	36
14	6	11	71	24	6	42
16	6	11	70	17	6	46
18	6	10	64	—	—	—
20	8	10	65	—	—	—

In the experiments of Garner and Allard, in which light and dark periods alternated (in the ratio of 1:2) in such a way that the total sum of light and darkness was 8 and 16 hr, respectively, *R. bicolor* plants flowered under all short cycles, but flowering was progressively delayed with increase of the length of the cycles.[63] In the experiments of Chailakhyan and Rupcheva,[60] intermittent light (the ratio of light to darkness being 1:4.5) given for 11 hr, in addition to 6 hr of continuous light, exerted the same effect on *R. bicolor* plants as continuous 17 hr of light daily. Development in this case was more rapid in those treatments where the dark periods were shorter, with corresponding decrease of the light periods.

When the light periods were 2 times as long as the dark periods (16 hr of light and 8 hr of darkness in a 24-hr cycle), *R. bicolor* plants formed flower buds as rapidly as under continuous illumination.[63]

Continuous Darkness

Stem elongation and formation of flower primordia in large rosette plants of *R. bicolor* could be obtained by long exposures (23 to 31 days) to continuous darkness.[18,61] In later experiments, full flowering was obtained even when the exposure to continuous darkness was reduced to 13 days, by successive exposure of the plants to short days, then to continuous darkness, and again to short days.[47]

The use of a wider range of treatments made it possible to characterize the inductive action of continuous darkness (Table 8) and of a long day (Table 7) in a quantitative manner. Induction by LD causes coneflower plants to form tall stems (100 to 120 cm) and to start flowering rapidly (Figure 4). After staying in continuous darkness, the plants develop dwarf stems (10 to 20 cm) and begin to flower with a delay of 15 to 20 days as compared to exposure to the same number of long days.[82]

The minimum exposure to continuous darkness sufficient for the initiation of stem growth was 2 to 6 days, and for the onset of flowering 8 to 14 days in 5- to 7-month-old plants, depending on plant age.[82]

A 14-day exposure of *R. bicolor* plants to continuous darkness considerably enhances the effect of subsequent induction by LD, the plants starting to flower earlier and having taller stems.[48]

The effect of continuous darkness may be explained if we assume that two factors are responsible for the control of flowering, namely age and environment. Elimination of one of them, i.e., of the photoperiodic effect, leads to the predominant influence of age control over the realization of the genetic program of ontogenesis. The potential readiness of adult plants for flowering and transition to a ripeness to flower state, which is suppressed by noninductive SD conditions, comes into play here.

Spectral Dependence and Threshold Light Intensity

Under SD conditions, the quality of light was found to exert no effect upon the development of *R. bicolor* plants.[86] Under LD conditions, there was a favorable influence of green light on stem elongation and flowering. The first visible and irreversible sign of transition to reproductive development in *Rudbeckia* appears to be stem growth. The most rapid rates of stem formation and elongation were found under exposure to green light. Stem elongation under exposure to white, red, and blue illumination was somewhat slower. The plants also formed flower buds most early under exposure to green light, 2 weeks earlier than under blue light. Plants which were exposed to red and white illumination occupied the intermediary position.[86]

Preliminary culture (2,5 months, SD) of *Rudbeckia* plants under differential spectral illumination exerted no considerable influence upon the rate of their subsequent development under conditions of LD provided either by white and colored light.[86]

The minimum intensity of supplementary illumination, for providing a 20-hr light period in winter and causing stem formation of *R. bicolor superba* plants was between 4 to 20 lx.[8] When the light intensity was lower, the plants did not form stems, whereas under exposure to higher light intensity they formed stems and flower buds and flowered, with flower bud formation starting under a supplementary light intensity of 80 to 600 lx, after 12 weeks, 3 weeks earlier than under supplementary illumination intensity of 20 lx.

The Role of Leaves and Roots

In LDP with the rosette type of development under SD conditions (as in *R. bicolor* plants), the formation of flower stalks and flowers results from the interaction of the three plant organs leaf, root, and shoot apex. Investigations by Chailakhyan[22] showed that exposure of a leaf or even only a small part of it (as little as 1/8 of the leaf blade), to LD was sufficient for the plant to form a stem and flowers.

Defoliation of plants exposed to LD prevented stem formation. After defoliation was terminated and leaves with a very small area had developed, plants exposed to LD formed stems, flower buds, and flowers on dwarf stalks.[28,61]

The role of roots in the growth of stems and leaves as well as in the formation of flower organs was studied in experiments with derooted *R. bicolor* plants. If rosette coneflower plants with ample leaves were exposed to SD, then completely derooted and transferred to LD conditions with any newly appearing roots continuously removed, the plants did not flower at all, since while formation of flower stems started, their growth soon stopped.[19,30,38] Thus, the ability to form flower organs in plants deprived of roots was found to be dependent on the presence of stems. *Rudbeckia* rosette plants without roots were unable to form stems and subsequently to flower.

However, rosette plants without roots are capable to perceive photoperiodic induction by LD. Upon transfer to SD conditions and after regeneration of roots with plants begin to develop stems and form flower organs with only a slight delay as compared to plants which had been held under inductive LD conditions in the presence of roots.[30,32,41] Since the flowering of rosette types of LDP, to which *R. bicolor* belongs, consists of two phases, the phase of stem formation and that of flower formation [31] the phase of stem formation must always precede that of flower formation. Thus, the presence of roots, with their specific

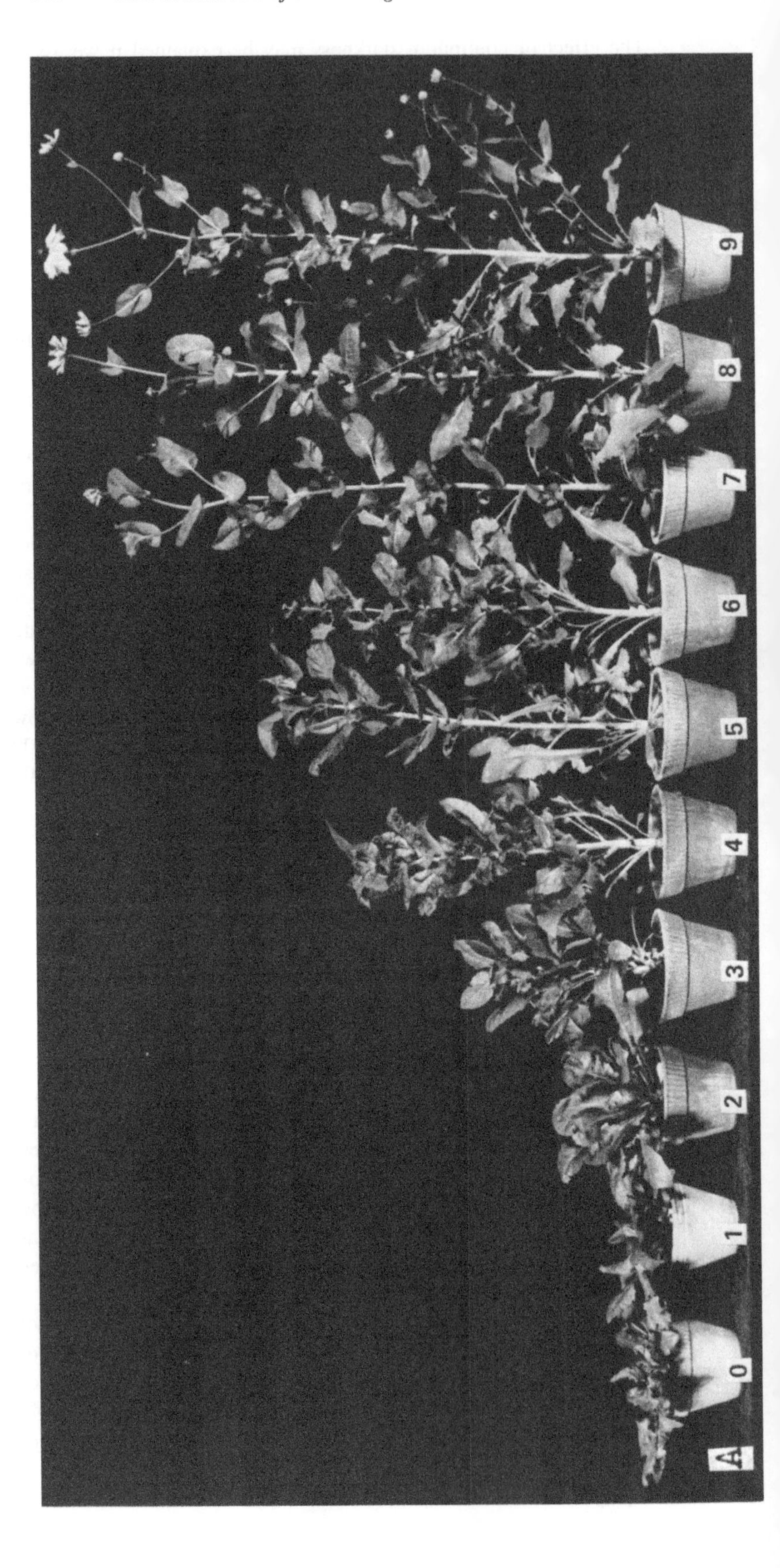
A
0
1
2
3
4
5
6
7
8
9

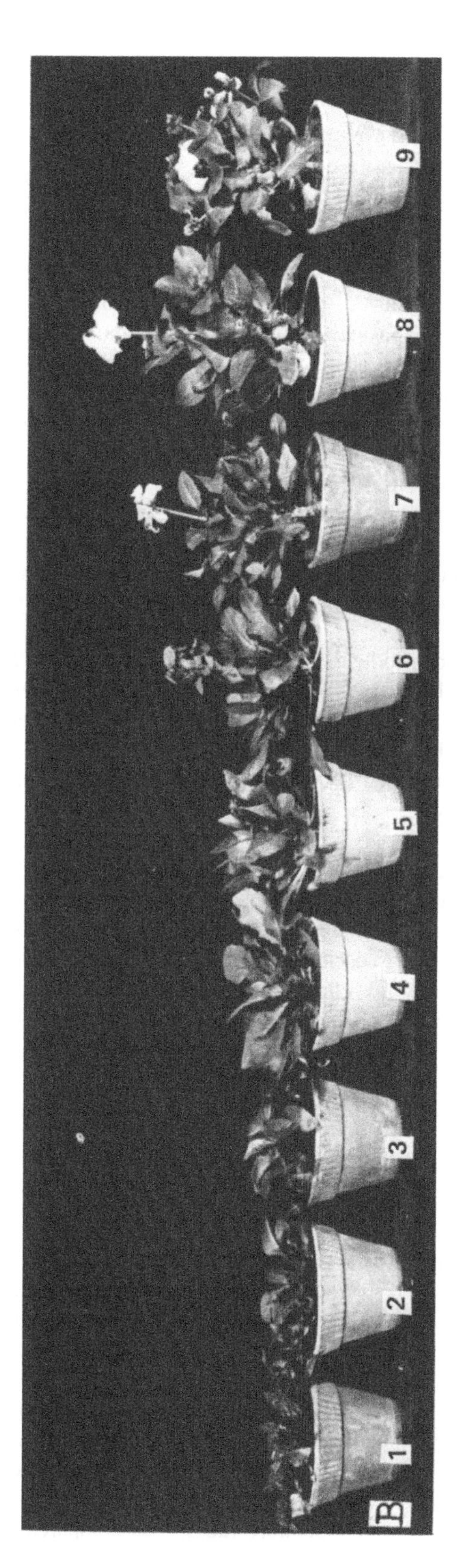

FIGURE 4B

metabolic pattern, directly determines in coneflower the completion of the first phase of flowering — i.e., stem formation — and thereby is reflected in the flowering process as a whole.[32]

Photoperiodic Inhibition

The effect of photoperiod to shorten *Rudbeckia* stems was called, by the American plant physiologist A. E. Murneek,[96] photoperiodic inhibition. Murneek regarded the influence of short photoperiods of less than 12 hr in which *R. bicolor* plants remain as rosette and do not form flowers as the most prominent manifestation of photoperiodic inhibition. Reproductive processes may be induced (photoperiodic induction) to a different degree by LD depending on the time of exposure. Flower organs continue to develop and can normally function when the plants are transferred to SD conditions. In Murneek's opinion, stem elongation cannot be induced but occurs and is maintained only under long photoperiods, since it rapidly ceased when the plants are transferred from long to short photoperiodic conditions. The effects of photoperiodic induction and photoperiodic inhibition may vary depending on the combinations of long and short days. This provides a means for changing the size and shape of vegetative and reproductive organs in the plants.[97]

Murneek was the first to show that in *Rudbeckia* "the mechanism of photoperiodism is not trigger-like in its action that possibly a certain substance or substances are produced gradually, the quantitative accumulation of which results eventually in the development of reproductive organs."[96]

Another aspect of photoperiodic inhibition, the inhibiting effect of SD leaves on flowering, was investigated by Chailakhyan.[17] When in *R. bicolor* plants kept under SD conditions and growing as rosettes, the main bud and 2 to 6 large leaves were retained, with one of the leaves being exposed to LD while the other(s) were kept under SD, the inhibitory effect of the SD leaves was insignificant: stem formation was delayed by only 6 days as compared to plants without SD leaves.

Effects of Anaerobiosis.

Under completely anaerobic conditions (that is, in pure nitrogen atmosphere) *R. bicolor* plants die rapidly.[50] Under partially anaerobic conditions (0.6 to 0.8% oxygen), plants growing under 10-hr SD were accelerated in their development. If plants were exposed to anaerobiosis during the first 5 to 8 hr of the dark period, they formed only stems, but when grown under anaerobic conditions during the last 5 to 8 hr of darkness they produced flowers (Figure 3) on dwarf stems, while control plants grown in normal air remained in the rosette stage.[49,50]

The effect of anaerobiosis during the last 7 hr of a 17-hr day retarded the flowering of *R. bicolor*.[4,50,85]

EFFECTS OF TEMPERATURE

Studies on temperature effects on the transition to flowering were carried out with *R. laciniata*,[111-113] *R. bicolor*,[115] cvs of *R. bicolor* var. *superba*, and *R. speciosa newmanii*.[97] Temperature was found to modify or change the response of some *Rudbeckia* species to daylength, this effect being species specific and apparently connected with the geographic area of their origin.

Roberts and Struckmeyer found[112,113] that *R. laciniata* plants under 9.5- to 10-hr SD started to flower at a mean daily temperature of 12.7°C, forming tall stems, and did not flower at 17 to 18°C or 21 to 24°C, remaining as rosettes. Under LD conditions, the plants flowered in all three temperature ranges.

On the other hand, a higher temperature (28.3°C as compared to 25°C) contributed to earlier flowering (9 to 14 days, earlier depending upon the duration of the photoperiod) and to stem elongation in *R. bicolor* plants grown under long (14 to 24 hr) photoperiods from the emergence of the seedlings on. Under 6-, 9-, or 12-hr SD the plants did not start flowering either at 25 or at 28.3°C.[115] In older plants (8 months) of *R. bicolor* var. *superba* exposure to high daytime temperatures (32.2 to 37.7°C) and correspondingly higher night temperatures in summer (when growing in a greenhouse without cooling) resulted in the emergence of normal-sized and completely developed flowers on dwarf stems under conditions of 7-hr SD. Under similar conditions 8-month-old plants of *R. speciosa* did not form flowers.[97]

Altogether, modifying the night temperature (low for *R. laciniata,*[113] high for *R. bicolor* var. *superba*[97]) appears to be more effective in changing the influence of photoperiod on the plants then changing the day temperature.

EFFECTS OF GROWTH REGULATORS

The pattern of photoperiodic response of *Rudbeckia* plants (formation of rosettes under SD conditions, formation of stems, as a first stage of transition to flowering, under LD conditions) made it possible to use them as materials for investigating the influence of various growth regulators (both stimulators and inhibitors of flowering). Of the greatest interest in this regard were annual species, such as *R. bicolor,* being (generally) the most suitable ones for experimental purposes.

Rudbeckia plants, mainly *R. bicolor,* were used to study the effect of a variety of chemical compounds on stem growth and flowering, including phytohormones (gibberellins, auxins, cytokinis, abscisic acid) natural and synthetic growth retardants and inhibitors, nucleic metabolites, vitamins, phenolic compounds, organic acids, and antibiotics.

The only compounds which were capable to induce stem growth and formation of flowers in *R. bicolor* plants under SD conditions were gibberellins.[35,42,58]

However, when given in addition to induction by LD, treatment of *R. bicolor* plants with a number of physiologically active substances may exert either a stimulating effect on stem growth and flowering (some auxins,[3-7,40,42,58] polyphenols and phenolic acids,[40,42,58] vitamins,[24,37,42,58] and nucleic metabolites[42]), or an inhibiting one, e.g., retardants,[44,46,75,78,104] morphactins,[45,46,78] and maleic hydrazide.[78]

Gibberellins

The response of *R. bicolor* plants to gibberellins under SD conditions is specific[42] and may be employed as a bioassay to identify gibberellins and GA-like substance (GLS) in microbial exudates[39,51,88,89] and in tissues of higher plants.[29,52-56,59,90,91] Apart from *R. bicolor,* rosette plants of *R. speciosa*[67] and *R. grandiflora*[15] were used as test material to determine GLS in plant tissues.

Among *R. hirta,*[116] *R. speciosa,*[99] and *R. bicolor,*[16,26] the latter proved to be most sensitive to GA_3.

Not all gibberellins were active to the same extent on *R. bicolor* (Table 9). Among the seven gibberellins studied, GA_3 and GA_1 were the most active and GA_8[102] was the least active one in causing stem elongation and flowering.

The most effective method of treatment with GA is direct administration to the plant (spraying the leaf surface or applying drops of aqueous solutions containing surfactant to the shoot apex).[26,27,43] Daily application fo 24 μg of GA_3 per apex by a drop of a solution of 50 mg/ℓ for 12 days under SD conditions had the same effect on stem growth and flowering in 4-month old plants as 12 days of induction by LD (Figure 1).[78] Feeding GA through roots is less effective.[43]

Stem growth and flowering in *R. bicolor* plants induced by GA_3 under SD conditions were inhibited by simultaneous action both of some natural growth retardants and inhibitors

Table 9
THE EFFECT OF VARIOUS GIBBERELLINS ON STEM ELONGATION AND FLOWERING OF *R. BICOLOR* UNDER SD CONDITIONS

Gibberellins	Stem elongation[a]	Flowering[a]
A_3	+ + + 16,26,102[b]	+ + + 16,26,102
A_1	+ + + 102	+ + + 102
A_7	+ + 102	+ + 102
A_4	+ + 102	+ 102
A_5	− 102	+ 102
A_8	− 102	− 102
A_9	− 102	+ 102

[a] Activity of the gibberellins: + + + highest
+ + mean
+ lowest
− inactive

[b] Figures designate numbers of references to literature

(abscisic and *p*-coumaric acid, coumarin, and phloridzin) and some synthetic ones (chlormequat, morphactin, and MH).[45,46,77,78]

Since, as mentioned above, the action of GA in imitating the LD effect is specific, it was expedient to study the effect of other phytohormones and growth regulators simultaneously with induction by a LD, to see whether they accelerate or inhibit stem formation and transition to flowering.

Auxins

Under inductive LD conditions IAA (100 mg/ℓ) accelerates stem growth and transition to flowering.[31] Treatment with 2,4-D (100 mg/ℓ under SD conditions causes modifications of stem formation and of the shape and size of apices.[58]

Cytokinins

Kinetin (1 g/ℓ) inhibits stem growth and flowering.[31]

Abscisic Acid

Abscisic acid (1 to 100 mg/ℓ) exerts no effect on stem growth and flowering in *R. bicolor* plants induced by LD.[45,46,78]

Growth Retardants

The retardants chlormequat, (1 to 10 g/ℓ)[44,46,75,77,78,104] and chlorophonium chloride (20 to 60 g/ℓ)[104] considerably (up to 80%) inhibit stem elongation and to a lesser extent, but rather effectively (by 10 to 15 days), the onset of flowering. The action of these retardants is similar to that of SD.

Growth Inhibitors

At high concentrations, synthetic growth inhibitors — MH (1 to 5 g/ℓ) and morphactin IT 3456 (1 to 10 mg/ℓ) inhibited the growth of the main stem, while growth inhibitors of natural origin (*p*-coumaric acid, 100 to 200 mg/ℓ; coumarin, 0.1 to 1 g/ℓ; and phloridzin, 0.1 to 1 g/ℓ) had no effect on stem growth and flowering.[78]

CHEMICAL CHANGES DURING TRANSITION TO FLOWERING

Hormonal Changes

The leaves of *R. bicolor* and *R. speciosa* plants were found to contain gibberellin-, auxin-, and abscisic acid-like substances. *Rudbeckia* plants grown on LD are suitable for studying correlations between the content of some growth regulators and stem elongation, as the first stage of transition from the vegetative state to flowering. Various external conditions make it possible to grow these plants in the rosette form (SD or treatment with retardants), in a caulescent form (by induction with LD or treatment with GA), or with a dwarfed stem (exposure to continuous darkness).

Gibberellins

Various authors have employed different methods for extraction and determination of gibberellins either in young leaves and shoot apices,[66,67] or in leaves of different levels,[52,78] or in all leaves[78—84,98,100,101,103,104] in several *Rudbeckia* species and varieties, namely, *R. speciosa,*[66,67] *R. bicolor,*[52,78,81—84,98] and *R. bicolor* cv Etoile de Kelvedon.[100,101,103,104] Despite different ages of the plants used, some common features have been found.

In plants having a stem and forming flower buds, the activity of gibberellins in the leaves is higher than in rosette plants.[52] The content of gibberellins in leaves during transition to flowering under LD conditions changes, exhibiting several peaks. The first maximum corresponds to or precedes the onset of stem growth and is related to the initial differentiation of the apical meristem.[66,67,83,93] The second maximum corresponds to the period of intensive elongation prior to the onset of flower bud formation.[66,67,83,101,103] Exposure to one long photoperiod in this case was sufficient for a marked increase in the level of GA (Figure 5). The leaves of *R. bicolor* were found to contain several gibberellins and, along with the increase in the total content of gibberellins after exposure to LD, the activity of various individual gibberellins as well as their ratios changed.[78,81,98,100,101]

In continuous darkness, *R. bicolor* plants showed, prior to differentiation of the apices and stem growth (during the first 3 to 6 days of exposure), a rise in the activity of GA which, unlike that after exposure to LD, decreased toward the onset of flower bud formation.[79,83]

Under SD conditions, leaves of rosette plants were found to contain a low level of GA which, however, increased with age.[78,83,101]

Use of the growth retardants chlormequat and chlorophonium chloride, which strongly inhibit stem growth and flowering, caused the activity of gibberellins in the plants to disappear.[98,104]

Auxins

IAA in *R. bicolor* leaves predominantly occurs in the free form.[83] The rosette form of *R. bicolor* plants is characterized by a high content of free IAA,[83,84] which sharply declines after exposure to 1 to 3 LD or to 3 days of continuous darkness and again increases at the onset and the following period of intensive stem growth (11 to 28 LD or 10 to 20 days of continuous darkness).[83,84]

Abscisic Acid

ABA is present in *R. bicolor* leaves both in free[76,77,83,84,103] and bound[84] forms, the bulk being in the free form.

Under SD conditions, there is a steady high level of free and bound ABA in the plants. Under the influence of long photoperiods the activity of ABA falls before the start of the differentiation of the apices and stem growth (1 to 6 LD in *R. bicolor*[83,84] or 15 LD in *R. bicolor* cv Etoile de Kelvedon[103]). After this, the activity of ABA increases, and a second decrease in its level is observed at the period of intensive stem growth prior to flower bud formation.[76]

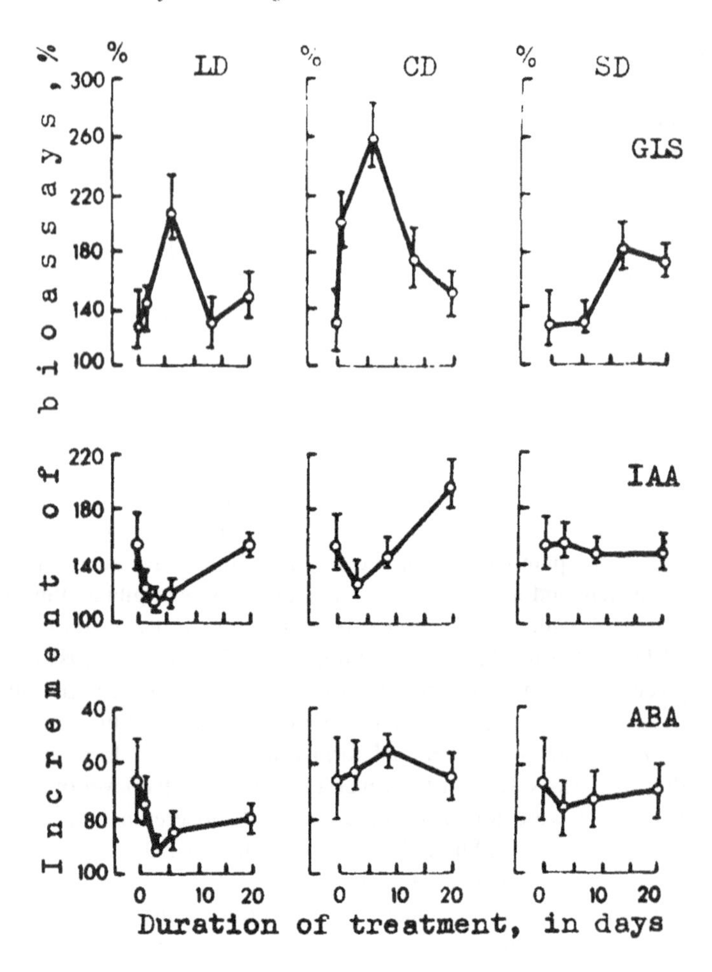

FIGURE 5. Comparative effects of LD, continuous darkness (CD), and SD on changes in the activities of gibberellin (GLS), auxin (IA), and abscisic acid (ABA)-like substances in leaves of *R. bicolor* plants. Results of biotests on lettuce seedlings (GLS) and segments of wheat coleoptiles (IA and ABA). Vertical bars designate the product of standard error in the Student's *t* test under 5% level of significant difference. (From Kochankov, V. G., Muzafarov, B. M., and Chailakhyan, M. Kh., in *Metabolism and Mode of Action of Phytohormones,* Salyaev, R. K. and Gamburg, K. Z., Eds., Siberian Institute of Plant Physiology and Biochemistry, Irkutsk, 67—70, 1979. With permission.)

Under conditions of continuous darkness the activity of ABA in *R. bicolor* remained at a high level.[83]

Thus, the activities of free gibberellins, IAA, and ABA in *R. bicolor* leaves exposed to LD, SD, and continuous darkness correlated with the different morphogenetic patterns observed under these conditions. Absence of flower stalks or their slow growth depend not only upon a low content of gibberellins and a high content of IAA, but also upon the high level of ABA activity.[76,78,83] Apart from this, the changes in the content of phytohormones in *R. bicolor* show that "the triggering mechanism" for transition from the rosette state to stem growth under LD is similar to that under continuous darkness. The further changes in the content of these phytohormones, which are responsible for the rate of flower stem elongation and the onset of flowering, are different after treatment with LD or continuous darkness.[80]

Metabolic Changes

The first attempts at studying the metabolism of *Rudbeckia* plants during transition from vegetative growth to flowering were undertaken soon after the discovery of the photoperiodic phenomenon. Thus, a study carried out by Garner et al.[64] dealt with changes in the concentration of H-ions in the cell sap of *R. bicolor.* These authors assumed that pH values indicated metabolic activity and that major changes in the concentration of H-ions were intimately associated with changes in the development of daylength-induced plants. They established that the acidity of the cell sap of leaves under LD conditions decreased as compared to leaves of rosette plants grown under SD conditions, while during stem formation the acidity of cell sap in the upper part of the stem increased up to flower bud formation; the highest acidity was found in flower buds.

Interest in studies of particular metabolic processes in *Rudbeckia* plants has varied depending on which class of substances was thought more important in relation to flowering. Following the discovery of the perceptive role of the leaves in the photoperiodic response, metabolic changes have been studied mainly in the leaves. It was found that during transition to flowering of *R. laciniata*[74] and *R. bicolor*[20,21,23,25] plants maintained under LD conditions their leaves contained more carbohydrates (sugars and starch) and less proteins and other nitrogenous compounds than leaves of plants grown on SD. Studies of diurnal changes in the starch content in *R. bicolor* leaves showed that after the short night (in plants kept under LD conditions) a low amount of starch remained, and by the end of a long night (under SD conditions) no starch was found at all.[20]

Under LD conditions, *R. bicolor* leaves showed a decrease in peroxidase[34] activity, an increase in polyphenoloxidase[2] activity, and a lower rate of cyanide-sensitive respiration[1,2,31,33] as compared to leaves of plants kept on SD.

The coupling of oxidation and phosphorylation in the leaves of *R. bicolor* plants increased under light and decreased when the plants were placed in darkness.[3]

The length of day caused changes in the activity of the chloroplasts. The Hill reaction in chloroplasts from leaves of *R. bicolor* plants grown under LD conditions was higher than in chloroplasts isolated from leaves of SD grown plants;[10] this indicates that the flowering of *R. bicolor* depends to a great extent upon chloroplast activities.[11]

The fact that photoperiodic induction causes changes in the apical meristems of *Rudbeckia* has drawn an increasing attention to metabolic changes in the apices.

A very rapid increase (after 10 to 18 hr) of DNA synthesis in the shoot apices of *R. bicolor* cv Etoile de Kelvedon was found by Jacqmard under conditions of LD[69] or treatment with GA.[71]

Shoot apices of *Rudbeckia* plants in the first stage of reproductive development (evocation phase) were found, as early as 2 days after the beginning of exposure to LD, to contain a new protein that had been absent in the apices of the vegetative plants. At later stages of reproductive morphogenesis more new proteins appeared.[94,95]

CONCLUSION

The rosette form of *Rudbeckia* plants growing under SD conditions makes them a suitable tool for studying the mechanism of the transition from the vegetative state to flowering. Studies on the physiology of flowering in *Rudbeckia* formed the basis for putting forward the concept that flowering proceeds in two phases; (1) the phase of flower stem formation, and (2) the phase of flower formation.

The first phase of flowering in *Rudbeckia,* flower stem formation, is critical and occurs only under LD conditions. Its regulation depends mainly upon the ratio between phytohormones, both promoters and inhibitors synthesized in the leaves and transported to the shoot apex. Flower stem growth and flower formation in *Rudbeckia* are distinct processes, occurring successively but closely interrelated. The data obtained on the growth of flower stems in *Rudbeckia* plants may be of great theoretical and practical importance for the regulation of flower formation in other cultivated plants. The possibility to delay the transition to flowering in many rosette, shrub-like, or low-growing vegetable and other crops or, on the contrary, promotion of their transition to flowering can be important for enhancing their productivity or improving the quality of their yield, as well as for solving the "rosette problem" in some cultivated plants.

REFERENCES

1. **Aksyonova, N. P.,** Effect of day length on the activity of glycolysis and on the cycle of tricarboxylic acids, *Fiziol.* Rast. 8, 3, 338—344, 1961 (in Russian).
2. **Aksyonova, N. P.,** Effect of day length on the activity of oxidases in plants, *Fiziol. Rast.*, 10, 2, 166—175, 1963 (in Russian).
3. **Aksyonova, N. P., Konstantinova, T. N., and Nikitina, A. A.,** Effect of light, darkness and length of day on the conjugation of oxidation and phosphorylation in the leaves of long- and short-day plants, *Fiziol. Rast.*, 15, 5, 859—864, 1968 (in Russian).
4. **Allard, H. A.,** Daylight factor in flowering, *U.S. Dep. Agric. Yearb. Agric.*, 306—309, 1926.
5. **Allard, H. A.,** Length of day in relation to the natural and artificial distribution of plants, Ecology, 13, 3, 221—234, 1932.
6. **Allard, H. A. and Garner, W. W.,** Further observations of the response of various species of plants to length of day. *U.S. Dep. Agric. Tech. Bull.*, 727, 1—64, 1940.
7. **Astvatsatryan, Z.,** *Rudbeckia, Tsvetovodstvo,* 4, 7—8, 1968 (in Russian).
8. **Austin, J. P.,** Minimum intensity of artifical illumination effective in supplementing the normal photoperiod, *Pap. Mich. Acad. Sci., Arts Lett.*, 22, 25—26, 1936.
9. **Austin, J. P.,** The influence of the length of the photoperiod on the vegetative and reproductive development of *Rudbeckia superba, Delphinium ajacis, Cosmos sulphureus,* and *Impatiens balsamine, Am. J. Bot.*, 28, 3, 244—250, 1941.
10. **Bavrina, T. V.,** The effect of day length on the photochemical activity of chloroplasts in the leaves of plants of various photoperiodical groups, *Dokl. Akad., Nauk S.S.S.R.*, 192, 2, 451—454, 1970 (in Russian).
11. **Bavrina, T. V., Aksyonova, N. P., and Konstantinova, T. N.,** On the participation of photosynthesis in photoperiodism, *Fiziol. Rast.*, 163, 381—391, 1969.
12. **Bernier, G., Bronchart, R., and Jacqmard, A.,** Action of gibberellic acid on the mitotic activity of the different zones of shoot apex of *Rudbeckia bicolor* and *Perilla nankinensis, Planta,* 61, 3, 236—244, 1964.
13. **Bernier, G., Bronchard, R., Jacqmard, A., and, Sylvestre, G.,** Acide gibbèrellique et morphogénèse cauilinaire, *Bull. Soc. Roy. Bot. Belg.*, 100 1, 51—71, 1967.
14. **Bouillenne, R.,** Recherche de la photopèriode critique chez diverses espèees de jours-longs et de jours-courts cultivées milieu conditionné, *Bull. Acad. Roy. Belg., Classe Soc.*, 49, 4, 337—345, 1963.
15. **Budagyan, E. G., Lozhnikova, V. N., and Goldin, M. I.,** Some data on the presence of gibberellin-like substances in vital diseases of the jaundice type, *Biol. Zhu. Armen.*, XX, 3, 56—60, 1967 (in Russian).
16. **Bünsow, R. and Harder, R.,** Blütenbildung von *Adonis* und *Rudbeckia* durch Gibberellin, *Naturwissenschaften,* 41, 16, 453—454, 1957.
17. **Chailakhyan, M. Kh.,** The nature of substances accelerating and inhibiting the flowering of plants and Klebs theory. *Uzb. Sovremennoi Biol.*, 26 (1), 515—530, 1948 (in Russian).
18. **Chailakhyan, M. Kh.,** Photoperiodism and the flowering capacity in plants, *Dokl. Akad. Nauk S.S.S.R.*, 59, 5, 1003—1006, 1948 (in Russian).
19. **Chailakhyan, M. Kh.,** On the role of roots in the photoperiodic response of plants, *Dokl. Akad. Nauk S.S.S.R.*, 72, 1, 201—204, 1950 (in Russian).
20. **Chailakhyan, M. Kh.,** On the relation of photoperiodism to the basic physiological processes in plants, *Dokl. Akad. Nauk Arm. S.S.R.*, 16(4), 109—115, 1953 (in Russian).
21. **Chailakhyan, M. Kh.,** The effect of day length on the character of carbohydrate-protein metabolism in plant leaves, *Dokl. Akad. Nauk S.S.S.R.*, 100 (2), 373—376, 1955 (in russian).
22. **Chailakhyan, M. Kh.,** The Integrity of Organism in the Plant World, Publication of Academy of Science, Armenian S.S.R., Yerevan, 1955, 1—57 (in Russian).
23. **Chailakhyan, M. Kh.,** On localization of starch and protein in plant leaves in relation to differentiated photoperiodic exposure, *Dokl. Akad. Nauk Arm.*, S.S.R., 21 (1), 37—42, 1955 (in Russian).
24. **Chailakhyan, M. Kh.,** The effect of vitamins on the growth and development of plants, *Dokl. Akad. Nauk S.S.S.R.*, 111 (4), 894—897, 1956 (in Russian).

25. **Chailakhyan, M. Kh.,** Photoperiodism and basic physiological processes in plants, *Zh. Obshch. Biol.*, 17 (2), 121—141, 1956 (in Russian).
26. **Chailakhyan, M. Kh.,** The effect of gibberellins on the growth and flowering of plants, *Dokl. Akad. Nauk S.S.S.R.*, 117 (6), 1077—1080, 1957 (in Russian).
27. **Chailakhyan, M. Kh.,** Hormonal factors in flowering of plants, *Fiziol. Rast.*, 5 (6), 541—560, 1958 (in Russian).
28. **Chailakhyan, M. Kh.,** *Basic Regularities in Ontogenesis of Higher Plants*, U.S.S.R. Academy of Science Publ., Moscow, 1958, 1—78 (in Russian).
29. **Chailakhyan, M. Kh.,** The effect of gibberellins and substances of nucleic nature on the growth and flowering of plants, *Izv. Akad. Nauk Arm. S.S.R.*, 12, 11, 3—13, 1959 (in Russian).
30. **Chailakhyan, M. Kh.,** Photoperiodic sensitivity of plants deprived of stems and roots, *Dokl. Akad. Nauk S.S.S.R.*, 135 (1), 213—216, 1960 (in Russian).
31. **Chailakhyan, M. Kh.,** *Factors of generative development of plants. The XXV Timiryazev Lecture*, Nauka Publ., Moscow, 1964, 1—57 (in Russian).
32. **Chailakhyan, M. Kh.,** The integrity and differentiated models of plants flowering, in *Biology of Plant Development*, Chailakhyan, M. Kh., Ed., Nauka Publ., Moscow, 1975, 24—47. (in Russian).
33. **Chailakhyan, M. Kh. and Aksyonova, N. P.,** On the relation between photoperiodism and respiration of plants, *Fiziol. Rast.*, 6 (6), 699—708, 1959 (in Russian).
34. **Chailakhyan, M. Kh. and Boyarkin, A. N.,** The effect of day length on the activity of oxidative enzymes in plants, *Dokl. Akad. Nauk S.S.S.R.*, 105 (3), 592—595, 1955 (in Russian).
35. **Chailakhyan, M. Kh., Ilgach, G. V., Yegorova, T. A., and Yanina, L. I.,** The action of regulatory substances on the growth and forming of *Rudbeckia*, *Dokl. Akad. Nauk Arm. S.S.R.*, 51 (5), 298—303, 1970 (in Russian).
36. **Chailakhyan, M. Kh., Kakhidze, N. T., Milyaeva, E. L., Gukasyan, I. A., and Yanina, L. I.,** The effect of day length and gibberellins on the growth rate, flowering and differentiation of apices in *Rudbeckia bicolor*, *Fiziol. Rast.*, 16 (3), 392—399, 1969 (in Russian).
37. **Chailakhyan, M. Kh. and Khlopenkova, L. P.,** The effect of auxins and vitamins on the growth and development of gibberellin-treated plants, *Dokl. Akad. Nauk S.S.S.R.*, 129 (2), 454—457, 1959 (in Russian).
38. **Chailakhyan, M. Kh. and Khlopenkova, L. P.,** On the factors of stem growth in rosette plants of long-day species, *Dokl. Akad. Nauk S.S.S.R.*, 135 (2), 482—485, 1960 (in Russian).
39. **Chailakhyan, M. Kh. and Khlopenkova, L. P.,** Comparative data on the physiological activity of various gibberellin preparations, *Izv. Akad. Nauk S.S.S.R., Biol. Ser.*, 1, 87—92, 1961 (in Russian).
40. **Chailakhyan, M. Kh. and Khlopenkova, L. P.,** The effect of growth preparations and derivatives of nucleic metabolism on the growth and flowering of photoperiodically induced plants, *Dokl. Akad. Nauk S. S. S. R.*, 141 (6), 1497—1500, 1961 (in Russian).
41. **Chailakhyan, M. Kh. and Khlopenkova, L. P.,** The photoperiodic sensitivity of rosette leaves in *Rudbeckia* plants deprived of roots, *Dokl. Akad. Nauk S.S.S.R.*, 189 (6), 1400—1403, 1969 (in Russian).
42. **Chailakhyan, M. Kh., Khlopenkova, L. P., and Lozhnikova, V. N.,** On the specificity of responses of rosette plants to the action of gibberellin, *Dokl. Akad. Nauk Arm. S.S.R.*, 38 (1), 45—51, 1964 (in Russian).
43. **Chailakhyan, M. Kh. and Kochankov, V. G.,** The effect of gibberellin on the growth and flowering of ornamental plants, *Izv. Akad. Nauk S.S.S.R. Biol. Ser.*, 1, 3—12, 1961 (in Russian).
44. **Chailakhyan, M. Kh. and Kochankov, V. G.,** Effect of retardants on the growth and flowering of plants, *Fiziol. Rast.*, 14 (5), 773—784, 1967 (in Russian).
45. **Chailakhyan, M. Kh. and Kochankov, V. G.,** Growth and flowering plants in relation to the influence of retardants and inhibitors, in Stimulants of Organisms' Growth, Proc. Conf. Baltic Republics on the Problems of Stimulation of Plants, Animals and Microorganisms, Vilnius, 1969, 76—77 (in Russian).
46. **Chailakhyan, M. Kh. and Kochankov, V. G.,** The effect of CCC retardant, morphactin and abscisin on the growth and flowering of *Rudbeckia* plants. *Dokl. Akad. Nauk S.S.S.R.*, 188 (2), 477—480, 1969 (in Russian).
47. **Chailakhyan, M. Kh., Kochankov, V. G., and Muzafarov, B. M.,** Induction of stem elongation and flowering of *Rudbeckia* rosette plants by means of continuous darkness, *Dokl. Akad. Nauk S.S.S.R.*, 231 (6), 1497—1501, 1976 (in Russian).
48. **Chailakhyan, M. Kh., Kochankov, V. G., and Muzafarov, B. M.,** The effect of continuous darkness on bolting and flowering of *Rudbeckia* rosette plants in combination with long-day induction, *Dokl. Akad. Nauk S.S.S.R.*, 234 (1), 252—255, 1977 (in Russian).
49. **Chailakhyan, M. Kh. and Konstantinova, T. N.,** Influence of anaerobiosis on the photoperiodism of plants, *Dokl. Akad. Nauk S.S.S.R.*, 135 (6), 1539-1542, 1960 (in Russian).
50. **Chailakhyan, M. Kh. and Konstantinova, T. N.,** The effect of aeration conditions on the photoperiodic response of plants, *Fiziol. Rast.*, 9 (6), 693—701, 1962 (in Russian).
51. **Chailakhyan, M. Kh., Krasilnikov, N. A., Kuchaeva, A. G., Ivanov, K. I., Khlopenkova, L. P., Aseyeva, I. V., and Kravchenko, B. F.,** Production of gibberellin and determination of its physiological activity in connection with its use in plant growing, *Fiziol. Rast.*, 7 (1), 112—119, 1960 (in Russian).

52. **Chailakhyan, M. Kh. and Lozhnikova, V. N.,** The effect of gibberellin-like substances derived from leaves of various plants on the growth and flowering of coneflower, *Dokl. Akad. Nauk S.S.S.R.*, 128 (6), 1309—1312, 1959 (in Russian).
53. **Chailakhyan, M. Kh. and Lozhnikova, V. N.,** Gibberellin-like substances in higher plants and their influence on growth and flowering, *Fiziol. Rast.*, 7 (5), 521—530, 1960 (in Russian).
54. **Chailakhyan, M. Kh. and Lozhnikova, V. N.,** Gibberellin-like substances and vernalization of plants, *Fiziol. Rast.*, 9 (1), 21—31, 1962 (in Russian).
55. **Chailakhyan, M. Kh., and Lozhnikova, V. N.,** Photoperiodism and dynamics of gibberellins in plants, *Fiziol. Rast.*, 11 (6), 1006—1014, 1964 (in Russian).
56. **Chailakhyan, M. Kh. and Lozhnikova, V. N.,** Photoperiodic action and dynamics of gibberellin-like substances in plants, *Dokl. Akad. Nauk S.S.S.R.*, 157 (2), 482—485, 1964 (in Russian).
57. **Chailakhyan, M. Kh. and Lozhnikova, V. N.,** Response to interruption of darkness by light and gibberellins of plants. *Fiziol. Rast.*, 13 (5), 833—841, 1966 (in Russian).
58. **Chailakhyan, M. Kh., Milyaeva, E. L., and Yanina, L. I.,** Influence of physiologically active substances on the growth, development and differentiation of *Rudbeckia bicolor* apices, *Biol. Zh. Arm.*, 23 (11), 13—23, 1970 (in Russian).
59. **Chailakhyan, M. Kh., Nekrasova, T. V., Khlopenkova, L. P., and Lozhnikova, V. N.,** The role of gibberellins in the processes of photoperiodism, vernalization and stratification of plants, 10 (4), 465—476, 1963 (in Russian).
60. **Chailakhyan, M. Kh. and Rupcheva, I. A.,** Influence of intermittent light on the generative development of plants, *Dokl. Akad. Nauk S.S.S.R.*, 61 (3), 565—568, 1948 (in Russian).
61. **Chailakhyan, M. Kh. and Samygin, G. A.,** On the role of leaves in photoperiodic responses of long-day plants, *Dokl. Akad. Nauk S.S.S.R.*, 59 (1), 187—190, 1948 (in Russian).
62. **Garner, W. W.,** Plant growth by artificial light has possibilities, *U.S. Dep. Agric. Yearb. Agric.*, 436—439, 1931.
63. **Garner, W. W. and Allard, H. A.,** Effect of abnormally long and short alterations of light and darkness on growth and development of plants, *J. Agric. Res.*, 42 (10), 629—651, 1931.
64. **Garner, W. W., Bacon, C. W., and Allard, H. A.,** Photoperiodism in relation to hydrogen-ion concentration of the cell sap and carbohydrate content of the plant, *J. Agric. Res.*, 27 (3), 119—156, 1924.
65. **Gleason, H. A.,** *The New Britton and Brown Illustrated Flora of the Northeastern United States and Adjacent Canada,* Volume 3, Macmillan, New York, 1974, 346—352.
66. **Harada, H. and Nitsch, J. P.,** Changes in endogenous growth substances during flower development, *Plant Physiol.*, 34 (4), 409—415, 1959.
67. **Harada, H. and Nitsch, J. P.,** Extraction d'une substance provoquant la floraison chez *Rudbeckia speciosa* Wend., *Bull. Soc. Bot. France*, 106 (9), 451—454, 1959.
68. **Jacqmard, A.,** Action comparée de la photopériode et al'acide gibbérellique sur méristèeme caulinaire de *Rudbeckia bicolor* Nutt., *Bull. Acad. Roy. Belg., Classe Sci.*, 50 (2), 174—188, 1964.
69. **Jacqmard, A.,** Comparaison des actions de la photopériode et de l'acid gibbérellique sur le méristèeme caulinaire de *Rudbeckia bicolor* Nutt., *Bull. Soc. France Physiol. Veg.*, 11, 165—171, 1965.
70. **Jacqmard, A.,** Etude cinétique de la stimulation de l'activité mitotique dans le bourgeon terminal de *Rudbeckia bicolor* traité par l'acide gibbérellique, *C.R. Acad. Sci. Paris*, 264, 1282—1285, 1967.
71. **Jacqmard, A.,** Early effect of gibberellic acid on mitotic activity and DNA synthesis in the apical bud of *Rudbeckia bicolor*, *Physiol. Veg.*, 6 (4), 409—416, 1968.
72. **Khudzhaniyazova, S.,** Morphobiological Characteristics and Biology of *Rudbeckia* and *Echinaceae* Flowering under Conditions of Tashkent, abstr. thesis for a Master of Biology, Academy of Science of Uzbek S.S.R., Tashkent, 1964 (in Russian).
73. **Khudzhaniyazova, S.,** On the morphology of seedlings of some species in genus *Rudbeckia* L. and *Echinaceae* Moench, in *Introduction and Acclimatization of Plants*, Vol. 14, Rusanov, F. N., Ed., Publ. Academy of Science of Uzbek S.S.R., Tashkent, 1977, 121—129 (in Russian).
74. **Knodel, H.,** Läst sich die klebssche Ansicht über die Abhängigkeit der Blütenbildung von der chemischen Zusammensetzung der Pflanze aufrechterhalten, *Zeitschr. Bot.*, 29, 449—501, 1936.
75. **Kochankov, V. G.,** Effect of various concentrations of CCC retardant on growth and development of annual plants, in *Organogenesis of Higher Plants*, Kazaryan, V. O., Ed., Publ. Academy of Science of the Armenian S.S.R., Yerevan, 1970, 184—192 (in Russian).
76. **Kochankov, V. G.,** Effect of day length on the level of the activity of abscisin-like inhibitor in *Rudbeckia* plants, *Dokl. Akad. Nauk S.S.S.R.*, 198 (4), 959—962, 1971 (in Russian).
77. **Kochankov, V. G.,** Effect of CCC and gibberellin on the growth and flowering of long-day species in separate and joint treatment, *Dokl. Akad. Nauk S.S.S.R.*, 199, 2, 485—488, 1971 (in Russian).
78. **Kochankov, V. G.,** Regulation of Plant Flowering by Means of Gibberellins, Growth Retardant CCC and Other Inhibitors, dissertation for Master of Biology, Institute of Plant Physiology, Moscow, 1972, 145.

79. **Kochankov, V. G.,** Alternative ways of growth and development of rosette plants and the activity of natural growth regulators, in *Proc. 3rd Symp. Regulation of Plant Growth (October 18-22, Varna, Bulgaria)*, Bulgarian Academy of Sciences, 1981, 57.
80. **Kochankov, V. G.,** Phytohormone content in coneflower *(Rudbeckia bicolor)* ontogenesis, in *Abstr. 11th International Conf. Plant Growth Substances, Aberystwyth, Wales, 12—16 July, 1982*, University College of Wales and I.P.G.S.A., Aberystwyth, 1982, 44.
81. **Kochankov, V. G., Muzafarov, B. M., Andreyev, L. V., and Balakhinin, V. V.,** Changes in the activity of gibberellins in alternative ways of transition to flowering of *Rudbeckia* plants, approaches to identification of gibberellin by the GC-MS method, in *Regulators of Plant Growth and Development*, Proc. 1st All-Union Conf., Moscow, Chailakhyan, M. Kh., Ed., Nauka Publ., Moscow, 1981, 111.
82. **Kochankov, V. G., Muzafarov, B. M., and Chailakhyan, M. Kh.,** Influence of continuous darkness, length of day and treatment with gibberellin on bolting and flowering of *Rudbeckia* of rosette plants, *Fiziol. Rast.*, 25 (1), 145—154, 1978 (in Russian).
83. **Kochankov, V. G., Muzafarov, B. M., and Chailakhyan, M. Kh.,** Changes in hormonal-inhibitory balance during transition of plants to flowering, in *Metabolism and Mode of Action of Phytohormones*, Salyaev, R. K. and Gamburg, K. Z., Eds., Siberian Institute of Plant Physiology and Biochemistry, Siberian Branch of the U.S.S.R. Academy of Sciences, Irkutsk, 1979, 67—70 (in Russian).
84. **Kochankov, V. G., Muzafarov, B. M., and Chailakhyan, M. Kh.,** Dynamics in the activity of free and bound auxins and abscisic acid at the initial stage of *Rudbeckia* flower stem elongation, in *Regulators of Plant Growth and Development*, Proc. 1st All-Union conf., Moscow, Chailakhyan, M. Kh., Ed., Nauka Publ., Moscow, 1981, 112 (in Russian).
85. **Konstantinova, T. N.,** Influence of anaerobiosis on plant respiration in relation to photoperiodism, *Fiziol. Rast.*, 10 (4), 480—482, 1963 (in Russian).
86. **Konstantinova, T. N., Aksyonova, N. P., and Nikitina, A. A.,** Effect of a spectral light composition on the development of *Rudbeckia* and *Perilla* under long and short day conditions, *Fiziol. Rast.*, 15 (2), 363—366, 1968 (in Russian).
87. **Kotova, I.,** Hybrid *Rudbeckia, Tsvetovodstvo*, 1, 12, 1971 (in Russian).
88. **Krasilnikov, N. A., Chailakhyan, M. Kh., Aseyeva, I. V., and Khlopenkova, L. P.,** On gibberellin-like substances produced by soil yeasts, *Dokl. Akad. Nauk S.S.S.R.*, 123, 16, 1124—1127, 1958 (in Russian).
89. **Krasilnikov, N. A., Chailakhyan, M. Kh., Skryabin, G. K., Khokhlova, Yu. M., Ulezlo, I. V., and Konstantinova, T. N.,** On the stimulating action of gibberellins of various origin, *Dokl. Akad. Nauk S.S.S.R.*, 121 (4), 755—758, 1958 (in Russian).
90. **Lozhnikova, V. N.,** Dynamics of natural gibberellins under conditions of various photoperiodic cycles, *Dokl. Akad. Nauk S.S.S.R.*, 168 (1), 223—226, 1966 (in Russian).
91. **Lozhnikova, V. N., Khlopenkova, L. P., and Chailakhyan, M. Kh.,** A method of determining natural gibberellins in plant tissues, *Agrokhimiya*, 10, 132—139, 1967 (in Russian).
92. **Milyaeva, E. L., and Chailakhyan, M. Kh.,** Changes in shoot apices of plants during transition from their vegetative growth to flowering, *Izv. Akad. Nauk S.S.S.R., Biol. Ser.*, 3, 342—352, 1974 (in Russian).
93. **Milyaeva, E. L. and Chalakhyan, M. Kh.,** Daily fluctuations of mitotic activity in shoot apices of *Rudbeckia bicolor* during flowering evocation, *Fiziol. Rast.*, 28 (2), 302—306, 1981 (in Russian).
94. **Milyaeva, E. K., Kovalyova, L. V., and Chailakhyan, M. Kh.,** Formation of specific proteins in shoot apices during plant transition from vegetative growth to flowering, *Fiziol. Rast.*, 29 (2), 253—260, 1982 (in Russian).
95. **Milyaeva, E. L., Kovalyova, L. V., Lobova, N. V., and Chailakhyan, M. Kh.,** Changes in protein spectra of shoot apices in *Rudbeckia bicolor* during transition from vegetative to generative state, *Dokl. Akad. Nauk S.S.S.R.*, 245, (1), 269—272, 1979 (in Russian).
96. **Murneek, A. E.,** A separation of certain types of responses of plants to photoperiod, *Proc. Am. Soc. Hortic. Sci.*, 34, 507—509, 1936.
97. **Murneek, A. E.,** Length of day and temperature effects in *Rudbeckia, Bot. Gaz.*, 102 (2) 269—279, 1—40.
98. **Muzafarov, B. M., Mekhti-zade, R. M., and Kochankov, V. G.,** Gibberellin-like substances (GLS), bolting and flowering of plants, in *Regulators of Plant Growth and Development*, Proc. 1st All-Union Conf., Moscow, Chailakhyan, M. Kh., Ed., Nauka Publ., Moscow, 1981, 122 (in Russian).
99. **Nitsch, J. P. and Harada, H.,** Production de fleurs en jours courts par l'alcool furfurylique chez le *Rudbeckia speciosa, Bull. Soc. Bot. France*, 105, 319—322, 1958.
100. **Pont Lezica, R. F.,** Gibberellins in *Rudbeckia bicolor* grown under different photoperiods. *Bull. Soc. Roy. Sci. Liège*, 34 (1—2), 49—55, 1965.
101. **Pont Lezica, R. F.,** Les gibberellins naturelles dans la floraison d'une plante de jours-longs, in *Les Phytohormones et l'Organogenèse*, Cong. et Colloq. Univ. Liège, No. 38, Universite de Liège, 401—409, 1966.
102. **Pont Lezica, R. F.,** Effect of seven different gibberellins on stem elongation and flower formation in *Rudbeckia bicolor* Nutt. grown under noninductive conditions, *Fyton*, 26, (2), 185—190, 1969.

103. **Pont Lezica, R. F.,** Relacion entre periodo de induction fotoperiodica y gibberellinas en *Rudbeckia bicolor* Nutt., *Rev. Fac. Cien. Agrar. (Mendoza, Argentina),* 15 (2), 203—209, 1969.
104. **Pont Lezica, R. F., Jacqmard, A., and Deltour, R.,** Accion del Phosfon-D y CCC en *Rudbeckia bicolor* Nutt., *Rev. Fac. Cien. Agrar. (Mendoza, Argentina),* 14 (1—2), 27—38, 1968.
105. **Rickett, H. W.,** *Wild Flowers of the United States,* McGraw-Hill, Vol. 3, New York, 1970, 556.
106. **Rickett, H. W.,** *Wild Flowers of the United States,* Vol. 4, McGraw-Hill, New York, 1970, 802.
107. **Rickett, H. W.,** *Wild flowers of the United States,* Vol. 5, McGraw-Hill, New York, 1971, 668.
108. **Rickett, H. W.,** *Wild Flowers of the United States,* Vol. 1, McGraw-Hill, New York, 1973, 560.
109. **Rickett, H. W.,** *Wild Flowers of the United States,* Vol. 6, 3, McGraw-Hill, New York, 1963, 786.
110. **Rickett, H. W.,** *Wild Flowers of the United States,* Vol. 2, McGraw-Hill, New York, 1975, 690.
111. **Roberts, R. H. and Struckmeyer, B. E.,** Photoperiod, temperature and some hereditary responses of plants, *J. Hered.,* 29 (3), 94—98, 1938.
112. **Roberts, R. H. and Struckmeyer, B. E.,** The effects of temperature and other environmental factors upon the photoperiodic responses of some of the higher plants, *J. Agric. Res.,* 56 (9), 633—677, 1938.
113. **Roberts, R. H. and Struckmeyer, B. E.,** Further studies of the effects of temperature and other environmental factors upon the photoperiodic responses of plants, *J. Agric. Res.,* 59 (9), 699-709, 1939.
114. **Samygin, G. A.,** Photoperiodism of plants, *Tr. Inst. Fisiol. Rast. K.A. Timiriazeva (Moscow),* 3, 2, 129—262, 1946 (in Russian).
115. **Steinberg, R. G. and Garner, W. W.,** Response of the certain plants to length of day and temperature under controlled conditions, *J. Agric. Res.,* 52 (12), 943—960, 1936.
116. **Wittwer, S. H. and Bukovac, M. J.,** Effects of gibberellin on the photoperiodic responses of some higher plants, in *Photoperiodism in Plants and Animals,* Withrow, R. B., Ed., Publ. No. 55, American Association for the Association of Science, Washington, D.C., 1959, 373—379.
117. **Zhdanova, L. P.,** Photoperiodic response of short- and long-day plants in relation to their age, *Dokl. Akad. Nauk S.S.S.R.,* 58 (3), 485—488, 1947 (in Russian).
118. **Zhdanova, L. P.,** On the analysis of vegetative growth phenomenon, *Dokl. Akad. Nauk S.S.S.R.,* 60 (8), 1421—1424, 1948 (in Russian).
119. **Zhdanova, L. P.,** The rate of ouflow of flowering hormones under photoperiodic induction, *Dokl. Akad. Nauk S.S.S.R.,* 61, 3, 553—556, 1948 (in Russian).
120. **Zhdanova, L. P.,** Comparative analysis of photoperiodic induction in short- and long-day plants, *Tr. Inst. Fisiol. Rast. K.A. Timiriazeva (Moscow),* 6, 1, 69—84, 1948 (in Russian).

SAUROMATUM GUTTATUM

En. Voodoo lily, Leopard lily, Monarch of the East

B.J.D. Meeuse

INTRODUCTION

Sauromatum guttatum Schott (Araceae) is usually assumed to be native to northwestern India and Pakistan. However, there is evidence that its natural area of distribution includes Mesopotamia, Jordan, and Egypt. Both in Europe and the U.S., where this plant is propagated in greenhouses, it has become popular as a botanical curiosity; the large, smelly inflorescences are produced from a dry corm without benefit of water or soil (hence the Dutch name "stinking dry-bloomer").

MORPHOLOGY AND GROWTH HABIT

Almost without exception, voodoo lily individuals are monoecious; the single inflorescence produced by each mature corm always displays a group of well-developed female flowers at the base of the spadix and a group of equally well-developed male ones much higher up (Figure 1). However, on the basis of 26 years of experience in growing large numbers of *Sauromatum* plants in the greenhouse, we can state that unisexual individuals (more often female than male) are occasionally found. In addition, shoots programmed to turn into normal (bisexual) inflorescences may turn into vegetative shoots, depending on circumstances, as we will explain.

In common greenhouse practice, a voodoo lily plant is often grown from a small "cormlet", produced vegetatively by a large corm. The cormlet develops into a vegetative (i.e., nonflowering) plant. When, after some 6 or 7 months, its foliage dies down, it turns out that the initial cormlet has been replaced by a larger corm. If circumstances were extremely favorable, as can be judged by the size of the corm, the new corm will (after a certain period of dormancy) produce a shoot which will turn into a normal inflorescence. When flowering is over, this same corm (after it has been replanted) will produce another vegetative plant. After several months, when *its* foliage has died down, the new corm it has produced develops into another sexual phase in the form of an inflorescence, and so on. The life-cycle of *Sauromatum* is thus characterized by a regular alternation of sexual and vegetative phases. When applied to the normal course of events in the greenhouse, this statement is too sweeping only in the sense that the very first vegetative phase, produced by the cormlet, is followed by a second one when conditions are not ideal. It is only after this phase that the regular alternation will manifest itself. Obviously, blooming will not occur until a corm of sufficient size has been built up. However, duplication of the vegetative phase in 2 successive years may, under special conditions, occur even if flowering is not limited by corm size. Such conditions obtain when corms reach the blooming stage in late spring rather than in late fall (which would be normal). The available evidence indicates that prolonged exposure of the corms to fairly high temperatures is responsible for the switch from "prospective sexual" to "prospective vegetative".[42a]

Production of Subterranean Infructescence

Recognizing that even very exceptional events may have profound evolutionary consequences, we wish to report here on two cases in which a large ready-to-flower corm, inadvertently left underground, produced a perfect infructescence with a large number of

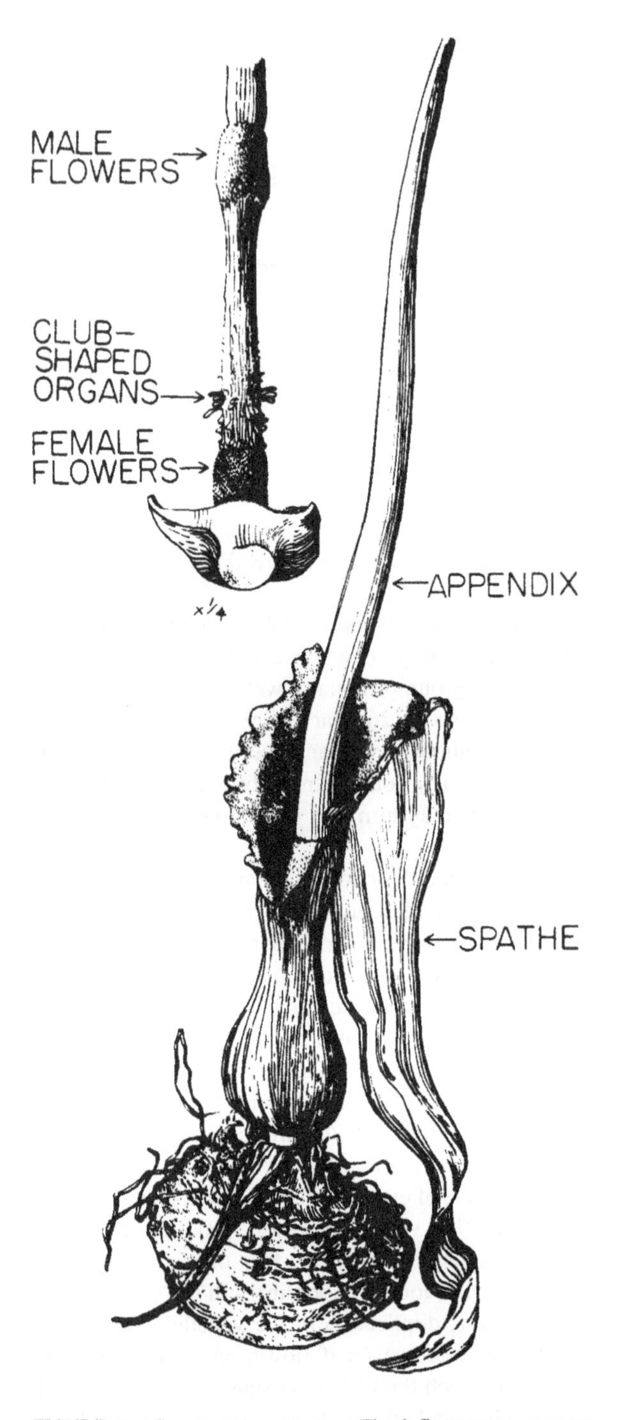

FIGURE 1. *Sauromatum guttatum*. The inflorescence emerges from the tuber or corm without any water of soil. Above, arrangement of the various organs on the base of the appendix in the floral chamber.

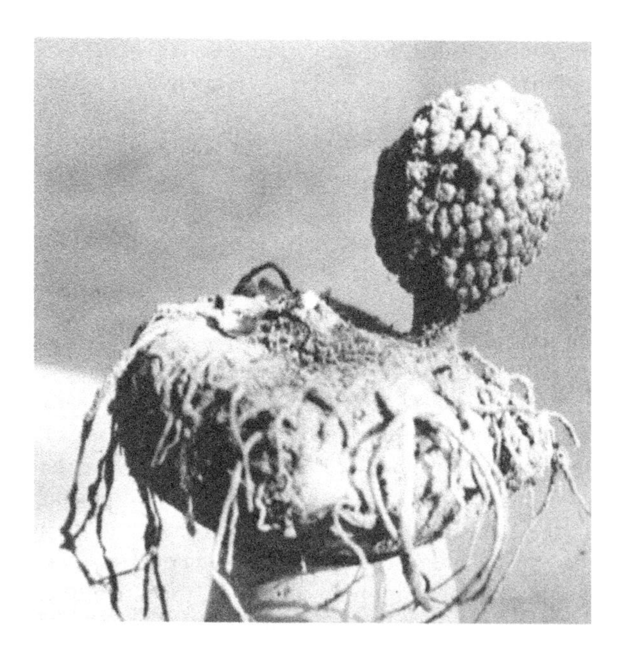

FIGURE 2. An abnormal corm of *Sauromatum guttatum* with infructescence formed in sites where normally a cormlet is produced.

pinkish fruits that, when ripe, contained viable seeds (Figure 2). The structures were already mature when found. In contrast to normal inflorescences, which always emerge from the very center of a corm, the underground infructescences were attached more laterally on the upper surface of the corm, several inches away from the center. Such a position is normal for a cormlet, and this makes it conceivable that these infructescences each developed from a cormlet primordium. Covered with soil as they were and not being accompanied by male flowers or any remnants of a spathe, there can be little doubt that they were *not* the product of some sexual event. The seeds yielded plants that were apparently normal. When the first vegetative phase of these was over, some of the newly produced corms were allowed to form roots. In the tips of these, the ploidy level of the cells was checked with the aid of a Feulgen staining reaction. It was found to be normal (diploid). In subsequent years no further formation of underground infructescences occurred in the (vegetatively produced) descendants of the seedlings that grew from the seeds of the original underground infructescences. For the time being, the situation can only be called enigmatic. However, it is clear that voodoo lilies are capable of producing apparently normal offspring in a number of ways — including unexpected ones!

FLOWERING

It has been demonstrated that anthesis in *Sauromatum* is controlled by the particular light/dark regime under which the inflorescence is allowed to develop.[2,4,37,39—42] Constant darkness prevents completely the unfolding of the spathe and the normally occurring production of heat and smell in the appendix (see below). Constant light leads to slow and incomplete unfolding of the spathe and (more importantly!) to the loss of the synchronization that normally can be observed in inflorescences developing from a given group of corms. Inflorescences allowed to grow under constant light up to the point where, under a normal day and night regime, they would have opened, and then given a single "dark shot" of at least 6 hr duration, will display normal and synchronized anthesis, but only after a lag period of almost 2 days (45 hr). The length of this lag period, initially surprising, becomes acceptable as a reality when one considers that in evening primrose *(Oenothera biennis)* the shift from a normal day and night regime to one in which day and night are reversed will not manifest itself in a 12-hr shift in the opening hour of flower-ripe buds (from, e.g., 6 p.m. to 6 a.m.) before 2 days have passed.[1] (See also chapter in this volume.) Obviously, the unfolding of flowers and inflorescences, far from being a "simple" happening, is (in certain cases, at least) a precisely programmed event, perhaps requiring the synthesis of new enzymatic proteins. Once a bud is set upon a certain course (i.e., programmed for opening), it cannot be deflected from it.

Thermogenic Respiration during Anthesis

On the first day of flowering, the so-called appendix of the inflorescence goes through a truly sensational, cyanide-insensitive and thermogenic (heat-producing) *respiratory explosion;* when this process reaches its peak, the metabolism of the organ compares favorably with that of a flying hummingbird.[10,37-39,41-43]The thermogenicity is based on the presence of a dual pathway for the respiratory transport of electrons from the "fuel" (starch) to atmospheric oxygen: the "classical" one, which traps energy in the form of high-energy phosphate, and the so-called "alternate" pathway, which does not generate much ATP, so that the energy originally locked up in the fuel appears rapidly in the form of heat.[8—10,17,19,31-34,45,46,56] As a result, the temperature of a *Sauromatum* appendix may reach a level about 8 to 10°C higher than that of the environment (in *Arum,* a very close relative, the difference is even higher, namely, 15°C[27a]). The heat has survival value (for the species) in that it serves as a volatilizer for the odoriferous compounds (indole, ammonia, and various amines[11-14,53]) which attract the pollinating carrion and dung flies (or beetles). No heat would simply mean no attraction of pollinators, no fertilization, and thus no sexual reproduction with the reshuffling of the genetical cards that is of such paramount importance in evolution.

The alternate pathway has also been found in other plant materials such as bean seedlings and potatoes,[34] but nowhere does it have the enormous capacity one observes in voodoo lilies. Thus, a mere 15 g of *Sauromatum* appendix tissue has the same capacity as 7500 g of potatoes!

The respiratory explosion is triggered by the hormone *calorigen,* which is produced by the buds of the male flowers.[3,13,14,20-22,24,26,27,35,36] Its function is at least threefold, for it also controls the unfolding of the spathe and the induction of mild climacteric, characterized by sweet fragrance, in the yellow, club-shaped organs that are situated just above the female flowers[42] (see Figure 1). Proof for this contention can easily be obtained by cutting through the spadix of the inflorescence just below the zone of the male flowers; this will completely prevent spathe opening and fragrance development, at least if the operation is performed more than 24 hr before normal opening time; if it is carried out later, the hormone has already begun to leave the male flower buds. Aqueous extracts of the latter, when injected into the spongy central cavity of "virgin" appendices, will lead to the development of heat and smell, after a lag time of about a day (22 hr).[21,22] The fragrance of the club-shaped organs is reminiscent of papaya and lemon; lemonene and alpha-pinene are indeed major odor components here. The contrast between the sweet fragrance of the yellow organs and the powerful aminoid odor of the appendix, at first sight paradoxical, makes it tempting to assume that the two types of odor appeal to different instincts of the insect visitors. There is at least circumstantial evidence that the fragrance induces mating in visiting beetles;[15,18,42,47] this tends to prolong the sojourn of these animals in the strongly protogynous inflorescences, giving the pollen the opportunity to fall down on them so that later it can be carried away to other inflorescences, still in the female stage, where the female flowers will be subjected to beneficial cross-pollination.

Calorigen has now been purified to a considerable extent,[14] and has been used in "push-button" systems in which virgin appendix sections are forced to go through the metabolic explosion by confronting them with calorigen at a precisely defined moment;[35,36] the resulting biochemical changes can then be monitored with great accuracy over a period of, for example, 24 hr.

The respiratory explosion can be understood by recognizing that under the influence of calorigen there is a spectacular boost in the activity of phosphofructokinase, a key glycolytic enzyme.[25-28] As a result, glycolysis becomes a mighty stream; the classical pathway of electron transport becomes saturated, and the overflow has to go through the alternate system. Since this does not lead to much ATP production (as pointed out earlier), there is no significant depletion of ADP (the compound that is needed to keep glycolysis going full tilt) and, for this reason, the metabolic explosion does not "pinch itself off" at the base, but keeps going on and on until essentially all the starch is gone. The important question under investigation in this author's Seattle laboratory at present is how the boost in phosphofructokinase activity is brought about? An investigation carried out a few years ago[34,35] permitted the conclusion that application of calorigen leads to the synthesis of proteins, and it is not unlikely that the enzymatic protein of phosphofructokinase is among those generated in this fashion. However, it is also conceivable that there is a synthesis of other enzymes and metabolites which strongly influence the *activity* of phosphofructokinase. A number of revolutionary discoveries in glycolysis research made in the last 5 years can potentially shed light on the situation. It is now known that in addition to the "classical", ATP-requiring phosphofructokinase there is another one, which is dependent upon pyrophosphate.[1a,5-7,29,30,44,48,49,52,54,55,58,59] The latter, which in some plant materials can be present in two interconvertible forms,[59] can be boosted in spectacular fashion by a newly discovered metabolite, fructose-2,6-bisphosphate.[16,23,50,51,57] There is thus a veritable network of interacting enzymes and metabolites, and this offers a multitude of possibilities for the control and modulation of glycolysis. There can be no doubt that the study of these in *Sauromatum* will, in the very near future, make it possible to understand the metabolic explosion more fully.

REFERENCES

1. **Arnold, C.-G.,** die Blutenöffnung bei *Oenothera* in Abhängigkeit vom Lichtdunkelrhythmus, *Planta*, 53, 198—211, 1959.

1a. **Balogh, A., Wong, J. H., and Buchanan, B. B.,** Metabolite mediated interconversion of PFP/PFK: a regulatory mechanism to direct cytosolic carbon flux, *Plant Physiol.*, 75(1), Suppl. 53, 1984.

2. **Buggeln, R. G.,** Control of Blooming in *Sauromatum gluttatum* Schott (Araceae), Ph.D. thesis, University of Washington, Seattle, 1969.

3. **Buggeln, R. G. and Meeuse, B. J. D.,** Hormonal control of the respiratory climacteric in *Sauromatum guttatum* (Araceae), *Can. J. Bot.*, 49, 1373—1377, 1971.

4. **Buggeln, R. G., Meeuse, B. J. D., and Klima, J. R.,** Control of blooming in *Sauromatum guttatum* Schott (Araceae) by darkness, *Can. J. Bot.*, 49, 1025—1031, 1971.

5. **Carnal, N. W. and Black, C. C.,** Pyrophosphate-dependent 6-phosphofructokinase. A new glycolytic enzyme in pineapple leaves, *Biochem. Biophys. Res. Commun.*, 86, 20—26, 1979.

6. **Carnal, N. W. and Black, C. C.,** Phosphofructokinase activities in photosynthetic organisms. The occurrence of pyrophosphate-dependent 6-phosphofructokinase in plants, *Plant Physiol.*, 71, 150—155, 1983.

7. **Carnal, N. W. and Black, C. C.,** Kinetic characteristics of pineapple pyrophosphate-dependent phosphofructokinase, *Plant Physiol.*, 72, 5—124, 1983.

8. **Chauveau, M.,** La chaine respiratoire des mitochondries d'*Arum*. I. Inhibition de la voie d'oxydation cytochromique, *Physiol. Veg.*, 14, 309—323, 1976.

9. **Chauveau, M.,** La chaine respiratoire des mitochondries d'*Arum*. II. Inhibition de la voie d'oxydation insensible au cyanure, *Physiol. Veg.*, 14, 325—337, 1961.

10. **Chauveau, M. and Lance, C.,** Respiration et thermogénèse chez les Aracées, *Bull. Soc. Bot. Fr.*, 129, Actual. Bot. 1982(2), 123—134.

11. **Chen, J. and Meeuse, B. J. D.,** Production of free indole by some aroids, *Acta Bot. Neerl.*, 20, 627—635, 1971.
12. **Chen, N. and Meeuse, B. J. D.,** Production of free indole by the appendix of *Sauromatum* (Araceae), *Am. J. Bot.*, 58, 478, 1971.
13. **Chen, J. and Meeuse, B. J. D.,** Induction of indole synthesis in the appendix of *Sauromatum guttatum* Schott, *Plant Cell Physiol.*, 13, 831—841, 1972.
14. **Chen, J. and Meeuse, B. J. D.,** Purification and partial characterization of two biologically active compounds from the inflorescence of *Sauromatum guttatum* Schott., *Plant Cell Physiol.*, 16, 1—11, 1975.
15. **Croat, T. B.,** *Dieffenbachia* (Loterias, Dub Cane), in *Costa Rican Natural History*, Janzen, D. H., Ed., University of Chicago Press, Chicago, 1983, 234—236.
16. **Cséke, C., Weenden, N. F., Buchanan, B. B., and Uyeda, K.,** A special fructose bisphosphate functions as a cytoplasmic regulatory metabolite in green leaves, *Proc. Natl. Acad. Sci. U.S.A.*, 79, 4322—4326, 1982.
17. **Dizengremel, P. and Lance, C.,** La respiration insensible an cyanure chez les végétaux, *Bull. Soc. Bot. Fr.*, 129, Actual Bot. 2, 19—36, 1982.
18. **Gottsberger, G.,** Pollination strategies in Brazilian *Philodendron* species, *Ber. Dtsch. Bot. Ges.*, 97, 391—410, 1984.
19. **Henry, M. F. and Nyns, E. G.,** Cyanide-insensitive respiration. An alternate mitochondrial pathway, *Sub-Cell Biochem.*, 4, 1—65, 1975.
20. **Herk, A. W. H. van,** Die chemischen Vorgänge im *Sauromatum* -Kolben, *Rec. Rav. Bot. Néerl.*, 34, 69—156, 1937.
21. **Herk, A. W. H. van,** Die chemischen Vorgange im *Sauromatum* -Kolben, II. *Proc. Ned. Akad. Wet. Ser. C*, 40, 607—614, 1937.
22. **Herk, A. W. H. van,** Die chemischen Vorgänge im *Sauromatum* -Kolben, III, *Proc. K. Ned. Akad. Wet. Ser. C*, 40, 709—719, 1937.
23. **Hers, M. G., Hue, L., and Schaftingen, E. van,** Fructose 2,6 bisphosphate, *Trends Biochem. Sci.*, 7, 329—331, 1982.
24. **Hess, C. M.,** A Preliminary Study of the Respiratory Metabolism in the Spadix of the Arum Lily *Sauromatum guttatum* Schott, M.Sc. thesis, University of Washington, Seattle, 1961.
25. **Hess, C. M.,** Glycolysis with its Associated Enzymes and Intermediates in the Appendix of *Sauromatum guttatum* Schott and Other Aroids, Ph.D. thesis, University of Washington, Seattle, 1964.
26. **Hess, C. M. and Meeuse, B. J. D.,** Factors contributing to the respiratory flare-up in the appendix of *Sauromatum* (Araceae). I, *Proc. K. Ned. Akad. Wet. Ser. C*, 74, 443—455, 1968.
27. **Hess, C. M. and Meeuse, B. J. D.,** Factors contributing to the respiratory flare-up in the appendix of *Sauromatum* (Araceae). II, *Proc. K. Ned. Akad. Wet. Ser. C*, 74, 456—471, 1968.
27a. **James, W. O. and Beevers, H.,** The respiration of *Arum* spadix. A rapid respiration resistant to cyanide, *New Phytol.*, 49, 353—357, 1950.
28. **Johnson, T. F. and Meeuse, B. J. D.,** The phosphofructokinase of the *Sauromatum* -appendix (Araceae). Purification and activity-regulation *in vitro*, *Proc. K. Ned. Akad. Wet. Ser. C*, 75, 1—19, 1972.
29. **Kombrink, E., Kruger, N. J., and Beevers, H.,** Kinetic properties of pyrophosphate: fructose-6-phosphate phosphotransferase from germinating castor bean endosperm, *Plant Physiol.*, 74, 395—401, 1984.
30. **Kruger, N. J., Kombrink, E., and Beevers, H.,** Pyrophosphate: fructose-6-phosphate phosphotransferase in germinating castor bean seedlings, *FEBS Lett.*, 153, 409—412, 1983.
31. **Lambers, H.,** The physiological significance of cyanide-resistant respiration, *Plant Cell Environ.*, 3, 293—302, 1980.
32. **Lambers, H.,** Cyanide-resistant respiration: a non-phosphorylating electron transport pathway acting as an energy overflow, *Physiol. Plant.*, 55, 478—485, 1982.
33. **Lance, C.,** Cyanide-insensitive respiration in fruits and vegetables, in *Recent Advances in the Biochemistry of Fruits and Vegetables*, Academic Press, London, 1981, 63—87.
34. **Laties, G. G.,** The cyanide-resistant, alternative path in higher plant respiration, *Annu. Rev. Plant Physiol.*, 33, 549—555, 1982.
35. **McIntosh, L.,** A Developmental Analysis of Anthesis in *Sauromatum guttatum* Schott, Ph.D. thesis, University of Washington, Seattle, 1977.
36. **McIntosh, L. and Meeuse, B. J. D.,** Control of the development of cyanide-resistant respiration in *Sauromatum guttatum* (Araceae), in *Plant Mitochondria*, Ducet, G. and Lance, C., Eds., Elsevier/North Holland Biomedical Press, Amsterdam, 1978, 339—345.
37. **Meeuse, B. J. D.,** Van Herk's "botanical liver", or the appendix of the arum lily inflorescence, in *What's New in Plant Physiology*, 4, 1—4, 1972.
38. **Meeuse, B. J. D.,** Films of liquid crystals as an aid in pollination studies, in *Pollination and Dispersal* , Brantjes, M. B. M., Ed., University of Nijmegen, Nijmegen Netherlands, 1973, 19—20.
39. **Meeuse, B. J. D.,** Thermogenic respiration in aroids, *Annu. Rev. Plant Physiol.*, 26, 117—126, 1975.

39a. **Meeuse, B. J. D.,** The physiology of some sapromyophilous flowers, in *The Pollination of Flowers by Insects,* Richards, A. J., Ed., Linnean Soc. Symp. Ser. 6, Academic Press, London, 1978, 97—104.

40. **Meeuse, B. J. D.,** Regulation of anthesis phenomena in the voodoo lily, *Sauromatum guttatum* Schott, *Plant Physiol.,* 75(1), Suppl. 135, 1984.

41. **Meeuse, B. J. D., and Buggeln, R. G.,** Time, space, light and darkness in the metabolic flare-up of the *Sauromatum* -appendix, *Acta Bot. Neerl.,* 18, 159-172, 1969.

42. **Meeuse, B. J. D., Schneider, E. L., Hess, C. M., Kirkwood, K., and Patt, J.,** Activation and possible role of the "food-bodies" of *Sauromatum* (Araceae), *Araceae), Acta Bot. Neerl.,* 33(4), 483-496, 1984.

42a. **Meeuse, B. J. D.,** unpublished.

43. **Nagy, K. A., Odell, D. K., and Seymour, R. S.,** Temperature regulation by the inflorescence of *Philodendron, Science,* 178, 1195—1197, 1972.

44. **O'Brien, W. E., Bowien, S., and Wood, H. G.,** Isolation and characterization of a pyrophosphate-dependent phosphofructokinase from *Propionibacterium shermanii, J. Biol. Chem.,* 250, 8690—8695, 1975.

45. **Palmer, J. M.,** The organization and regulation of electron transport in plant mitochondria, *Annu. Rev. Plant Physiol.,* 27, 133—157, 1976.

46. **Plas, L. H. W. van der,** CN-resistente ademhaling by planten: energie-verspilling of nuttige aanpassing?, *Vakbl. Biol.,* 61(22), 478—482, 1981.

47. **Ray, T.,** *Syngonium triphyllum* (Mano de Tigre), in *Costa Rican Natural History,* Janzen, D. H., Ed., University of Chicago Press, Chicago, 1983, 333—335.

48. **Reeves, R. E., South, D. J., Blytt, H. J., and Warren, L. G.,** Pyrophosphate: D-fructose 6-phosphate 1-phosphotransferase, *J. Biol. Chem.,* 249, 7737—7741, 1974.

49. **Reeves, R. E., Serrano, R., and South, D. J.,** 6-Phosphofructokinase (pyrophosphate). Properties of the enzyme from *Entamoeba histolytica* and its reaction mechanism, *J. Biol. Chem.,* 251, 2958—2962, 1976.

50. **Sabularse, D. C. and Anderson, R. L.,** D-fructose, 2,6-bisphosphate: a naturally occurring activator for inorganic pyrophosphate: D-fructose-6-phosphate 1-phosphotransferase in plants, *Biochem. Biophys. Res. Commun.,* 103, 848—855, 1981.

51. **Schaftingen, E. van and Hers, M.-G.,** Inhibition of fructose 1,6-bisphosphatase by fructose-2,6 bisphosphate, *Proc. Natl. Acad. Sci. U.S.A.,* 78, 2861—2863, 1981.

52. **Schaftingen, E. van, Lederer, B., Bartrons, R., and Hers, H. G.,** A kinetic study of pyrophosphate: fructose-6-phosphate phosphotransferase from potato tubers, *Eur. J. Biochem.,* 129, 191—195, 1982.

53. **Smith, B. M. and Meeuse, B. J. D.,** Production of volatile amines an skatole at anthesis in some arum lily species, *Plant Physiol.,* 41, 343—347, 1966.

54. **Smyth, D. A., Wu, M.-X., and Black, C. C.,** Phosphofructokinase and fructose 2,6-bisphosphatase activities in developing corn seedlings (*Zea Mays* L.), *Plant. Sci. Lett.,* 33, 61—70, 1984.

55. **Smyth, D. A., Wu, M.-X., and Black, C. C.,** Pyrophosphate and fructose 2,6-bisphosphate effects on pea seed glycolysis, *Plant Physiol.,* 75(1), Suppl. 52, 1984.

56. **Solomos, T.,** Cyanide-resistant respiration in higher plants, *Annu. Rev. Plant Physiol.,* 28, 279—297, 1977.

57. **Stitt, M., Kurzel, B., and Heldt, H. W.,** Control of photosynthetic sucrose synthesis by fructose 2,6-bisphosphate. II. Partitioning between sucrose and starch, *Plant Physiol.,* 75, 554—560, 1984.

58. **Wu, M.-X, Smyth, D. A., and Black, C. C.,** Fructose 2,6-bisphosphate and the regulation of pyrophosphate-dependent phosphofructokinase activity in germinating pea seeds, *Plant Physiol.,* 73, 188—191, 1983.

59. **Wu, M.-X, Smyth, D. A., and Black, C. B., Jr.,** Regulation of pea seed pyrophosphate-dependent phosphofructokinase: evidence for interconversion of two molecular forms as a glycolytic regulatory mechanism, *Proc. Natl. Acad. Sci. U.S.A.,* 81, 5051—5055, 1984.

SCABIOSA

En. Scabiosa, Scabious, Mourning bride, Pincushion flower; Fr. Scabieuse; Ge. Grundkraut, Skabiose

H. F. Wilkins and A. H. Halevy

INTRODUCTION

The genus *Scabiosa* L. belongs to the Dipsacaceae family. There are some 80 or more species in this genus, native mainly to the Mediterranean area.[1,11]

Leaves are opposite, entire to dissected. Flowers are long-stalked with an involucrate head subtended by nonspiny receptacular bracts. Colors range from white to yellowish, rose, blue, dark purple, and lilac. The calyx is cup-shaped with 5 bristly teeth and is enveloped by a cup-shaped involucel. The corolla is 4- to 5-lobed, and lobes are nearly equal or, sometimes, 2-lipped; marginal corollas are usually larger. There are 4 stamens.[1]

There are three species of *Scabiosa* L. which are commonly seen in commercial floriculture as cut flowers: *S. atropurpurea, S. caucasica,* and *S. japonica.*[11]

Scabiosa atropurpurea L. is an erect annual growing over 1 m and originating from southern Europe, but it has become naturalized in California. Flowers are 5 cm across and are dark purple, rose, lilac, or white. *S. caucasica* Bieb. is a perennial plant originating in the Caucasus Mountains and grows to 60 cm or more in height. The foliage is gray-tomentose; the flowers are blue and are 7.5 cm across. *S. japonica* Miq. is a biennial plant originating in Japan and growing to 76 cm, with blue flowers which are some 5 cm across.[1,11]

Scabiosas are popular for flower gardens and thrive in good soil with a pH of 6.0 to 7.0, moist conditions, and sunny locations. Commercially, they are mostly a field crop, as under greenhouse conditions they respond slowly to forcing.[1,11]

Researchers have used the *Scabiosa* species *S. canescens, S. succisa,* and *S. ukranica* to study vernalization and photoperiod interactions.[3—7] Their responses have been cited in several reviews as classic examples of reponses of these species to photoperiod and temperature interactions.[7,9,15]

CULTURE

Perennial and biennial forms are seed propagated in the spring and summer. These plants flower the following year.[10,11]

The annual forms are seed propagated under greenhouse conditions in early spring, transplanted into cold frames or fields after frost danger has passed, and flower in the summer.[10,11]

Photoperiod and Temperature

Under winter or SD conditions the annual form *(S. atropurpurea)* does not elongate stems at 18°C. If 5 or 6 hr of supplementary incandescent lighting are given, they flower as if it were summer. The best greenhouse temperature is 13°C, under full light conditions.[10,11]

Shillo and Shillo,[14] working with several cultivars of *S. atropurpurea,* found that holding seedlings for 6 to 8 weeks at 13°C under 6 klx of continuous fluorescent night lighting advanced flowering by some 6 months but greatly reduced flower numbers. With larger plants, a 3-hr incandescent NB of 100 to 200 lux from October to mid-April also advanced flowering 2 to 4 months, with acceptable flower production in the field under natural mild winter temperatures in Israel.

Biran,[2] also working with the annual *S. atropurpurea,* used a 4-hr NB with an incandescent light source to dramatically hasten flowering. Flowering commenced 2 to 3 months after the NB commenced, regardless of plant size; however, large rosetted plants at the beginning of NB resulted in maximum flower production. These results with this annual species indicate that the main floral induction factor is LD; but again there was no doubt that some cold promotion was acquired during the mild winter in Israel where these plants were grown in the field.

S. caucasia, a perennial, has been forced into early spring flowering under unheated plastic tunnels or in greenhouses with little heat.[8] Plömacher[12] employed daytime temperatures of 12°C and 4 to 6°C at night. Plants were field planted in April and May, overwintered, and flowered in the spring. Protective covering was provided, and forcing began when the snow melted, resulting in flowering in mid-April.

With *S. canescens (S. suaveolens),* flowering is rapid when plants are cold treated and grown under LD; under SD, flowering is slowed.[4,6,7] Thus *S. canescens* has a partial vernalization and partial LD requirement for flowering.[7] *S. succisa (S. pratensis* and *S. praemorsa)* has been reported to have an obligate vernalization requirement with a partial LD requirement.[7] However, *S. succisa* has been classed by Lang[9] as a SLDP flowering in LD only if previously under SD.

Growth Regulators

GA has been reported to promote stem elongation with *S. ukrainica* in SD, but does not induce flowering;[4] with *S. succisa,* GA promotes initiation in LD and can substitute for SD in this SLDP.[5]

Sachs and Thimann[13] working with intact plants of *Scabiosa* have demonstrated that bud release was possible by kinetin applications.

REFERENCES

1. **Bailey, L. H.,** *Hortus Third,* Macmillan, New York, 1979, 1013—1015.
2. **Biran, I.,** Flowering Control of *Scabiosa atropurpurea* for the Winter Months (in Hebrew), Biannual Rep. Dept. Orn. Hortic., The Hebrew University, Rehovot, Israel, 1978—1980, 1980, 68—70.
3. **Chouard, P.,** Expériences de longue durée sur la photopériodisme: lecons que en découlent, *Mem. Soc. Bot. Fr.,* 96, 106—146, 1949.
4. **Chouard, P.,** Diversité des mechanismes des dormances, de la vernalization et du photopériodisme reveleé notamment par l'action de l'acide gibberellique, *Mem. Soc. Bot. Fr.,* 51—64, 1956—57.
5. **Chouard, P.,** La journée courte ou l'acide gibbérellique comme succedanées du froid pour la vernalisation d'une plante vivace en rosette, le *Scabiosa succisa* L., *C. R. Seances Acad. Sci. Paris,* 245, 2520—2522, 1957.
6. **Chouard, P.,** Quelques problèmes évogués par la diversité des réactions des plants à fleurs au photopériodisme, *Colloq. Int. Photo-Thermopériodisme,* Union Internationale des Sciences Biologiques, Laboratoire de Physiologie Vegétale, Paris, 1957, 7—23.
7. **Chouard, P.,** Vernalization and its relations to dormancy, *Annu. Rev. Plant. Physiol.,* 11, 191—238, 1960.
8. **Kneissl, P. and Söllner, V.,** Forcing of cut flowers from perennials under plastic (in German), *Dsch. Gartenbau,* 32, 1946—1947, 1978.
9. **Lang, A.,** Physiology of flower initiation, in *Encyclopedia of Plant Physiology,* Vol. 15 (Part 1) Ruhland, W., Ed., Springer-Verlag, Berlin, 1965, 1380—1536.
10. **Post, K.,** Effects of Daylength and Temperature on Growth and Flowering of Some Florist Crops, Cornell University Agricultural Experiment Station Bulletin, No. 787, Cornell University, Ithaca, N.Y., 1942, 1—70.
11. **Post, K.,** *florist Crop Production,* Orange Judd, New York, 1952, 808—809.
12. **Plömacher, H.,** Plants grown under plastic for cutting, *Scabiosa caucasica, Zierpflanzenbau,* 80, 17, 1980.
13. **Sachs, T. and Thimann, K. V.,** Release of lateral buds from apical dominance, *Nature, (London),* 201, 939—940, 1964.
14. **Shillo, G. and Shillo, R.,** Experiments in flowering control of *Scabiosa atropurpurea* (in Hebrew), Biannual Rep. Dept. Orn. Hortic., The Hebrew University, Rehevot, Israel, 1975—1977, 1977, 52—53.
15. **Vince-Prue, D.,** *Photoperiodism in Plants,* McGraw Hill, London, 1975, 11—29, 46, 219—245.

SETARIA ITALICA

En. Foxtail-, Italian-, Siberian-, German-, Hungarian-, Hay-millet; Ge. Kolbenhirse; Fr. Millet des oiseaux; Sp. Mijo

Jerry D. Eastin and Lenis A. Nelson

INTRODUCTION

Setaria is a large genus in the Gramineae family consisting of domestic and wild species, both of which are harvested for grain. De Wet et al.[2] recently summarized the origins and evolutions of foxtail millets. Malm and Rachie[3] reviewed world literature earlier. Currently, foxtail millet (*Setaria italica* [L.] Beauv.) is generally considered a minor or relict crop but it was a staple in early agriculture.[2,4] Williams[5] and Werth[6] consider China or Central Asia to be the center of cultivated varieties. Foxtail millet was moved from China to Japan[7] and occurs in the wild state there.[5] Green foxtail, *S. viridis*, is wild foxtail millet and is recognized as a subspecies of S. *italica*.[2] *S. viride* grades morphologically into cultivated *S. italica* and the two cross to produce fertile hybrids.[8] The subspecies *S. italica* subsp. *viridis* (L.) is a weedy annual native to temperate Eurasia[2] introduced to temperate areas all over the world. De Wet et al.[2] consider it unlikely that *S. italica* was domesticated in the Loes highlands of China and transported to Europe or vice versa. Possibly the species was domesticated in more than one area simultaneously or at different times.[9,10] Domestication dates nearly to 5000 B.C. in Shenshi, China.[11,11a] *Setaria* was the most important food in the Neolithic culture in China.[12] Foxtail millet provided 17% of the total food consumed in China as recently as 1949[13] and remains important today on poor agricultural lands.[2] It was cultivated in Russia about 1500 years ago.[3] The review of De Wet et al.[2] refers to *Setaria* species (domestic and wild) used for grain in Australia, the Philippines, New Guinea, Mexico, Hungary, Czechoslovakia, and Afghanistan.

INFLORESCENCE MORPHOLOGY

Robbins[14] refers to the inflorescence as a modified spike, but others consider it to be a panicle.[2,15,16] The spikelet contains an upper fertile (bisexual) floret and a lower sterile floret. A lemma and palea enclose 2 lodicules, 3 anthers, and a pistil (twin stigmas). Glumes enclose the fertile and sterile floret pair. The main axis of the inflorescence also bears bristles (modified spikes).[14]

Cultivated races are morphologically variable and two races are commonly recognized on the basis of differences in panicle structure. The two races are *moharia* (from Afghanistan) and *maxima*.[15,16] *Maxima* is advanced phylogenetically and was derived from more primitive cultivars in race *moharia*.[17] *Moharia* panicles are 4 to 15 cm long, are erect to slightly nodding, have branches 1 to 3 cm long with clusters of spikelets, bristles are well developed, panicles are straw to purple colored, and grain ranges from white to black. By contrast *maxima* panicles are large (up to 35 cm long and 8 cm wide) causing them to be pendulous at maturity. Panicle branches are long (up to 6 cm) with clusters of spikelets on secondary branches. Bristles tend to be shorter than in race *moharia* with straw to dark-purple colors.[17,18] The race *maxima* has been classified into three complexes, the Manchurian complex, the Mongolian complex, and the Korean complex, which is high yielding.[18]

Table 1
EFFECT OF PHOTOPERIOD AND LIGHT QUALITY ON STEM AND PANICLE LENGTH IN FOXTAIL MILLET

	Photoperiod[a] (12 hr)		Photoperiod[a] (16 hr)	
	Incandescent	Fluorescent	Incandescent	Fluorescent
Stem length (mm)	420	308	222	82
Panicle length[b] (mm)	21.6	16.9	1.4	0.6

[a] A total of 8 hr of natural light plus the duration of supplemental light necessary to obtain these photoperiods.
[b] A measure of the rate of flowering.

FLORAL INDUCTION AND DEVELOPMENT

Malm and Rachie[3] reviewed literature indicating *Setaria* is a quantitative SDP. A broad analysis of flowering reaction in 50 varieties coming from 31 to 42°N latitude in Japan was made by Kokubu and Miyaji.[7] They were seeded 6 times at monthly intervals from April 10 to September 10. Growth periods were gradually reduced to the shortest point in August and were longer again in the September planting. Genotypes from higher latitudes had shorter growth periods and were less sensitive to daylength. They also had longer critical daylengths than genotypes from lower latitudes. Presumably they were more temperature dependent for flowering.

Kokubu et al.[19] later did field and pot plantings of 126 varieties from 26 countries. European varieties (German millets) had short growth periods, a relatively long critical daylength for heading, and flowered even under relatively low temperatures. Japanese varieties (Italian millets) had longer growth periods which were heavily influenced by daylength. The required daylengths for flowering were shorter and required temperatures were higher. Obviously flowering in the German millets is more temperature regulated and flowering in the Italian millets is more daylength regulated.

Kokubu and Nagakura[20] subsequently subjected 20 Italian millets to different temperatures under different daylengths. Critical temperature required to initiate flowering rose as daylength increased in some varieties. Some varieties decreased leaf numbers as conditions shortened the period to flowering, while others did not. Clearly the diversity of environments in which foxtail millet has been domesticated and is grown has resulted in a wide range of temperature and daylength responses related to control of flowering in specific environments.

In a review of photocontrol of growth in plants, Downs[21] considered the effect of light on flowering in *Setaria italica* (L.) Beauv. Downs generally categorized light effects in terms of (1) photoperiod which influences whether or not stem elongation will occur and (2) light quality (R and FR ratios) which relates to the level of elongation. The ratio of R to FR presumably shifts the phytochrome system to either the R or the FR absorbing form. Incandescent lamps tend to induce the R form while fluorescent lights induce the FR form. Table 1 shows the influence of photoperiod and light quality on stem and panicle length in foxtail millet.[21]

Downs[21] emphasized the interdependence between flowering and stem elongation. Stem length in 12-hr days was two to four times longer than in 16-hr days, depending on light quality. Both absolute and relative differences induced in panicle length were even greater. Incandescent light (R form) increased panicle length 22% under 12-hr days and more than doubled panicle length under 16-hr days, emphasizing the R-form effect under a photoperiod unfavorable to flowering. He further showed that only 5 min of FR radiation at the end of 16 hr of fluorescent light could induce flowering in this SD plant.

There appear to be no investigations which key on the potential relationship between variation in daylength and temperature after floral initiation and before anthesis on grain yield and its seed-number and seed-size components. Surprisingly, there are two reports[22,23] which show little or no positive correlation between days to flowering and grain yield. This may relate to the finding that foxtail millet is almost equally sensitive to damage from water stress at all stages from the primordial (boot leaf stage) clear through the hard dough stage. This seems to be a more extended range of sensitivity to stress than in many other cereals. Also, the lack of correlation between yield and flowering data may relate to the generally meager production resources characteristic of many areas where the crop is grown.

GROWTH REGULATORS

Kumar et al.[24] treated plants with GA_3 every other day after the 4- to 7-leaf stage under both 8- and 24-hr days. Flowering occurred in 53.7 days under continuous light, confirming that *Setaria* is a quantitative SDP. GA_3 hastened flowering by 13.2 days (24%) under continuous light. The 8-hr plants flowered in 29.7 days; this was hastened 11.8 days (40%) by GA_3.

Other chemicals are generally not known to be involved in control of flowering in *Setaria*. Ascorbic acid is mentioned in the literature. Tayal[19] planted one variety in January, and after 68 days of growth exposed each of three sets to 8-hr days, normal days, and continuous light. The 8-hr-day plants flowered in 79 days compared to 87 days for the normal-day plants, while the continuous-light plants did not flower. Absorbic acid increased simultaneously with the change to reproductive status, but a cause and effect relationship was not demonstrated.

SUMMARY

Historically, the *Setarias* provided grain for consumption in several cultures in precultivation eras. Domestication appears to have occurred independently in more than one center. As a consequence, a wide range of plant reaction to temperature and daylength control of flowering exists. Basically *Setaria* is a quantitative SDP. GA_3 hastens flowering. The level of importance of *Setaria* has declined considerably in deference to corn, sorghum, and wheat. However, it remains important in several areas and may increase where agricultural lands are poor, rainfall is low, and the growing season is short.[2]

REFERENCES

1. **Rominger, J. J. M.,** *Taxonomy of Setaria (Gramineae) in North America,* University of Illinois Press, Urbana, 1962.
2. **De Wet, J. M. J., Oestrey-Sidd, L. L., and Cubero, J. I.,** Origins and evolution of foxtail millets *(Setaria italica),* J. Agric. Trad. et de Bota. Appl XXVI (1), 53—64, 1979.
3. **Malm, N. R. and Rachie, K. O.,** The *Setaria Millets.* A Review of World Literature, University of Nebraska SB513, 1971, 13 pp.
4. **Kawase, M. and Sakamoto, S.,** *Setaria* is a relict crop in Eurasia but was a staple in early agriculture, *Theor. Appl. Genet.,* 63, 117—119, 1982.
5. **Williams, T. A.,** Millets, *USDA Yearbook,* 1898, 267—290.
6. **Werth, E.,** Geography and history of millets, *Ang. Bot.,* 19, 41, 1937.
7. **Kokubu, T. and Miyaji, Y.,** Variations of growth-period of Italian millet strains, *Setaria italica* Beauv. and their response to day-length and temperature. I. Changes of main standard varieties in Japan due to the different seeding dates, *Mem. Fac. Agric. Kagoshima Univ.,* 12, 77—86, 1976.
8. **Li, C. H., Pao, W. K., and Li, M. W.,** Interspecific crosses in *Setaria, J. Hered.,* 33, 351—355, 1942.

9. **De Wet, J. M. J. and Harlan, J. R.,** Weeds and domesticates: evolution in the man-made habitat, *Econ. Bot.*, 29, 99—107.
10. **Harlan, J. R.,** The possible role of weed races in the evolution of cultivated plants, *Euphytica*, 14, 173—176, 1965.
11. **Ho, Ping-ti,** *The Cradle of the East,* University of Chicago Press, Chicago, 1968.
11a. **Bray, F.,** Millet cultivation in China: a historical survey, *J. Agric. Trad. Bot. Appl.*, 28, 291—307, 1981.
12. **Chang, K.,** Ch., Archeology of ancient China, *Science*, 162(3853) 519—526, 1968.
13. **Anderson, E. and Martin, J. H.,** World production and consumption of millet and sorghum, *Econ. Bot.*, 3, 265—288, 1949.
14. **Robbins, W. W.,** *Botany of Crop Plants*, P. Blakiston's Son & Co., 1931.
15. **Dekaprelevich, L. L. and Kasparian, A. S.,** A contribution to the study of foxtail millet *(Setaria italica* P. B. *maxima* ALF.) cultivated in Georgia (western Transcaucasia), *Bull. Appl. Bot. Plant Breed.*, 19, 533—572, 1928.
16. **Scheibe, A.,** Die Hirsen in Hindukusch, *Z. Pflanzensucht* 25, 392—436, 1943.
17. **Kornicke, F. and Werner, H.,** *Handbuch des Getreidebaues, Vol. 2, Die Sorten und der Anbau des Getreides,* Verlag Paul Parey, Berlin, 1885.
18. **Gritzenko, R. J.,** Chumiza (Italian millet) taxonomy *(Setaria italica* [L.] P. B. subsp. *maxima* Alef.), *Bull. Appl. Bot. Genet. Plant Breed.*, 32, 145—182, 1960.
19. **Kokubu, T., Ishmine, Y., and Miyaji, Y.,** Variations of growth period of Italian millet strains, *Setaria italica* Beauv. and their response to daylength and temperature. II. Changes of growth period of strains gathered from different districts, both native and foreign, due to the different seeding dates, *Mem. Fac. Agric. Kagoshima Univ.*, 12, 55—75, 1977.
20. **Kokubu, T. and Nagakura, T.,** Variations of growth period of Italian millet strains. *Setaria italica* Beauv. and their responses to daylength and temperature. III. Changes in growth-period due to temperature under the different daylengths, *Mem. Fac. Agric. Kagoshima Univ.*, 17, 53—68, 1981.
21. **Downes, R. J.,** Photocontrol of vegetative growth, in Photoperiodism and Related Phenomena in Plants and Animals, Withrow, R. B., Ed., Publication No. 55, American Association for the Advancement of Science, Washington, D.C., 1957, 129—135.
22. **Sandhu, T. S., Arora, B. S., and Singh, Y.,** Interrelationships between yield and yield components in foxtail millet, *Indian J. Agric. Sci.*, 44(9), 563—566, 1974.
23. **Vishwanatha, J. K.,** Studies on the genetic variability in the world germplasm collection of foxtail millet, *Setaria italica* Beauv., *Mysore J. Agric. Sci.*, 19, 519—520, 1978.
24. **Kumar, S., Datta, K. S., and Nanda, K. K.,** Gibberellic-acid causes flowering in the short-day plants *Panicum milliaceum* L., *P. milliare* Lamk., and *Setaria italica* (L.) P. Beauv. *Planta*, 134, 95—96, 1977.
25. **Tayal, M. S.,** Effect of photoperiod on ascorbic acid contents and flowering in *Setaria italica, J. Indian Bot. Soc.*, 51, 223—226.

SMYRNIUM OLUSATRUM

En. Alexanders; Ge. Gelbdolde

Jon Lovett Doust

INTRODUCTION

This monocarpic (characteristically biennial) species occurs naturally in southern Europe, and extends northward to France. It is extensively naturalized in Britain and Ireland, in hedges, waste places, and on cliffs, in particular near the sea. It is a large herbaceous species, 50 to 150 cm high, often attaining mean dry weights of about 100 g in the field. Flowering occurs from April to June. In Roman times the plant was used both as a pot herb and vegetable.[1]

Like many other Umbelliferae, *Smyrnium* is andromonoecious, that is, on a single plant there are produced both hermaphrodite (perfect) flowers and male (staminate) flowers. At flowering, the main shoot of the rosette develops as a leafy stem terminating in a primary (first-order) umbel. Lateral shoots are produced from leaf axils on this stem and these terminate also in inflorescences (second-order umbels). Further umbels may arise on higher-order shoots and, in *S. olusatrum*, L. fourth-order umbels typically are produced. Nutrient availability seems to be a major factor determining final plant size and hence total production of flowers.[2] In most plants, the number of modular units (including, for example, flowers and leaves) will be strongly influenced by the extent of competition with other plants and local nutrient availability.

INFLORESCENCE

Inflorescences are compound umbels, each bearing large numbers of small, closely packed flowers. Sepals are absent. Bracts and bracteoles are few, small, and sometimes absent altogether. The five yellow-green petals are lanceolate to obcordate, with an inflexed apex. An umbellifer inflorescence has a hierarchic organization. The separate rays of an umbel each bear an umbellet (or simple umbel) containing a number of pedicels, each in turn bearing either a staminate or hermaphrodite flower. Thus, the single flowers are arranged within simple umbels, and these are ordered within compound umbels.

FLORAL COMPOSITION AND NUTRIENT ALLOCATION

Table 1 shows aspects of sexual composition in *S. olusatrum*. Third-order umbels are most numerous and contribute some 60% of the total flowers of a plant. However, of these third-order flowers, only 11.8 to 18.6% are perfect flowers (and thus capable of setting seed). In fact, there occurs a gradual increase in the relative maleness of an umbel, along the hierarchy of inflorescences, from first- to fourth-order (these latter umbels being composed almost entirely of male flowers). Table 1 also provides an index of the success at producing mature fruit (mericarps) of hermaphrodite flowers of a particular umbel order. This index associates the probability of a hermaphrodite flower of a particular inflorescence becoming a ripe fruit with the relative contribution of the hermaphrodite flowers of that umbel to the total output of hermaphrodite flowers of the plant.

Smyrnium has been studied in terms of patterns of flower number and resource allocation to pollen/ovules and to seed.[3,4] The overall floral ratio (staminate to hermaphrodite) of individual plants has been found to remain within quite narrow limits, despite variation in

Table 1
ASPECTS OF MEAN SEXUAL COMPOSITION IN *SMYRNIUM OLUSATRUM* (SE IN PARENTHESES)

		Primary umbels	Secondary umbels	Tertiary umbels	Quaternary umbels
Number of umbels per plant	Site 1	1	7.9 (0.4)	21.9 (2.3)	16.7 (6.3)
Sum of flowers of an umbel, as a proportion of total flowers of a plant	Site 1	1.1 (0.3)	33.9 (3.9)	59.6 (2.7)	5.2 (2.2)
Ratio of mature fruit of an umbel to hermaphrodite flowers of a plant	Site 1	3.0 (0.9)	34.8 (6.8)	17.2 (7.7)	0
	Site 2	0.7 (0.4)	16.3 (4.6)	4.6 (1.9)	0
Percentage perfect flowers	Site 1	67.6 (2.6)	30.3 (3.3)	18.6 (4.9)	0
	Site 2	50.1 (3.0)	34.0 (3.6)	11.8 (4.1)	3.5 (2.5)

sexual composition between the four orders of inflorescence, and between the same order of inflorescence in different populations. Typically, four male flowers are present on a plant, for each hermaphrodite flower. (This ratio has been found to obtain in at least eight other umbellifer species.[5-7]) Lovett Doust and Harper[4] reported that the ratio of one hermaphrodite to four staminate flowers was maintained in a greenhouse experiment that produced plants with a fivefold range in both total flower number and total plant dry weight. These authors also described in *S. olusatrum* the allocation of nitrogen, phosphorus, and potassium in floral organs of plants that had received various experimental treatments. Within the flower, the ratio of resource allocated to sporophylls, as opposed to ancillary organs of attraction, was also found to remain fairly constant. The relative allocation of nutrient to male as opposed to female activities was quite distinct for each of the resources,[4] but the ratio was scarcely altered by treatments (imposed to manipulate the internal balance of resources of the plant). Of floral N, P, and K, 41 to 47% was found in the stamens and only 7 to 11% was in pistils. In contrast, the plants allocated 32% of floral dry matter to stamens and 18% to pistils. Thus, it was estimated that this plant at flowering expends about five times as much of its mineral resources and nearly twice as much of its dry matter on male as on female sporangia. The constancy of the sexual expression of whole plants over a wide range of plant size suggests that the ratio of ovulate to staminate activity is tightly canalized, and hence the result of strong selection pressures.

DICHOGAMY

Many Umbelliferae, including *S. olusatrum,* are protandrous, wherein stamens mature and become functional in a flower before pistils.[8] This protandry may regulate the extent of outcrossing and favor outcrossing while not precluding selfing. Flowers are tightly clustered, so pollen might be transferred readily between flowers within an umbel, depending upon the extent of protandry. As Bell[8] has described, if protandry is only weakly developed, then geitonogamous self-pollination (literally pollination of "sister" flowers on the same plant) can occur; if very strongly expressed, it will enforce cross-pollination. In *Smyrnium*, it would be misleading to assume protandry is simply an outbreeding device.[7] The individual flowers are protandrous, but individual umbellets (the simple umbels comprising a compound umbel) are in effect *protogynous.* This is so because flowers from the predominantly male, central section of any umbel mature later than those of the periphery, which are predominantly hermaphrodite flowers. Moreover, lower-order umbels are in effect protogynous with respect to higher-order (later) umbels, because they bear a greater proportion of hermaphrodite flowers. Lovett Doust[7] interpreted the apparently opposing forces of floral protandry and umbel protogyny as a possible solution to the problem of allocating limited resources and minimizing resource competition between male and female activities, pollen and ovule production, and nurturing of the seed.

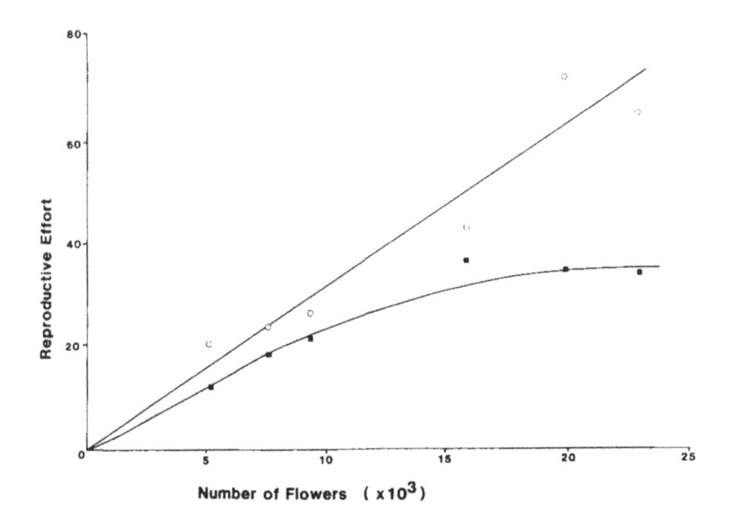

FIGURE 1. The relation between allocation of phosphorus and dry matter to reproduction, and the total flower production of *S. olusatrum*. Reproductive effort was calculated as the proportion of total plant phosphorus or biomass accounted for by flowers, schizocarps, pedicels, and rays. Phosphorus, open circles; dry matter, closed squares. (See Reference 2 for further details.)

FACTORS AFFECTING FLOWERING

Plants grown experimentally in a heated greenhouse did not differ in total accumulated biomass from controls grown in an unheated coldframe, by the end of the first year of growth. The hothouse plants did, however, have a significantly greater mean number of leaves at that time (November) — 7.4 leaves compared to 4.9 in coldframe plants.[8a] Over the first year, plants in the heated greenhouse also had undergone more rapid turnover of leaves, having borne a mean total of 23.4 leaves in their first year of growth (as vegetative rosettes), compared to 11.1 leaves in plants grown in the coldframe.

The greater cumulative foliage, however, did not serve to improve the likelihood of flowering in plants grown in the heated greenhouse; on the contrary, only 27% of the hothouse plants flowered (by July of their second year), compared to 67% of control plants grown in the coldframe. Of those that flowered, the usual 4:1 ratio of male to hermaphrodite flowers obtained, although the mean total number of flowers was about 7500, in contrast to control plants, which produced over 22,500 flowers.

It does not seem that the plants which flowered were simply those having greater reserves of biomass. Indeed, the mean total dry matter of flowering plants grown in the heated glasshouse was 62.4 g (SE 2.4), compared to 105.7 g (SE 7.9) for those which remained vegetative. In this species it appears that there is not a critical accumulation of biomass needed to stimulate flowering, as has been found for other weedy biennial species (see Reference 9). In *Smyrnium*, the total number of leaves borne by hothouse plants was greater, yet they flowered rather less frequently, suggesting that there is not a critical number of nodes that get produced before flowering will occur. It is likely that a period of cold plays a role in the life history strategy of *Smyrnium*. The greater rate of flowering of plants grown in the coldhouse suggests that vernalization may enhance the probability that a flowering apex will be produced in the second year of growth. This needs further study.

The allocation of dry matter may be compared with the allocation of phosphorus, and both of these resources related to flower and seed production (see Reference 2). In Figure 1 a linear relationship is apparent between allocation of phosphorus to reproduction and the mean total number of flowers of a plant. This is in contrast to the relationship between reproductive dry matter and flower number. Progressive increase in flower number was accompanied in a linear fashion by an increase in phosphorus reproductive effort. This suggests that phosphorus is a limiting factor to plant growth. Even in large plants, an increase in flower production requires a corresponding increase in phosporus utilization. Carbon (dry matter), on the other hand, seems less of a limiting resource for large plants; the carbon reproductive effort of large plants levels off while seed production continues to increase. Werner[10] pointed out that future studies of the evolution of resource allocation should consider the fact that selection of any characteristic ultimately is related to the number of descendants produced. An equal reduction in reproductive effort, in plants having different numbers of seeds, may have very different evolutionary consequences. The point at which any resource becomes limiting is critical.

REFERENCES

1. **Salisbury, E. J.,** *Weeds and Aliens,* Collins, London, 1961.
2. **Lovett Doust, J.,** Strategies of Resource Allocation and Sexuality in the Umbelliferae, with Special Reference to *Smyrnium olusatrum* L., Ph.D. thesis, University of Wales, Bangor, 1978.
3. **Lovett Doust, J.,** Experimental manipulation of patterns of resource allocation in the growth cycle and reproduction of *Smyrnium olusatrum* L., *Biol. J. Linn. Soc.,* 13, 155—166, 1980.
4. **Lovett Doust, J. and Harper, J. L.,** The resource costs of gender and maternal support in an andromonoecious umbellifer, *Smyrnium olusatrum* L., *New Phytol.,* 85, 251—264, 1980.
5. **Cruden, R. W.,** Intraspecific variation in pollen-ovule ratios and nectar secretion: preliminary evidence of ecotypic adaptation, *Ann. Mo. Bot. Garden,* 63, 277—289, 1976.
6. **Lloyd, D. G.,** The distributions of gender in four angiosperm species, illustrating two evolutionary pathways to dioecy, *Evolution,* 34, 123—134, 1980.
7. **Lovett Doust, J.,** Floral sex ratios in andromonoecious Umbelliferae, *New Phytol.,* 85, 265—273, 1980.
8. **Bell, C. R.,** Breeding systems and floral biology of the Umbelliferae, in *The Biology and Chemistry of the Umbelliferae,* Heywood, V. H., Ed., Academic Press, London, 1971, 93—108.

8a. **Lovett Doust, J.,** unpublished, 1978.

9. **Werner, P. A.,** Predictions of fate from rosette size in teasel *(Dipsacus fullonum* L.), *Oecologia,* 20, 197—201, 1975.
10. **Werner, P. A.,** Ecology of plant populations in successional environments, *Syst. Bot.,* 1, 246—268, 1976.

SOLIDAGO

En. Goldenrod; Fr. Solidage; Ge. Goldrute, Steingünsel

W. W. Schwabe

Solidago — the goldenrod — belongs to the same tribe of the Compositae as *Aster novi-belgii*, the *Astereae*.[1] While there is only one native European species, *S. virgaurea*, the goldenrod of gardens (*S. canadensis* L.) is a North American plant and the experiments described below were carried out with this species.

There have been only a few published investigations on the flowering of *Solidago* species. Allard and Garner[2] studied nine different North American species of *Solidago*, some of which were classified as SD requiring, some as intermediate, some as indeterminate, and one species and one strain of another were classified as LD requiring. These were as follows:

1. SD — *S. altissima, S. fistulosa, S. graminifolia, S. rugosa*, and *S. nemoralis*
2. Indeterminate — *S. juncea, S. sempervirens*, and *S. ulmifolia*
3. LD — *S. cutleri* and a strain of *S. sempervirens*

Goodwin[3] has shown that *S. sempervirens* L. is a SD-requiring plant for flowering and cites Turesson[4] as having found the same to be true for *S. virgaurea* L. *S. sempervirens*, being self-sterile, clearly depends on effective synchronization of flowering and his observations indicate a spread of flower initiation over only 2 weeks in September in natural conditions in Rochester, N.Y. Hurlbert[5] compared ten species of *Solidago* in relation to flowering time, all being SD requiring but differing in what amounts to their critical day-lengths; while van de Krogt[6] induced earlier flowering in strains of *S. virgaurea* by shortening the photoperiod so as to produce stems suitable as cut flowers. He found that in his cultivars only continuous light prevented flowering altogether.

The goldenrod is a rhizomatous perennial, and single shoots taken in the spring were used in the present experiments. No vernalization requirement was observed and plants given the appropriate daylength were capable of flowering. A comparison of the heights attained confirms this (Table 1). It requires no statistics to see that only the final daylength had any effect on height. None of the plants flowered normally, with those in LD continuing vegetative growth and those in SD ceasing to grow and becoming dormant. However, a few plants from LD stock, given chilling followed by SD, produced abortive inflorescences. Clearly, in a 16-hr LD, flowering is prevented by too long a photoperiod, while in an 8-hr day flowering is also prevented; the initial conclusion was that the goldenrod is a good example of a species with an intermediate daylength requirement — a category originally reported by Allard[7] and Allard and Garner.[2]

In a follow-up experiment, a series of plants were exposed to three daylengths: 12, 14, and 16 hr. The result is perhaps most simply demonstrated by Figure 1. Clearly a 16-hr day again proved too long to permit any flowering at all; normal flowering occurred in 14-hr days, but in a 12-hr photoperiod only some plants flowered while others aborted after reaching an early flower-initiation stage.

Briefly, the final results indicate that *S. canadensis* is a SD-requiring plant for flower initiation, but also responds to this daylength in becoming dormant. In fact, the dormancy inducing response is more rapid and predominates when the daylength is *very short;* i.e., in an 8-hr SD the onset of dormancy is so rapid that in most instances, flower formation cannot even be initiated. In a 12-hr SD there is almost a balance between the two responses, while in a 14-hr photoperiod normal flowering occurs. Thus, in the normal annual cycle of

Table 1
HEIGHT GROWTH IN *SOLIDAGO CANADENSIS*

	Plants from 24-hr LD stock				Plants from 8-hr SD stock			
	6 weeks chilling		No chilling		6 weeks chilling		No chilling	
Final daylength	**16-hr LD**	**8-hr SD**	**16-hr LD**	**8-hr SD**	**16-hr LD**	**8-hr SD**	**16-hr LD**	**8-hr SD**
Height (cm)	28.6	2.93	27.3	2.17	32.7	2.60	31.2	1.80

FIGURE 1. Plants of *S. canadensis* exposed to different daylengths. Left to right: plant in 8-hr SD (dormant and aborted inflorescence); two plants in 12-hr SD (poorly developed inflorescences); plant in 14-hr SD (normal inflorescence); and plant in 16-hr LD (vegetative).

changing daylengths, vegetative growth and shoot elongation are promoted by a LD; then as days shorten in late summer, flowering occurs; and finally as the equinox is approached and passed, the dormancy stimulus predominates and dormancy of basal shoots ensues, incidentally preserving these shoots for the next season. Clearly the ''intermediate'' photoperiodic behavior results from the competition of two SD responses; it then depends on the actual length of the photoperiod which of these predominates — a nice example of environmental adaptation to secure reproduction and survival by this perennial plant.

REFERENCES

1. **Clapham, A. R., Tutin, T. G., and Warburg, E. F.,** *Flora of the British Isles*, 2nd ed., Cambridge University Press, Cambridge, 1962.
2. **Allard, H. A. and Gardner, W. W.,** Further observations on the response of various species of plants to length of day, *U.S. Dep. Agric. Tech. Bull.*, No. 727, 1940, 1—64.
3. **Goodwin, R. H.,** The inheritance of flowering time in a short-day species, *Solidago sempervirens* L., *Genetics*, 29, 503—519, 1944.
4. **Turesson, G.,** The selective effect of climate upon the plant species, *Hereditas*, 14, 99—152, 1930.
5. **Hurlbert, S. H.,** Flower number, flowering time, and reproductive isolation among ten species of *Solidago* (Compositae), *Bull. Torrey Bot. Club*, 97, 189—195, 1970.
6. **van de Krogt, Th. M.,** *Solidago* als snijbloem de hele zomer beschikbaar, *Vakbl. Bloemisterij*, 29, 28—29, 1981.
7. **Allard, H. A.,** Complete or partial inhibition of flowering in certain plants when days are too short or too long, *J. Agric. Res.*, 57, 775—789, 1938.

SPIRAEA

En. Bridal wreath, Meadowsweet, Queen of the meadows, Spiraea; Fr. Reine des pres, Spairelle; Ge. Geissbart, Spierstaude

Dennis P. Stimart

INTRODUCTION

The genus *Spiraea,* a member of the Rosaceae, is composed of some 100 species. Plants are deciduous shrubs native to the northern hemisphere of mainly China, the Himalayan regions, Russia, India, North America, Europe, and Japan. As ornamentals in the garden they flower with great freedom and are often very graceful in growth habit. Many make excellent low hedges. In some places, the reddish or rich brown stems of many spiraeas make excellent winter color.

The flowering season of spiraeas can be divided into two groups.[2] The first group (Group A) contains species that flower from April to June with all white flowers. Flowers are usually produced all at once on wood from the previous season. Some of the most important species of this group arranged by relative flowering time are *S. thunbergii* Siebold ex Blume., *S.* × *arguta* Zab., *S. hypericifolia* L., *S. prunifolia* Siebold & Zucc., *S. media* Francz Schmidt, *S. pikoviensis* Bess., *S. pubescens* Turcz., *S. chamaedryfolia* L., *S. trilobata* L., *S.* × *vanhouttei* Zab., and *S. cantoniensis* Lour.[1]

The second flowering group (Group B) contains species that flower from late June to September with white and pink to crimson flowers which are produced on the current season's shoots. Flowers are not produced all at once but continue blooming for a longer time than those of the first flowering group. The flowering period of this group may be lengthened by removing spent flower clusters. Some of the most important species of Group B arranged by flowering time are *S. bella* Sims., *S. corymbosa* Raf., *S. densiflora* Nutt. ex Rydb., *S. japonica* L.f., *S. albiflora* (Mig.) Zab., *S. salicifolia* L., *S. alba* Du Roi (Meadowsweet), *S. menziesii* Hook., *S. douglasii* Hook., and *S. tomentosa* L.[1]

MORPHOLOGY

There are five short sepals, five petals, and numerous stamens attached to the rim of the cup-shaped receptacle of the flower. In some species, the stamen length exceeds that of the petals.[1,4] The inflorescence is a condensed raceme or rounded corymb produced from buds on the wood of the previous season or it terminates the growth of the current season in the form of a panicle or a complex corymb. The flowers are very uniform in size, and rarely more than 1 cm wide and 1.3 cm across in the doubled-flowered forms of *S. prunifolia* and *S. cantoniensis.* They are hermaphroditic. The species of Group A have flowers in umbels or umbel-like racemes at the end of short leafy twigs while those of Group B have flowers in corymbs or panicles produced at the end of a long shoot. The leaves are alternate, simple, without stipules, sometimes pinnately lobed, toothed, or more rarely entire.[4]

FLOWER INITIATION AND DEVELOPMENT

Although there have been extensive studies on floral initiation and development on other rosaceous plants, mainly the fruit plants, relatively few studies have been conducted on *Spiraea.* In *S. thunbergii,* normally an early-spring-flowering species, flower initiation does not begin until late summer or early fall of the previous season. During early September at East Lansing, Mich., flower formation first became apparent by the presence of several rounded masses of meristematic cells. By mid-September individual flower primorida could be distinguished as rounded masses with a flattened surface. Within the next 5 days the surface flattened and sepal primordia appeared. By early October, petal differentiation occurred and was followed 3 days later by stamen development. In late October, minute ovules were visible and all flower parts were differentiated. Microsporogenesis was not completed and megasporogenesis was not started until growth began the following spring. Anthesis started after several days of warm temperatures and lasted over a 1- to 2-month period.[3]

Similar results of floral initiation were found on *S. thunbergii* observed in Tokyo and Kagawa, Japan.[7] Flower bud cluster formation occurred in lateral buds of the current season's growth in early October. Pistils developed 10 days later and 7 days after the stamens formed. However, a few flowers began to bloom by mid-November with most of the them blooming at the end of March.

In studies with *S. cantoniensis* over 2 years, flower buds formed in the lateral buds of the current season's growth by October 17 in 1950, but on September 28 in 1951 in Tokyo.[7] However, sepal formation and differentiation of individual flower buds occurred on November 14, 1950 and November 9, 1951, respectively. The initial formation of petals, pistils, pollen, and ovules occurred from February 2 through April 17, 1951. Flowers bloomed on May 7. This development is slower in contrast to *S. thunbergii* and shows yearly variation.

Temperature can affect flower initiation and development of spireas. In controlled-temperature experiments with *S. cantoniensis,* flower initiation and development were not observed at 20 to 25°C, but were observed at 5 to 15°C.[6] The optimum temperature for floral initiation was about 10°C. However, further development of these flower buds was not observed for 5 months under 5 to 25°C. Only when plants were treated at 0°C for 6 to 8 weeks after initiation and then grown at 15°C or above in the greenhouse did further development of the flower bud and anthesis occur.

Under natural temperatures in Japan, the flower formation process was not affected by photoperiods.[6]

In summary, there appears to be considerable variation within a species as well as between species as to time of year at which the various sequences of flower development occur. Differences in seasonal temperatures may account for some of this variation.

DORMANCY

Photoperiod was found to influence vegetative bud growth of spiraeas. A light break in the middle of the night successfully overcame the effect of long nights in inducing dormancy of vegetative buds in *S.* × *bumalda* Burv. cv Froebelli.[8]

Temperature has been reported to influence the breaking of flower buds in the spring. In studies with *S. cantoniensis,* a period of low temperature followed by warm conditions was essential for flowering.[6] The range of low temperatures effective for flowering was −5 to −10°C, with the optimum temperature at 0°C. If a low-temperature treatment of 0°C for 6 weeks was given to plants before floral initiation, emergence and elongation of axillary buds occurred. If the treatment was given after flower initiation, emergence and elongation of axillary buds bearing the inflorescence occurred.

The forcing of spring flowers to bloom on *S. prunifolia* 'Plena' was found to require 19 days after the rest period was satisfied, provided the plants were grown at 13.0 to 16.0°C.[5]

PROPAGATION

Spiraeas can be propagated from newly ripened seed or after seed stratification.[2] Propagation can be by semihardwood stem cuttings taken in July and August or by hardwood stem cuttings taken in September.[2] Some species spread by underground sucker growths and are easily increased by dividing. Spiraeas cross fertilize freely. Therefore, seed can only be depended on to come true when plants are isolated.

CULTURE

Spiraeas prefer a loamy soil, abundant moisture, and full sunlight for optimum growth. The time of flowering affects the time of year in which plants are optimally pruned. Those species which are early blooming in Group A must be pruned immediately after flowering by thinning out and cutting away the older and weaker shoots. Unnecessary shortening back of the shoots will mean a reduction in the next crop of blossoms. The later flowering species of Group B benefit by shoot pruning. This should be done in late winter or early spring and at the same time superfluous old shoots should be removed. Unless pruning of either group is done, many spiraeas become weedy with thin growth and their blossoms will not compare in quantity and quality to pruned plants.

REFERENCES

1. **Bailey, L. H.,** *Cyclopedia of American Horticulture,* Vol. II, Doubleday, Page and Company, New York, 1906, 595.
2. **Bean, W. J.,** *Trees and Shrubs,* Vol. IV, St. Martins, London, 1980, 808.
3. **Carpenter, E. D. and Watson, D. P.,** The initiation and differentiation of flower buds of four rosaceous shrubs, *J. Am. Soc. Hortic. Sci.,* 86, 806—808, 1965.
4. **Chittenden, F. J.,** *Dictionary of Gardening,* Vol. IV, 2nd ed., Synge, P. M., Ed., Clarendon Press, Oxford, 1977, 1088.
5. **Coggeshall, R. G.,** Forcing woody plants for flower shows, *Am. Nurseryman,* 130, 7—63, 1956.
6. **Goi, M., Kawanishi, T., and Ihara, Y.,** Studies on the acceleration of flowering in woody ornamentals by low temperature treatments. V. On the flowering behaviour of *Spiraea cantoniensis* Lour., *Tech. Bull. Fac. Agric. Kagawa Univ.,* 26, 84—92, 1975.
7. **Kosugi, K. and Miyoshi, Y.,** Studies on the flower bud differentiation and development in some ornamental trees and shrubs. IV. On the time of flower bud differentiation and flower bud development in *Spiraea thunbergii* Sieb. and *S. cantoniensis* Lour., *J. Jpn. Soc. Hortic. Sci.,* 23, 172—176, 1954.
8. **Mahlstede, J. P.,** Report on field trial committee, *Proc. Int. Plant Prop. Soc.,* 6, 130—134, 1956.

TABEBUIA

En. Trumpet Tree

Rolf Borchert

INTRODUCTION

Leafless trees of *Tabebuia* (Bignoniaceae), covered with masses of large, trumpet-shaped, pink or yellow flowers (Figure 1) present one of the most spectacular sights encountered in tropical and subtropical countries during the dry season. Trumpet trees of the genus *Tabebuia* together with *Jacaranda mimosifolia* and *Spathodea campanulata,* also members of the Bignoniaceae, are among the most common flowering trees planted in warm climates.[11,13,16,17] Several Latin American countries have designated species of *Tabebuia* as their national tree or flower: Brazil *(T. serratifolia),* Ecuador *(T. chrysantha),* El Salvador *(T. rosea),* Paraguay *(Tabebuia* sp.), and Venezuela *(T. billbergii).*[10]

CLASSIFICATION, DISTRIBUTION AND ECOLOGICAL ADAPTATIONS OF *TABEBUIA* SPECIES

The genus *Tabebuia* comprises approximately 110 exclusively Neotropical species, ranging from northern Mexico and the West Indies to northern Argentina.[11] Synonymous names formerly applied to trees of the genus are *Bignonia, Cybistax, Couralia, Stenolobium,* and *Tecoma.* Species cultivated in the U.S. are listed in Table 1. Excellent color photographs of *Tabebuia* species are to be found in Menninger's *Flowering Trees for Warm Climates.*[17] Gentry distinguishes three major groups within the genus:[11]

1. Yellow-flowering, large (>30 m) trees occurring mostly in seasonally dry forests of continental tropical America from Mexico to Argentina, where isolated rainfalls during the dry season induce spectacular mass flowering of leafless trees[11] (Table 1). These slow-growing species produce some of the hardest, heaviest, and most durable woods among Neotropical trees, e.g., "guayacan", the wood of *T. guayacan.*[11]
2. Species range from microphyllous shrubs to large trees. Most have lavender or magenta flowers, but there are also species with red, white, and yellow flowers. This group includes all the West Indian species, which account for 80 of the 110 species of *Tabebuia* including over 50 species on Cuba and 30 on Hispaniola. These species are mostly distinguished by vegetative characters and are often adapted to specific substrates such as limestone or serpentine and a variety of other habitats. A Central American species of this group, *T. rosea,* is one of the principal timber trees of the region ("roble" [Span.] = oak).[11]
3. Two species, *T. donnell-smithii* and *T. chrysea,* were formerly segregated from the genus *Tabebuia* into the genera *Roseodendron* or *Cybistax.*[11]

Most species of *Tabebuia* are native to seasonally dry tropical and subtropical forests and flower during the dry season when leafless. However, several species are adapted to special habitats, which might make them useful for special horticultural situations.[11] Two subspecies of *T. chrysantha* occurring at higher elevations in the Colombian Andes (ssp. *pluvicola)* and up to 28°N in Sonora and Chihuahua, Mexico (ssp. *chrysantha)* should be fairly cold hardy, as is *T. umbellata,* which is widely cultivated in central Florida. Four species, all with white flowers and light wood, are salt tolerant and live in or just behind mangrove

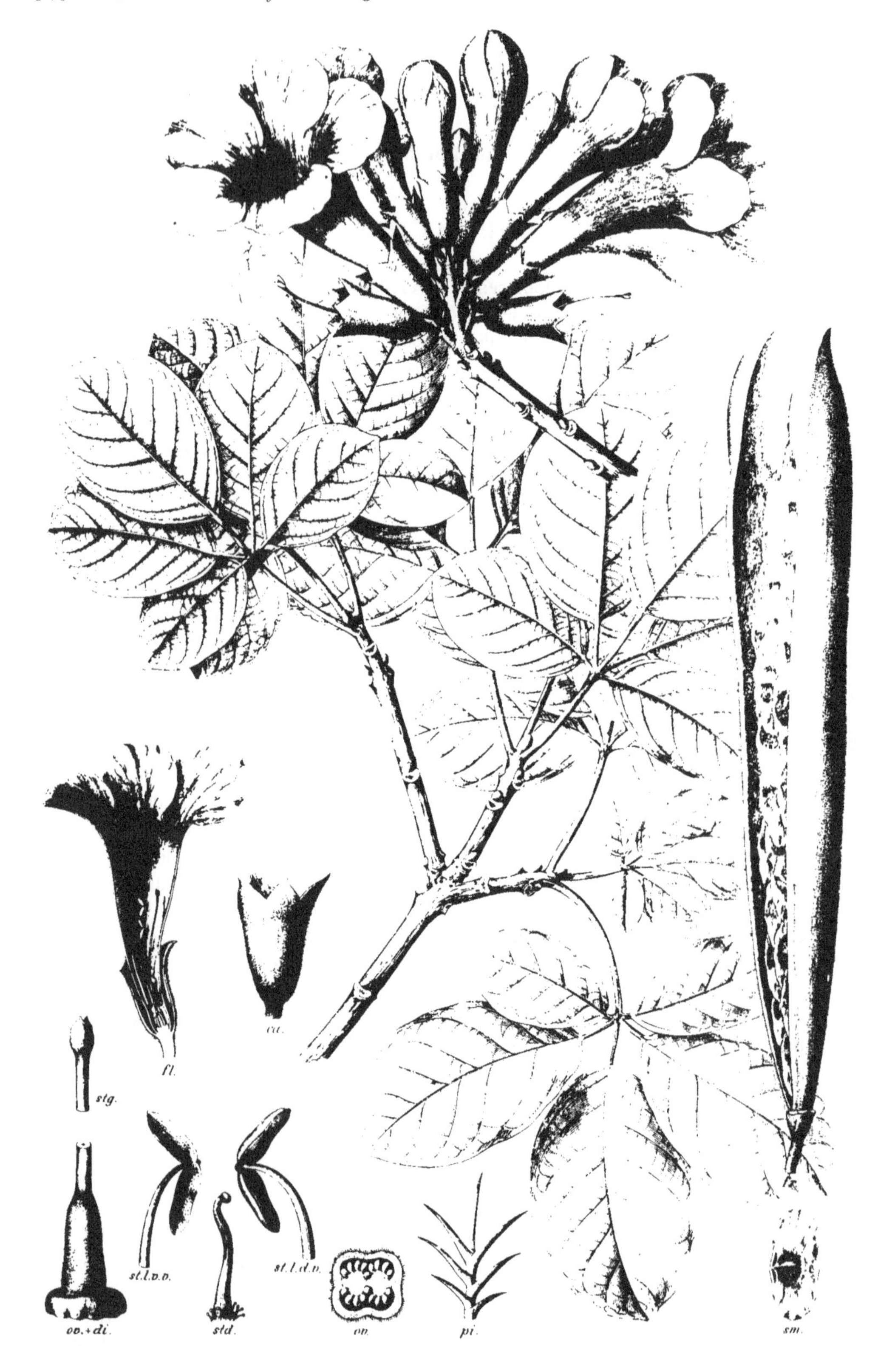

FIGURE 1. *Tabebuia ochracea:* inflorescence, vegetative shoot, flower parts, and fruit. (Shown as *Tecoma ochracea* in Table 112 of Reference 15.)

Table 1
HORTICULTURALLY IMPORTANT SPECIES OF *TABEBUIA* (AFTER GENTRY[10])

Species	Color	Origin
Group I. Large, Yellow-Flowered Trees of Dense Wood		
T. chrysantha	Yellow	High altitude, S.W. South America
T. chrysotricha	Yellow	Coastal Brazil
T. pulcherrima	Yellow	N. Argentina, Paraguay, S. Brazil
T. ochracea	Yellow	El Salvador to Argentina
T. guayacan	Yellow	Mexico to Colombia, Venezuela
T. serratifolia	Yellow	South America
T. umbellata	Yellow	Coastal Brazil
T. impetiginosa	Magenta	Mexico to Brazil
T. heptaphylla	Yellow	S. Brazil, Argentina, Paraguay
Group II. Microphyllous Shrubs to Large Trees of Mostly Lavender or Magenta Flowers		
T. caraiba (= *T. argenta*)	Yellow	Brazil, Paraguay, N. Argentina
T. roseo-alba	White-pink	Brazil, Paraguay, Peru
T. heterophylla	Pink	West Indies
T. rosea	Pink-magenta	Mexico to Ecuador, Venezuela
T. riparia	White	
T. dubia	Magenta	Cuba
T. haemantha	Red	Puerto Rico
T. lepidota		Cuba, Bahamas
T. bahamensis	Red	Cuba, Bahamas
Group III. Yellow-Flowered Trees of Lighter Wood		
T. donnelli-smithii	Yellow	Mexico to El Salvador
T. chrysea	Yellow	N. Colombia, Venezuela

swamps *(T. palustris, T. aquatilis, T. obtusifolia,* and *T. cassinoides).* The white-flowered *T. insignis* is restricted to freshwater swamps. Species like *T. pilosa* and *T. orinocensis* grow on outcrops of pure rock in the Amazonian basin, and others are adapted to special substrates such as limestone or serpentine (see the second item in the listing above). As virtually all species have the same number of chromosomes ($2n = 40$) and most, if not all, are interfertile, it should be possible to develop many useful hybrids combining desired properties of habitat, foliage, flower form and color, flowering behavior, and resistance to drought, cold, and salt.[11]

FLOWER MORPHOLOGY AND POLLEN VECTORS

The sympetalous flowers are borne in terminal panicles or racemes (Figure 1). Color and shape of the tubular corolla vary widely with the principal pollinators (see below). Flowers possess four didynamous stamens with bithecate anthers and a bilamellate stigma with elongate style. Fruits are elongate-linear or short oblong dehiscent capsules. Seeds may be thin with two hyaline-membranaceous wings or thick and corky without wings[8] (Figure 1).

Tabebuia has radiated to make use of many pollinators.[10,11] Most species, including all the cultivated ones (Table 1), are bee pollinated. Bee-pollinated flowers are yellow, pink, purple, or white, and are delicately fragrant. They possess nectar guide lines, and anthers are held against the dorsal side. At least two species, *T. stenocalyx* and *T. striata*, are hawkmoth pollinated; their white flowers open at night, produce exceedingly strong, sweet fragrances and abundant nectar, and have long, narrow corolla tubes. A few West Indian species (e.g., *T. haemantha),* lack fragrance, have relatively small, narrowly tubular red flowers, and are hummingbird pollinated. *T. platyantha* and *T. obtusifolia* are bat pollinated; their whitish, open, campanulate flowers are rather musky smelling, produce copious nectar, and bloom at night.[11]

PHENOLOGY AND CONTROL OF FLOWERING

Vegetative and reproductive development are strongly interrelated in all plants. In trees these relationships are considerably more complex than in herbaceous plants because of the structural complexity of the shoot system,[12] the life-long alteration between periods of vegetative growth and periods of rest, and repeated flowering.[3]

In *T. ochracea* ssp. *neochrysantha* (henceforth *T. neochrysantha),*[19] *T. rosea,*[3] and most likely all other species inhabiting areas with a pronounced dry season, the principal flush of vegetative shoot growth occurs during the early rainy season and is completed within at most 2 to 3 months. Shoot growth ceases under climatic conditions favorable for plant growth and must therefore be caused by internal, correlative inhibition. The small, naked, terminal buds of newly formed shoots remain dormant for several months and eventually develop into the terminal inflorescence, usually after leaves have fallen.[3,5,19] Observations with *T. rosea*[3,5] and published drawings (e.g., Figure 1) and photographs of other *Tabebuia* species[17] suggest that terminal buds do not always develop into an inflorescence, but may also form the next season's vegetative shoot, or abscise.[5] This implies, first, that internal factors responsible for shoot growth cessation are not identical with those inducing the apical meristem to become a flower bud, and, second, the transition from vegetative to reproductive mode must take place within the resting terminal bud, as it does in most deciduous temperate trees.[3]

This mode of flower development has several implications:

1. After a period of vegetative growth, the apical meristem may become wholly transformed into a flowering axis (hapaxanthy[3,12]) and vegetative growth is continued during the following growing season by the outgrowth of a pair of subtending, opposite, lateral buds.[5] Formation of terminal infloresceness is thus intimately related to the distinct, bifurcating branching pattern of *Tabebuia.*[5]
2. In *Tabebuia* as in many other trees, particularly those adapted to a seasonal climate, flower development is not continuous from flower induction to anthesis, as is common in annual, herbaceous plants, but is temporarily arrested at some intermediate state. Final development of the inflorescence and anthesis occur several months after formation of the putative flower bud (proleptic flowering[3]). Blooming of leafless trees, as is prevalent in *Tabebuia,* is thus indicative of proleptic flowering.
3. In proleptic trees like *Tabebuia,* flower induction and anthesis must be controlled by two different sets of factors, which must be considered separately.

Available evidence supports the hypothesis that in trees the transition of the apical meristem from the vegetative to the reproductive mode is determined mainly by internal, correlative factors, which are only poorly understood.[3,14] Flower induction and determination of flower buds appear to be stimulated by conditions which inhibit vegetative growth at or near the potential flower bud and favor the establishment of high carbohydrate levels.[3] These conditions prevail in the terminal bud of *Tabebuia* after shoot growth cessation, when shoot growth has been eliminated as a sink for assimilates, while mature leaves continue to produce assimilates. Contrary to repeated claims in the ecological literature, there is no good evidence for the induction of flowering in tropical trees such as *Tabebuia* by short photoperiods.[3]

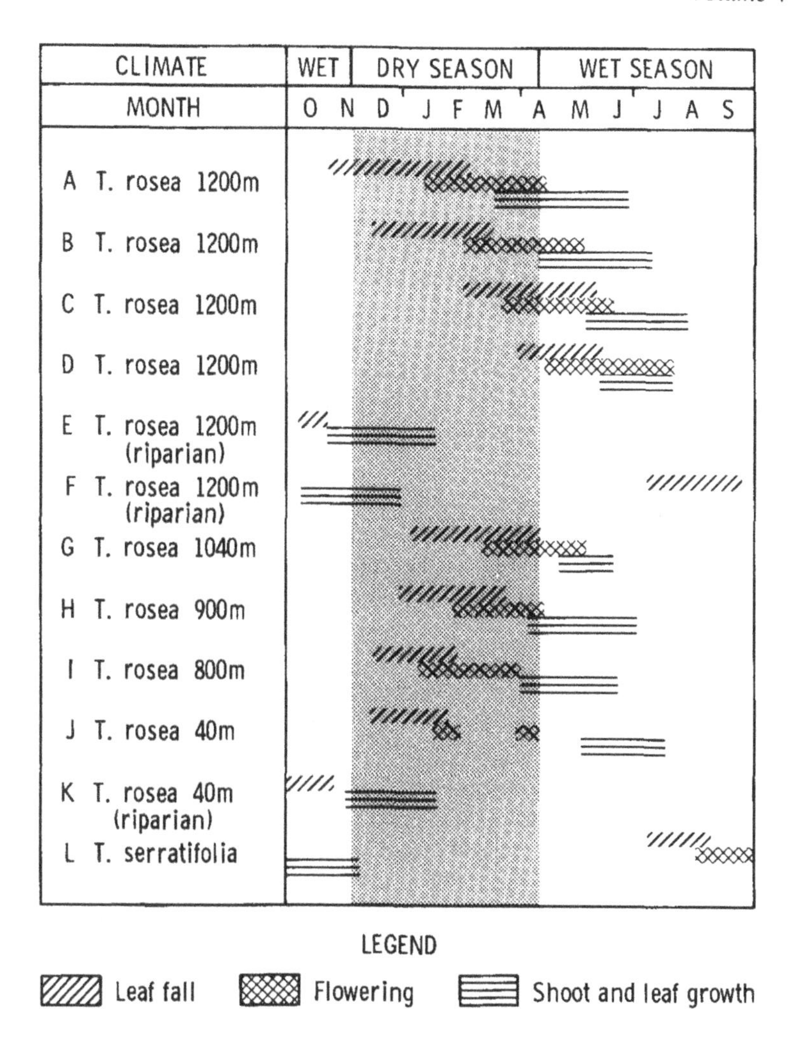

FIGURE 2. Periodicity of leaf fall, flowering, and shoot growth in *Tabebuia rosea* (A through K) and *T. serratifolia* (L). (Data from References 3, 20, and 21.)

have been shed (Figure 2). Increasing drought during the early dry season causes progressively greater tree water deficits, as indicated by shrinkage of trunks, and then leaf shedding.[20] Leaf fall occurring in moderately water-stressed trees eliminates the transpiring surface, thus reduces water loss, and permits rehydration of trees, as indicated by growth in girth during continuing drought.[2,20,21] As leaves are shed, flower buds begin to expand and flowers open gradually during a flowering period of several weeks (Figure 2). In *T. rosea,* leaves in the upper part of the crown often are shed earlier than those on the lower branches, and flowers open first on the bare, upper branches. This phenomenon clearly illustrates the correlative inhibition of inflorescence development and anthesis by senescent leaves, as discussed above. In areas of moderate drought, new shoots emerge shortly after flowers have opened (Figure 2A through D).

Along a descending, altitudinal gradient on the Pacific slope of Costa Rica ranging from 1200 to 40 m, mean temperature, evaporative demand of the atmosphere and the severity of the dry season increase. At lower altitudes, leaf fall and, hence, flowering occurred earlier, but shoot growth began later than at higher altitude; flowering and shoot growth became thus progressively more separated in time (Figure 2C, D, and G through J). In the tropical lowland deciduous forest, two flowering episodes were observed in *T. rosea:* the

The control of anthesis is basically different from that of flower induction. While the latter implies a switch from vegetative to reproductive development (i.e., a basic redirection of morphogenesis in the apical meristem), the triggering of anthesis mainly involves the release from an inhibitory, endogenous or environmental condition, usually water stress or low temperature, which prevents flower buds from completing their development. Virtually all successful attempts to establish good temporal correlations between environmental changes and phenological data on synchronous flowering in tree populations, as observed in *Tabebuia*,[11,19] deal with the triggering of anthesis in proleptic species, not with flower induction.

For temperate trees, and probably for most proleptic tropical trees, anthesis, like expansion of early leaves in a flush of shoot growth, consists essentially in the expansion of flower parts preformed in the resting bud. Cell expansion is known to be suppressed by even moderate water stress, and in fully foliated *Tabebuia* daily water deficits resulting from transpiration of old leaves may suffice to suppress opening of the large flowers.[1,19,20]

While the large majority of *Tabebuias* flower when leafless, two distinct, major patterns concerning the timing of anthesis can be distinguished. In some of the best known species of *Tabebuia*, mainly those adapted to tropical deciduous forests, all or most of the trees in a population flower in synchrony, and bare trees are then covered with a profusion of flowers which lasts only a few days ("big-bang" species[9,11,18,19]). In other species, such as *T. rosea*, flowers open more gradually and mass-flowering extends over a period of several weeks ("cornucopia" species[3,9], Figure 2). In view of the sensitivity of cell expansion and hence flower opening to water stress, mass flowering during the dry season raises the following basic question: how can trees in which increasing drought has caused leaf shedding eliminate water stress and flower during the dry season?

In leafless *T. neochrysantha* growing at dry sites, isolated rainfalls of >25 to 30 mm, as they occur at irregular intervals during the late dry season in Central America, caused rapid expansion of the trunk,[6,19] indicative of rehydration of the tree, and highly synchronous opening of flowers within 5 or 6 days after rainfall.[6,18,19] Flowering lasted for 4 days. Later rainfalls caused either a second, less intensive flowering episode or the emergence of young vegetative shoots.[18,19] If the first rainfall exceed 50 mm, flowers and young leaves appeared in rapid succession.[19] Similarly, watering of individual trees for 12 hr caused flowering only, while flowering and shoot growth were induced by continuous irrigation.[19] Trees at less-dry sites produced few flowers after leaf shedding and a few more after the first rains (compare Figure 2J). At wet sites trees retained some foliage until new shoot growth began and did not flower[19] (compare Figure 2E, F, and K).

Two prerequisites must thus be met for mass flowering in *T. neochrysantha* to occur: (1) trees must be exposed to drought for complete shedding of leaves, and (2) leafless trees must receive sufficient water for complete rehydration. Depending on the amount of water supplied to desiccated trees, they may (1) undergo partial rehydration without flowering (<20 mm), (2) open part of their flowers (~20 to 30 mm), or (3) flower and initiate shoot growth successively (>50 mm).[18,19] Lack of synchronized flowering, as observed in cultivated "big-bang" species,[11] is most likely the consequence of incomplete leaf fall due to exposure to only moderate drought.

It is not known, but should be investigated, whether trees at wet sites do not flower because (1) their terminal buds have not differentiated into flower buds or (2) existing flower buds fail to expand due to correlative inhibition by persisting foliage. Old leaves of *Tabebuia* are known to have lost stomatal control entirely and to experience severe daily water deficits,[1] which might prevent expansion of flower buds.

In contrast to the "big-bang" species, in *T. rosea*, *T. serratifolia*, and probably other species subject to only moderate drought, flowering is not synchronous among trees of a population (Figure 2) and hence cannot be caused by a well-defined environmental trigger such as a single, heavy rainfall. In these species, flowers always open shortly after leaves first, brief flowering episode after leaf fall was probably limited by rapidly declining soil moisture,[19] while the second was induced by a heavy rainfall during the late dry season, as described above for *T. neochrysantha;* shoot growth began during the early wet season after the soil moisture had been repleted (Figure 2J[20]). This pattern of vegetative growth and flowering in *T. rosea* is thus identical to that of *T. neochrysantha* growing at moderately dry habitats in the same area.

T. serratifolia is one of the exceptional deciduous tree species in the rain forest of Suriname (northern South America).[21] The tree normally sheds its leaves and then flowers at the beginning of the "long dry season" in August. This timing of leaf fall and flowering, about 6 months out of phase with flowering of *T. rosea* in Central America, clearly demonstrates the timing of the vegetative and reproductive cycle by seasonal drought, and makes control by photoperiod unlikely.[4] Trees of *T. serratifolia* growing on sand shed their leaves and came into flower one or more weeks before trees at nearby mesic sites. During an unusually dry "short dry season" (February to April), trees at sandy sites shed their leaves and blossomed while those at mesic sites did not.[21]

In moderately water-stressed *Tabebuia* the timing of flowering is thus a function of the timing of leaf fall, which varies with leaf age, leaf structure (mesomorphic vs. xeromorphic leaves), soil moisture availability, evaporative demand of the atmosphere, and other factors. Various combinations of factors determining leaf fall may cause variations of 4 to 5 months in the timing of leaf shedding and hence anthesis, within a population of the same species (Figure 2A through F) or among sympatric species in the same habitat. Similar correlated variation in the timing of leaf fall and flowering has been observed in the tropical tree *Erythrina poeppigiana* (Fabaceae) and in other tree species.[2,4,20]

Based on the above observations, it should be possible to induce prolonged mass flowering in cultivated *Tabebuia* by first withholding moisture until all leaves have fallen and then watering sparsely and repeatedly to cause gradual opening of flowers. In regions without prolonged, severe drought, chemical defoliation might be necessary to induce mass flowering in bare trees.

REFERENCES

1. **Borchert, R.,** Complete loss of stomatal functioning in aging leaves of tropical broadleaved trees, *Plant Physiol.*, 60(5), S60, 1979.
2. **Borchert, R.,** Phenology and ecophysiology of a tropical tree, *Erythrina poeppigiana* O. F. Cook, *Ecology*, 61, 1065—1074, 1980.
3. **Borchert, R.,** Phenology and control of flowering in tropical trees, *Biotropica*, 15, 81—89, 1983.
4. **Borchert, R.,** *Erythrina* in *Handbook on Flowering*, Vol. 5, Halevy, A. H., Ed., CRC Press, Boca Raton, Fla., 1986.
5. **Borchert, R. and Tomlinson, P. B.,** Architecture and crown geometry in *Tabebuia rosea*, *Am. J. Bot.*, 71, 958—969, 1984.
6. **Daubenmire, R.,** Phenology and other characteristics of tropical semi-deciduous forest in northwestern Costa Rica, *J. Ecol.*, 60, 147—170, 1972.
7. **Fournier, L. A.,** Estudio preliminar sobre la floración en el Roble de Sabana, *Tabebuia pentaphylla* (L.) Hemsl., *Rev. Biol. Trop.*, 15, 259—267, 1969.
8. **Gentry, A. H.,** Flora of Panama: Bignoniaceae, *Ann. Mo. Bot. Gard.*, 60, 781—977, 1974.
9. **Gentry, A. H.,** Bignoniaceae of southern Central America: distribution and ecological specificity, *Biotropica*, 8, 117—131, 1976.
10. **Gentry, A. H.,** Bignoniaceae. I. Crescentieae and Tourrettieae, *Flora Neotrop. Monogr.*, 25, 1—130, 1980.
11. **Gentry, A. H.,** The cultivated species of *Tabebuia* with notes on other cultivated Bignoniaceae, in *Proc. Menninger Flowering Tree Conf: Third Annu. Conf. 1982*, Burch, D., Ed., Florida Nurserymen and Growers Association, Tampa, 1982, 52—79.

12. **Hallé, F., Oldeman, R. A. A., and Tomlinson, P. B.,** *Tropical Trees and Forests. An Architectural Analysis,* Springer-Verlag, Berlin, 1978.
13. **Kunkel, G.,** *Flowering Trees in Subtropical Gardens,* W. Junk, The Hague, 1978.
14. **Longman, K. A.** Tropical forest trees, in *Handbook on Flowering,* Vol. 1, Halevy, A. H., Ed., CRC Press, Boca Raton, Fla., 1985, 23—39.
15. **Martius, C. F. P. and Eichler, A. W.,** *Flora Brasiliensis,* Vol. 8/1 Fleischer, Munich, 1856—1872.
16. **Menninger, E. A.,** *Tabebuia,* our best yard trees, *Proc. Fla. State Hortic. Soc.,* 73, 366—373, 1960.
17. **Menninger, E. A.,** *Flowering Trees of the World for Tropics and Warm Climates,* Heartside Press, New York, 1962.
18. **Opler, P. A., Frankie, G. W., and Baker, H. G.,** Rainfall as a factor in the release, timing and synchronization of anthesis by tropical trees and shrubs, *J. Biogeogr.,* 3, 231—236, 1976.
19. **Reich, P. B. and Borchert, R.,** Phenology and ecophysiology of the tropical tree, *Tabebuia neochrysantha* (Bignoniaceae), *Ecology,* 63, 294—299, 1982.
20. **Reich, P. B. and Borchert, R.,** Water stress and the phenology of trees in a tropical lowland dry forest of Costa Rica, *J. Ecol.,* 72, 61—74, 1984.
21. **Schulz, J. P.,** Ecological studies on rain forest in northern Suriname, *Verh. K. Ned. Akad. Wet. Afd. Natuurkd. Reeks 2,* 53, 3—268, 1960.

TAGETES

Tagetes patula **L.**
En. French marigold; Fr. Petit oeillet d'Inde; Ge. Ausgebreitete Studentenblume

Tagetes erecta **L.**
En. African or Aztec marigold; Fr. Rose d'Inde, Ge. Aufrechte Studentenblume

Tagetes lucida **Cav.**
En. Sweet-scented marigold

Tagetes tenuifolia **Cav.**
En. Signet marigold

Allan M. Armitage

INTRODUCTION

The genus *Tagetes* is a member of the Asteraceae (Compositae) family and consists of over 30 species native from Arizona and New Mexico to Argentina.[3,4] The garden marigold in North America is usually *T. patula, T. erecta*, or hybrids between the two resulting in triploid marigolds.[5] *T. lucida* and *T. tenuifolia* are used much more in Europe than in North America as bedding plants but not to the extent of *T. patula* and *T. erecta*.

T. patula is generally 30 to 45 cm tall, while *T. erecta* is usually >60 cm in height with large flowers. Triploid marigolds have been selected with the large flower properties of *T. erecta* combined with the branching properties of *T. patula*. Their height is intermediate between the two parents. Most marigold cultivars have flowers which are double, but single flower types are also available.

MORPHOLOGY

Tagetes are mostly annual plants with opposite pinnately dissected leaves. Leaves of *T. patula* are 1.5 to 2.5 cm long and are pinnately divided into about 12 lanceolate or oblong serrated segments, each serration usually with a large gland at its base. Flower heads are solitary and about 2 to 4 cm across. The involucre is thick and oblong with acute teeth, and is dotted with glands. There are numerous rays and the pappus is made up of one to two long-awned scales and two to three shorter blunt ones.[3,4]

The leaves of *T. erecta* are similar to *T. patula* except for their larger size (being 1.5 to 5 cm long) and the presence of a few large glands near the leaf margins. The peduncle of the solitary flower is swollen at the flower base and the teeth of the involucre are often elongate or dentate. The pappus is similar to that of *T. patula*.

The leaves of T. *tenuifolia* are also compound, approximately 0.6 to 1.8 cm long, and each segment has two rows of glandular dots. The solitary flower head is approx. 2.5 cm across and has few ray flowers (approximately five). The involucre has acute teeth with large glandular dots, while the pappus is made up of several scales, one or two of which are sharp-pointed or awned.[3,4]

The leaves of *T. lucida* are simple and sessile, characteristics not found on the other species previously mentioned. The flowers heads are approximately 1.5 cm across and occur in dense terminal clusters.[3,4]

ENVIRONMENTAL EFFECTS ON FLOWERING

Most of the research on the flowering process of *Tagetes* involves *T. patula* and *T. erecta* and the bulk of this chapter will be concerned with those species. Flowering of *T. tenuifolia* and *T. lucida* will be discussed where sufficient data exist.

Light

Photoperiod

Marigolds are quantitative SDP, but there are differences in degree of response to photoperiod. Flower bud primordia initiate just after germination in marigold regardless of daylength, but LD appear to retard flowering in marigold due to LD retardation of floral bud development.[18] Tsukamato et al. suggest that the strongest requirement for SD is exhibited by *T. tenuifolia* and the weakest by *T. erecta*, while *T. patula* appears to be intermediate in response.[18] Carlson et al.,[5] however, suggested that *T. erecta* has a greater requirement for SD than *T. patula*. The intermediate response of the French marigold shown by Tsukamoto et al.[18] is likely the result of its being the offspring of *T. erecta* × *T. tenuifolia* parentage.[17] Tsukamato et al.[18] did not note any retardation of daylength in African or French marigolds, although there was a slight delay in initiation in 'Signet' marigold in LD. In *T. patula* the most responsive age for reception of SD is between 14 and 34 days after germination. After that time, daylength is unimportant with respect to flowering time.[18] SD not only enhanced flowering in the main shoot but also in the secondary and tertiary shoots.

Flower number was increased under SD, indicating an enhancement of lateral shoot formation by SD compared with LD, particularly in French and Signet marigolds. There also appears to be considerable variation even between cvs within one species in their response to photoperiod.[5,18] Carlson et al.[5] found that some French marigold cvs such as 'Petite Orange' showed no response to photoperiod while others showed the typical quantitative SD response. Most of the African and triploid types, however, were strongly photoperiodic and many had not flowered 85 days from sowing when grown under LD conditions.

The critical photoperiod for floral development of *T. erecta* appears to be between 12.5 and 13 hr,[18] but critical photoperiods for other species have yet to be determined.

Flowering in marigolds was delayed equally by light levels of 10 lux from different lamp sources for 16 hr.[11] At this illuminance, no differences in stem length due to lamp sources occurred. However, 300 lux of incandescent light significantly increased stem length.[11]

Light Intensity

Work with 'Petite Yellow', a nonphotoperiodic French marigold, indicated that high light intensity could overcome the inhibiting effects of cool temperatures on flowering time.[1] Armitage et al. also suggested that high light intensity may be important for transition from floral bud to anthesis especially under high night temperatures (>26°C).[1] The greatest number of days to flower occurred under low light conditions regardless of day or night temperatures. Using daylength-responsive cvs Tsukamato et al.[18,19] studied flowering response in both the low light (December to January) and high light (August to September) months in Kyoto, Japan. During the low light period at 20°C, most flowers were produced when plants received LD compared with SD, while during the high light period there was no difference in flower number. They suggested that lack of light during SD in winter resulted in poor flower production as well as the occurrence of some blind shoots.[18]

They also found that SD accelerated flowering regardless of light intensity but as light intensity decreased, the retarding effect of LD was accentuated.[19] The greater the light intensity, the greater the flower number. When light was reduced to 10% of ambient, no flowers were developed regardless of photoperiod. The time to flowering was not markedly different when plants were returned to full light after 25 days at 75% shade compared with continuous 75% shade treatment; however, the number of flowers was greater when plants were returned to full light. This suggests that light intensity affects flower development more than flower initiation.[19]

Supplemental Light

Marigold 'Moonshot' given 4 weeks of continuous high-pressure sodium (HPS) supplemental irradiation flowered significantly earlier than those given either natural SD or 16-hr LD provided by incandescent light.[6] Plant height was greater for incandescent-treated plants but fresh weight was highest for plants receiving high-intensity discharge lighting. They also noted that root and top growth were greater with HPS after 3 or 4 weeks of continuous irradiance compared with 0, 1, or 2 weeks.[6] Carpenter[7] summarized results of supplemental lighting of many herbaceous plants and found that only 1 week of HPS lighting resulted in greater growth than nonlighted controls. There also appeared to be no difference in flowering time between fluorescent lighting of 1.5 to 2 Wm^{-2} for 3 to 4 weeks and HPS lighting of 0.5 to 1 Wm^{-2} for 2 to 3 weeks.[7] These data suggest the major role of supplemental light intensity is additional growth through photosynthesis. Supplemental lighting applied when the plants are young will likely result in accelerated flowering time due to decrease in time for floral initiation.

Temperature

Temperature appears to interact with light intensity and photoperiod in the French marigold. Cool temperatures (10°C), whether applied during the day or night, resulted in slower time to reach visible bud stage, but high light intensity appeared to be more important for flower anthesis under high night (20°C) temperatures compared with low night temperature (10°C).[1] In general, low light and low day temperature resulted in slowest flowering regardless of night temperature.[1]

French marigolds grown under LD and 30°C did not flower, but those grown under SD and 30°C resulted in a high percentage of flowering.[19] The authors noted that flower buds were formed in both LD and SD at cool temperatures but only in SD at high temperatures (30°C).[19] Flower development (initiation to anthesis) was accelerated by warm temperatures in the French marigold,[1,19] but Tsukamoto et al.[19] found that various combinations of daylength and temperature caused no marked difference in flowering of African marigolds.[2] Summerfield et al. grew *T. patula* for 11 weeks after transplanting at different night temperatures during the first 6 weeks compared with the last 5 weeks.[15] Raising night temperatures during early stages (weeks 1 to 6) resulted in increased flower dry weight compared with elevating night temperatures at later growth stages. This would be expected due to the early flower initiation in *T. patula* noted by Tsukamoto et al.[18]

Growth Regulators

Cathey suggested that phosphon, ancymidol (A-Rest), chlormequat (Cycocel), and daminozide (SADH, B-Nine) may be used for height control in *T. erecta*.[10]

McConnell and Struckmeyer[13] noted that plants of *T. erecta* 'Sovereign' flowered later when treated with daminozide compared with control plants regardless of photoperiod. The delay in flowering depended on the concentration applied, and a maximum delay of 8 days occurred with 2000 ppm daminozide application. *T. patula* was not responsive to daminozide.[8] GA (25 ppm) promoted reproductive development in *T. patula* regardless of photoperiod;[14,19] however, TIBA had no significant effect.[19] The application of ABA delayed flowering of *T. erecta* 'Golden Eagle' regardless of photoperiod.[9]

Scheduling

Scheduling of *Tagetes* for flowering in the greenhouse is dependent on species, growth habit, and environmental conditions under which they are grown. *T. patula* will require approximately 6 to 10 weeks from sowing to bloom depending on cultivar, with most flowering in 6 to 7 weeks.[12] *T. erecta* requires 9 to 13 weeks for bloom under "normal" greenhouse practices. Scheduling data in the southeast U.S. indicate that *T. patula* requires 6 to 7 weeks,[2] while data from the midwest area of the U.S. report 6 to 13 weeks for *T. patula.*[16]

REFERENCES

1. **Armitage, A. M., Carlson, W. H., and Cress, C. E.,** Determination of flowering time and vegetative habit of *Tagetes patula* through response surface techniques, *J. Am. Soc. Hortic. Sci.*, 106, 632—638, 1981.
2. **Armitage, A. M.,** Determining optimum sowing time of bedding plants for extended marketing period, *Acta Hortic.*, 147, 143—152, 1983.
3. **Bailey, L. H.,** *Manual of Cultivated Plants*, Macmillan, New York, 1951.
4. **Bailey, L. H.,** Staff of Hortorium, *Hortus Third, a Concise Dictionary of Plants Cultivated in the United States and Canada*, Macmillan, New York, 1976.
5. **Carlson, W. H., Armitage, A., and Mischel, J.,** Producing Marigolds for Profit, a Commercial Grower's Guide, Michigan State University Extension, Bulletin E-1443, Michigan State University, East Lansing, 1982, 4 pp.
6. **Carpenter, W. J. and Beck, G. R.,** High intensity supplementary lighting of bedding plants after transplanting, *HortScience*, 8, 482—483, 1973.
7. **Carpenter, W. J.,** High intensity lighting in the greenhouse, *Mich. Agric. Exp. Stn. Res. Rep. No.*, 255, 1974.
8. **Cathey, H. M.,** Prescriptions for annual plants — light and chemicals. II, *Flor. Rev.*, 134(3459), 23—25, 1964.
9. **Cathey, H. M.,** Response of some ornamental plants to synthetic abscisic acid, *Proc. Am. Soc. Hortic. Sci.*, 93, 693—698, 1968.
10. **Cathey, H. M.,** Growth regulators. I, in *Bedding Plants, A Manual on the Culture of Bedding Plants as a Greenhouse Crop*, Mastalerz, J. W., Ed., Pennsylvania Flower Growers Association, University Park, 1976, 177—189.
11. **Cathey, H. M. and Campbell, L. E.,** Effectiveness of five vision lighting sources on photo-regulation of 22 species of ornamental plants, *J. Am. Soc. Hortic. Sci.*, 100, 65—71, 1975.
12. **Goldsmith, G. A. and Wilson, A.,** Sun plants, in *Bedding Plants, A Manual on the Culture of Bedding Plants as a Greenhouse Crop*, Mastalerz, J. W., Ed., Pennsylvania Flower Growers Association, University Park, 1976.
13. **McConnell, D. B. and Struckmeyer, B. E.,** Effect of succinic acid 2, 2-dimethylhydrazide on the growth of marigold in long and short photoperiods, *HortScience*, 5, 391—392, 1970.
14. **Peat, J. R. and Summerfield, R. J.,** Environmental and cultural effects on vegetative growth and flowering of selected "bedding" ornamentals. II. Night-break lighting and gibberellic acid applications, *Scientia Hortic.*, 7, 81—89, 1977.
15. **Summerfield, R. J., Dawson, E. M., and Peat, J. R.,** Environmental and cultural effects on vegetative growth and flowering of selected "bedding" ornamentals night. I. Night temperature, *Scientia Hortic.*, 7, 67—79, 1977.
16. **Tayama, H. K., Kiplinger, D. C., Brooks, W. M., Staby, G. L., Powell, C., Lindquist, R. K., Poole, H. A., Farley, J. D., and Roberston, J. L.,** Tips on Growing Ornamental and Vegetative Bedding Plants, Cooperative Extension Service, Ohio State University, Columbus, 1976.
17. **Towner, J. W.,** Cytogenetic studies on the origin of *Tagetes patula*. I. Meiosis and morphology of diploid and allotetraploid *T. erecta* × *T. tenuifolia*, *Am. J. Bot.*, 48, 743—751, 1961.
18. **Tsukamoto, Y., Imanishi, H., and Yahara, H.,** Studies on the flowering of marigold. I. Photoperiodic response and its differences among strains, *J. Jpn. Soc. Hortic. Sci.*, 37, 47—55, 1968.
19. **Tsukamoto, Y., Imanishi, Y., and Yahara, H.,** Studies on the flowering of marigold. II. Interactions among day-length, temperature, light intensity and plant regulators, *J. Jpn. Soc. Hortic. Sci.*, 40, 65—70, 1971.
20. **Youngsman, J. E.,** Response of Marigolds, *Tagetes erecta* Linn., to Photoperiod, Temperature and Phosphonium, MS thesis, Pennsylvania State University, State College, 1962.

THEOBROMA CACAO

En. Cacao; Fr. Cacao; Ger., Kakao; Sp. Cacao

Paulo de T. Alvim

INTRODUCTION

Cacao *(Theobroma cacao* L.) is a native species of the rain forest region of tropical South and Central America.[18,27] In the English language the name "cocoa" is commonly used as a synonym of cacao, but most authors employ "cacao" to designate the plant, while "cocoa" is used more frequently to refer to its product, as cocoa bean, cocoa powder, cocoa butter, etc.

Cacao belongs to the family Sterculiacee and its center of origin is believed to be the headwaters of the Amazon basin, where *Theobroma* and related genera are found in greater proportion than elsewhere.[18] From its center of origin the species spread mostly in two directions, resulting in the two primary groups of cultivars known as "criollo" and "forastero". The former originated from dispersion across the Andes toward the lowlands of Venezuela, Colombia, and Ecuador, and northward to Central America and Mexico. The latter resulted from dispersion down the Amazon valley toward northern Brazil and the Guianas. Over 95% of the current world production of cacao is of forastero origin and comes mainly from West African countries and Brazil.

Cacao was first domesticated by the Mayas in Central America a long time before the Europeans arrived in the Western Hemisphere.[27] The name of the plant is derived from the Nahuatl word "cacahuatl", just as the name of the popular drink chocolate is derived from "xocoatl" or "chocoatl".[18]

As a typical crop of the humid tropics, cacao requires a relatively high rainfall, usually 1500 to 2500 mm/year, and a mean annual temperature above 23°C, normally 24 to 25°C.[8] Most producing areas are located within a fairly narrow belt about 15°N and 15°S of the equator. Exceptionally, some small plantings are found in subtropical latitudes as, for instance, in the lowlands of the State of São Paulo, Brazil, at 21 to 23°S.

As a rule, cacao is grown at a relatively low altitude, usually below 300 m, but in places near the equator plantings are sometimes found at fairly high elevations, for example, in some regions of Colombia (1000 to 1200 m), in the Chama valley of Venezuela (900 m), and in Uganda (1100 to 1300 m).

Cacao production started to expand during the last century, first in the Western Hemisphere (Ecuador, Venezuela, and in the State of Bahia, Brazil) in response to the development of the chocolate industries in Europe and in the U.S. Following the introduction of the crop about a century ago in West Africa, world production increased very rapidly from about 100,000 tons at the beginning of the 20th century to the present production of 1.5 to 1.7 million metric tons per year. About 80% of this production come from the following countries: Ivory Coast (24%), Brazil (22%), Ghana (14%), Nigeria (10%), Cameroon (6%), and Ecuador (4%). The remaining 20% comes from about 30 other countries.

PLANT AND FLOWER MORPHOLOGY

Cacao is a small tree, normally 5 to 8 m high, with a dimorphic growth habit. Its main stem is orthotropic with long-petiolated leaves arranged in double spiral phyllotaxis.[55] After growing for about 1 year to a height of 1.0 to 1.5 m, the terminal bud of the orthotropic trunk is replaced by a cluster of secondary buds which produces three to five plagiotropic

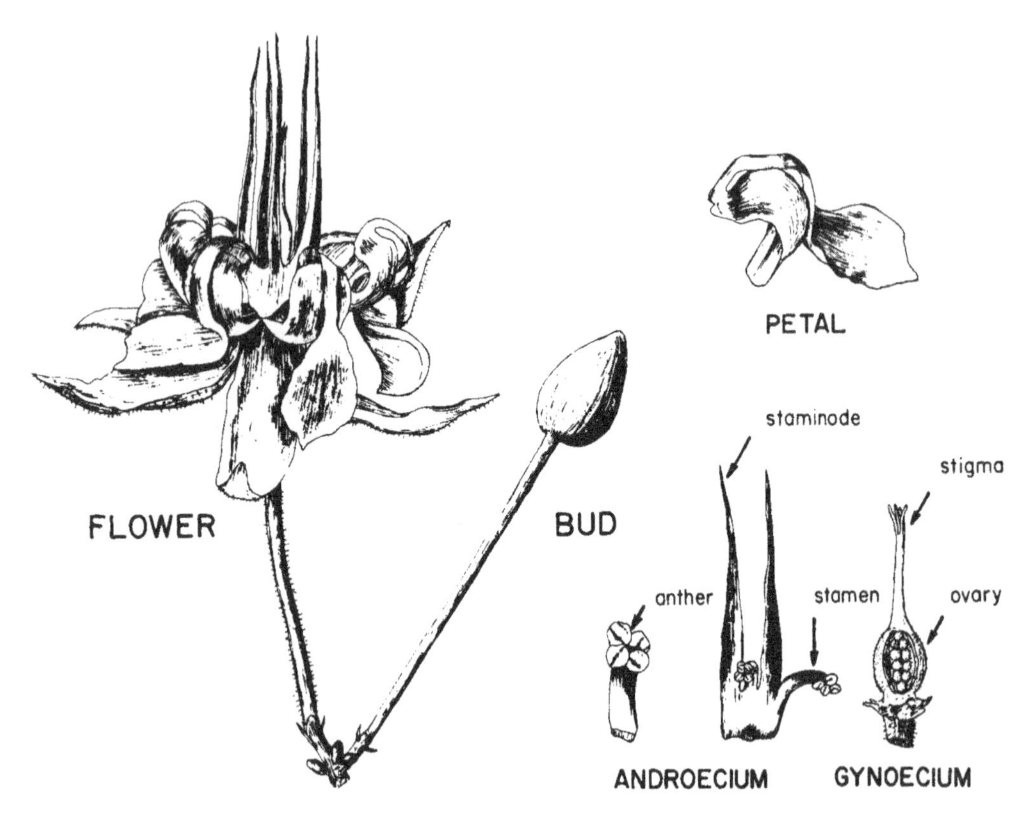

FIGURE 1. The flower of cacao and its parts.

or "fan" branches arranged in a terminal whorl known as "jorquette". The leaves of these fan branches are short-petiolated and arranged in a distichous phyllotaxis. Further growth of the plant may take place from axillary buds in orthotropic or plagiotropic branches, the former producing new vertical shoots, known as "chupons", with the same architecture of the main trunk, and the latter giving origin to new fan branches. In mature trees orthotropic branches sometimes originate from plagiotropic branches of thicker diameter, but in most trees the upright growth comes usually from chupons of the main trunk, and this leads to the formation of a sympodial main trunk with alternating internodes and irregular nodes.

Cacao is a typical cauliflorous species, the flowers giving the impression of originating directly from the bark of the plant. Actually they are formed in minute inflorescences with short internodes and branchlets, originating from the axil of the leaves.

Cacao has two buds in the axils of most of its leaves, a larger one known as the "principal" bud, visible to the naked eye, and a smaller one, the "subordinate" bud, often obscured by the dense indumentum of the stem and located in a depression in the stem between the principal bud and the base of the petiole.[55] The principal bud grows only as a vegetative branch, while the subordinate bud usually forms an inflorescence.[55] Some authors were able to cause a subordinate bud to grow out as a vegetative branch by decapitating the shoot and destroying the principal bud.[26,53]

The individual cacao flower (Figure 1) has a thin and relatively long (1 to 1.5 cm) pedicel and a calyx with 3 to 7 (normally 5) sepals measuring 6 to 8 mm long. Prior to anthesis these sepals form an oval body, measuring 5 × 3 mm, at the tip of the pedicel (Figure 1). The corola has 5 petals about 6 to 8 mm long which are divided in two sections: (1) a lower one corresponding to the claw, which is strongly veined and in the shape of a hood *(cucullus)*, normally with two red lines, and (2) an upper white blade *(lamina)* which is flat, flexible, and articulated to the apex of the claw. The androecium has an outer whorl of 5 sterile and rather long (5 to 7 mm) staminodes, opposite to the sepals, usually with red pigmentation, and an inner whorl of 5 fertile and relatively short stamens opposite to the petals. The anthers are hidden inside the petal hoods and have two unilocular thecae which open by a longitudinal cleft. The gynoecium has an ovoid 5-celled ovary with 40 to 60 ovules (two rows of 8 to 12 in each cell) and 5 apical stigmas.

The term "flower cushion" is commonly used to refer to the site where the cacao inflorescences appear. Each cushion remains active (i.e., produces flowers) for many years, apparently only becoming sterile when accidentally injured by cuts (as may happen when the fruits are harvested) or because of bark diseases (canker).

In young plants the flowers are produced mainly on the trunk, but in mature trees they normally appear all over the plant or in most of the branches thicker than about 1 cm in diameter. As the branches get thicker, the flower cushions usually become swollen and the number of flower buds per cushion increases. As a rule, no more than one or two flowers open per cushion on the same day, but under favorable conditions for flowering it is common to find as many as five to ten open flowers in older cushions and a much greater number of flower buds at different stages of development. An adult cacao plant may produce several thousand flowers per year, sometimes more than 50,000, of which only a small proportion (usually less than 5%) are pollinated, and an even smaller fraction (0.5 to 2.0%) lead to the final crop. From pollination to harvesting, the cacao fruit, which weights from 400 to 800 g and is usually referred to as a "pod", takes from 5 to 6 months to mature, depending on the temperature.[8,10]

FLOWER BIOLOGY

The anthesis of the cacao flower follows a typical nyctitropic pattern which has been described by several authors[21,34,35] and also recorded by the author through slow-motion pictures.[9] Normally, the process begins late in the afternoon, when the sepals start to separate from each other, and is completed early in the morning of the next day, usually between 4 to 6 a.m. This pattern has been observed also with leafless branches, detached from the plant, supplied with water, and exposed to different treatments of temperature and photoperiodism.[9] The nyctitropic movement remained unalterable under all conditions, suggesting an endogenous stimulus for anthesis.

Cacao is an entomophilous species with a flower structure adapted for cross-fertilization. Flowers which are not pollinated within the first 8 to 10 hr after anthesis will normally fall in the next 24 to 48 hr. It is now well established that the most effective and important natural pollinators are the tiny midges of the genus *Forcipomyia* (Diptera: Ceratopogonidae) of which there are innumerable species.[17,46,56] A few other small insects, such as thrips, ants, and aphids, may also pollinate cacao flowers, but *Forcipomyia*,[17] because of their size and shape, are more efficient in collecting pollen (when crawling under the petal hoods which cover the anthers) and in transferring pollen to the stigma (when crawling down the staminodes of a next flower).[17] This is regarded as a case of "mutual adaptation" between plant and insect.[51] The midges are believed to visit the flowers in search of food and/or perfume, but this has not been experimentally demonstrated. Some authors suggested that the red pigmentation of the staminodes and petal hoods would be stimulus to attract the insects,[32-34] but this hypothesis also is not generally accepted, as flowers of unpigmented cultivars are equally attractive to the midges.[48]

A considerable amount of research has been done on the biology and taxonomy of *Forcipomyia*.[39-41,49,50] Of the various species reported as important cacao pollinators in different regions, the most frequently found belong to the subgenus *Euprojoannisia*, formerly named *Proforcipomyia*.[39,41,56]

Recent experiments have shown that cacao flowers may also be "artificially" pollinated by mechanical means, such as, for instance, by agitating the flowers with a hand brush or by blowing air directly on the flowers with a hand brush or by blowing air directly on the flowers with a motorized blower.[45,47] These treatments have been reported to increase fruit setting and yield when applied to self-compatible cultivars in areas where shortage of natural pollinators is a common problem.[11,29,47]

FACTORS AFFECTING FLOWERING

Comparative studies on the seasonal flowering pattern in the various cacao-growing countries show that flower production is primarily controlled, either directly or indirectly, by climatic factors. As a rule, in regions with well-defined seasonal variability in rainfall and/or temperature, flowering is much reduced or may be completely inhibited during excessively dry or relatively cold periods.[2,5,6,12,13,23] In places with well-distributed rainfall and no marked seasonal differences in temperature, cacao usually produces flowers and fruits throughout the year. Even in such regions, however, flowering is never uniform during the year but varies in intensity under the influence of different factors, some of which are not completely understood. Since different factors may simultaneously affect the flowering process, it is often difficult to establish causal relationships for the variability of flowering.

In the author's opinion, fluctuations in flowering intensity in cacao are linked to the growth rhythms of the inflorescences themselves, which apparently follow a pattern similar to, but not necessarily synchronous with, the growth rhythms or flushing cycles of the vegetative buds of the plant. Both phenomena are affected not only by climatic factors but also by internal plant factors. Once an inflorescence starts its own "flushing cycle", the young flower buds begin to expand in sequence (as in the case of leaf flushes), this leading to a "wave" of flowering which may last for several days or weeks. The period of fastest growth of the individual buds apparently occurs during the last 10 to 12 days prior to anthesis, when they expand from the size of a pinhead to fully grown.[21] The buds apparently do not have a rest period (as in the case of coffee) and anthesis occurs as soon as they reach full size.[37]

In order to review our present state of knowledge on factors controlling cacao flowering, a separate analysis will be made of internal and external factors. Regrettably, because of limited research on flower-bud initiation in cacao, it is not always possible to separate the influences of these factors on bud differentiation and bud growth.

Internal Factors

Our knowledge about internal or plant factors is based on studies related to the effect of age, girdling, internal correlations, and cambium activity. No attempt has been made to identify the chemical nature of flowering stimulus in cacao.

Age

The age at which cacao starts producing flowers varies greatly with cultural practices and genetic origin of the plant. As a rule, high-yielding varieties under good management start flowering at the age of about 2 years in the field, while unselected varieties under traditional cultural practices start at the age of 3 to 4 years. Plants growing in nutrient solutions under greenhouse conditions may start flowering before 1 year of age. An extreme case of genetically controlled premature flowering has been reported for some "criollo" seedlings from Mexico, which began to produce flowers at the age of only 3 months.[44]

In places where there is a marked seasonal fluctuation in flowering it has been observed that this fluctuation is much more pronounced in mature plants than in young ones. This was demonstrated in an experiment comparing plants aged 5, 12, and 23 years.[8] The 23-year-old plants produced about ten times as many flowers as the 5-year-old plants during the peak of flowering (February to April), and about ten times less during the low-flowering period (July to September). The data for the 12-year-old-plants fell between the other two age groups. These differences may be due not to age itself, but to plant size or to the microclimate around the plants. Smaller plants, because of reduced leaf canopy or less self-shading, are more exposed to solar radiation than larger plants, and this may be an important factor in the difference in their response to variations in climate. Older plants, on the other hand, because of their larger proportion of nonphotosynthetic tissues which act as a sink for photosynthates, may also be more readily depleted of flower-promoting substances than younger plants.[8]

Girdling

As reported for several other species,[19,24] the removal of a ring of bark from the trunk of mature cacao trees has the effect of increasing flowering on the portion of the trunk above the ring and decreasing or completely arresting flowering below the ring.[10] Approximately 45 days elapse between girdling and full bloom.[52] Some studies suggest the possibility of using ringing as a method to induce out-of-season flowering or to change the cropping pattern of cacao.[52] In Ecuador, where fruit losses carried by the diseases "witches' broom" *(Crinipellis perniciosa)* and "monilia pod-rot" *(Monilia roreri)* are strongly seasonal, the method is claimed to be relatively efficient in reducing the disease incidence in some regions.[52]

Internal Correlations

Among some cvs of cacao it is common to find a few plants which bear no fruit or have a limited crop because of a genetically controlled flower sterility or incompatibility. Such plants, which are usually referred to as "male cacao", invariably show very profuse flowering. It has also been observed that in most regions where there is a pronounced seasonal variation in flowering, the period of low flowering invariably occurs when the plants have the highest fruit load.[8,10] These are clear indications of a competitive correlation or a depressing effect of fruits on flowering.[10,12,23]

To study interactions between flowering and fruiting of cacao an experiment was conducted in Brazil in which fruits were manually removed twice a week for a period of 2 years.[54] Flowering intensity which was measured in comparison with untreated plants growing in the same area showed marked increases following fruit removal. Partial fruit removal also stimulated flowering, although not as markedly as when all fruits were removed. The study indicated that fruit size or age appeared to make little difference, as flowering intensity increased following the removal of either young (between 1 to 6 cm in length) or of nearly mature fruits. In other species, this depressing effect of fruits on flowering has been attributed to growth substances produced by the seeds of the growing fruits, as GA, which is known to inhibit flower initiation.[16,20,22]

Cambium Activity

With the help of a sensitive liquid-displacement dendrometer, cambium growth was measured at weekly intervals in Bahia, Brazil from 1975 to 1978.[7,9] Simultaneous studies on the flowering pattern of the same plants showed that the period of minimum cambium activity, normally from July to October, coincided every year with the period of minimum flowering. Resumption of cambium growth took place in October or mid-November and was followed by resumption of flowering about 5 to 6 weeks later. It was concluded that two factors probably contributed to the seasonal decrease in flowering and cambium growth in this study: the relatively low temperature from June to September and the inhibiting effect of the heavy load of fruits in the plants as previously discussed.

External Factors

The main environmental factors affecting cacao flowering are shading, temperature, and rainfall distribution or moisture availability. Photoperiodism has apparently no effect on cacao flowering.[6]

Shading

Because cacao is usually grown in association with larger trees, much attention has been given to research on its response to shading. It is now well established that the benefit of shading is not to provide a low light intensity considered optimal for growth and yield of cacao, but mainly to counteract unfavorable factors such as moisture stress, nutrient deficiencies, wind damages, etc., which are usually more severe when the plants are grown without shade.[6,8,31] When such unfavorable conditions are not present or are controlled by methods other than the use of overhead shade, yield is usually higher when the plants are grown with little or no shade.[8,12,31]

Several studies have demonstrated that plants grown without shade flower earlier and more profusely than shaded plants.[12,13,25,31] An experiment in Ghana comparing three shade regimes (heavy, light, and no shade) showed that flowering followed the same seasonal pattern in all regimes but the total number of flowers produced decreased significantly with increasing shade.[12] Conditions for effective pollination were more favorable under light shade than under no shade and fruit set was highest in that regime. However the final yield was higher under no shade because of reduced loss of fruits from the physiological disturbance known as "cherelle wilt".

Relatively young plants appear to be more sensitive to shade than old ones. This was found in a study in Brazil comparing 12- and about-50-year-old plants, in which shading reduced flowering by two thirds in the former and about one half in the latter.[25]

Temperature

The effect of temperature on flowering of cacao has been studied both under field conditions and in controlled environment rooms.[5,6,8,36] Field studies in Bahia, Brazil and in Cameroon[6,13] suggested that flowering was inhibited when the mean monthly temperature dropped below 23°C and that this effect of low temperature appeared to be "indirect" in the sense that it first affected vegetative growth which in turn would affect flowering about 2 months later.[5,6] This "indirect" influence was not supported by research under controlled temperature conditions which showed that temperature affected directly the process of flower-bud differentiation.[36] The 2 months of time lag observed under field conditions may probably be attributed to the fact that flower counts were not based on bud differentiation but on flower dropping. As suggested by the work on the effect of girdling on flowering, the time lag between differentiation and anthesis appears to be about 45 days.[52]

The work under controlled temperature conditions also demonstrated that flowering was greater at day temperatures of 26 and 30°C than at 23°C and at each level of day temperature, flowering was greater at a night temperature of 26°C than at one of 23 or 30°C. The relative effects of temperature were similar on numbers of flowering cushions per plant and of flowers per cushion.[36]

Some recent studies carried on also in Bahia revealed that low temperature alone cannot explain the reduced flowering observed in that region from July to October. It has been found that the period of low flowering from July to August resulted mainly from relatively low temperature but in September to October the heavy load of fruits on the plants appears to be the main cause of low flowering.[8,10] As mentioned in our previous discussion on internal factors, fruit removal greatly increased flowering during the period of relatively low temperature, although not as much as during warmer periods.[54]

Moisture Availability

Among perennial crops grown in the wet tropics, cacao is regarded as one of the most sensitive to moisture stress. Changes in soil moisture availability during the year are generally recognized as the most important climatic factor controlling the physiological processes of the plant, including flowering. In all producing regions variation in yield from year to year or changes in the cropping pattern during the year are caused primarily by irregularities in the distribution and total volume of rainfall.

Because of the great commercial interest in developing methods to forecast the cacao crop, considerable attention has been given to research on the relationship between rainfall and yield. Good correlations have been shown to exist between monthly yield and the rainfall 5 to 7 months before, when flowering and fruit setting take place.[1,14,30,43] In practically every cacao region good crops have long been known to be associated with favorable rainfall distribution during the year.

The effect of moisture availability on flowering of cacao has been the subject of much investigation in several countries.[5,13,28] Results from field studies in places with well-defined dry seasons have clearly demonstrated the depressing effect of moisture stress on flowering.[2,13,28] Research with plants grown in containers under controlled soil-moisture conditions also demonstrated that flowering rapidly diminishes, then ceases altogether, when the plants are submitted to moisture stress.[37] This research has also shown that when stressed plants are rewatered or submitted to a "wet regime" following a fairly long (6 months) "dry regime", flowering starts again and becomes exceptionally profuse a few weeks later. This has been interpreted as an indication that flower initiation had been enhanced during the dry regime and only bud development was inhibited by moisture stress.[37] Research with potted plants under controlled atmospheric relative humidity has also shown that flowering of cacao can be induced by transferring plants from a low (50 to 60%) or medium (70 to 80%) relative humidity to a high one (90 to 95%).[38]

Results from phenological studies in Brazil also demonstrated that flowering is greatly stimulated when rain follows a period of moisture deficiency.[8,10,42] A similar type of response to a transition from dry to wet conditions has been clearly demonstrated in the case of coffee flowering[3] (and see that chapter in Volume 2[3a]), but here the flower buds remain in a state of complete rest for as long as the plants are frequently watered and kept under high atmospheric humidity, a "dry shock" apparently being necessary to break the resting period. In the case of cacao this does not appear to be the case, as the flower buds do not go through a rest period imposed by high water potential and flowers are produced even under conditions of frequent watering or high humidity,[37,38] although not as profusely as when watering alternates with dry periods. This has been demonstrated in an experiment carried out in a very dry region (about 400 mm of annual rainfall) where cacao was submitted to different irrigation regimes. Plants which received frequent irrigation so as to avoid alternate periods of moisture stress produced less than half the number of flowers of periodically stressed plants. The difference between treatments became particularly evident about 7 or 8 weeks after the drier plots were wetted either by irrigation or rain.[42] It is worth pointing out that in cacao a transition from dry to wet periods has a stimulating effect not only on flowering but also on the flushing cycles of vegetative buds.[8,10] A similar type of response to a relatively dry period followed by a wet one has been observed in citrus in connection with flowering and flushing cycles.[15] (See also Chapter on *Citrus*, Volume 2.[29a]) The term "hydroperiodicity" has been suggested by the author to refer to this relationship between growth and flowering pattern and the exogenous conditions affecting the internal moisture status of the plant.[3,4]

REFERENCES

1. **Alvim, P. de T.,** Correlação entre chuva, temperature e produção do cacaueiro, in Reunião do Comitê Técnico Interamericano de Cacau, Salvador, Brasil, 1956, Instituto de Cacau de Bahia, Salvador, 1956, 133—137.
2. **Alvim, P. de T.,** Las necesidades de agua del cacao, *Turrialba,* 10, 6—16, 1960.
3. **Alvim, P. de T.,** Moisture stress as a requirement for flowering of coffee, *Science,* 132(3423), 354, 1960.

3a. **Alvim, P. de T.,** *Coffea,* in *CRC Handbook of Flowering,* Vol. 2, Halevy, A. H., Ed., CRC Press, Boca Raton, Fla., 1985, 308—316.

4. **Alvim, P. de T.,** Tree growth periodicity in tropical climates, in *The Formation of Wood in Forest Trees,* Zimmermann, M. H., Ed., Academic Press, New York, 1964, 479—495.
5. **Alvim, P. de T.,** Factors affecting the flowering of the cocoa tree, *Cocoa Growers Bull.,* 7, 15—19, 1966.
6. **Alvim, P. de T.,** Eco-physiology of the cacao tree, Cónference Internationale sur les Recherches Agronomiques Cacaoyères, 1967, Abidjan, Côte d'Ivoire, 1965; Institut Français du Café et du Cacao, Paris, 1967, 23—25.
7. **Alvim, P. de T.,** A new dendrometer for monitoring cambium activity and changes in the internal water status of plants, *Turrialba,* 25, 445—447, 1975.
8. **Alvim, P. de T.,** *Cacao, in Ecophysiology of Tropical Crops,* Alvim, P. de T. and Kozlowski, T. T., Eds., Academic Press, New York, 1977, 279—313.
9. **Alvim, P. de T.,** unpublished results.
10. **Alvim, P. de T., Machado, A. D., and Vello, F.,** Physiological responses of cacao to environmental factors, in 4th Int. Cocoa Res. Conf., St. Augustine, Trinidad, 1972, Government of Trinidad and Tobago, 1972, 210—225.
11. **Arevalo, A. R. and Soria, S. de J.,** Evaluación de cuatro metodos de polinización artificial en el aumento de producción de cacao *(Theobroma cacao* L.), 5th Int. Cocoa Res. Conf., Ibadan, 1975, Cocoa Research Institute of Nigeria, Ibadan, 1977, 78—84.
12. **Asomaning, E. J. A., Kwakwa, R. S., and Hutcheon, W. V.,** Physiological studies on an Amazon shade and fertilizer trial at the Cocoa Research Institute, *Ghana J. Agric. Sci.,* 4(1), 47—64, 1971.
13. **Boyer, J.,** Influence des régimes hydrique, radiatif et thermique du climat sur l'activité végétative et la floraison de cacaoyers cultivés au Cameroun, *Café Cacao Thé,* 14(3), 189—201, 1970.
14. **Bridgland, L. A.,** Study of the relationship between cacao yield and rainfall, *Papua New Guinea Agric. Gaz.,* 8(2), 7—14, 1953.
15. **Cassin, J.,** The influence of climate upon the blooming of citrus in tropical areas, *Proc. Int. Citrus Symp. 1st,* 1969, Vol. 1, Sect. 1, 315—323.
16. **Chan, B. G. and Cain, J. C.,** The effect of seed formation on subsequent flowering in apple, *Proc. Am. Soc. Hortic. Sci.,* 91, 63—68, 1967.
17. **Chapman, R. K. and Soria, S. de J.,** Comparative *Forcipomyia* (Diptera, Ceratopogonidae) pollination of cacao in Central America and Southern Mexico, *Rev. Theobroma,* 13(2), 129—139, 1983.
18. **Cuatrecasas, J.,** *Cacao and Its Allies: A Taxonomic Revision of the Genus Theobroma,* Smithsonian Institute, Washington, D.C., 1964.
19. **Furr, J. R., Cooper, W. C., and Reece, P. C.,** An investigation of flower formation in adult and juvenile citrus trees, *Am. J. Bot.,* 34, 1—4, 1947.
20. **Gil, G. F., Martin, G. C., and Griggs, W. H.,** Fruit-set and development in the pear: diffusible growth substances from seeded and seedless fruit, *J. Am. Soc. Hortic. Sci.,* 98, 51—54, 1975.
21. **Gorrez, D. D.,** The flower biology, morphology and pollination and cross habits of cacao, *Phillipp. Agric.,* 46(4), 288—302, 1962.
22. **Hoad, G. V. and Ramirez, H.,** La función de las giberelinas sintetizadas en las semillas del fruto para el control de la floración en manzanos, *Turrialba,* 30(3), 284—288, 1980.
23. **Hutcheon, W. V., Smith, R. W., and Asomaning, E. J. A.,** Effect of irrigation on the yield and physiological behaviour of mature Amelonado cocoa in Ghana, *Trop. Agric. (Trinidad),* 50(4), 261—272, 1972.
24. **Kozlowski, T. T.,** *Growth and Development of Trees,* Vol. 2, Academic Press, New York, 1971.
25. **Leite, R. M., de O. and Müller, M. W.,** Estudos fenológicos em cacaueiros expostos ao sol e à sombra, Informe Técnico do CEPEC 1981 (Brasil), 1981, 105.
26. **Lent, R.,** The origin of the cauliflorous inflorescence of *Theobroma cacao, Turrialba* 16(4), 352—358, 1966.
27. **Leon, J.,** Fundamentos botánicos de los cultivos tropicales, Instituto Interamericano de Ciencias Agricolas de la O.E.A., San José, Costa Rica, 1968.
28. **Machado, R. C. R., Alvim, P. de T., and Milde, L. E.,** Efeito da deficiência de água no solo sobre o comportamento fisiológico do cacaueiro, CEPLAC-CEPEC, Informe Técnico 1981, Ilhéus, 1982.

29. **Martin, E. J.,** Efectos de polinización controlada en cacao, *Proc. 8th Int. Cocoa Res. Conf.*, Carragena, Colombia, Cocoa Producers' Alliance, Lagos, Nigeria, 1982, 57—60.

29a. **Monselise, S. P.,** *Citrus* and related genera, in *CRC Handbook of Flowering*, Vol. 2, Halevy, A. H., Ed., CRC Press, Boca Raton, Fla., 1985, 275—294.

30. **Monti, J. R.,** La périodicité des pluies au Mayumbe et leur relation avec la production de cacao, *Bull. Agric. Congo Belge*, 44(3), 493—510, 1953.

31. **Murray, D. B.,** The use of shade for cacao, 6th Conf. Interam. Cacau, Salvador, Brazil, Instituto de Cacau de Bahia, Salvador, 1956, 57—60.

32. **Posnette, A. F.,** Pollination of cacao in Trinidad, *Trop. Agric. (Trinidad and Tobago)*, 21, 115—118, 1944.

33. **Posnette, A. F.,** The pollination of cacao in the Gold Coast, *J. Hortic. Sci.*, 25, 155—163, 1950.

34. **Posnette, A. F. and Entwistle, H. M.,** The Pollination of Cocoa Flowers, 1957 Report, Cocoa, Chocolate and Confectionery Alliance, London, 1958, 66—68.

35. **Pound, F. J.,** A note on a method of controlled pollination of cacao, 4th Annu. Rep. Cacao Research, 1934, Imperial College of Tropical Agriculture, Trinidad, 1935, 15—16.

36. **Sale, P. J. M.,** Flowering of cacao under controlled temperature conditions, *J. Hortic. Sci.*, 44, 163—173, 1969.

37. **Sale, P. J. M.,** Growth, flowering and fruiting of cacao under controlled soil moisture conditions, *J. Hortic. Sci.*, 45, 99—118, 1970.

38. **Sale, P. J. M.,** Growth and flowering of cacao under controlled atmospheric relative humidities, *J. Hortic. Sci.*, 45, 129—132, 1970.

39. **Saunders, L. G.,** Revision of the genus *Forcipomyia* based on characters of all stages, *Can. J. Zool.*, 34, 657—705, 1956.

40. **Saunders, L. G.,** Report to A.C.R.I. on the cacao pollination study conducted in Trinidad and Costa Rica, June-September, 1957.

41. **Saunders, L. G.,** Methods for studying *Forcipomyia* midges, with special reference to cacao pollinating species (Diptera, Ceratopogonidae), *Can. J. Zool.*, 37, 33—51, 1959.

42. **Silva, W. S. da, Alvim, P. de T., and Aragão, O. P.,** Fenologia do cacaueiro sob regime de irrigação diferencial, Informe Técnico do CEPEC 1981 (Brasil), 1981, 35—37.

43. **Skidmore, C. L.,** Indications of existing correlation between the rainfall and the number of pods harvested at Aburi and Asuansi, *Gold Coast Dep. Agric. Bull.*, 16, 114—120, 1929.

44. **Soria, J.,** A note on cacao plants which flowered at the age of three months, *Cacao*, 6, 11—12, 1961.

45. **Soria, J. and Cerdas, M.,** Rendimiento del cacao en relación con polinización artificial y abonamiento, *Cacao*, 15(3), 11—12, 1970.

46. **Soria, S. de J.,** Studies on *Forcipomyia* spp. Midges (Diptera, Ceratopogonidae) Related to the Pollination of *Theobroma cacao* L., Ph.D. thesis, University of Wisconsin, Madison, 1970.

47. **Soria, S. de J.,** Indução da produção em cacaueiro com o uso de atomizador portátil na Bahia, Brasil, *Rev. Theobroma*, 4(2), 3—13, 1974.

48. **Soria, S. de J., Silva, P., and Chapman, R. K.,** Influence of floral pigmentation on field pollination rates in some cultivated varieties of *Theobroma cacao* L.: some effects on yield, *Rev. Theobroma*, 13(2), 141—149, 1983.

49. **Soria, S. de J. and Wirth, W.,** Ciclos de vida dos polinizadores do cacaueiro *Forcipomya* spp. (Diptera, Ceratopogonidae) e algumas anotações sobre o comportamento das larvas no laboratório, *Rev. Theobroma*, 5(4), 3—22, 1975.

50. **Soria, S. de J., Wirth, W., and Besemer, H. A.,** Breeding places and sites of collection of adults of *Forcipomyia* spp. midges (Diptera, Ceratopogonidae) in cacao plantations in Bahia, Brazil: a progress report, *Rev. Theobroma*, 8, 21—29, 1978.

51. **Takhitajian, A. L.,** *Flowering Plants & Origin and Dispersal*, Edinburg, Oliver & Boid, 1969, 310 pp.

52. **Vera, B. J., Mogrovejo, J. E., and Moreira, D. M.,** La polinización suplementaria y cortes en el tallo como practicas simples para incrementar los rendiminetos en cacao, *Proc. 8th Int. Cocoa Res. Conf.*, Lagos, Nigeria, Cocoa Producers' Alliance, 1982, 31—35.

53. **Vogel, M.,** Existence d'un gradient morphogène au niveau des burgeons végétatifs supérieurs des axes orthotropes chez le cacaoyer (*Theobroma cacao* L. Sterculiaceae). Apparition de la plagiotropie, *Cafe Cacao The*, 22, 13—30, 1978.

54. **Vogel, M., Machado, R. C. R., and Alvim, P. de T.,** Remoção de orgãos jóvens como método de avaliação das interações fisiológicas no crescimento, floração e frutificação do cacaueiro, *Proc. 8th Int. Cocoa Res. Conf.*, Lagos, Nigeria, Cocoa Producers' Alliance, 1982, 31—35.

55. **Wheat, D.,** Branch formation in cocoa (*Theobroma cacao* L., Sterculiaceae), *Turrialba*, 29(4), 275—284, 1979.

56. **Winder, J. A.,** The role of non-dipterous insects in the pollination of cocoa in Brazil, *Bull. Entomol. Res.*, 68, 559—574, 1978.

VIBURNUM

En. Arrowwood, Guelder Rose, Snowball; Fr. Viorne; Ge. Brandzweig, Schlingbaum, Schneeball, Schwelke

Dennis P. Stimart

INTRODUCTION

Viburnum L., a member of the Caprifoliaceae, is a genus of some 225 species. Plants are upright shrubs or small trees. Viburnums are widely distributed in North America and Europe, and are most abundant and diverse in eastern Asia. Their distribution also extends into the Andes of South America, the mountains of Central America, and Malaysia. The genus is generally divided into nine sections based principally on fruit and leaf characteristics. These sections are *Viburnum* (Lantana), *Pseudotinus, Pseudopulus, Lentago, Odontotinus, Thyrsoma, Tinus, Opulus*, and *Megalotinus*.[1,2,13]

In horticulture, viburnums are among the most popular ornamental shrubs. They are grown in the landscape mostly for their attractive flowers, autumn leaf color, and autumn and winter fruits. A few of the snowball types (inflorescence with only sterile flowers) are forced in the greenhouse as potted plants.[1]

MORPHOLOGY

The flowers of *Viburnum* are white or pinkish, and borne in terminal panicles or umbel-like cymes. In most species the inflorescence has a stalk, but some (i.e., *V. furcatum* Blume ex Maxim) produce sessile inflorescences (Figure 1). Some Asian species (i.e., *V. farreri* Stearn and *V. sieboldii* Miq.) produce paniculate inflorescences with opposite branches (Figure 1). The inflorescences of a few species are overwintered in naked buds (i.e., *V. carlesii* Hemsl.). However, in most species the next year's leaves and inflorescences overwinter enclosed by bud scales. There may be one or two bud scales present.

The number of flowers in each inflorescence can range from 15 to 500, depending on the species. The flowers of viburnums are usually small (4 to 10 mm across) and have a uniform morphology. There are five small, green calyx lobes (Figure 1). The corolla consists of five fused petals. In most species, the corolla tube is shallow. However, a few of the Asian species (i.e., *V. carlesii* and *V. farreri)* have a more elongate (15 mm long) and tubular corolla. There are five stamens attached near the base of the corolla. Anthers usually extend out of the throat beyond the corolla. The anthers split open longitudinally when releasing pollen. The pistil is of three carpels. The ovary is mostly inferior, and the style is cone-shaped and short. The stigma is slightly trilobed and fairly broad. Most viburnums are protandrous. In some species (i.e., *V. furcatum* and *V. opulus* L.) sterile flowers are produced around the periphery of the inflorescence. The corollas of these flowers can be 30 mm across. The function of these large sterile ray-flowers is to attract insects to the inflorescence. The fruit is a one-seeded drupe often persisting into winter. The leaves of the viburnums are opposite, simple, toothed, sometimes lobed and deciduous or persistent. Stipules are sometimes present but are small and adnate to the petiole.[1,2,4,13]

VEGETATIVE AND REPRODUCTIVE DEVELOPMENT

The vegetative growth on viburnums appears to be controlled by photoperiod and temperature. The main shoot axis and lateral branches on *V.* × *burkwoodii* Hort. Borkw. &

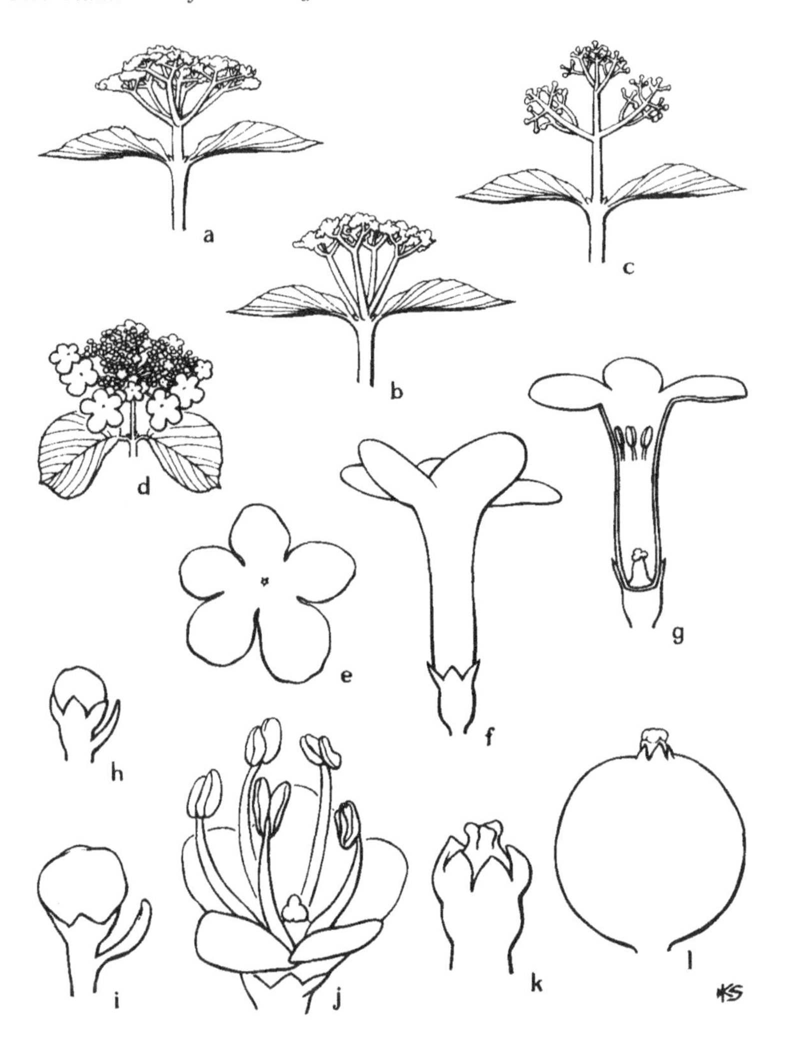

FIGURE 1. Viburnum. Inflorescence types: (a) stalked umbel-like inflorescence with all its branches originating at one level; (b) stalkless umbel-like inflorescence; (c) paniculate inflorescence with its branches originating at different levels; (d) inflorescence with sterile marginal flowers. Flower types: (e) close-up of a sterile marginal flower; (f) a tubular flower with its corolla tube much longer than its corolla lobes; (g) the inside of a tubular flower showing the attachment of the stamens to the corolla tube, the inferior ovary, and the short style. Development of a flower from young bud through fruit: (h) very young flower bud; (i) older flower bud just prior to opening; (j) an open flower with a short corolla tube showing the five stamens which alternate with the five corolla lobes; (k) a flower from which the corolla and its attached stamens have fallen, leaving the sepals and the style; (l) the fruit (ripened ovary), which contains one seed. (From Donoghue, M., *Arnoldia,* 40, 2—22, 1980. With permission.)

Skipw., *V. carlesii, V.* × *chenaultii, V.* × *juddii* Rehd., and *V. plicatum* Thunb. f. *tomentosum* made relatively little growth under 8-hr photoperiods. Growth of both the main axis and lateral branches increased under treatments of 12-, 14-, and 16-hr photoperiods (day continuation), except for lateral growth on *V. juddii.* The sum of growth made by the main axis and lateral branches of *V. burkwoodii* and *V. chenaultii* over a 24-week period was twice as great under 16-hr days as under 8-hr days. Similarly, *V. carlesii* produced three times the growth, *V. plicatum* f. *tomentosum* four times the growth, and *V. juddii* five times the amount of shoot growth under 16-hr days compared to 8-hr days.[5,6] Similar results were found with *V. trilobum* in that 8-hr photoperiods suppressed shoot growth while 16-hr treatment promoted continuous growth.[3]

The critical length of the night rather than the day appears to regulate the shoot growth on some viburnums. In *V. opulus,* a 1-hr NB in the middle of a 15-hr night kept shoots actively growing. If plants were grown in 10 consecutive hours of light they became dormant.[14] However, *V. carlesii* showed less of a response and *V. prunifolium* L. showed no response to the same light treatments.

In some viburnums, photoperiod and temperature may interact to regulate shoot-growth response. Dormant plants of *V. sieboldii* which set terminal buds resumed uninterrupted growth under 10- to 15.5-hr photoperiods when at continuous 22°C. However, shoot growth was inhibited without the formation of terminal buds when plants were transferred to 10-hr photoperiods at cyclic temperatures of 28°C days and 18°C nights.[9] Similarly, Morita found a quantitative requirement for chilling and photoperiod in promoting shoot growth of *V. awabuki.*[12]

Limited information is available regarding flower initiation and development of viburnums. Downs and Piringer[5,6] found that *V. burkwoodii, V. chenaultii,* and *V. carlesii* formed flower buds under 8-, 12-, 14-, or 16-hr photoperiods given as a day continuation. They concluded that photoperiod does not seem to control flower initiation. However, marked differences in the extent of flower bud development occurred among species. Under all photoperiod treatments (8 to 16 hr), (1) at least one visible flower bud per plant formed on *V. burkwoodii* and *V. chenaultii,* (2) one fourth of the plants of *V. carlesii* formed flower buds, and (3) no visible flower buds appeared on *V. juddii* and *V. plicatum* f. *tomentosum.* Observation of axillary buds on those plants without visible flower buds showed minute flower buds were present, but remained compressed and inconspicuous.

The time of seasonal flowering varies among *Viburnum* species. Some species such as *V. samucinum* and *V. hartwegii* in tropical latitudes flower nearly continuously. Often matured fruits and blooming flowers coexist on plants. However, some other tropical species flower in a given session, but may have two or more flowering flushes within this period. Most of these species flower during the wet period of the summer months, but others such as *V. blandum* and *V. venustum* flower during the drier months of winter. In contrast, most viburnums of temperate regions have well-defined flowering periods, usually restricted to the spring or early summer. Flowering on these species is synchronized and generally limited to one flowering period yearly. However, under periods of severe stress resulting from drought or unseasonably warm temperature, fall flowering can result.[4]

The length of flowering of temperate zone viburnums lasts some 7 to 12 days. Inflorescences on the upper part of the plant tend to bloom slightly in advance of those lower on the plant. The flowers in an inflorescence open within 2 to 4 days of each other. Sterile marginal flowers, if present, tend to open 1 to 4 days before perfect flowers.

The flower in the spring begins its development as a compact bud. It expands and the corolla turns white and inflates. The flower opens, and remains open for 4 to 5 days then abscises. After the corolla falls, the calyx lobes and style attached to the ovary persist. The ovary then enlarges if fertilization has occurred.[4]

DORMANCY

The onset of summer bud dormancy in *V. lantana* L. was studied from 1976 to 1979 at Lednice, Czechoslovakia. Plants were hand defoliated from June 1 to October 31 and then scored for bud break and stem growth. Buds on plants defoliated up to the end of July to the beginning of August were not in summer dormancy, since regrowth occurred. Those defoliated beyond the beginning of August failed to regrow and were categorized as being in summer dormancy.[10]

Bud break on dormant *V. trilobum* Marsh. stored at 1°C for 3 or 10 weeks began 3 to 6 weeks earlier on plants stored for 10 weeks.[3] Photoperiods of 8 or 16 hr (day continuation) given to chilled plants did not affect the rate of bud break. However, bud break on dormant

V. lantana chilled (1°C) for 3 weeks broke dormancy more quickly under the 16-hr photoperiod than under the shorter one. In contrast, flower buds of *V. sieboldii* which had initiated and developed under 10- or 15.5-hr photoperiods flowered without a requirement for a cold exposure.[9]

Development of cold hardiness of *V. plicatum tomentosum* was independent of the induction of bud dormancy. Plants given 10°C under LD as a day continuation (1700 to 2300 HR) forced plants to harden as if they were under SD, even though they were not dormant. Increasing the number of weeks of SD brought about a progressive increase in hardiness. The SD stimulus could be reversed by LD. Following 6 weeks of SD, the rate of hardening was increased twofold if plants were put in darkness at 5°C compared to those previously exposed to LD. Thus, development of hardiness is a photoperiodic- and temperature-dependent response.[8]

The loss of cold hardiness from equally hardy dormant and nondormant plants artificially hardened showed that dormancy depressed the rate of cold hardiness loss when plants were given 21°C conditions.[7]

The breaking of flower buds and flowering times of 21 *Viburnum* taxa from the sections *Lantana, Lentago, Odontotinus, Opulus, Pseudopulus, Pseudotinus,* and *Thyrsoma* were observed over 2 years at the Arnold Arobretum, Jamaica Plain, Mass. Results there demonstrated that viburnums flower over a 2- to 3-month period, depending on the weather during a particular year. The order of flowering time of the species did not vary between years, but the exact dates of flowering fluctuated. It was concluded that there is a genetic basis for the timing of flowering.[4]

Photoperiod may be involved in initiating spring growth on some viburnums. Species with exposed naked buds are invariably among the early flowering viburnums, while those with two pairs of bud scales flower during the latter half of the flowering season.[4] However, plants with naked buds begin their inflorescence development earlier the previous season and overwinter their inflorescences in a more fully developed state than plants with one or two pairs of bud scales. As a consequence they require less time to mature their flowers during the following season. Therefore, the role of photoperiod and physiological development in time of spring flowering is not well understood.

GROWTH REGULATORS

The breaking of bud dormancy on *V. plicatum* f. *tomentosum* can be accelerated by treatment with GA_3 at 1000 ppm. This treatment did not cause a loss of plant cold hardiness. Also, the shoot growth response of *V. opulus* cv 'Nanum' was promoted by GA_3 treatment at 500 ppm. Histological observations showed that the increase in shoot growth resulted from an increase in cell division.[11]

CULTURE AND PROPAGATION

Viburnums are generally of easy cultivation but there are exceptions. They thrive best in moist conditions and in a deep rich sandy loam type of soil. Most are considered as being cold hardy.[1]

Propagation is by stem cuttings taken in early summer. Seeds are often slow to germinate and may be the result of double dormancy. Layering is another means of propagation and is often used on *V. plicatum, V. lantanoides,* and *V. furcatum.*

REFERENCES

1. **Bailey, L. H.,** *Hortus Third,* Macmillan, New York, 1976, 1290.
2. **Bean, W. J.,** *Trees and Shrubs,* Vol. IV, John Murray, London, 1970.
3. **Benjamin, L. P.,** Responses of woody plants to photoperiod and chilling, *Greenhouse Garden Grass,* 3, 32—33, 1963.
4. **Donoghue, M.,** Flowering times in *Viburnum, Arnoldia,* 40, 2—22, 1980.
5. **Downs, R. J. and Piringer, A. A.,** Responses of several *Viburnum* species to day-lengths (Part II in "Report of field committee for 1956-photoperiod studies", by John Mahlstede, Chairman), *Proc. Int. Plant Prop. Soc.,* 6, 134—136, 1956.
6. **Downs, R. J. and Piringer, A. A.,** Growth and flowering responses of five *Viburnum* species to various photoperiod, *J. Am. Soc. Hortic. Sci.,* 72, 511—513, 1958.
7. **Irving, R. M. and Lanphear, F. O.,** Dehardening and the dormant condition in *Acer* and *Viburnum, J. Am. Soc. Hortic. Sci.,* 91, 699—705, 1967.
8. **Irving, R. M. and Lanphear, F. O.,** Environmental control of cold hardiness in woody plants, *Plant Physiol.,* 42, 1191—1196, 1967.
9. **Jennings, P. H. and Timmerman, J. C.,** Effects of daylength and temperature on growth of *Viburnum sieboldii, J. Am. Soc. Hortic. Sci.,* 51, 467—471, 1976.
10. **Kralik, J. and Sebanek, J.,** A contribution to the study on the onset of endogenous dormancy in some decorative coniferous and deciduous woody species, *Acta Univ. Agric.,* 29(1725), 55—64, 1981.
11. **McCarthy, D. and Bunemann, G.,** The use of growth regulators in the production of dwarf ornamental shrubs. III. Histological observations, *Gartenbauwissenschaft,* 47, 33—36, 1982.
12. **Morita, M., Iwanoto, S., and Higuchi, H.,** Interrelated effects of thermo- and photo-periodism on growth and development of ornamental woody plants. IV. The effects of chilling in winter and photoperiod on growth and development of several ornamental woody plants, *J. Jpn. Soc. Hortic. Sci.,* 48, 205—212, 1979.
13. **Rehder, A.,** *Manual of Cultivated Trees and Shrubs,* 2nd ed., Macmillan, New York, 1976, 1290.
14. **Waxman, S. and Nitsch, J. P.,** Influence of light on plant growth, *Am. Nurseryman,* 104, 11—12, 1956.
15. **Wyman, D.,** *Wyman's Gardening Encyclopedia,* Macmillan, New York, 1977, 1221.

VIOLA ODORATA

En. Violet; Fr. Violette; Ge. Veilchen; Sp. Viola

Anna M. Mayers*

INTRODUCTION

The genus *Viola* (Violaceae), with over 400 species, is native to both north and south temperate zones. Fifty-six *Viola* species are known to produce two distinct floral forms: (1) a showy, open or chasmogamous (CH) flower, and (2) a smaller, closed or cleistogamous (CL) flower. These morphological differences are environmentally induced. Cleistogamous flowers are produced under LD (14 to 17 hr) in all species studied.[1-5] Chasmogamous flowers are produced under SD (8 to 10 hr) in seven *Viola* species (Table 1). *Viola sylvestris, V. arenaria,* and *V. lancifolia* produce CH flowers only after a short period of low temperature.[3] No flowers are formed in SD alone, and a burst of CH flowering occurs after cold treatment even under LD. *V. palustris* produces occasional CH flowers under SD, but CH flowering is enhanced considerably by 8 weeks of vernalization at 4°C.[5] Under LD, this species responds like *V. sylvestris,* producing a burst of CH flowers after vernalization, followed by continuous CL flowering. Both cold treatment and SD are required for CH flowering in *V. hirta.*[3] The two flower types of *Viola* also represent differences in breeding systems.[6] The zygomorphic CH flower has several adaptations of form associated with insect pollination, and thus functions as an outcrosser.[7,8] All floral organs are reduced in the CL flower which selfs in the bud without opening.

Viola odorata L., an acaulescent herbaceous perennial, is native to Europe.[9] The leaves are cordate with crenate margins and are covered with a fine pubescence. Leaf size and petiole length, as well as flower form, vary in response to seasonally varying photoperiod.[10,11] This species has been cultivated for centuries for its sweetly scented flowers, as a source of violet perfume, and for herbal remedies. Large-scale commercial production reached its peak in Europe and America in the early 1900s. However, discovery of the chemical composition of its scent and rising costs associated with intensive greenhouse cultivation have resulted in a decline of this industry since World War II.[12]

FLORAL MORPHOLOGY

Flowers of *V. odorata* vary in size and form, ranging from the large blue-violet CH flower, to the much-reduced CL flower. The mature CH flower is approximately 20 mm long, compared to 5 mm for the mature CL flower (Figure 1). Each flower is 5-merous, perfect, and is subtended by 2 bracteoles. In the CH flower, the anterior petal is spurred and encloses 2 club-shaped nectaries produced from anther connective tissue (Figures 2 and 3). Petals are small, white, and membranous in the CL flower. These petals do not expand beyond the sepals. The petal spur and nectaries also show repressed development in this floral form (Figures 2 and 4). In both flowers, ovule placentation is parietal and styles are hollow throughout their lengths (Figures 3 and 4). However, the CH style elongates beyond 5 interlocking anther caps and bends 90° toward the anterior petal (Figures 2 and 3), while the CL style recurves 180° toward the ovary and is enclosed by the ring of anther caps (Figures 2 and 4). This stylar modification facilitates self-pollination. The CL flower selfs within the bud without undergoing anthesis.[10,13]

* This represents work done for the Ph.D. at the University of California Riverside, with Dr. E. M. Lord.

Table 1
REQUIREMENTS FOR CH AND CL FLOWERING IN *VIOLA* SPP.

Species	CL	Ch
V. arenaria	LD	Cold temperature
V. biflora	LD	SD
V. canina	LD	SD
V. fimbriatula	LD	SD
V. hirta	LD	SD + cold temperature
V. lancifolia	LD	Cold temperature
V. mirabilis	LD	SD
V. odorata	LD	SD
V. palustris	LD	Cold temperature
V. papilionacea	LD	SD
V. riviniana	LD	SD
V. sylvestris	LD	Cold temperature

The fruit is a many-seeded unilocular capsule that splits open along 3 sutures, spilling the seeds passively rather than explosively as in some *Viola* species.[14] Seeds are cream colored, producing lipid-rich elaisomes from integumental tissue at the micropylar end. The elaisomes are attractive as food to ants, a phenomenon known as myrmecochory. For several *Viola* species, dispersal of seeds is achieved by ant-seed interactions.[15]

PHOTOPERIODIC EFFECTS

Flowers of *V. odorata* show a gradual, seasonal transition in floral form (Table 2). CH flowers are produced in response to 11 hr or less of daylight; CL flowers in response to 14 hr or more of daylight. Flowers intermediate in form between CH and CL types show a progressive reduction in size of all floral organs (Figure 5), repressed growth of the anterior petal spur and staminal nectaries, and morphological modification of the elongate CH style which gradually recurves and is enclosed by the circle of anther caps in the fully CL flower. The reverse process occurs in floral forms in transition between CL and CH types (Figures 1 to 4 and 9).

Under both inductive daylengths, seedling plants produce serial vegetative buds in the axil of each leaf of the first nine nodes (Figure 6). These buds elongate into stolons with long internodes. Each primary stolon branches several times producing a network of runners that encircle the mother plant. Frequent nodal rooting results in ramet formation. Stolon production from these lower nodes continues throughout the life of the plant. Solitary axillary flowers replace the vegetative buds beginning typically at node 10 of a seedling plant, and flowers are produced indefinitely thereafter, at the axils of each leaf.[11]

Dimorphic leaf production associated with CH and CL flowering is correlated with daylength.[1,4,11] In greenhouse-grown plants, SD induce CH flowers and short-petioled leaves. These leaves also have leaf blade areas that are significantly smaller than those of the long-petioled leaves produced by CL flowering plants under LD (Table 3, Figure 7). In addition, CH flower peduncle lengths are approximately five times longer than those of CL flowers. These long CH peduncles exceed the total leaf length of the small CH leaves, while CL flower peduncles are markedly shorter than the CL leaves (Figure 7).

The early phase of leaf elongation is exponential under both LD and SD. CL leaves grow at a significantly faster rate (0.0821) relative to CH leaves (0.0611). CL leaves also attain a maximum length of 10 cm, compared to 7 cm for CH leaves under these same conditions (Figure 8). The CH plastochron under 8-hr days is 4.3 days, while the CL plastochron under 16-hr days is 3.4 days (Figure 8). Meiosis within the anthers occurs 9 plastochrons after

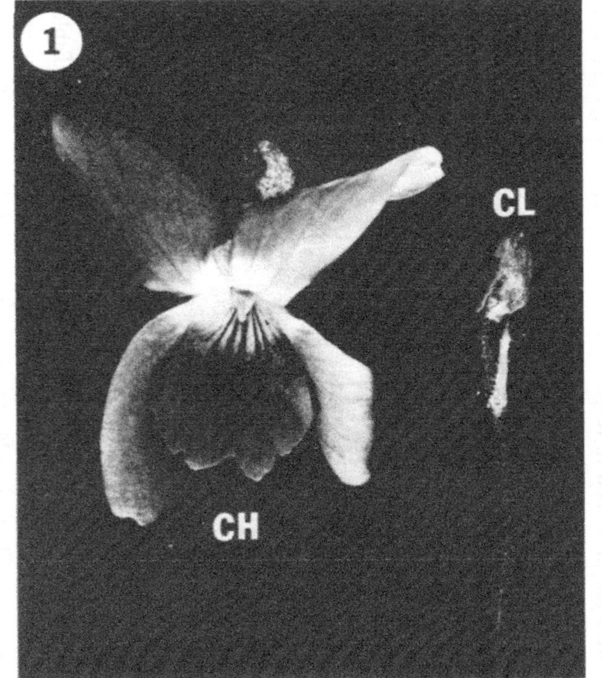

FIGURE 1. Mature CH and CL flower of *V. odorata*. (Magnification × 2.5.) (From Mayers, A. M. and Lord, E. M., *Am. J. Bot.*, 70, 1548—1555, 1983. With permission.)

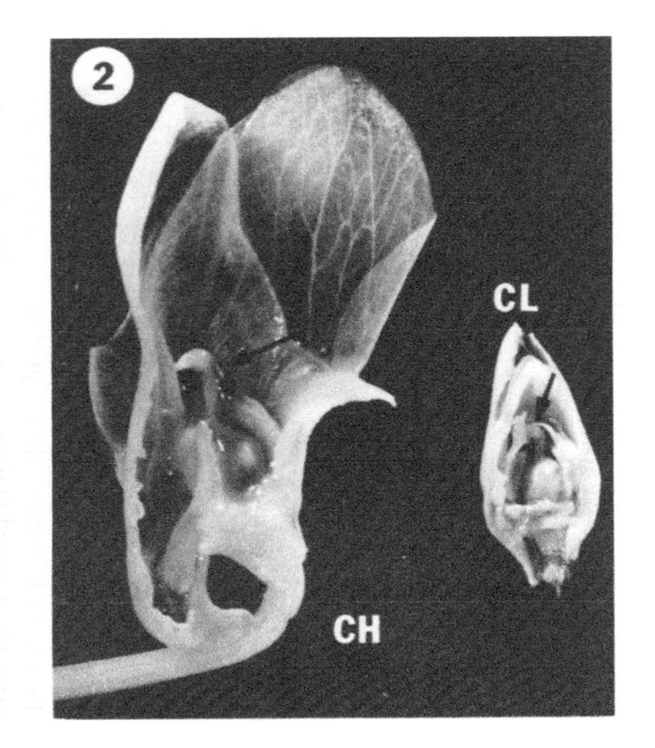

FIGURE 2. Mature CH and CL flowers with lateral sepals and petals removed to reveal stylar morphology (arrows). (Magnification × 4.) (From Mayers, A. M. and Lord, E. M., *Am. J. Bot.*,

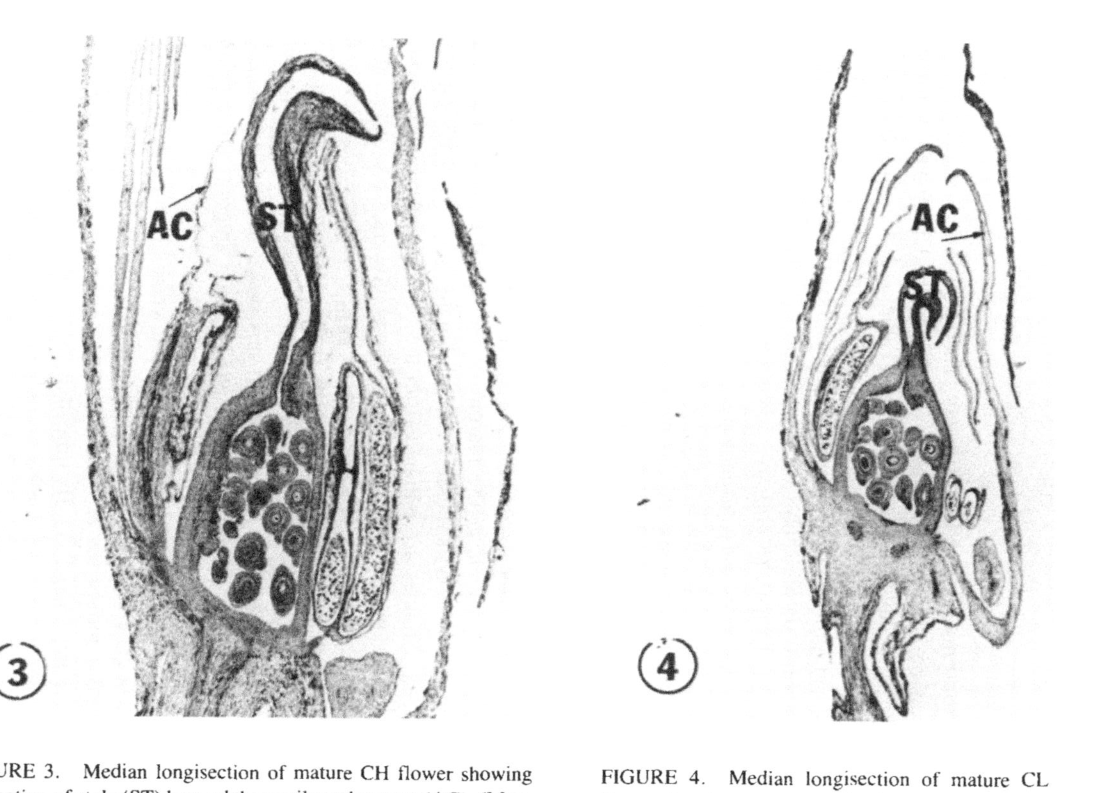

FIGURE 3. Median longisection of mature CH flower showing elongation of style (ST) beyond the sterile anther caps (AC). (Magnification × 15.) (From Mayers, A. M. and Lord, E. M., *Bot. Gaz. (Chicago)*, 145, 83—91, 1984. With permission.)

FIGURE 4. Median longisection of mature CL flower showing recurved style (ST) enclosed by anther caps (AC). (Magnification × 15.) S = sepal, O = ovary, AS = anther sac, N = nectary. (From

Table 2
SEASONAL FLOWERING CYCLE OF *VIOLA ODORATA*

Type of flower	Flowering period[a]	Daylength (hr)
CH	Nov. 1979—Feb. 1980	9.5—11
Intermediates	Mar. 1980—May 1980	11.5—14.5
CL	June 1980—Aug. 1980	15—13
Intermediates	Sept. 1980—Oct. 1980	13—10.5
CH	Nov. 1980—Feb. 1981	9.5—11

[a] Greenhouse conditions, temperature ranges: June to October, 12 to 40°C; November to May, 2 to 30°C.

floral initiation in both flowers. Therefore, time from floral initiation until time of meiosis can be indirectly calculated. The age of the CH flower at this stage is 38.7 days compared to 30.6 days for the CL flower (Table 4).[11] CH flowers mature 21.2 days after meiosis, while CL flowers mature 13.8 days after meiosis (Table 4). When the times of the two developmental periods are combined, the estimated period from floral initiation until flower maturity is 59.9 days for the CH flower and 44.4 days for the CL flower, a difference of about 15 days (Figure 9).

Seasonally varying photoperiod influences the rate of both leaf and flower development in *V. odorata*. This rate difference may in part determine mature flower size and morphology of the two extreme floral forms.[11,16]

COMPARATIVE FLOWER DEVELOPMENT

Shoot apices from CH (Figure 10) and CL (Figure 11) flowering plants show similarities in early ontogeny of both leaves and flowers. The youngest leaf primordia (P_1) arise as circular mounds at the margin of the apical domes. The second leaf primordia (P_2) are crescent-shaped structures which compose approximately one third of the perimeter of the apex. Stipules are initiated from the base of the primordia while the upper part elongates into the midrib of the laminar portion. Flowers are initiated in the axils of the third oldest leaf (P_3) in both CH and CL shoot apices (Figures 10 and 11). CH apices are significantly larger than CL apices.[13,16] Chasmogamous flower primordia are also larger than CL primordia from inception, although the two floral types are morphologically identical until initiation of petals (Figures 12 and 13). At this stage, in the CH flower, the newly initiated anterior petal causes the separation of two adjacent anterior sepals (Figure 12), a condition that is not observed in the CL flower (Figure 13).

There are ontogenetic parallels between CH and CL forms during subsequent floral organogenesis. Five stamens are initiated alternate to five petals (Figures 14 and 15). The gynoecium arises as a circular structure from the center of the remaining floral meristem (Figures 16 and 17). By time of meiosis within the anthers, the CH floral form is considerably larger in size. CH anthers sampled at pollen mother cell stage have a nine times greater volume than comparable CL anthers (Figures 18 and 19). The reduced anther volume of the CL flower results in approximately one seventh the amount of pollen produced in the CH anther.[16]

ANTHESIS AND POLLINATION

At anthesis, the CH flower is four times the size of the CL flower which fails to open. Aside from this obvious difference between the flowers, there are other morphological

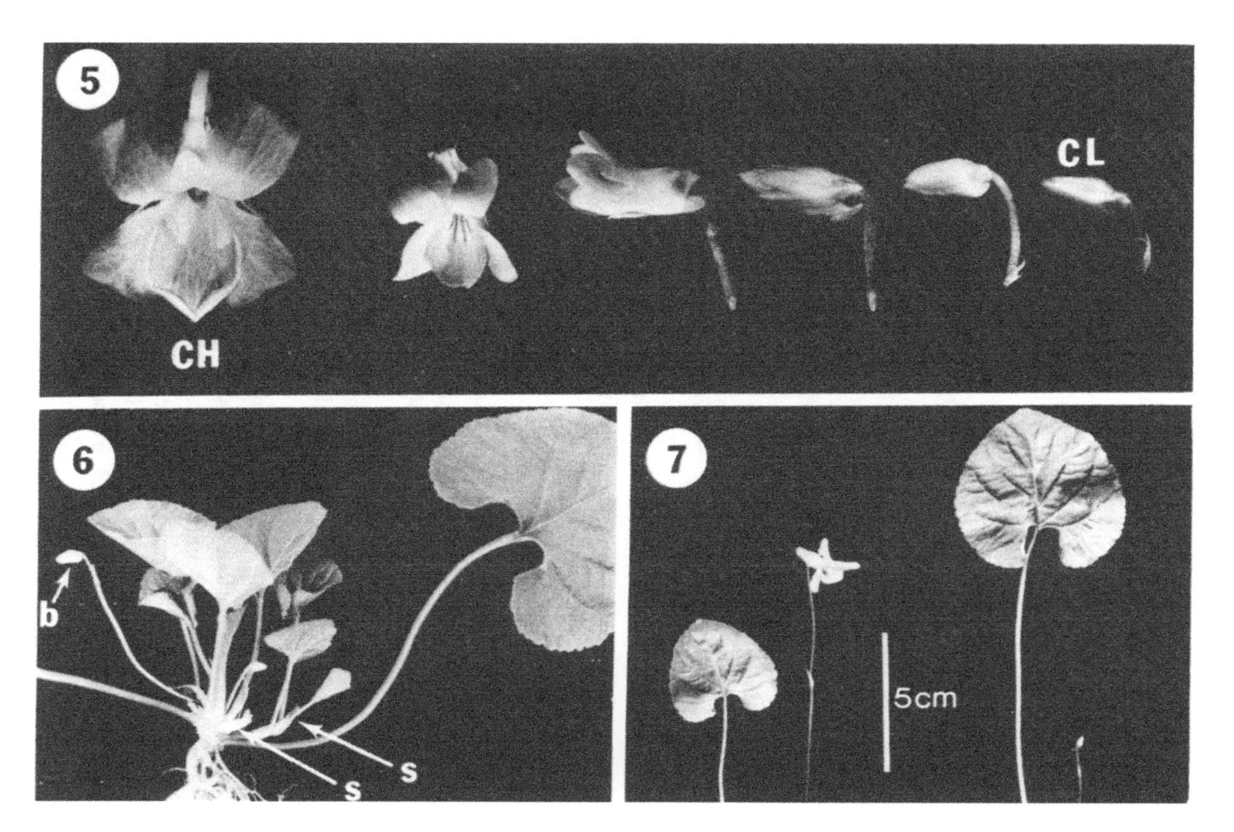

FIGURE 5. Mature flowers of *V. odorata:* CH, intermediate series illustrating progressive modification of form, and mature CL flower. (Magnification × 2.)

Table 3
COMPARISON OF *VIOLA ODORATA* CH AND CL LEAF DIMENSIONS[a]

Plant part	Greenhouse	
	8 hr[b]	16 hr[b]
Total leaf length (cm)	8.2 ± 0.87	14.8 ± 2.3*[c]
Petiole length (cm)	5.4 ± 1.2	11.0 ± 2.1*
Lamina area (cm^2)	11.9 ± 2.6	19.2 ± 3.6*

[a] Sample size = 30 leaves.
[b] Daylength at time of harvest.
[c] Significant difference at 1% level indicated by *.

Table 4
***VIOLA* FLOWER GROWTH**

Stage	CH	CL
Initiation to meiosis[a] (days)	38.7 (9 plastochrons)	30.6 (9 plastochrons)
Meiosis to flower maturity[b] (days)	21.2 ± 2.5	13.8 ± 3.1
Total (days)	59.9	44.4

[a] Meiosis in the anthers.
[b] Significant difference between CH and CL at 1% level, N = 30 flowers.

modifications, particularly in the gynoecia. Both forms have hollow styles with a stigmatic cavity at their tips. These are lined with a secretory epidermis. In the CH flower, the style elongates beyond the ring formed by five sterile anther caps and bends down toward the anterior petal (Figure 3). This bent style acts as a hindrance to a pollinating insect and must be forced upward. Beattie[7] has described the pivotal mechanism of the CH style in *Viola*, which results in the opening of the anther ring and dispersal of pollen onto the insect vector. A pollination droplet is formed at the stigma tip when the style is triggered. This droplet, carrying any pollen that was deposited, is drawn back into the stigmatic cavity when the style springs back into place. Thus, the CH flower normally functions as an outcrossing floral form. However, in several *Viola* species, CH pollen germinates prematurely within the anther sacs, prohibiting pollen dispersal.[10,11,17,18]

The CL floral form appears adapted for self-pollination, representing an inherited alternative breeding system according to Darwin.[6] Selfing occurs as a result of pollen germination inside anther locules before tapetal degeneration (Figure 20). Pollen tubes then grow through the undehisced anther sac wall at its apex where endothecial thickenings are lacking (Figures 21 and 22). The short, recurved form of the CL style facilitates pollination by positioning the stigmatic cavity entirely within the ring of interlocking anther caps (Figures 4 and 23). This enclosure of the CL gynoecium within the closed bud also provides a humid chamber that prevents dessication of the developing gametophytes.[16] Fertilization proceeds normally as described by Madge.[10]

The production of distinctly different flower forms within a single individual in *V. odorata* provides an experimental system useful in further investigation into the physiological changes associated with dimorphic flowering.

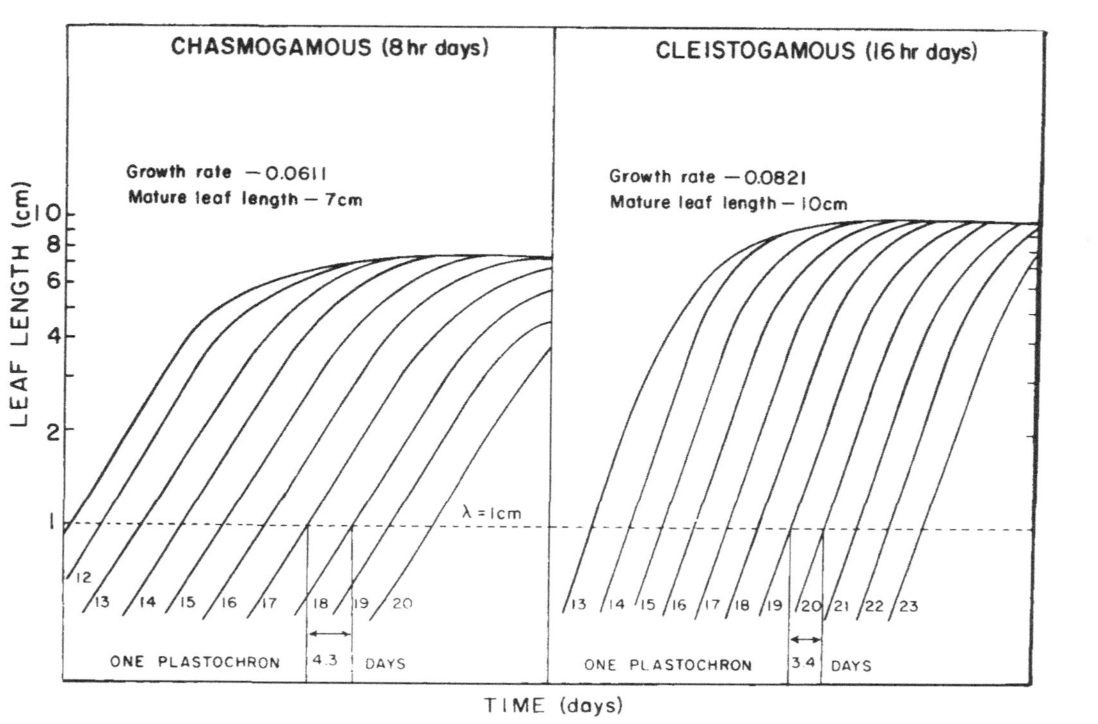

FIGURE 8. Lengths of successive leaves of *V. odorata* plants plotted logarithmically against time. Each growth curve represents the mean of five leaves sampled from two populations of five plants each, grown in growth chambers under 8- and 16-hr days, and identical temperature and light treatments. Growth rates were determined by regression analysis for the linear portion of the curves. (From Mayers, A. M. and Lord, E. M., *Am. J. Bot.*, 70, 1548—1555, 1983. With permission.)

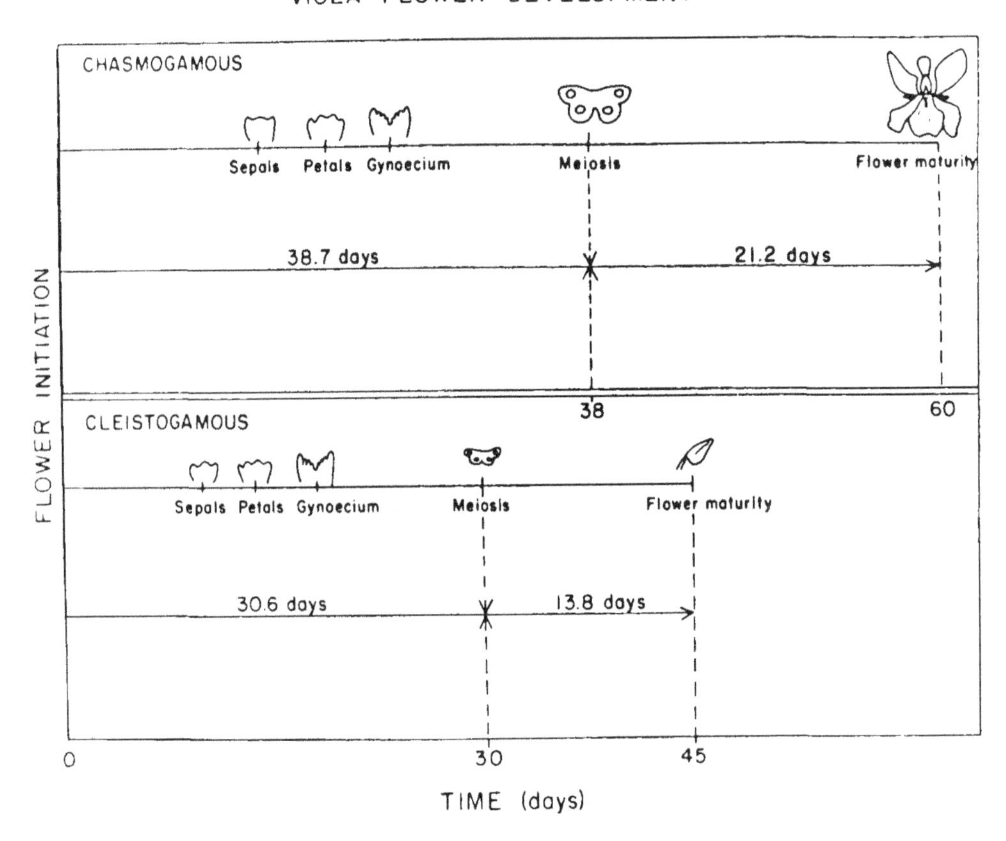

FIGURE 9. Comparative developmental rates of CH and CL floral forms of *V. odorata*. Meiosis occurs 8 days earlier in CL flowers. CL flowers mature in approximately 45 days vs. 60 days for the CH flowers. (From Mayers, A. M. and Lord, E. M., *Bot. Gaz. (Chicago)*, 145, 83—91, 1984. With permission.)

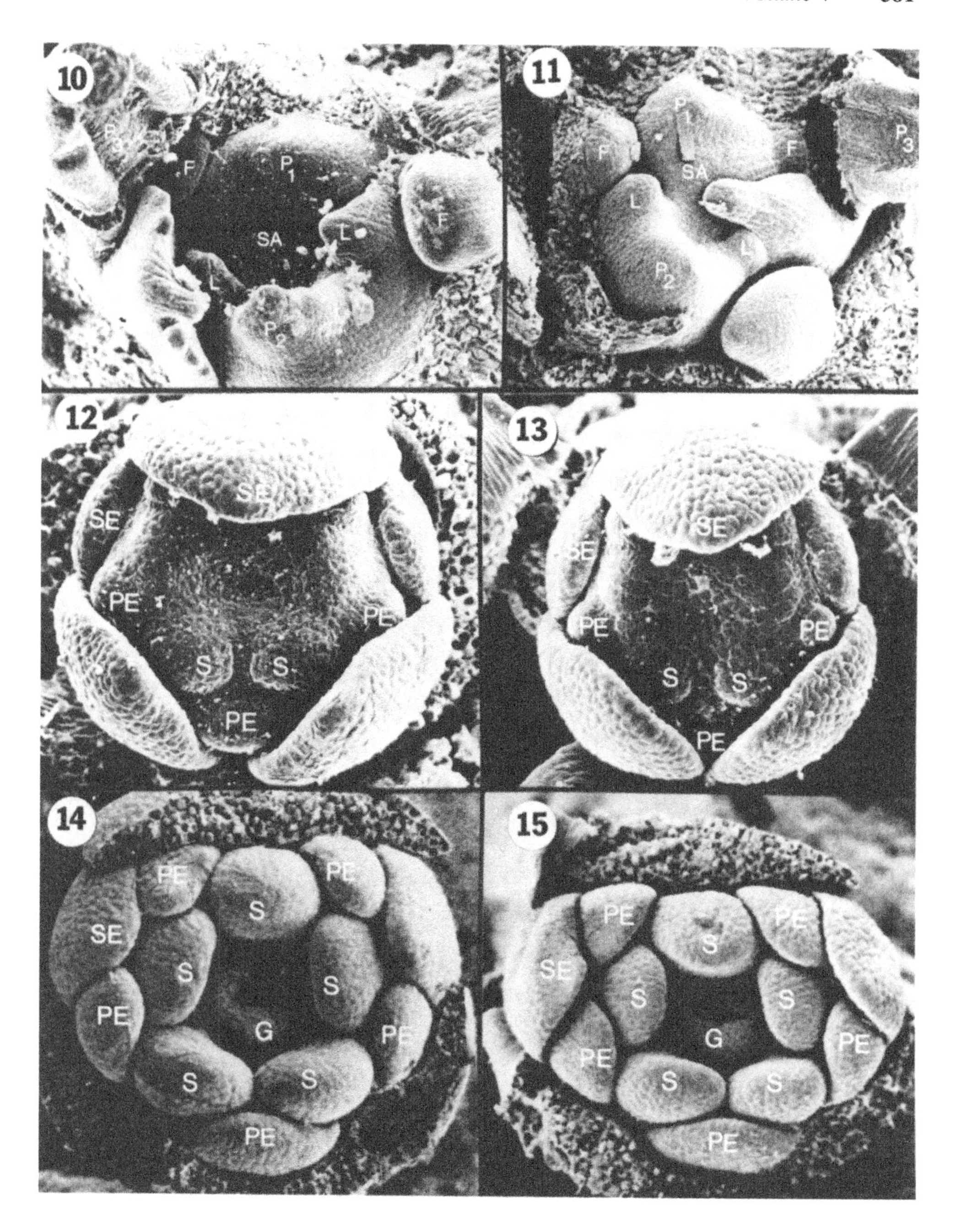

FIGURES 10 to 15: Scanning electron micrographs of shoot apices and floral primordia of *V. odorata*. CH, left; CL, right. FIGURES 10 and 11, youngest leaf primordia (P_1) just after inception from shoot apical meristem (SA). Initiation of floral primordia (F) occurs in the axil of P_3 in both cases. (Magnification × 135.). FIGURES 12 and 13. CH and CL floral primordia with anterior stamens (S) being initiated. (Magnification × 180.). FIGURES 14 and 15. CH and CL floral primordia with three advanced sepals removed to reveal floral organs. (Magnification × 135.) SE = sepals, PE = petals, S = stamens, G = gynoecium. (From Mayers, A. M. and Lord, E. M., *Am. J. Bot.*, 70, 1556—1563, 1983. With permission.)

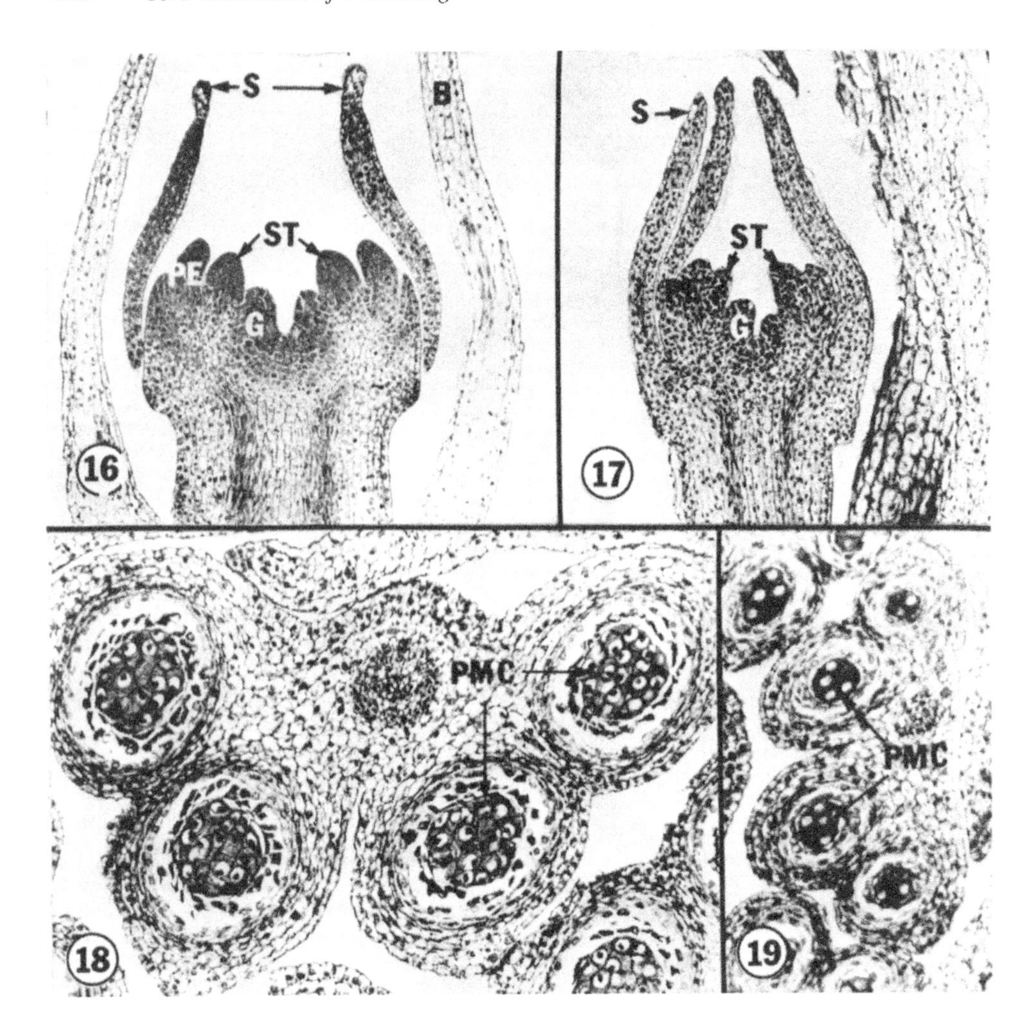

FIGURES 16 and 17. Median longisections of CH (left) and CL (right) floral primordia with all organs initiated. (Magnification × 180.) S = sepals, B = bracts, PE = petals, ST = stamens, G = gynoecia. FIGURES 18 and 19. Cross sections of CH (left) and CL (right) anthers in pollen mother cell (PMC) stage just prior to meiosis. (Magnification × 180.) (From Mayers, A. M. and Lord, E. M., *Bot. Gaz. (Chicago)*, 145, 83—91, 1984. With permission.)

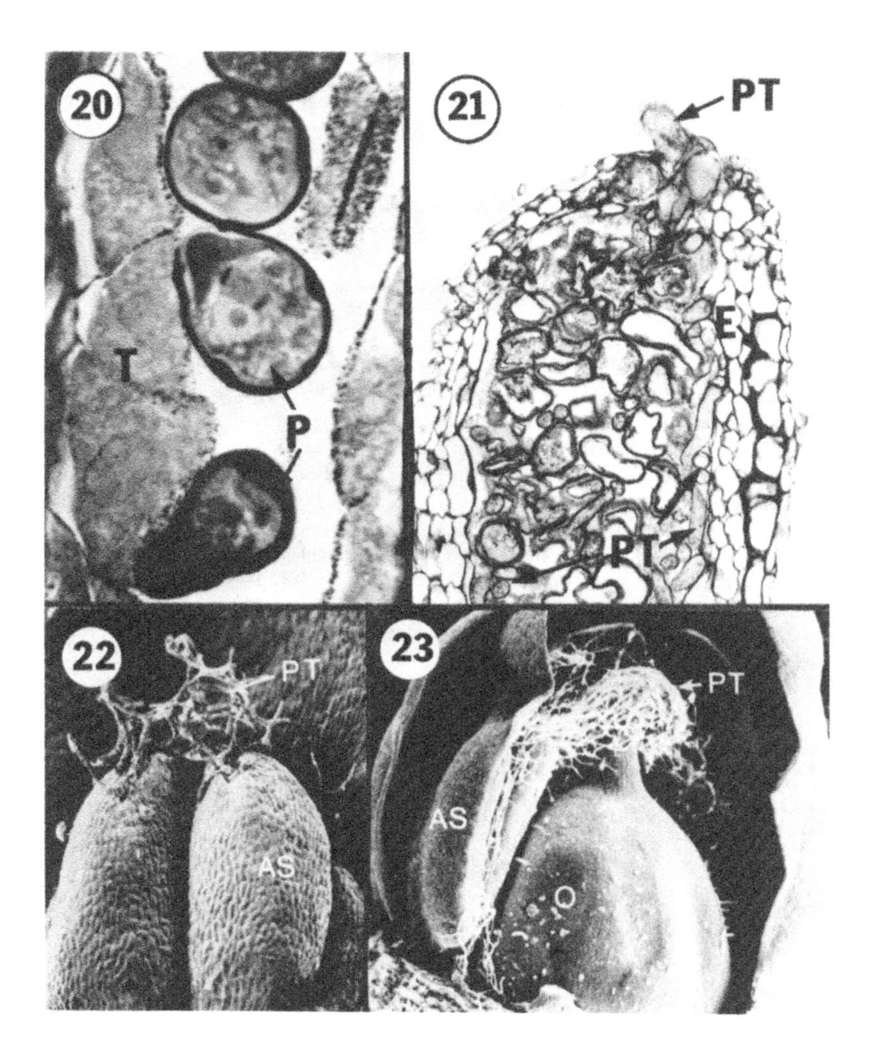

FIGURE 20. Mature CL pollen (P) germinating precociously within the anther locule before degeneration of the tapetum (T). (Magnification × 400.) (From Mayers, A. M. and Lord, E. M., *Bot. Gaz. (Chicago)*, 145, 83—91, 1984. With permission.) FIGURE 21. Longisection of CL anther showing germinated pollen inside locule and pollen tube (PT) penetrating anther at the apex. Note undeveloped endothecial thickening (E) at the apex. (Magnification × 180.) (From Mayers, A. M. and Lord, E. M., *Bot. Gaz. (Chicago)*, 145, 83—91, 1984. With permission.) FIGURE 22. Undehisced CL anther with pollen tubes (PT) growing from upper anther sac. (Magnification × 72.) (From Mayers, A. M. and Lord, E. M., *Am. J. Bot.*, 70, 1556—1563, 1983. With permission. FIGURE 23. Pollination of CL flower. Pollen tubes (PT) are growing over the recurved style into the stigmatic cavity; As, anther sac; O, ovary. (Magnification × 30.) (From Mayers, A. M. and Lord, E. M., *Am. J. Bot.*, 70, 1556—1563, 1983. With permission.)

REFERENCES

1. **Allard, H. A. and Garner, W. W.,** Further observations of the response of various species of plants to length of day, *U.S. Dep. Agric. Tech. Bull. 727*, 1—64, 1940.
2. **Borgstrom, G.,** Formation of cleistogamic and chasmogamic flowers in wild violets as a photoperiodic response, *Nature (London)*, 144, 514—515, 1939.
3. **Chouard, P.,** Diversité des types de compartement au photo-et au thermoperiodisme dans le genre *Viola* (Violettes et Pensees), *C. R. Acad. Sci.*, 226, 1831—1833, 1948.

4. **Cooper, C. C. and Watson, D. P.,** Influence of daylength and temperature on the growth of greenhouse violets, *Proc. Am. Soc. Hortic. Sci.*, 59, 549—553, 1952.
5. **Evans, L. T.,** Chasmogamous flowering in *Viola palustris, Nature (London)*, 178, 1301, 1956.
6. **Darwin, C.,** *The Different Forms of Flowers on Plants of the Same Species*, Appleton, New York, 1877, 352.
7. **Beattie, A. J.,** The floral biology of three species of *Viola, New Phytol.*, 68, 1187—1201, 1969.
8. **Beattie, A. J.,** Pollination mechanisms in *Viola, New Phytol.*, 70, 343—360, 1971.
9. **Clausen, J.,** Chromosome number and the relationship of the species in the genus *Viola, Ann. Bot.*, 41, 678—714, 1927.
10. **Madge, M. A. P.,** Spermatogenesis and fertilization in the cleistogamous flower of *Viola odorata*, var. *praecox*, Gregory., *Ann. Bot.*, 43, 545—577, 1929.
11. **Mayers, A. M. and Lord, E. M.,** Comparative floral development in the cleistogamous species *Viola odorata*. I. A growth rate study, *Am. J. Bot.*, 70, 1548—1555, 1983.
12. **Coombs, R. E.,** *Violets*, Croom Helm, London, 1981, 11—23.
13. **Mayers, A. M. and Lord, E. M.,** Comparative floral development in the cleistogamous species *Viola odorata*. II. An organographic study, *Am. J. Bot.*, 70, 1556—1563, 1983.
14. **Valentine, D. H.,** Variation and evolution in the genus *Viola, Preslia*, 34, 190, 1962.
15. **Culver, J. and Beattie, A. J.,** Myrmecochory in *Viola:* dynamics of seedant interactions in some West Virginia species, *J. Ecol.*, 66, 53—72, 1978.
16. **Mayers, A. M. and Lord, E. M.,** Comparative floral development in the cleistogamous species *Viola odorata*. III. A histological study, *Bot. Gaz.* (Chicago), 145, 83—91, 1984.
17. **Ritzerow, H.,** Über Bau and Befruchtung kleistogamer Blüten, *Flora*, 98, 163—212, 1908.
18. **West, G.,** Cleistogamy in *Viola riviniana* with special reference to the cytological aspects, *Ann. Bot.*, 44, 85—109, 1930.
19. **Goebel, K.,** Chasmogamie und kleistogamie Blüten bei *Viola, Allg. Bot. Z. Syst.*, 95, 234—239, 1905.

WEIGELA

En. Weigela; Fr. Diervillée; Gr. Weigelie

Dennis P. Stimart

INTRODUCTION

Weigela (family Caprifoliaceae) is a genus of some 12 species of deciduous shrubs. They are native to the temperate zones of China, Korea, Japan, and Russia. In horticulture, the plants are grown in the landscape as specimen plants or as hedges for their flowers which are produced during the summer.[1,2] Due to the ease of hybridization within and among species, numerous cultivars are now available which surpass the original species. The species which have gone into producing some of the most popular hybrids of white, pink, carmine, or crimson flower types and hybrids with colored foliage are *W. coraeenis* Thunb., *W. floribunda* C. Koch, *W. florida* A. DC., and *W. japonica* Thunb.[9] Until a few years ago there were at least 150 named cultivars.[9]

MORPHOLOGY

The flowers of *Weigela* are borne on short lateral twigs, usually on 1-year-old branches. The petals are showy and colored pink, purplish, carmine, or white. They are about 3.8 cm long and often in more-or-less compound clusters. The corolla is tubular-companulate or funnel-shaped and more-or-less regular. There are five stamens which are shorter than the corolla. The ovary is inferior, slender, and elongate. One style is present with a capitate or hood-like stigma. The fruit is a capsule and opens by two valves. The seeds are numerous and very small. The leaves are serrated and attached oppositely. The species range in height from 1 to 4 m.[1]

VEGETATIVE AND REPRODUCTIVE DEVELOPMENT

A critical photoperiod appears to determine if plants of *W. florida* will stop or continue growing vegetatively as measured by increases in stem length. Growth entirely stopped under 8- or 12-hr photoperiods (SD) forcing plants into setting terminal buds and premature dormancy.[5,6,11] LD, on the contrary, of 14- or 16-hr photoperiods, prevented the plants from becoming dormant, resulting in considerable softwood production and terminal shoot tips that remained vegetative.[5,6,11]

The SD photoperiodic inhibition of vegetative growth can be reversed by LD treatment. In switching experiments of dormant 8-hr grown plants into 16-hr photoperiods, it was found that buds began to swell within a week and within 12 days new vegetative growth was apparent.[5,6] Thus, vegetative growth appears to be controlled easily by photoperiodic manipulation.

The cessation and initiation of cold acclimation in *W. florida* appears to be phytochrome mediated. In experiments where dark period interruptions of R irradiation were given, the cold acclimation process was suppressed, which consequently decreased plant cold hardiness.[13] However, cold acclimation and cold hardiness suppression was relieved when R light treatment was followed by FR irradiation.

The flowering response of *Weigela* has been described as being typical of many woody plants in that floral initiation occurs the previous season to floral expression.[4] According to Nitsch,[8] *Weigela* flower initiation occurs at the same period in spring when the flowers initiated last year are in bloom. Another author states that some *Weigelas* bear occasional flowers during summer months in addition to the spring flush.[14] Howard recognized 117 cvs of *Weigela* but mentions only 4 cvs that flower continuously throughout the summer.[7]

Limited information is available regarding the control of flowering in *Weigela*. However, photoperiod is one environmental factor which has been associated with the flowering response. Lateral buds of *W. florida* were formed under 18-hr photoperiods, but only the older buds differentiated into flower buds.[4] These later buds failed to bloom and remained dormant under LD. However, flowering occurred when LD plants were given 8 weeks of SD and returned to LD. Flowering of apical nodes on plants grown under LD was also obtained after the terminal portion of the shoot was removed. Lateral buds in the median and basal portions of the shoot remained dormant.[4]

Even though flower buds initiated and developed under LD conditions, it is not clear if the stimulus of floral development is photoperiodic dependent. It has been suggested that physiological age of dormant lateral buds is involved in the differentiation of flower buds.[4] Also, conditions which inhibit vegetative growth appear promotive to floral development.[4]

The genetic composition of seedling progeny appears to influence the flowering of *Weigela*. The repeat flowering response within a season of a progeny occurred regardless of the parent habit.[10] However, flowering intensity of seedlings has been found to vary from none to complete coverage of a bush.[12] Also, a positive relationship exists between the floriferousness of a parent and the percentage of progeny that flowered the first growing season.[12]

CULTURE AND PROPAGATION

Weigelas are considered as cold hardy shrubs, although they benefit by winter protection of the ground. *W. hortensis* (Siebold and Zucc.) C. Koch and *W. middendorfiana* (Trautv. and C. A. Mey.) C. Koch are the most cold susceptible. Plants are easily cultivated, but they are gross feeders and require rich soils which are not too dry.

Pruning should be done as soon as flowering is completed, by entirely removing the old shoots. Young shoots of the current season must not be pruned since they produce their crop the following year.

Propagation is done mainly by semihardwood cuttings.

GROWTH REGULATORS

Weigela subjected to 9-hr photoperiods and 50 ppm of GA as a foliar spray continued vegetative shoot elongation while control plants under the same photoperiod without GA ceased vegetative extension and became dormant. Dormancy induced by a short photoperiod (9 hr) was broken by GA applications.[3]

Flowering of lateral buds of *Weigela* was obtained on plants growing under LD following exogenous application of MH at 2500 ppm.[4] This response was on developed lateral buds at treatment time. The apical meristem remained vegetative, and lateral buds which formed after treatment remained dormant.

REFERENCES

1. **Bailey, L. H.,** *Hortus Third*, Macmillan, New York, 1976, 1290.
2. **Bean, W. J.,** *Trees and Shrubs*, Vol. I, John Murray, London, 1970, 845.
3. **Bukovac, M. J. and Davidson, H.,** Gibberellin effects on photoperiod controlled growth of *Weigela*, *Nature (London)*, 183, 59—60, 1959.
4. **Davidson, H., Bukovac, M. J., and MacLean, D. C.,** Photoperiodic and chemical control of vegetative growth and flowering in *Weigela*, *J. Am. Soc. Hortic. Sci.*, 82, 589—595, 1963.

5. **Downs, R. J. and Borthwick, H. A.,** Effects of photoperiod on growth of trees, *Bot. Gaz.*, 117, 310—326, 1956.
6. **Downs, R. J. and Borthwick, H. A.,** Effects of photoperiod on the vegetative growth of *Weigela florida* var *variegata*, *J. Am. Soc. Hortic. Sci.*, 68, 518—521, 1956.
7. **Howard, R. A.,** A check-list of cultivar names in *Weigela*, *Arnoldia*, 25, 49—69, 1965.
8. **Nitsch, J. P.,** Photoperiodism in woody plants, *Proc. Am. Soc. Hortic. Sci.*, 70, 526—544, 1957.
9. **Pizetti, I. and Cocker, H.,** *Flowers*, Vol. II, Harry N. Abrams, New York, 1975, 1477.
10. **Poszwinska, J.,** Preliminary progeny analysis of some *Weigla* crosses, *Arbor. Kornickie*, 6, 143—167, 1961.
11. **Waxman, S.,** The Development of Woody Plants as Affected by Photoperiodic Treatments, Ph.D. thesis, Cornell University, Ithaca, N.Y., 1957, 193.
12. **Weigle, J. L. and Beck, A. R.,** Flowering characteristics of *Weigela* and their relationship to progeny age at first bloom, *J. Am. Soc. Hortic. Sci.*, 96, 685—686, 1971.
13. **Williams, B. J., Pellet, N. E., and Klein, R. M.,** Phytochrome control of growth cessation and initiation of cold acclimation in selected woody plants, *Plant Physiol.*, 50, 262—265, 1972.
14. **Zucker, I.,** *Flowering Shrubs*, D. Van Nostrand, Princeton, N.J., 1966, 380.

WISTERIA

En. Wisteria: Fr. Gycine; Gr. Glyzine

Dennis P. Stimart

INTRODUCTION

Wisteria is a genus of seven to ten species of deciduous twining climbers. Two species (*W. frutescens* L. and *W. macrostachya* [Torr. and A. Gray] Nutt. ex Torr. and A. Gray) are native to North America, while the others are from China or Japan. *Wisteria* belongs to the family Leguminosae.[2] In horticulture, the plants are grown for their long racemes of attractive flowers which bloom in the spring. Frequently, the vines are allowed to grow on walls or on a trellis support over a walkway for the display of the long, colorful racemes. The twisting of branches around supports provides a dense shade during the summer and an attractive effect when the leaves are not on the vines during the winter. Two species, *W. floribunda* (Willd.) DC. and *W. sinensis* (Sims) Sweet are the most commonly grown.

MORPHOLOGY

The flowers of *Wisteria* are in pendant and usually terminal racemes. The petals are colored blue, violet, purple, rose, or white. The standard is reflexed. There are ten stamens: nine are united and one is separate. The Chinese wisteria *(W. sinensis)* produces flower clusters ranging from 18 to 36 cm long and all flowers open simultaneously. The flower clusters on the Japanese wisteria *(W. floribunda)* vines range from 20 to 132 cm in length and the individual flowers do not open at once. Instead, the flowers open at the base of the cluster first and progress towards the tip. Not all of the flowers in a cluster are in a good condition at one time.[4]

The fruit is an elongate, thick, knobby legume which dehisces late in the growing season. On the Asiatic species (except *W. japonica* Siebold and Zucc.), the outer surface of the legume is velvety, while on the American species it is glaborous.

The leaves of wisteria are alternate, deciduous, and unequally pinnate with 9 to 13 leaflets.

VEGETATIVE AND REPRODUCTIVE DEVELOPMENT

There are two kinds of shoots that produce flower buds on wisteria vines. The first type is a shoot terminated by a flower cluster containing a number of leaves in their axils. Flower buds normally develop, except at very small basal leaves where the buds are vegetative and remain dormant for one to several years. The second type are shoots originating from dormant basal buds on shoots terminated by a flower cluster. The shoots can become 3 to 6 m long and will produce rudimentary flower clusters if exposed to a sunny location.[1,3] Flower clusters on these long shoots tend to open later in the spring.

The time of flower bud initiation of *Wisteria* begins during late spring and progresses to the pistil stage during mid summer. The progress of this development, however, is species and cultivar dependent. The time of flower bud differentiation and development in *W. brachybotrys* 'Arkebon', 'Blue Capitan', and 'White Capitan' and *W. floribunda* 'Honbeni', 'Kuchibeni', and 'Yatsubusa' were studied at Matsudo, Japan.[3] Flower-bud samples were taken from May 25 to July 14, 1965. Nine stages of flower bud development were identified:

1. Vegetative stage. The growing point was low and small.
2. Predifferentiation stage. The growing point became swollen, larger, and thicker.
3. Scale and bract differentiation stage. Scale formation began on the basal parts of the enlarged meristem and almost concurrently bracts appeared on the upper portion where the scales were formed; thus differentiation occurs in an acropetal succession.

4. Floret primordium differentiation stage. Formation of individual floret primordia occurred at the basal and upper parts of the domed apex in an acropetal order.
5. Bractlet differentiation stage. Two bractlets were produced on the sides of each floret primordium.
6. Sepal formation stage. Five sepal protuberances appeared at the basal end of each floret primordium.
7. Petal formation stage. Five petals were formed in the inner parts of each floret primordium.
8. Stamen formation stage. Pistil formation occurred within the center of the floret primordium.

In 'Blue Capitan' and 'White Capitan' Stage 4 was reached by May 25, 1964 and Stage 9 was reached on June 15 in 'Blue Capitan' and on July 6 in 'White Capitan'. In 'Akenbono', Stage 2 was observed on May 15, 1965 and Stage 9 on July 2. In 'Kuchibeni' Stage 3 was seen on May 27, 1965 and Stage 9 on July 2. In 'Honbeni' Stages 4 through 8 were found on June 22, 1964 and Stage 9 on June 29. 'Yatsubusa' Stage 3 was observed on June 8, and Stage 9 on July 6.

Pruning long shoots of *W. floribunda* 'Royal Purple', *W. sinensis, W. sinensis alba*, and *W. venusla violacea* in late summer or late winter caused the flower clusters produced the following spring at the base of the pruned stems to be longer and to contain more flowers than unpruned shoots. Pruning also reduced the abscission of flower initials and dying of protruding inflorescences at the base and apex of the long shoots.[1]

When the shoots of *Wisteria* were cut back in the early stages of floral development, the development of flower buds was stimulated and flowering occurred rapidly and prematurely in the summer.[3]

CULTURE AND PROPAGATION

Wisterias are easy to cultivate. They thrive best in deep rich soils and in sunny locations. The total amount of light energy seems to influence the flowering response since shade-grown wisterias do not flower as freely as full-sunlight-grown ones.[4]

Heavy fertilizing of wisterias, especially excess amounts of nitrogen, result in heavy stem and leaf growth which discourages flower-bud formation. Application of phosphorous may perhaps hasten flowering.[4]

Propagation is by seeds, division layering, hardwood stem cuttings, root cuttings, and grafting. Wisterias propagated by seed have a juvenility period often lasting some 7 years to reach the flowering phase. However, flowering of the Japanese wisteria *(W. floribunda)* may take up to 15 years.[4]

REFERENCES

1. **Chandler, W. H.,** Pruning trials on *Wisteria* vines, *J. Am. Soc. Hortic. Sci.*, 54, 482—484, 1949.
2. **Chittenden, F. J.,** *Dictionary of Gardening*, Vol. IV, 2nd ed., Synge, P. M., Ed., Clarendon Press, Oxford, 1977, 1088.
3. **Kosugi, K., Yokoi, M., Inaba, T., and Kusajima, I.,** Studies on the flower bud differentiation and development in ornamental trees and shrubs. VIII. On the time of flower bud differentiation and development in *Wisteria, Tech. Bull. Fac. Hortic. Chiba Univ.*, 14, 13—18, 1966.
4. **Perkins, H. O. and Wyman, D.,** Flowering in *Wisteria, Bull. Brooklyn Bot. Gard.*, 34, 64—66, 1978.

ERRATA: VOLUMES I THROUGH IV

Page	Line	Reads	Correction
		Volume I	
425	8	Involucels	Involucra
		Volume II	
Table of contents (2nd page)	13	*Dactylus*	*Dactylis*
60	11	DWP at low temperatures	DNP at low temperatures
		Volume III	
66	37	. . . *which it becomes* . . .	. . . which becomes . . .
314	(last line)	. . . evolution.	. . . evocation.
438	19	Reception . . .	Perception . . .
	20	. . . received . . .	. . . perceived . . .
		Volume IV	
22	4, 7, 13, 16	spontaneous (forms)	wild forms
209	Figure 4	A strictly female inflorescence	A strictly male inflorescence
210	Figure 5	A strictly male inflorescence	A strictly female inflorescence
254	Last paragraph, line 1	. . . controvertible.	. . . controversial.
310	14 (of text)	. . . photoperiod is controlled . . .	. . . photoperiod response is controlled . . .
339	Third secondary head	**Concentraion**	**Concentration**
367	Table	Drumondii (Subseries I-A) *halapense, almun* (Subseries VII)	Drummondii *halepense, almum*
445	Heading, and 1st line	*Anthrescus*	*Anthriscus*
481	22	. . . were carratic . . .	. . . were erratic . . .
493	5 (text)	. . . given at a 7.5-hr . . .	. . . given a 7.5-hr . . .

ABBREVIATIONS AND ALTERNATIVE NAMES OF CHEMICALS

The following list of terms and chemicals are those used in several chapters. Abbreviations and names used in only one chapter are specified in the paper.

ABA	(±) Abscisic acid
ADP	Adenosine diphosphate
Alar	See Daminozide
AMO-1618	2′-Isopropyl-4′-(dimethylammoniumchloride)-5′-methylphenyl piperdine-1-carboxylate
AMP	Adenosine monophosphate
Ancymidol	α-Cyclopropyl-α-(*p*-methoxyphenyl)-5-pyrimedine methyl alcohol
ATP	Adenosine triphosphate
Atrinal	See Dikegulac
AVG	Aminoethoxyvinylglycine
B	Blue light
BA	6-Benzyladenine, 6-benzylaminopurine
B-nine, B-9	See daminozide
CAM	Crassulacean acid metabolism
CCC	See chlormequat
CEPA	See ethephon
Chlorflurenol	See Morphactin
Chlormequat	2-Chloroethyl-trimethyl ammonium chloride (CCC, Cycocel)
Chlorophonium chloride	2,4-Dichlorobenzyl-tributylphosphonium chloride (Phosphon, Phosfon)
c	Centimeter
4-CPA	4-Chlorophenoxyacetic acid
cpm	Counts per minute
cv, cvs	Cultivar, cultivars
Cycocel	See chlormequat
2,4-D	2,4-Dichlorophenoxyacetic acid
Daminozide	Succinic acid-2,2-dimethylhydrazide (B-nine, Alar, SADH)
DCMU	3(3,4-dichlorophenyl)-1,1-dimethylurea
Dikegulac	2,3:4,6-di-*O*-isopropylidene-L-xylo-2-hexulo-furanosonic acid
DNA	Deoxyribonucleic acid
DNP	Day neutral plant
2,4-DNP	2,4-Dinitrophenol
EDTA	Ethylenediamine tetraacetic acid
En.	English
Ethephon	2-Chloroethylphosphonic acid (CEPA, Ethrel)
Ethrel	See ethephon
FI	Flower (floral) initiation
FLDP	Facultative long-day plant
FSDP	Facultative short-day plant
5-FDU	5-Fluorodeoxyuridine
Fr.	French
FR	Far-red light
g	Gram
GA	Gibberellin

GA_3	Gibberellic acid
Ge.	German
GMP	Guanosine 5′-monophosphate
hr	Hour
HIR	High irradiance response
IAA	Indole-3-acetic acid
IBA	Indole-3-butyric acid
ID	Intermediate day
IDP	Intermediate day plant
IR	Infrared
k	Kilo
LD	Long day(s)
LDP	Long-day plant
LSDP	Long-short-day plant
LB	Light break (same as NB)
lx, klx	Lux, Kilo lux
m	Meter
mg	Milligram
MH	Maleic hydrazide, 6-hydroxy-3(2H)-piridazinone
min	Minute(s)
mm	Millimeter
Morphactins	Inhibitors, derivatives of fluorene-9-carboxylic acid (chlorflurenol)
NAA	1-Naphthaleneacetic acid
NAD, NADH	Nicotinamide adenine dinucleotide and its reduced form
NADP, NADPH	Nicotinamide adenine dinucleotide phosphate and its reduced form
NB	Night break
nm	Nanometer
NS	Not significant
PAR	Photosynthetically active radiation
Pfr	Phytochrome-far-red absorbing form
Pr	Phytochrome-red absorbing form
PEP	Phosphoenolpyruvate
Phosfon, Phosphon	2,4-Dichlorobenzyl-tributylphosphonium chloride
R	Red light
RH	Relative humidity
RNA	Ribonucleic acid
sec	Second
SADH	See daminozide
SD	Short day(s)
SDP	Short-day plant
SE	Standard error of the mean
SEM	Scanning electron microscope
SH	Sulfhydryl
SLDP	Short-long-day plant
Sp.	Spanish
TIBA	2,3,5-Triiodobenzoic acid
2,4,5 TP	2-(2,4,5-trichlorophenoxy)propionic acid (Silvex)
UV	Ultraviolet

INDEX

A

B

C

D

E

H

I

J

K

L

M

Q

R

S

W

X

Z